Experimental Glycoscience

Glycobiology

N. Taniguchi, A. Suzuki, Y. Ito, H. Narimatsu,
T. Kawasaki, S. Hase (Eds.)

Experimental Glycoscience
Glycobiology

Naoyuki Taniguchi, M.D., Ph.D.
Professor, Department of Disease Glycomics, Research Institute for Microbial
Diseases, Osaka University, 1-1 Yamadaoka, Suita, Osaka 565-0871, Japan;
Systems Glycobiology Group, Advanced Science Institute, RIKEN, Wako,
2-1 Hirosawa, Wako, Saitama 351-0198, Japan

Akemi Suzuki, M.D., Ph.D.
Professor, Institute of Glycotechnology, Future Science and Technology Joint Research
Center, Tokai University, Hiratsuka, Kanagawa 259-1292, Japan

Yukishige Ito, Ph.D.
Chief Researcher, RIKEN Discovery Research Institute, Synthetic Cellular Chemistry
Laboratory, 2-1 Hirosawa, Wako, Saitama 351-0198, Japan

Hisashi Narimatsu, M.D., Ph.D.
Research Center for Medical Glycoscience, Advanced Industrial Science and
Technology, AIST Tsukuba, Central 2, Tsukuba, Ibaraki 305-8568, Japan

Toshisuke Kawasaki, Ph.D.
Professor, Research Center for Glycobiotechnology, Ritsumeikan University, 1-1-1
Noji-Higashi, Kusatsu, Shiga 525-8577, Japan

Sumihiro Hase, Ph.D.
Professor, Graduate School of Science, Osaka University, 1-1 Machikaneyama,
Toyonaka, Osaka 560-0043, Japan

ISBN: 978-4-431-77921-6 Tokyo Berlin Heidelberg New York
eISBN: 978-4-431-77922-3

Library of Congress Control Number: 2008930135

Cover: A ribbon diagram of an influenza virus neuraminidase subtype N9. Image created by Dr. Yoshiki Yamaguchi

© Springer 2008
This work is subject to copyright. All rights are reserved, whether the whole or part of the material is concerned, specifically the rights of translation, reprinting, reuse of illustrations, recitation, broadcasting, reproduction on microfilms or in other ways, and storage in data banks.
The use of registered names, trademarks, etc. in this publication does not imply, even in the absence of a specific statement, that such names are exempt from the relevant protective laws and regulations and therefore free for general use.
Product liability: The publisher can give no guarantee for information about drug dosage and application thereof contained in this book. In every individual case the respective user must check its accuracy by consulting other pharmaceutical literature.

Springer is a part of Springer Science+Business Media
springer.com
Printed in Japan

Typesetting: SNP Best-set Typesetter Ltd., Hong Kong
Printing and binding: Nikkei Printing, Japan

Printed on acid-free paper

Preface

There is growing interest in glycoscience, which is known to be one of the most important research areas in the medical and life sciences. It is now clear that glycans are implicated in various diseases such as cancer, immune diseases, infectious diseases, muscle-degenerative diseases, neurodegenerative diseases, and diabetes. Glycans also play a pivotal role in fertilization, growth and development, regeneration, and the aging process. Moreover, glycans are likely targets for the discovery of biomarkers, vaccines, and drugs for cancer and inflammatory diseases.

For these reasons, scientists in every field of research have realized that the application of glycoscience contributes to the development of their research projects. They also have a feeling, however, that glycoscience is technically difficult because of the heterogeneity and complexity of glycans. It is evident that many researchers from different disciplines who are not familiar with the specialized techniques used in glycoscience require detailed advice.

Newly developed analytical techniques—namely, the development of mass spectrometry, capillary electrophoresis, and high-performance liquid chromatography—and the accumulated resources of cloned glycosyltransferase genes, lectins, chemically and/or enzymatically synthesized compounds, bioinformatics, and KO mice or gene-targeting animals have opened new avenues for glycoscience.

The Japan Consortium for Glycobiology and Glycotechnology (JCGG) published the Japanese monograph entitled *Glycoscience: A Door to Open the Future* in 2006 written by more than 150 glycoscientists to summarize recent advances in this field in Japan and to offer readers a broad introduction to glycoscience. This present book is the English version but is largely modified and updated. Many of the contributors are experts in their respective fields and have made an effort to present their material in a manner that is understandable to those with general knowledge but with a different background in the biological sciences. The book is presented in three ways: 1) as a "cookbook" with which researchers will be able to prepare the solutions described and follow "recipes" to complete their experiments without needing to consult original research papers; 2) by "concept," which will allow researchers to search out the discipline, research equipment, research device, or methodology required; and 3) as a "table" providing an overview of the research data, among other information. For those who wish to delve into these topics, we included several references at the end of each chapter for further reading. This book will therefore be of interest to many scientists both in glycoscience and in the broader fields of biology, chemistry, and medicine, as well as to postdoctoral fellows, students, and young scientists.

I would like to take this opportunity to introduce the JCGG which was launched 5 years ago (Yoshitaka Nagai, JCGG president, professor emeritus, The University of Tokyo). The organization of this consortium is slightly different from that of the Consortium for Functional Glycomics (CFG) in the United States. Instead of receiving official support from the Japanese government for its operation, scientists took the initiative for creating and supporting the consortium by using their existing government research grants. Scientists who have been funded by different ministries of the Japanese government, such as the Ministry of Education, Culture, Sports, Science and Technology

(MEXT); the Japan Society for the Promotion of Science (JSPS); the Japan Science and Technology Agency (JST); the Ministry of Economy, Trade and Industry (METI); the New Energy and Industrial Technology Development Organization (NEDO); and the Ministry of Health, Labour and Welfare, are involved in this effort. They have joined together to form this consortium and to provide support through their individual research grants.

The JCGG aims to facilitate the exchange of scientific information; the sharing of resources, equipment, and facilities; the fostering of young scientists; and the construction of a database and infrastructures, among other goals. It also aims to launch national research centers in Japan, such as a Systems Glycobiology Center, where glycobiology can merge with such areas as nanotechnology, bioinformatics, and chemical biology to pursue the goals mentioned above, including medical applications. The JCGG holds an annual symposium in which more than 600 people usually participate.

On behalf of the editors, I gratefully acknowledge all those who contributed to this volume despite their busy schedules. Special thanks are due to Mr. Keiichi Yoshida, the secretary general of JCGG, for his skillful assistance in editing. I also thank Kinpodo, Kyoto, who published the original Japanese version, for agreeing to publication of the English version.

Thanks also go to the editorial staff members at Springer Japan, for their constant support, patience regarding the deadline for the manuscript, and skill in directing the production of this book.

I hope this publication will provide an impetus for future research.
On behalf of the editors,

Naoyuki Taniguchi, MD, PhD
February 10, 2008

Editorial Board

Editors:
Naoyuki Taniguchi, M.D., Ph.D.
Department of Disease Glycomics, Research Institute for Microbial Diseases, Osaka University, 1-1 Yamadaoka, Suita, Osaka 565-0871, Japan;
Systems Glycobiology Group, Advanced Science Institute, RIKEN, Wako, 2-1 Hirosawa, Wako, Saitama 351-0198, Japan

Akemi Suzuki, M.D., Ph.D.
Institute of Glycotechnology. Future Science and Technology Joint Research Center, Tokai University, Hiratsuka, Kanagawa 259-1292, Japan

Yukishige Ito, Ph.D.
RIKEN Discovery Research Institute, Synthetic Cellular Chemistry Laboratory. 2-1 Hirosawa, Wako, Saitama 351-0198, Japan

Hisashi Narimatsu, M.D., Ph.D.
Research Center for Medical Glycoscience, Advanced Industrial Science and Technology, AIST Tsukuba, Central 2, Tsukuba, Ibaraki 305-8568, Japan

Toshisuke Kawasaki, Ph.D.
Research Center for Glycobiotechnology, Ritsumeikan University, 1-1-1 Noji-Higashi, Kusatsu, Shiga 525-8577, Japan

Sumihiro Hase, Ph.D.
Graduate School of Science. Osaka University, 1-1 Machikaneyama, Toyonaka, Osaka 560-0043, Japan

Co-editors:
Koichi Furukawa, M.D., Ph.D., School of Medicine, Nagoya University, Japan
Makoto Kiso, Ph.D., Faculty of Applied Biological Sciences, Gifu University, Japan
Tomoya Ogawa, Ph.D., RIKEN Yokohama Institute, Japan
Shoichi Kusumoto, Ph.D., Suntory Institute For Bioorganic Research, Japan

Contents

Preface V
Editorial Board VII
Contents of *Experimental Glycoscince: Glycochemistry* XVI
List of Contributors XVII

Part 1 Glycosyltransferase Genes

Section I Glycosyltransferases

Convenient Synthesis of Glycan-Related Oligosaccharides and Their Transformation into Neoglycoconjugates Using Enzymatic Methods
  T. Murata, T. Usui 5

Comparison of Glycosyltransferase Families Using the Profile Hidden Markov Model
  N. Kikuchi 9

N-Acetylglucosaminyltransferases Involved in N-Glycan Biosynthesis
  K. Inamori, N. Taniguchi 13

UDP-N-Acetyl-D-Galactosamine:polypeptide N-Acetylgalactosaminyltransferases (ppGaNTases, pp-GalNAc-Ts, EC 2.4.1.41)
  K. Tachibana 16

Glycosyltransferase Family with β4GT Motif (β4Gal-T and β4GalNAc-T Family)
  T. Sato, H. Narimatsu 19

β1,3-glycosyltransferase Gene Family and IGnT Gene Family
  A. Togayachi, T. Sato, H. Narimatsu 24

O-Mannosyltransferase and POMGnT1
  T. Endo, H. Manya 30

Glycosyltransferases Involved in O-fucose Glycan Synthesis
  T. Okajima, T. Matsuda 33

Fucosyltransferase (α1,2/ α1,3/ α1,4-fucosyltransferases)
  T. Kudo, H. Narimatsu 36

FUT8 Assay
  H. Ihara, H. Korekane, A. Matsumoto, N. Taniguchi 39

Sialyltransferases
S. Tsuji 42

Glucuronyltransferases Involved in the HNK-1 Biosynthesis
S. Kakuda, T. Kawasaki, S. Oka 47

α1,4-Linkage Glycosyltransferases
J. Nakayama, K. Furukawa 50

Glycosyltransferases Involved in the Biosynthesis of Glycolipids
K. Furukawa, A. Tsuchida, K. Furukawa 53

Sialyltransferases and Other Enzymes Involved in the Biosynthesis of Gangliosides
K. Furukawa, K. Furukawa 56

Heparan Sulfate Synthases and Related Genes
S. Mizumoto, H. Kitagawa 59

Chondroitin Sulfate Biosynthesis and Related Genes
H. Watanabe, K. Kimata 64

Biosynthesis of Keratan Sulfate
A. Seko, K. Yamashita 67

Hyaluronan Synthase Assay
N. Itano 70

Glycosylphosphatidylinositol-Anchored Protein Biosynthesis and Related Genes in Mammalian Cells
Y. Maeda, T. Kinoshita 72

Cloning of Genes for N-linked Oligosaccharide Synthesis in Endoplasmic Reticulum
K. Nakayama, Y. Jigami 76

Section II Sugar-Modified Enzymes

N-acetylglucosamine-6-O-sulfotransferases
K. Uchimura, T. Muramatsu 83

Sulfotransferases Involved in Sulfation of Glycosaminoglycans
O. Habuchi 87

Sulfatide Synthase
K. Honke 94

CMP-N-acetylneuraminic Acid Hydroxylase
A. Suzuki 97

Section III Transporters

Nucleotide Sugar Transporter Genes and Their Functional Analysis
S. Nishihara 103

Section IV Glycosidases

Processing Enzymes Involved in *N*-glycan Biosynthesis and Related Genes: the Golgi *N*-glycan Processing α-mannosidase II and α-mannosidase IIx
T.O. Akama, M.N. Fukuda 111

Sialidase Genes
T. Miyagi 115

Release of Sugar Chains from Glycosphingolipids
M. Ito 119

Heparan Sulfate Endosulfatase Assay
K. Keino-Masu, M. Masu 123

Sphingolipid Activator Proteins
J. Matsuda 125

Section V Animal Lectins

Affinity Purification of Recombinant Galectins
J. Iwaki, J. Hirabayashi 133

Siglec Family
T. Angata 136

Part 2 Functional Analyses of Sugar Chains

Section VI Immunology

NKT Cells and Their Recognition of Glycolipids
K. Seino, M. Taniguchi 145

Carbohydrate Ligands for Selectins in Immune Cell Trafficking
R. Kannagi, K. Ohmori 148

Carbohydrate Recognition by Cytokines and Its Relevance to Their Physiological Activities
K. Fukushima, K. Yamashita 154

Detection of Mouse MGL1/CD301a by a Specific Monoclonal Antibody
K. Denda-Nagai, K. Suzuki, M. Goda, A. Kudo, H. Kawakami, T. Irimura .. 158

Roles of Serum Lectins in Host Defense
N. Kawasaki, B.Y. Ma, T. Kawasaki 162

Siglec-2 Is a Key Molecule for Immune Response
T. Tsubata 167

Galectin and Immune System: Toward Clinical Applications
M. Hirashima 171

Section VII Brains and Nerves

Novel Glycolipids in Developing Brain: Determination by Nano-LC/MS/MS
Y. Hirabayashi, Y. Nagatsuka, S. Ito 177

Functional Roles of the HNK-1 Carbohydrate and Polysialic Acid in the
Nervous System
  Y. Kizuka, S. Oka 180
Neuronal Function of Sulfatide
  K. Honke 182
Accumulation of Alzheimer β-amyloid via Ganglioside Clusters
  K. Matsuzaki 185
Chondroitin Sulfate Proteoglycans and Neuronal Network Formation
  N. Maeda 188
Processing of Glycosyltransferases by Alzheimer's β-secretase (BACE1)
  S. Kitazume, S. Takashima, Y. Hashimoto 192
Chondroitin Sulfate Proteoglycans: A Key Substance for Central Nervous
System Injury Repair
  A. Oohira 195

Section VIII Quality Control of Proteins

A Cytoplasmic Peptide: *N*-Glycanase and ER-Associated Degradation
  T. Suzuki 201
Ubiquitination of Glycoproteins in the Cytosol
  T. Tai 204
Degradation of Misfolded Glycoproteins in the Endoplasmic Reticulum
  N. Hosokawa 207

Section IX Golgi and Lysosomal Diseases

Retrograde Transport of Glycolipid-Bound Toxins
  R. Misaki, T. Taguchi 213
Degradation of Hyaluronan and Its Disorder
  M. Yanagishita, K.A. Podyma-Inoue 216
Recent Advances in Enzyme Replacement Therapy for Lysosomal
Diseases
  K. Itoh 219

Section X Infections

Developing an Assay System for the Hemagglutinin Mutations
Responsible for the Binding of Highly Pathogenic Avian Influenza
A Viruses to Human-Type Receptors
  Y. Suzuki 225
Helicobacter pylori Growth Assay for Glycans
  J. Nakayama 227
Binding Properties of *Clostridium botulinum* Type C Progenitor Toxin
  A. Nishikawa 230

Section XI Cancer

Fucosylation and Cancer
E. Miyoshi 235

Biological Significance of Mucins Produced by Epithelial Cancer Cells
H. Nakada 238

Regulation of Glycosyltransferases for O-Glycan Core Structure in Cancer and Their Relations to Metastasis
T. Iwai, H. Narimatsu 243

Roles of Carbohydrate-Mediated Cell Adhesion in Cancer Progression
R. Kannagi, A. Seko 246

Conditional Transgenic Mouse of Hyaluronan Synthase: A Potential Model of Advanced Breast Cancer
N. Itano 252

Section XII Regeneration Medicine and Transplantation

Proteoglycans in Tissue Regeneration
K. Kimata 257

Stem Cell Glycobiology
M. Nakamura 262

Human Embryonic Stem (ES) Cells and Induced Pluripotent Stem (iPS) Cells Are Defined by Carbohydrate Markers
A. Umezawa, M. Toyoda 265

Glycoantigens in Xenotransplantation
S. Miyagawa 267

Section XIII Fertilization

Glycoconjugates in Spermatogenesis
K. Honke 275

Recognition Mechanism of Egg and Sperm Based on Sugar Chains
M. Matsumoto 278

Section XIV Development, Differentiation, and Morphogenesis

Functional Analysis of Sugar Chains Using a Genome-Wide RNAi System in *Drosophila*
S. Nishihara 285

Functional Glycomics at the Level of Single Cells: Studying Roles of Sugars in Cell Division, Differentiation, and Morphogenesis with 4D Microscopy
S. Mizuguchi, K. Dejima, K.H. Nomura, D. Murata, K. Nomura 290

Studies on Functions of Notch *O*-Fucosylation in *Drosophila*
T. Ayukawa, K. Matsuno 295

Membrane Microdomain as a Platform of Carbohydrate-Mediated Interactions During Early Development of Medaka Fish
T. Adachi, C. Sato, K. Kitajima 299

Functions of Heparan Sulfate Proteoglycans in Morphogenesis
H. Habuchi, K. Kimata 303

Fukutin and Fukuyama Congenital Muscular Dystrophy
M. Kanagawa, T. Toda 309

Section XV Muscle Dystrophy and Carbohydrate Disorders of Glycosylation

Congenital Muscular Dystrophies Due to the Glycosylation Defect
T. Endo 315

Molecular Diagnosis of Congenital Disorders of Glycosylation
Y. Wada 319

Research in Japan Has Contributed to the Understanding of GPI Anchor Deficiency
Y. Murakami, T. Kinoshita 323

Section XVI Lifestyle-Related Diseases

Aberrant Expression of Sialidase in Cancer and Diabetes
T. Miyagi 329

Insulin Resistance and Type 2 Diabetes as Microdomain Disease: Implication of Ganglioside GM3
J. Inokuchi, K. Kabayama 333

Section XVII IgA Nephropathy

Incompleteness of O-glycans at the Hinge Region of IgA1 and IgA Nephropathy
H. Iwase, Y. Hiki 339

β4-Galactosyltransferase Deficiency and IgA Nephropathy
M. Asano 343

Glycosyltransferases Involved in the Biosynthesis of IgA O-Glycans
Y. Narimatsu, H. Narimatsu 346

Section XVIII Growth Factor Receptors

Importance of Sugar Chains in the Function of Growth Factor Receptors
Y. Ikeda 349

Analysis of N-glycan of Growth Factor Receptors
M. Takahashi 351

The Midkine Family and Its Receptors
K. Kadomatsu 355

N-glycans Regulate Integrin α5β1 Functions
  J. Gu, Y. Sato, T. Isaji 358

Section XIX Blood Groups and Blood Cells

Blood Group Antigens: Blood Group Carbohydrate Antigens
  K. Furukawa, K. Iwamura, K. Furukawa 363

Bombay, Lewis and Ii Blood Group Antigens on Human Erythrocytes
  H. Narimatsu 366

Blood Cells in Glycobiology
  M. Nakamura 369

Part 3 Glycosyltransferase Gene KO Mice

Section XX Glycosyltransferase Gene KO Mice

Fucosyltransferase-9 Knockout Mouse
  T. Kudo, H. Narimatsu 377

Knockout Mice of α1,6 Fucosyltransferase (Fut 8)
  N. Taniguchi 379

Glucuronyltransferase (GlcAT-P) Gene-Deficient Mice
  I. Morita, S. Oka 381

KO Mice of β1,4-galactosyltransferase-I
  M. Asano 383

Sulfotransferases
  T. Muramatsu, K. Uchimura 386

KO Mice of Cerebroside Sulfotransferase
  K. Honke 389

KO Mice of β-1,4-*N*-acetylgalactosaminyltransferase (GM2/GD2 Synthase)
  K. Furukawa, K. Takamiya, K. Furukawa 391

KO Mice of α-2,8-sialyltransferase (GD3 Synthase)
  K. Furukawa, M. Okada, K. Furukawa 394

Double KO Mice of β1,4-*N*-acetylgalactosaminyltransferase (GM2/GD2 Synthase) and α-2,8-sialyltransferase (GD3 Synthase)
  K. Furukawa, O. Tajima, Y. Ohmi, K. Furukawa 396

Part 4 Infrastructures and Research Resources

Section XXI Glycosyltransferases and Related Genes, and Useful Cell Lines

Cell Lines for Glycobiology
  M. Nakamura 403

Sugar Chain-Related Genes: Glycosyltransferases Expression System
T. Sato, H. Narimatsu **407**

Section XXII Sugar Library

GALAXY Database and Pyridylaminated Oligosaccharide Library
K. Kato, N. Takahashi **413**

Mass Production of Sugar Intermediates for the Synthesis of Functional Oligosaccharides
Y. Matsuzaki **417**

Section XXIII Database

A Database System for Glycogenes (GGDB)
A. Togayachi, K.-Y. Dae, T. Shikanai, H. Narimatsu **423**

CabosDB: Carbohydrate Sequencing Database
N. Kikuchi **426**

GlycoEpitope: A Database of Carbohydrate Epitopes and Antibodies
T. Kawasaki, H. Nakao, T. Tominaga **429**

A Novel Lectin-Affinity Database for Structural Glycomics
Y. Takahashi, J. Hirabayashi **432**

Glycoconjugate Data Bank
N. Miura, S.-I. Nishimura **435**

KEGG GLYCAN for Integrated Analysis of Pathways, Genes, and Structures
K. Hashimoto, M. Kanehisa **441**

Section XXIV Microarray

DNA Microarray in Glycobiology
H. Yamamoto, H. Takematsu, Y. Kozutsumi **447**

Lectin Microarray
A. Kuno, J. Hirabayashi **451**

Carbohydrate Microarray for Deciphering the Information Embedded in Oligosaccharide Structures
S. Fukui **455**

Glossary **459**

Index **491**

Contents of *Experimental Glycoscince: Glycochemistry*

Part 1 Structural Analysis of Sugar Chains
Section I Release of Sugar Chains and Labeling
Section II Sequence Analysis
Section III Sugar Chain Analysis by Mass Spectrometry
Section IV Analysis of the Three-dimensional Structure of Sugar Chains
Section V Analysis of Sugar–Protein Interactions

Part 2 Chemical Synthesis of Sugar Chains
Section VI Chemical Synthesis of Sugar Chains

List of Contributors

Adachi, T. 299
Akama, T.O. 111
Angata, T. 136
Asano, M. 343, 383
Ayukawa, T. 295
Dae, K.-Y. 423
Dejima, K. 290
Denda-Nagai, K. 158
Endo, T. 30, 315
Fukuda, M.N. 111
Fukui, S. 455
Fukushima, K. 154
Furukawa, Keiko 53, 56, 363, 391, 394, 396
Furukawa, Koichi 50, 53, 56, 363, 391, 394, 396
Goda, M. 158
Gu, J. 358
Habuchi, H. 303
Habuchi, O. 87
Hashimoto, K. 441
Hashimoto, Y. 192
Hiki, Y. 339
Hirabayashi, J. 133, 432, 451
Hirabayashi, Y. 177
Hirashima, M. 171
Honke, K. 94, 182, 275, 389
Hosokawa, N. 207
Ihara, H. 39
Ikeda, Y. 349
Inamori, K. 13
Inokuchi, J. 333
Irimura, T. 158
Isaji, T. 358
Itano, N. 70, 252
Ito, M. 119
Ito, S. 177
Itoh, K. 219
Iwai, T. 243
Iwaki, J. 133
Iwamura, K. 363
Iwase, H. 339
Jigami, Y. 76

Kabayama, K. 333
Kadomatsu, K. 355
Kakuda, S. 47
Kanagawa, M. 309
Kanehisa, M. 441
Kannagi, R. 148, 246
Kato, K. 413
Kawakami, H. 158
Kawasaki, N. 162
Kawasaki, T. 47, 162, 429
Keino-Masu, K. 123
Kikuchi, N. 9, 426
Kimata, K. 64, 257, 303
Kinoshita, T. 72, 323
Kitagawa, H. 59
Kitajima, K. 299
Kitazume, S. 192
Kizuka, Y. 180
Korekane, H. 39
Kozutsumi, Y. 447
Kudo, A. 158
Kudo, T. 36, 377
Kuno, A. 451
Ma, B.Y. 162
Maeda, N. 188
Maeda, Y. 72
Manya, H. 30
Masu, M. 123
Matsuda, J. 125
Matsuda, T. 33
Matsumoto, A. 39
Matsumoto, M. 278
Matsuno, K. 295
Matsuzaki, K. 185
Matsuzaki, Y. 417
Misaki, R. 213
Miura, N. 435
Miyagawa, S. 267
Miyagi, T. 115, 329
Miyoshi, E. 235
Mizuguchi, S. 290
Mizumoto, S. 59
Morita, I. 331

Murakami, Y. 323
Muramatsu, T. 83, 386
Murata, D. 290
Murata, T. 5
Nagatsuka, Y. 177
Nakada, H. 238
Nakamura, M. 262, 369, 403
Nakao, H. 429
Nakayama, J. 50, 227
Nakayama, K. 76
Narimatsu, H. 19, 24, 36, 243, 346, 366, 377, 407, 423
Narimatsu, Y. 346
Nishihara, S. 103, 285
Nishikawa, A. 230
Nishimura, S.-I. 435
Nomura, K. 290
Nomura, K.H. 290
Ohmi, Y. 396
Ohmori, K. 148
Oka, S. 47, 180, 381
Okada, M. 394
Okajima, T. 33
Oohira, A. 195
Podyma-Inoue, K.A. 216
Sato, C. 299
Sato, T. 19, 24, 407
Sato, Y. 358
Seino, K. 145
Seko, A. 67, 246
Shikanai, T. 423

Suzuki, A. 97
Suzuki, K. 158
Suzuki, T. 201
Suzuki, Y. 225
Tachibana, K. 16
Taguchi, T. 213
Tai, T. 204
Tajima, O. 396
Takahashi, M. 351
Takahashi, N. 413
Takahashi, Y. 432
Takamiya, K. 391
Takashima, S. 192
Takematsu, H. 447
Taniguchi, M. 145
Taniguchi, N. 13, 39, 379
Toda, T. 309
Togayachi, A. 24, 423
Tominaga, T. 429
Toyoda, M. 265
Tsubata, T. 167
Tsuchida, A. 53
Tsuji, S. 42
Uchimura, K. 83, 386
Umezawa, A. 265
Usui, T. 5
Wada, Y. 319
Watanabe, H. 64
Yamamoto, H. 447
Yamashita, K. 67, 154
Yanagishita, M. 216

Glossary
Contributed by:
Tadashi Suzuki, Yoshiki Yamaguchi, Shinobu Kitazume

In collaboration with:
Yoko Funakoshi, Kazuyuki Nakajima, Kenji Kanekiyo, Tetsuya Suetake, Mayumi Kanagawa, Masaki Kato, Kana Matsumoto

Part 1
Glycosyltransferase Genes

Section I
Glycosyltransferases

Convenient Synthesis of Glycan-Related Oligosaccharides and Their Transformation into Neoglycoconjugates Using Enzymatic Methods

Takeomi Murata[1], Taichi Usui[2]

Introduction

Glycan-related oligosaccharides involved in a number of biological events. However, there is a paucity of structurally defined oligosaccharides that have been isolated from biological material. Currently, there is a great interest in developing synthetic routes for the oligosaccharide portion of glycoconjugates. The development of simple and effective methods for synthesizing neoglycoconjugates as mimetics of glycoproteins and glycolipids is also essential for the understanding of the biological function of these molecules. Here, we review the recent developments in the synthesis of glycan-related oligosaccharides and neoglycoconjugates.

Consecutive Synthesis of Glycan-Related Oligosaccharides

We have already reported the synthesis of possible arrays of linkages (1–3), (1–4), and (1–6) of β-D-Gal-D-GlcNAc and β-D-Gal-D-GalNAc by using some readily available β-D-galactosidase. Starting with these disaccharide units, some glycan-related oligosaccharides were enzymaticaly synthesized by consecutive addition of GlcNAc or Gal residue to acceptor sugars.

Core Structure of O-linked Glycan

Majority of the mucin-type carbohydrates of serum and membrane glycoproteins are O-linked glycans of the core 1 structure Galβ1-3GalNAcα-Ser/Thr, to which an addition of GlcNAc gives the core 2 structure Galβ1-3(GlcNAcβ1-6)GalNAcα-Ser/Thr. Regioselective β-D-galactosyl transfer from Galβ-pNP to the O-3 position of GalNAcα-pNP, using recombinant β-D-galactosidase from *Bacillus circulans* ATCC 31382, gives Galβ1-3GalNAcα-pNP (1 Apply the reaction mixture to a Chromatorex-ODS DM 1020T eluted with 10% MeOH and separate Galβ1-3GalNAcα-pNP with a Toyopearl HW-40S column). With the resulting Galβ1-3GalNAcα-pNP acceptor, the desired compound Galβ1-3(GlcNAcβ1-6)GalNAcα-pNP was obtained through N-acetylglucosaminyl transfer from N,N'-diacetylchitobiose donor mediated by β-N-acetylhexosaminidase from *Nocardia orientalis* [2, Apply the reaction mixture to a Toyopearl HW-40S column eluted with 10% MeOH and separate Galβ1-3(GlcNAcβ1-6)GalNAcα-pNP with an ODS column] (Scheme 1A). As a result, a core 2-oligosaccharide unit was enzymatically synthesized

---

Department of Applied Biological Chemistry, Faculty of Agriculture, Shizuoka University, 836 Ohya, Suruga-ku, Shizuoka 422-8529, Japan
Phone: +81-54-238-4873, Fax: +81-54-237-3028
E-mail: [1]actmura@agr.shizuoka.ac.jp, [2]actusui@agr.shizuoka.ac.jp

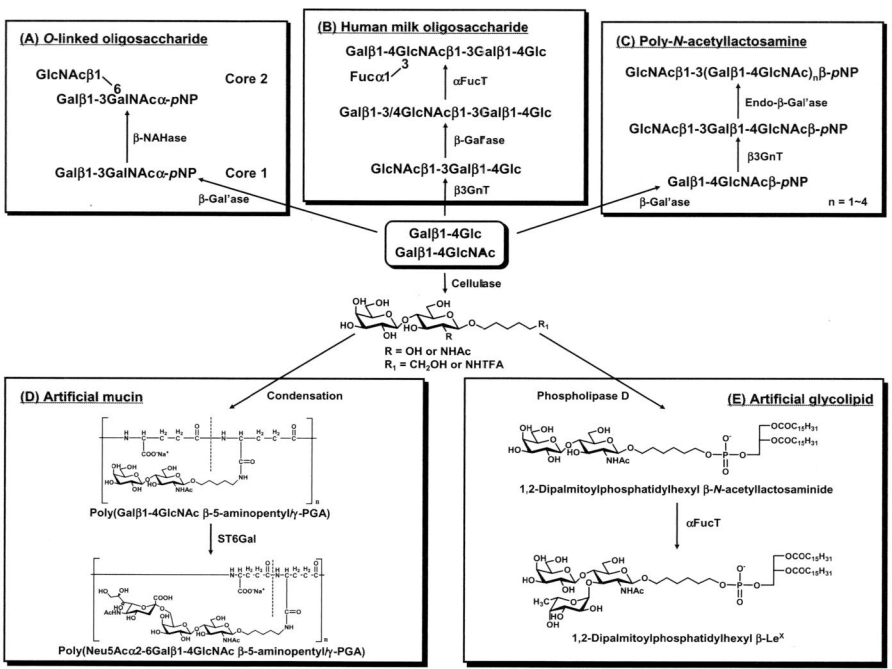

Scheme 1 Synthesis of glycan-related oligosaccharides and neoglycoconjugates using enzymatic methods

by consecutive additions of Gal and GlcNAc residues to GalNAcα-pNP (Murata et al. 1998).

Oligosaccharides in Human Milk

Lacto-N-tetraose (Galβ1-3GlcNAcβ1-3Galβ1-4Glc, LNT) and lacto-N-neotetraose (Galβ1-4GlcNAcβ1-3Galβ1-4Glc, LNnT) are human milk oligosaccharides that also form the core structure of glycolipids of the so-called lacto- and neolacto-series. GlcNAcβ1-3Galβ1-4Glc (lacto-N-triose II) was originally synthesized by the transfer of GlcNAc from UDP-GlcNAc to lactose using a crude β1,3-N-acetylglucosaminyltransferase (β3GnT) prepared from bovine serum (Scheme 1B, Murata et al. 1999). The availability of relatively large amounts of recombinant β3GnT has greatly improved the viability of this synthesis. The resulting lacto-N-triose II was readily converted into LNT (3, Apply the reaction mixture to a charcoal-Celite column eluted with 7.5–35% EtOH of linear gradient and purify LNT by rechromatography with the same column) and LNnT (4, Apply the reaction mixture to a charcoal-Celite column eluted with 20~40% EtOH of linear gradient and separate LNnT with a Bio-Gel P-2 column) utilizing two kinds of β-D-galactosidase-mediated transglycosylations. Specifically, recombinant β-D-galactosidase from *B. circulans* ATCC 31382 catalyzed the regioselective galactosyl transfer from Galβ-*o*NP to *O*-3″ of lacto-N-triose II. By contrast, β-D-galactosidase from *B. circulans* (Biolacta) catalyzed the transfer from lactose to *O*-4″ of lacto-N-triose II. We used α1,3fucosyltransferase (α3FucT) LNnT from chicken serum to convert into lacto-N-fucopentaose III, Galβ1-4(Fucα1-3)-GlcNAcβ1-3Galβ1-4Glc (5, Apply the reaction mixture to a charcoal-Celite column eluted with 0–30% EtOH of linear gradient and desalt Galβ1–4(Fucα1–3)–GlcNAcβ1–3Galβ1–4Glc with a mixed column of Dowex 50W–X2 and Bio-Rad AG-3-X4A), a human milk oligosaccharide that contains an Le^x antigen structure.

Poly-N-acetyllactosamine

Poly-*N*-acetyllactosamines are some of the most attractive structures in the field of chemical and enzymatic syntheses of glycans due to their numerous biological functions. Enzymatic synthesis of GlcNAc-terminated poly-*N*-acetyllactosamine was demonstrated using a transglycosylation reaction of *Escherichia freundii* endo-β-galactosidase, which was discovered as keratin sulfate-degrading enzyme, the so-called keratanase (Murata et al. 2003). The enzyme catalyzed a transglycosylation reaction on GlcNAcβ1-3Galβ1-4GlcNAc-*p*NP, which served both as a donor and as an acceptor, and converted the substrate into GlcNAc-terminated poly-*N*-acetyllactosamine β-glycosides.

GlcNAcβ1-3(Galβ1-4GlcNAcβ1-3)nGalβ1-4GlcNAcβ-*p*NP (n = 1–4) (6, Apply the reaction mixture to a Toyopearl HW-50S column eluted with H$_2$O and separate GlcNAcβ1-3(Galβ1-4GlcNAcβ1-3)nGalβ1-4GlcNAcβ-*p*NP (n = 1–4) with a Sep-pak Light C-18 column) through sequential transfer reaction of the disaccharide unit (Scheme 1C).

Molecular Design of Artificial Glycoconjugates

We have developed some facile synthetic methods for the preparation of biologically important glycan-related oligosaccharides. Well-defined oligosaccharides, such as recognition signals, were introduced in polypeptide or lipid for constructing neoglycoconjugates to act as glycomimetics.

Artificial Mucin

Mucins, highly glycosylated proteins that form the major component of mucosa, are known to play a physiological role as a barrier to biomolecules. Highly water soluble artificial glycopolypeptides with a polyglutamic acid backbone carrying multivalent sialyloligosaccharide units, termed as artificial mucins, have been chemoenzymatically synthesized as potential polymeric inhibitors of infection by bird and human influenza viruses (Totani et al. 2003; Ogata et al. 2007). Scheme 1D represents the pathway from *N*-acetyllactosamine or lactose via the following three steps: (1) enzymatic synthesis of a spacer-linked disaccharide glycosides; (2) coupling of the resulting glycoside to γ-polyglutamic acid (γ-PGA); and (3) sialylation of the resulting polymers to produce highly water-soluble glycopolymers carrying clustered identical sialyldisaccharide segments (7, Loaded the reaction mixture onto a Sephadex G-25 M PD-10 column to prepare artificial mucins). The sialylglycopolypeptides (artificial mucins) possessing Neu5Acα2-6LacNAc bound preferentially to human influenza A, whereas Neu5Acα2-3LacNAc/Lac bound to the avian influenza A The α-PGA-based sialylglycopolypeptides were useful probes for characterizing the sugar-binding specificity of hemagglutination (HA) of the Spanish influenza virus. H5N1 influenza A viruses have spread to numerous countries, infecting not only large numbers of poultry but also an increasing number of humans. The artificial glycopolymers were also helpful for assessing HA mutations responsible for the binding of H5N1 influenza A viruses to human-type receptors (Yamada et al. 2006).

Neoglycolipids

Neoglycolipids composed of a disaccharide glycoside and phospholipid were designed and prepared as mimetics of lactosylceramide (Scheme 1E, Harada et al. 2005). The lactosyl- and N-acetyl- lactosaminyl-phospholipids were enzymatically synthesized from lactose and LacNAc respectively using cellulose-mediated condensation with 1,6-hexanediol, followed by conjugation of the resulting glycosides and dipalmitoylphosphatidyl choline mediated by *Streptomyces* phospholipase D [8, Loaded the reaction mixture onto a Sep-pak Accell QMA column (CH_3Cl_3/MeOH/H_2O, 30:60:8) and separate lactosyl phospholipid or N-acetyllactosaminyl phospholipid by successive chromatographies on Sephadex LH-20 and Silica gel 60 N columns (CH_3Cl_3/MeOH/H_2O, 5:5:1)]. NMR spectroscopy showed that neoglycolipids in $CDCl_3$/CD_3OD (1/1) exist as a single molecule species, whereas in $CDCl_3$ they associate to form inverse micelles. X-ray diffraction showed that multilamellar vesicles (MLVs) of the neoglycolipids are in the bilayer gel phase at room temperature. These neoglycolipids have the ability to trap calcein, a chelating derivative of fluorecein, in MLVs and showed specific binding to lectin in plate assays using fluorescently labeled compounds. Such mimetics are especially attractive in the case of complex natural or synthetic glycolipids, which are expensive or difficult to obtain uniformly and /or in large amounts.

References

Harada Y, Murata T, Totani K, Kajimoto T, Masum S-D, Tamba Y, Yamazaki M, Usui T (2005) Design and facile synthesis of neoglycolipids as lactosylceramide and transformation into glycoliposome. Biosci Biotech Biochem 66:801–807

Murata T, Itho T, Usui T (1998) Enzymatic synthesis of β-D-Gal-(1-3)-[β-D-GlcNAc-(1-6)]-α-D-GalNAc-OC$_6$H$_4$NO$_2$-*p* as a carbohydrate unit of mucin-type 2 core. Glycoconjugate J 15:575–582

Murata T, Inukai T, Suzuki M, Yamagishi M, Usui T (1999) Facile enzymatic conversion of lactose into lacto-N-tetraose and lacto-N-neotetraose. Glycoconjugate J 16:189–195

Murata T, Hattori T, Amarume S, Koichi A, Usui T (2003) Kinetic studies on endo-β-galactosidase by a novel colorimetric assay and synthesis of N-acetyllactosamine-repeating oligosaccharide β-glycosides using its transglycosylation activity. Eur J Biochem 270:1–11

Ogata M, Murata T, Murakami K, Suzuki T, Hidari KI, Suzuki Y, Uusi T (2007) Chemoenzymatic synthesis of artificial glycopolypeptides containing multivalent sialyloligosaccharides with a γ-polyglutamic acid backbone and their effect for inhibition of infection by influenza viruses. Bioorg Med Chem 15:1383–1393

Totani K, Kubota T, Kuroda T, Murata T, Hidari I-PJK, Suzuki T, Suzuki Y, Kobayashi K, Ashida H, Yamamoto K, Usui T (2003) Chemoenzymatic synthesis and application of glycopolymers containing multivalent sialyloligosaccharides with a poly(L-glutamic acid) backbone for inhibition of infection by influenza viruses. Glycobiology 13:1–12

Yamada S, Suzuki Y, Suzuki T, Le MQ, Nidom CA, Sakai-Tgawa Y, Muramoto Y, Ito M, Kiso M, Horimoto T, Shinya K, Sawada T, Kiso M, Usui T, Murata T, Lin Y, Haire LF, Stevens DJ, Russell RJ, Gambin ST, Skehel JJ, and Kawaoka Y (2006) Hemagglutinin mutations responsible for the binding of H5N1 influenza A viruses to human-type receptor. Nature 444:378–382

Comparison of Glycosyltransferase Families Using the Profile Hidden Markov Model

Norihiro Kikuchi

Introduction

Several research groups have attempted to classify glycosyltransferases using sequence similarities, e.g., the CAZy (Carbohydrate-Active enZYmes) database (Coutinho et al. 2003). However, the structural, functional, and evolutionary relationships of glycosyltransferases are still unknown. We classified 47 of the 60 families from the CAZy database into four superfamilies using the profile hidden Markov model (HMM) method to reveal relationships between structure and function in glycosyltransferases.

Results

We named four superfamilies as GTS-A, -B, -C, and -D (Kikuchi et al. 2003). Glycosyltransferases with a GT-A or -B fold were classified as GTS-A or -B respectively; these glycosyltransferases had regions with similar sequences, including a donor/divalent cation-binding domain. These findings suggest that GTS-A and -B glycosyltransferases have a donor/divalent cation-binding domain with a similar structure. Glycosyltransferases that utilize dolichyl phosphate sugars as the donor substrate and have multi-spanning transmembrane domains fall into the GTS-C cluster; α1,2-/1,6-fucosyltransferases are classified as GTS-D.

We mapped glycosyltransferase superfamilies that are involved in four representative glycosylation pathways, namely, peptide glycan biosynthesis in bacteria, biosynthesis of the inner core region of bacterial lipopolysaccharides (LPS), biosynthesis of the glycosylphosphatidylinositol anchor of eukaryotes (GPI-anchor), and asparagine-linked glycan (N-glycan) modification in eukaryotes (Fig. 1). Our analysis indicates that there is conservation of glycosyltransferase superfamilies in the glycosylation pathways associated with the synthesis of N-glycans in eukaryotes and peptide glycans in bacteria. In both eukaryotes and bacteria, the initiation of glycosylation is catalyzed by sugar-1-phosphate transferase family enzymes [i.e., phospho-N-acetylmuramoyl-pentapeptide-transferase (mraY) and GlcNAc-1-phosphate transferase (GPT), respectively], followed by elongation catalyzed by GTS-B glycosyltransferases (Fig. 1). Thus, we propose the following hypothesis for glycosyltransferase evolution. Glycosyltransferases belonging to GTS-A and GTS-B are found in almost all life forms, suggesting that GTS-A- and GTS-B-like

---

Bioscience Group, Mitsui Knowledge Industry Co., Ltd., 7-14, Hitotsubashi SI bldg., 3-26, Kandanishiki-cho, Chiyoda-ku, Tokyo 101-0054, Japan
Phone: +81-3-5259-6570, Fax: +81-3-5280-2737
E-mail: kikuchi-norihiro@mki.co.jp

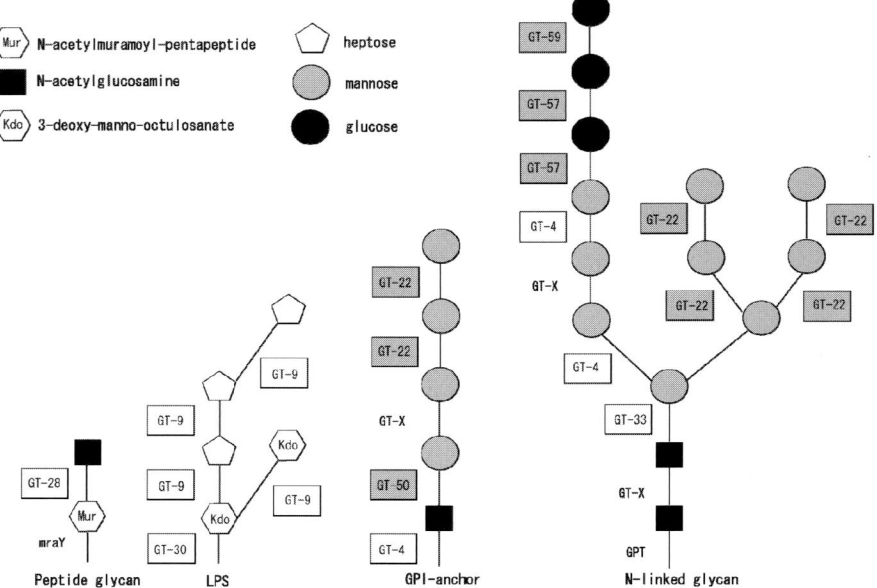

Fig. 1 Carbohydrate structures of peptide glycan, lipopolysaccharide (LPS), glycosylphosphatidylinositol anchor (GPI-anchor), and asparagines-linked glycan (N-glycan), together with the CAZy families involved in the synthesis. *Numbers in boxes* indicate the CAZy family number. The unknown family is represented by *GT-X*. Superfamilies GTS-B and -C are indicated in *white* and *gray boxes*, respectively. *mraY* represents phospho-N-acetylmuramoyl-pentapeptide-transferase and *GPT* represents GlcNAc-1-phosphate transferase. This figure is modified from *Biochem Biophys Res Commun*, 310, N. Kikuchi, Y.-D. Kwon, M. Gotoh, H. Narimatsu, Comparison of glycosyltransferase families using the profile hidden Markov model, 574-579, Copyright (2003), with permission from Elsevier

glycosyltransferases were present in the common ancestor of all modern life forms. Prior to the divergence of eukaryotes and prokaryotes, the initiation of glycosylation was catalyzed by a common ancestral sugar-1-phosphate transferase. Following initiation, elongation was catalyzed by an ancestor of the GTS-B family of enzymes. This indicates that the GTS-B family evolved first and is ancestral to other glycosyltransferase families. Although GTS-A and -B are present in both eukaryotes and prokaryotes, only GTS-B is involved in the synthesis of the root-side of glycans. The similarity between GTS-A and the nucleotidyltransferase family, which is involved in the synthesis of nucleotide sugars as sugar donors, suggests that the origin of GTS-A might be an enzyme that catalyzes the synthesis of this class of molecule. We speculate that an ancestral form of GTS-A supplied nucleotide sugars as donor substrates for the ancestral forms of nucleotide-1-phosphate transferase and GTS-B. GTS-A members have evolved into a large number of glycosyltransferases that catalyze the synthesis of diverse carbohydrate structures, such as the O-specific chains of LPS and the terminal structures of eukaryotic *N*- and *O*-glycans.

Most glycosyltransferases of the GTS-C superfamily are involved in the biosynthesis of dolichyl phosphate (Dol-P) glycan. The common ancestors of the GTS-C and GTS-A superfamilies and of the GTS-C and GTS-B superfamilies have yet to be determined. Thus, GTS-C may have first appeared during the evolution of eukaryotes.

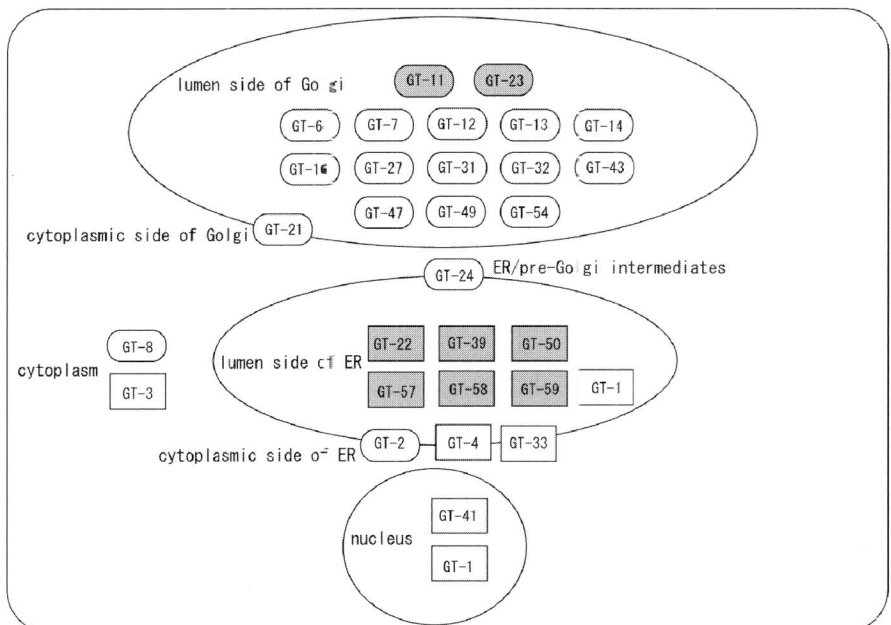

Fig. 2 Subcellular localizations of the superfamilies. *Numbers in boxes* indicate the CAZy family number. Superfamilies GTS-A, -B, -C, and -D are represented by *white ellipses, white rectangles, gray rectangles*, and *gray ellipses*, respectively. This figure is modified from *Biochem Biophys Res Commun*, 310, N. Kikuchi, Y.-D. Kwon, M. Gotoh, H. Narimatsu, Comparison of glycosyltransferase families using the profile hidden Markov model, 574-579, Copyright (2003), with permission from Elsevier

Some fucosyltransferases fall into the GTS-D superfamily. This might suggest that GTS-D is different from GTS-A and -B, although both GTS-A and -D are localized in the Golgi and are typical type II membrane proteins. Ihara et al. (2007) indicated that the catalytic region of FUT8, mammalian alpha1,6-fucosyltransferase, appeared to be similar to GT-B glycosyltransferases rather than GT-A. In eukaryotes, the glycosyltransferases of each superfamily show an extremely restricted pattern of subcellular localization (Fig. 2). Eukaryotic glycosyltransferases in each superfamily might have evolved from a common ancestor that functioned in a particular subcellular compartment.

The classification of glycosyltransferases into superfamilies has provided new insights into the evolution and function of these enzymes. Prior to 2003, the glycosyltransferase responsible for the synthesis of the second step glycan of Dol-P had not been cloned (Fig. 1). In our classification, this enzyme was predicted to be a member of the GTS-B family of enzymes. We used a profile HMM search to identify a candidate that was the product of two genes, one that encoded an N-terminal domain of the GT-B fold and the other that encoded a C-terminal domain (data not shown). In 2005, two enzymes homologous to the N- and C-terminal domains of a bacterial glycosyltransferase were reported to be responsible for the second step of dolichyl-linked oligosaccharide synthesis in *Saccharomyces cerevisiae* (Chantret et al. 2005).

Acknowledgment This work was supported by grants from the New Energy and Industrial Technology Development Organization (NEDO) of Japan.

References

Chantret I, Dancourt J, Barbat A, Moore SE (2005) Two proteins homologous to the N- and C-terminal domains of the bacterial glycosyltransferase Murg are required for the second step of dolichyl-linked oligosaccharide synthesis in *Saccharomyces cerevisiae*. J Biol Chem 280:9236–9242

Coutinho PM, Deleury E, Davies GJ, Henrissat B (2003) An evolving hierarchical family classification for glycosyltransferases. J Mol Biol 328:307–317

Ihara H, Ikeda Y, Toma S, Wang X, Suzuki T, Gu J, Miyoshi E, Tsukihara T, Honke K, Matsumoto A, Nakagawa A, Taniguchi N (2007) Crystal structure of mammalian {alpha}1,6-fucosyltransferase, FUT8. Glycobiology 17:455–466

Kikuchi N, Kwon Y-D, Gotoh M, Narimatsu H (2003) Comparison of glycosyltransferase families using the profile hidden Markov model. Biochem Biophys Res Commun 310:574–579

N-Acetylglucosaminyltransferases Involved in *N*-Glycan Biosynthesis

Kei-ichiro Inamori,[1] Naoyuki Taniguchi[2]

Introduction

We have succeeded in the purification and cDNA cloning of several enzymes involved in the biosynthesis of biologically important *N*-glycans of glycoproteins. These enzymes are responsible for the formation of branching structures of the core of *N*-glycans. To detect the activities with ease and sensitivity, we developed assay methods for these glycosyltransferases including *N*-acetylglucosaminyltransferases (GnTs) III, IV, V, VI, and IX using fluorescence-labeled oligosaccharides as acceptor substrates. The branching structures of *N*-glycans such as β1,6 GlcNAc branching and bisecting GlcNAc are formed by the actions of GnT-V and GnT-III, respectively. It has been reported that these *N*-glycans may play important roles in cell growth, cancer invasion, and metastasis. The assay methods for glycosyltransferases, including these enzymes involved in the branching formation of the core structure of *N*-glycan, are described (Fig. 1). The procedure described here is based on the methodology originally developed by Hase et al. (1978) and reported by our group earlier (Taniguchi et al. 1989). Each enzymatic product can be separated from the reaction mixture using reversed-phase HPLC. These methods allowed us to assay the activities with high sensitivity and substrate specificities of these glycosyltransferases.

Procedure

Acceptor Substrates

To prepare the fluorescence-labeled acceptor substrate for GnT-III, IV, V, and IX (Gn,Gn-bi-PA), the sialylglycopeptide from hen's egg yolk is subjected to hydrazinolysis, *N*-acetylation, and then pyridylamination, as described earlier by Hase et al. (1978). The resulting fluorescence-labeled sugar chains are digested sequentially with sialidase and β-galactosidase, and purified using a TSK-GEL ODS-80TM column (7.8 × 300 mm;

---

[1]Howard Hughes Medical Institute, Department of Physiology and Biophysics, Roy J. and Lucille A. Carver College of Medicine, University of Iowa, Iowa City, IA 52242, USA
Phone: +1-319-335-7823, Fax: +1-319-384-4793
E-mail: keiichiro-inamori@uiowa.edu

[2]Department of Disease Glycomics, Research Institute for Microbial Diseases, Osaka University, 2-1 Yamadaoka, Suita, Osaka 565-0871, Japan
Phone: +81-6-6879-4137, Fax: +81-6-6879-4137
E-mail: tani52@wd5.so-net.ne.jp

Fig. 1 Pathway for the synthesis of the core structure of complex-type *N*-glycans. From the biantennary structure as an initial substrate, various structures are formed by the actions of various *N*-acetylglucosaminyltransferases

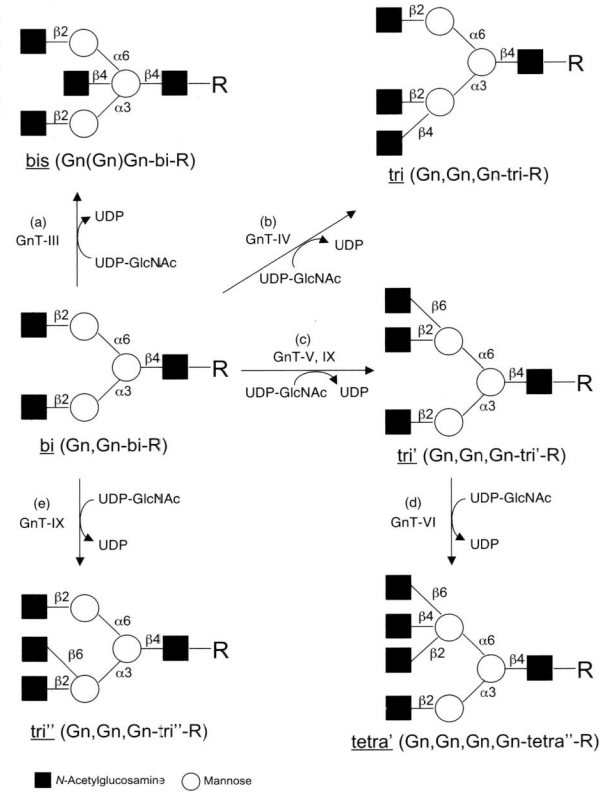

TOSOH) on a Shimadz LC-VP HPLC, which had been equilibrated with 0.1 M ammonium acetate (pH 4.0) containing 0.15% (v/v) *n*-butyl alcohol at a flow rate of 4.0 ml/min. Elution of the pyridylaminated sugar chain is performed with a linear concentration gradient from 0.15 to 0.45% of *n*-butyl alcohol for 30 min. Fluorescence is monitored with a fluorescence spectrophotometer (excitation 320 nm, emission 400 nm). Gn,Gn,Gn-tri′-PA (substrate for GnT-VI) is prepared by the hydrazinolysis of human α1-acid glycoprotein and labeled with 2-aminopyridine as described earlier (Taguchi et al. 2000).

Assay Premixtures

1. GnT-III assay premixture (2×): 250 mM MES (pH 6.5), 40 mM UDP-GlcNAc, 20 mM MnCl$_2$, 400 mM GlcNAc, and 1% Triton X-100 (w/v).
2. GnT-IV assay premixture (2×): 250 mM MOPS (pH 7.3), 80 mM UDP-GlcNAc, 15 mM MgCl$_2$, 400 mM GlcNAc, and 1% Triton X-100 (w/v).
3. GnT-V assay premixture (2×): 250 mM MES (pH 6.25), 80 mM UDP-GlcNAc, 20 mM EDTA, 400 mM GlcNAc, and 1% Triton X-100 (w/v).
4. GnT-VI assay premixture (2×): 250 mM Hepes (pH 8.0), 50 mM UDP-GlcNAc, 50 mM MgCl$_2$, 150 mM GlcNAc, and 1% Triton X-100 (w/v).
5. GnT-IX assay premixture (2×): 250 mM MOPS (pH 7.5), 80 mM UDP-GlcNAc, 20 mM EDTA, 400 mM GlcNAc, and 1% Triton X-100 (w/v).

Fig. 2 A typical elution pattern of the enzymatic products of GnT-III, GnT-IV, and GnT-V. A crude enzyme preparation from rat spleen was incubated in the reaction mixture, and an aliquot was applied to a TSK-GEL ODS-80TM column and eluted as described. The peak *S (bi)* represents the substrate, Gn,Gn-bi-PA. The peaks *tri'*, *tri*, and *bis* represent Gn,Gn,Gn-tri'-PA, Gn,Gn,Gn-tri-PA, and Gn(Gn)Gn-bi-PA, respectively

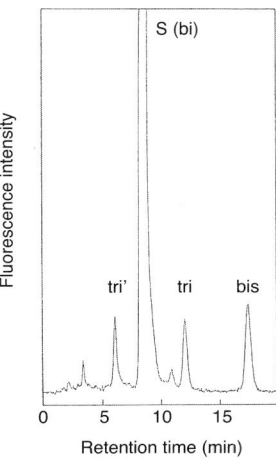

Preparation of Enzyme Extracts

To obtain the enzyme sources, various tissues or cells are homogenized in 10 mM Tris-HCl (pH 7.4) buffer containing 0.25 M sucrose. Following centrifugation at 900g for 10 min at 4°C, the supernatants are collected and used as a crude enzyme preparation.

Enzyme Reaction

1. Ten microliter of 3.85 mM Gn,Gn-bi-PA or 85 μM Gn,Gn,Gn-tri'-PA and 15 μl of enzyme solution are added to 25 μl of a 2× premixture solution, and the mixture is incubated at 37°C for 1 h or an appropriate time.

2. The reaction is stopped by heating at 100°C for 2 min, and then centrifuged at 15,000 rpm for 5 min.

3. An aliquot of the supernatant is subjected to an HPLC analysis. TSK-GEL ODS-80TM column (4.6 × 150 mm; TOSOH). Elution is performed at 50°C with 0.1 M ammonium acetate (pH 4.0) containing 0.1% *n*-butyl alcohol at a flow rate of 1.2 ml/min. Under these conditions, the products of GnT-V, GnT-IV, and GnT-III are eluted at 7, 12, and 17 min, respectively, as shown in Fig. 2. The products of GnT-VI and GnT-IX are separated as described elsewhere (Taguchi et al. 2000; Inamori et al. 2003).

References

Hase S, Ikenaka T, Matsushima Y (1978) Structure analyses of oligosaccharides by tagging of the reducing end sugars with a fluorescent compound. Biochem Biophys Res Commun 85:257–263

Inamori K, Endo T, Ide Y, Fujii S, Gu J, Honke K, Taniguchi N (2003) Molecular cloning and characterization of human GnT-IX, a novel β1,6-*N*-acetylglucosaminyltransferase that is specifically expressed in the brain. J Biol Chem 278:43102–43109

Taguchi T, Ogawa T, Inoue S, Inoue Y, Sakamoto Y, Korekane H, Taniguchi N (2000) Purification and characterization of UDP-GlcNAc: GlcNAcβ1-6(GlcNAcβ1-2)Manα1-R [GlcNAc to Man]-β1, 4-*N*-acetylglucosaminyltransferase VI from hen oviduct. J Biol Chem 275:32598–32602

Taniguchi N, Nishikawa A, Fujii S, Gu J (1989) Glycosyltransferase assays using pyridylaminated acceptors: *N*-acetylglucosaminyltransferase III, IV, and V. Methods Enzymol 179:397–408

UDP-*N*-Acetyl-D-Galactosamine:polypeptide *N*-Acetylgalactosaminyltransferases (ppGaNTases, pp-GalNAc-Ts, EC 2.4.1.41)

Kouichi Tachibana

Introduction

UDP-*N*-acetyl-D-galactosamine:polypeptide *N*-acetylgalactosaminyltransferases (ppGaNTases) are the glycosyltransferases that transfer GalNAc from the sugar donor UDP-GalNAc to serine and threonine residues on polypeptides (White et al. 1995; Ten Hagen et al. 2003). This is the initial step of mucin-type linkages (GalNAcα1-*O*-Ser/Thr); hence, this enzyme family is involved in the first step in mucin biosynthesis.

As most other glycosyltransferases, ppGaNTases are type II transmembrane proteins. Each ppGaNTase is composed of a short N-terminal cytoplasmic domain, a transmembrane domain, a stem domain, and a catalytic domain. In addition, a typical ppGaNTase has Ricin-like lectin domains at its C-terminal, which is specific for ppGaNTases (White et al. 1995).

The other characteristic of ppGaNTases is that they have a large number of family members (Ten Hagen et al. 2003; Chen et al. 2004). In human, there are 20 genes encoding a ppGaNTase or a ppGaNTase-like protein. These are almost one-tenth of all reported human glycosyltransferases. Although it is still unclear why ppGaNTases have so many members, possible explanations are (i) differences in tissue distributions and (ii) differences in catalytic specificities of ppGaNTases. Several ppGaNTases are specifically expressed in certain tissues, suggesting the tissue-specific function of such ppGaNTases. For catalytic specificity, each ppGaNTase shows a different preference for acceptor substrates.

One important point to note for enzyme assays of ppGaNTases is that the acceptor substrates for ppGaNTases are peptides, not sugars. To avoid proteolysis of acceptor substrates during enzyme reaction, purified ppGaNTases, not cellular lysates, are essential as enzyme sources. In addition, if synthetic peptides are used as acceptors, such peptides should be carefully chosen because of the preference of each ppGaNTase toward acceptor substrates.

Procedure

Production and Purification of Recombinant ppGaNTase

To obtain certain ppGaNTase for catalytic analysis, such ppGaNTase needs to be purified either by biochemical methods or by immunoprecipitation with specific antibodies.

Research Center for Medical Glycoscience, National Institute of Advanced Industrial Science and Technology (AIST), Umezono, Tsukuba, Ibaraki 305-8568, Japan
Phone: +81-29-861-3415, Fax: +81-29-861-3252
E-mail: kouichi-tachibana@aist.go.jp

Otherwise, recombinant ppGaNTase with an epitope peptide as an antibody tag is the easiest enzyme source to obtain. It is convenient to generate secretion-type recombinant ppGaNTase in culture cells. The procedure to generate such recombinant ppGaNTase is described below.

1. Obtain the cDNA encoding a ppGaNTase of your interest using RT-PCR or other methods.
2. Generate a ppGaNTase cDNA fragment that lacks N-terminal cytoplasmic and transmembrane domains using PCR or other methods, and subclone this cDNA fragment into an epitope-tagged secretion protein expression vector such as pFLAG-CMV-3 (Sigma) for the expression in mammalian cells, or BaculoDirect System (Invitrogen) for the expression with baculo virus.
3. Transfect the ppGaNTase expression vector in appropriate host cells, and collect culture supernatant after incubation for appropriate time.
4. Immunoprecipitate recombinant ppGaNTase with anti-epitope tag antibody. Analyze the amount of purified ppGaNTase by immunobloting with anti-tag antibody.

ppGaNTase Assay

There are two ways to analyze ppGaNTase activity. One method is to measure the incorporation of radiolabeled GalNAc into acceptor substrates. The other method is to detect the shift of the original peak in an HPLC profile as a result of the addition of GalNAcs to acceptor peptides. The latter method is described below.

Materials

1. Reaction buffer (final concentration): 25 mM Tris-HCl (pH 7.4), 5 mM MnCl, 0.2% Triton X-100, and 5 mM 2-mercaptoethanol
2. Donor substrate: UDP-GalNAc (Sigma)
3. Acceptor substrate: acceptor peptide labeled with 5-carboxyfluorescein succinimidyl ester (FAM) or the other fluorescent dye. Fluorescent dyes enable acceptor peptides for the retention on a ODS column and for the detection by a fluorescence detector in HPLC. There are several amino acid sequences frequently used in ppGaNTase assay. Most of them are derived from repeating units of mucins.
4. For HPLC: trifluoroacetic acid (TFA) and acetonitrile

Enzyme Reaction

1. Mix ppGaNTase in a total volume of 20 µl containing 0.25 mM UDP-GalNAc, 50 pmol acceptor substrate, and the reaction buffer.
2. Incubate reaction mixture at 37°C with agitation for an appropriate time.
3. Following incubation, the reaction is terminated by boiling reaction mixture for 3 min.
4. Add 80 µl distilled water to the boiled reaction mixture and centrifuge in a spin column (Ultrafree-MC 0.22 µm, Millipore) at 6,000 rpm for 5 min at 4°C to remove debris. Take filtrated solution for HPLC analysis.

HPLC Analysis

1. Set up a high performance liquid chromatography (HPLC, Shimadzu, Kyoto, Japan) equipped with a reverse C18 column (5C18-AR, 4.6 × 250 mm) and a fluorescence detector (model RF-10AXL, Shimadzu).
2. Equilibrate the column with 0.05% TFA at 40°C.
3. Apply 20 µl of the filtrated reaction mixture to HPLC.
4. Elute reaction products with a gradient (0–50%) of acetonitrile containing 0.05% TFA at a flow rate of 1.0 ml/min at 40°C.
5. Monitor the fluorescence of column elute with a fluorescence detector.

Further Analysis

1. To confirm that shifted peaks in HPLC represent GalNAc-attached products, products are fractionated and isolated by HPLC, and then subjected to mass spectrometry analysis to determine the molecular weights of the isolated products.
2. To determine the amino acid residue to which a GalNAc attached, products are isolated by HPLC, and then subjected to peptide sequencer analysis.

References

Cheng L, Tachibana K, Iwasaki H, Kameyama A, Zhang Y, Kubota T, Hiruma T, Tachibana K, Kudo T, Guo J-M, Narimatsu H (2004) Characterization of a novel human UDP-GalNAc transferase, pp-GalNAc-T15. FEBS Lett 566:17–24

Ten Hagen KG, Fritz TA, Tabak LA (2003) All in the family: the UDP-GalNAc:polypeptide N-acetylgalactosaminyltransferases. Glycobiology 13:1R–16R

White T, Bennett EP, Takio K, Sorensen T, Bonding N, Clausen H (1995) Purification and cDNA cloning of a human UDP-N-acetyl-α-D-galactosamine:polypeptide N-acetylgalactosaminyltransferas. J Biol Chem 270:24156–24165

Glycosyltransferase Family with β4GT Motif (β4Gal-T and β4GalNAc-T Family)

Takashi Sato, Hisashi Narimatsu

Introduction

Glycosyltransferase genes can be classified into several families, which have several characteristic motifs in the amino acid sequence. This section briefly describes a glycosyltransferase family having β4GT motif (WGxEDD/V/W). This family contains seven β1,4-galactosyltransferases, six chondroitin sulfate synthases (see a separate section for details), and two β1,4-N-acetylgalactosaminyltransferases. These enzymes have an activity to transfer carbohydrates via β1,4 linkage irrespective of donor and acceptor carbohydrates. The members of β4GT family in human and mouse are summarized in Table 1.

β1,4-Galactosyltransferase (β4Gal-T)

β4Gal-T1 is the first mammalian glycosyltransferase gene cloned in 1986. This is a unique enzyme having dual activities, which are galactosyltransferase activity to transfer Gal from the donor substrate, UDP-Gal, to the acceptor substrate, GlcNAcβ-, and lactose synthetase activity to synthesize lactose (Galβ1,4Glc) in cooperation with lactalbumin, and reaction mechanisms were clarified in detail by X-ray crystallography (Gastinel et al. 1999). The biological function of β4Gal-T1 has been studied using knockout mice. The β4Gal-T1$^{-/-}$ mice show growth retardation, reduced inflammatory responses and disease-like phenotypes of IgA nephropathy (Asano et al. 1997). With the completion of databases, homologous sequences to β4Gal-T1 were found in succession, where seven β4Gal-T genes had been reported in mammals to date (Taniguchi 2002). Similar to β4Gal-T1, β4Gal-T2, -T3, -T4, and -T5 synthesize Galβ1,4GlcNAc-, but the in vitro analysis revealed that these enzymes exhibited different specificity of acceptor substrate each other (Ito et al. 2007). β4Gal-T2 shows the similar substrate specificity to that of β4GalT1, but β4Gal-T3 prefers a glycolipid, Lc3Cer as a substrate. Since β4Gal-T4 uses GlcNAc-6-sulfate, a constituent of keratan sulfate, as a substrate, it is considered to be a keratan sulfate synthase. β4Gal-T6 uses Glc-Cer as an acceptor substrate and shows Lactosyl-Cer synthase activity. β4Gal-T7 uses Xyl-Ser as an acceptor substrate and synthesizes Galβ1,4Xyl-Ser which exists at the linkage region between chondroitin sulfate or heparan sulfate and core protein of proteoglycans. Furthermore, β4Gal-T7 was reported to be a responsible gene for Ehler–Danlos syndrome by Okajima et al. (1999). The above is a brief introduction about substrate specificity of enzymes in β4Gal-T family, and further studies are awaited to clarify differences in function and detail substrate specificity of these enzymes.

Research Center for Medical Glycoscience, National Institute of Advanced Industrial Science and Technology (AIST), 1-1-1 Umezono, Tsukuba, Ibaraki 305-8568, Japan
E-mail: takashi-sato@aist.go.jp

Table 1 β4GT family genes cloned to date

Gene	Human symbol	Other alias	CAZy	Chromosome	AccID (RefSeq)	Mouse symbol	Chromosome	AccID (RefSeq)	Glycan structure synthesized
β4Gal-T1	B4GALT1	GT1, GTB, GGTB2, B4GAL-T1, MGC50983, beta4Gal-T1	GT31	9p13	NM_001497	B4galt1	4	NM_022305	Galβ1-4GlcNAc
β4Gal-T2	B4GALT2	B4Gal-T2, beta4Gal-T2	GT7	1p34-p33	NM_001005417	B4galt2	4	NM_017377	Galβ1-4GlcNAc
β4Gal-T3	B4GALT3	beta4Gal-T3	GT7	1q21-q23	NM_003779	B4galt3	1	NM_020579	Galβ1-4GlcNAc
β4Gal-T4	B4GALT4	B4Gal-T4, beta4Gal-T4	GT7	3q13.3	NM_003778	B4galt4	16	NM_019804	Galβ1-4GlcNAc
β4Gal-T5	B4GALT5	gt-V, B4Gal-T5, beta4Gal-T5, beta4GalT-V	GT7	20q13.1–q13.2	NM_004776	B4galt5	2	NM_019835	Galβ1-4GlcNAc
β4Gal-T6	B4GALT6	beta3GalT6	GT7	18q11	NM_004775	B4galt6	18	NM_019737	Galβ1-4Glc
β4Gal-T7	B4GALT7	XGPT1, XGALT1, XGALT-1, B4GALT7, beta4Gal-T7	GT7	5q35.2–q35.3	NM_007255	B4galt7	13	NM_146045	Galβ1-4Xyl
β4GalNAc-T3	B4GALNT3	FLJ16224, FLJ40362, B4GalNac-T3	Not classified	12p13.33	NM_173593	B4galnt3	6	NM_198884	GalNAcβ1-4GlcNAc
β4GalNAc-T4	B4GALNT4	FLJ25045, NGalNAc-T1, Beta4GalNAc-T4	Not classified	11p15.5	NM_178537	B4galnt4	7	NM_177897	GalNAcβ1-4GlcNAc
CSGalNAc-T1	ChGn	FLJ11264, beta4GalNAcT	GT31	8p21.3	NM_018371	4732435N03Rik	8	NM_172753	Chondroitin
CSGalNAc-T2	GALNACT-2	PRO0082, MGC40204, CSGalNAcT-2, DKFZp686H13226	GT31	10q11.21	NM_018590	Galnact2	6	NM_030165	Chondroitin
ChSy	CHSY1	CSS1, KIAA0990	GT31	15q26.3	NM_014918	Chsy1	7	NM_001081163	Chondroitin
ChPF (CSS2)	CHPF	CSS2, FLJ22678	GT31	2q35	NM_024536	D1Bwg1363e	1	NM_001001566	Chondroitin
CSS3	CSS3	CHSY2	GT31	5q23.3	NM_175856	4833446K15Rik	18	NM_001081328	Chondroitin
CSGlcAT	CSGlcA-T	CSGlcAT	GT31	7q36.1	NM_019015	Not identified in MGI	5	BAC65786	Chondroitin

β1,4-Galactosyltransferase Assay

Enzyme Preparation

In this section, we describe the enzyme expression using mammalian cells; however, β4Gal-T1 can obtain as an active enzyme in bacterial expression systems.

Materials

pFLAG-CMV3 (Sigma),
  293T cells (available from ATCC), maintained in DMEM containing 10% FCS, penicillin, and streptomycin,
  LipofectAMINE 2000 (Invitrogen),
  Opti-MEM® I (Invitrogen),
  Anti-FLAG M2 Agarose Affinity Gel (Sigma-Aldrich),
  50 mM Tris-buffered saline (50 mM Tris–HCl, pH 7.4, and 150 mM NaCl).
  Recombinant enzymes of β4Gal-T1, -T2, -T3, -T4, -T5, -T6, and –T7 could be expressed in 293T cells as active soluble enzymes. A catalytic domain of β4Gal-Ts without transmembrane domain was amplified by PCR and DNA fragment was inserted into expression vector pFLAG-CMV3 (Sigma). Transfection procedures were described below.

1. One day before transfection, seed 2.0×10^6 of 293T cells in 10 ml of DMEM containing 10% FCS without antibiotics on 10 cm dish. Cells will be near confluent on the next day.
2. Dilute 30 μg of DNA in 1.5 ml of Opti-MEM®I and mix gently.
3. Dilute 30 μl of Lipofectamine 2000 in 1.5 ml of Opti-MEM®I and mix gently. Incubate the mixture for 5 min at room temperature.
4. Mix the diluted DNA solution with diluted Lipofectamine 2000 solution (total volume = 3 ml). Mix gently and incubate for 20 min at room temperature.
5. Drop the mixed solution on confluent 293T cells. Mix gently and incubate for 48–72 h at 37°C in a CO_2 incubator.
6. Recover the culture media and remove the cell debris by centrifugation. Mix with anti-FLAG M2 agarose affinity gel and incubate for O/N at 4°C with gentle agitation by rotary shaker.
7. Wash the enzyme–agarose gel mixtures three times with 50 mM Tris-buffered saline.
8. The enzyme–agarose gel mixtures can be suspended for enzyme reaction buffer and used for enzymatic glycan synthesis as enzyme sources.

β1,4-Galactosyltransferase Assay

1. The reaction mixture for the β1,4-galactosyltransferase assay contained the following components in a final volume of 20 μl.
   20 mM HEPES buffer (pH 7.0)
   10 mM $MnCl_2$
   2.5 μM UDP-Gal
   Various accepter substrates
   5 μl of enzyme suspension

2. Incubate at 37°C for various periods
3. Stop the reaction by heating at 100°C for 3 min
4. Centrifuge at 15,000 rpm for 5 min at 4°C and recover the supernatant.
5. Apply the supernatant to HPLC systems on the suitable columns to separate substrate and product.

β1,4-N-acetylgalactosaminyltransferase

The SO_4-4GalNAcβ1,4GlcNAcβ-structure was found on N-linked glycans of glycoprotein hormones secreted from the pituitary gland including luteinizing hormone (LH) and thyroid stimulating hormone (TSH). This structure has been reported to be involved in the metabolism of glycoprotein hormones recognized by R-type lectin domain of macrophage mannose receptor in the liver (Roseman and Baenziger 2000). Recently, two genes encoding β4GalNAc-T3 (Sato et al. 2003) and β4GalNAc-T4 (Gotoh et al. 2004), which exhibit β1,4 GalNAc-transferase activities and synthesize GalNAc β1,4GlcNAcβ1- (LacdiNAc, LDN) on both N- and O-glycans were cloned. β4GalNAc-T3 and β4GalNAc-T4 consist of 999 and 1,039 amino acid residues, respectively, and the predicted catalytic domain would lie in C-terminal region containing a β4GT motif and a DLH motif, which appeared to have a function similar to that of DXD. β4GalNAc-T3 and -T4 differed greatly in expression tissues except pituitary gland, in which both enzymes were detected by real-time PCR. β4GalNAc-T3 was expressed in the stomach and colon, whereas β4GalNAc-T4 was highly expressed in the brain and ovary suggesting that they would synthesize LDN structures on different carrier proteins in different tissues. An increasing number of glycoprotein carrying LDN structure has been found in human bodies (Fiete 2007) and the recognizing molecule for glycan structures containing LDN, which is mammalian lectins, i.e., DC-sign and MGL had been reported. The structure of LDN could be modified by sulfotransferases, α2,6 sialyltransferases and α1,3 fucosyltransferases which synthesize SO_4-4GalNAcβ1,4GlcNAc, Siaα2,6GalNAcβ1,4GlcNAc, and GalNAcβ1,4(Fucα1,3)GlcNAc, respectively. The differences of such glycan structures would be expected to give the differences of biological functions through the specific recognition by mammalian lectins. These would facilitate the discovery of the novel biological functions of LDN in the near future.

β1,4-N-acetylgalactosaminyltransferase Assay

Enzyme Preparation

Materials and procedures were described above.
Recombinant enzymes of β4GalNAc-T3 and β4GalNAc-T4 could be expressed in 293T cells as active soluble enzymes.

β1,4-N-acetylgalactosaminyltransferase Assay

1. The reaction mixture for the β1,4-N-acetylgalactosaminyltransferase assay contained the following components in a final volume of 20 μl.
   50 mM MES buffer (pH 6.5)
   0.1% triton X-100
   10 mM $MnCl_2$
   1 mM UDP-GalNAc

Various accepter substrates*
5 μl of enzyme suspension

*Following monosaccharide and oligosaccharides can be used for accepter substrates: GlcNAcβ-Bz, GlcNAcβ1,6(Galβ1,3)GalNAcα-pNp (core 2), GlcNAcβ1,3GalNAcα-pNp (core 3), GlcNAcβ1,6GalNAcα-pNp (core 6) (Calbiochem), biantenary structure of N-glycan with GlcNAc at non-reducing terminal and its derivative (TaKaRa).

2. Incubate at 37°C for various periods
3. Stop the reaction by heating at 100°C for 3 min.
4. Centrifuge at 15,000 rpm for 5 min at 4°C and recover the supernatant.
5. Apply the supernatant to HPLC systems on the suitable columns to separate substrate and product.

References

Asano M, Furukawa K, Kido M, Matsumoto S, Umesaki Y, Kochibe N, Iwakura Y (1997) Growth retardation and early death of beta-1,4-galactosyltransferase knockout mice with augmented proliferation and abnormal differentiation of epithelial cells. Embo J 16(8):1850–1857

Fiete D, Mi Y, Oats EL, Beranek MC, Baenziger JU (2007) N-linked oligosaccharides on the low density lipoprotein receptor homolog SorLA/LR11 are modified with terminal GalNAc-4-SO4 in kidney and brain. J Biol Chem 282(3):1873–1881

Gastinel LN, Cambillau C, Bourne Y (1999) Crystal structures of the bovine beta4galactosyltransferase catalytic domain and its complex with uridine diphosphogalactose. Embo J 18(13):3546–3557

Gotoh M, Sato T, Kiyohara K, Kameyama A, Kikuchi N, Kwon YD, Ishizuka Y, Iwai T, Nakanishi H, Narimatsu H (2004) Molecular cloning and characterization of beta1,4-N-acetylgalactosaminyltransferases IV synthesizing N,N'-diacetyllactosediamine. FEBS Lett 562(1–3):134–140

Ito H, Kameyama A, Sato T, Sukegawa M, Ishida H, Narimatsu H (2007) Strategy for the fine characterization of glycosyltransferase specificity using isotopomer assembly. Nat Methods 4(7):577–582

Okajima T, Fukumoto S, Furukawa K, Urano T (1999) Molecular basis for the progeroid variant of Ehlers-Danlos syndrome. Identification and characterization of two mutations in galactosyltransferase I gene. J Biol Chem 274(41):23841–28844

Roseman DS, Baenziger JU (2000) Molecular basis of lutropin recognition by the mannose/GalNAc-4-SO4 receptor. Proc Natl Acad Sci USA 97(18):9949–9954

Sato T, Gotoh M, Kiyohara K, Kameyama A, Kubota T, Kikuchi N, Ishizuka Y, Iwasaki H, Togayachi A, Kudo T, Ohkura T, Nakanishi H, Narimatsu H (2003) Molecular cloning and characterization of a novel human beta 1,4-N-acetylgalactosaminyltransferase, beta 4GalNAc-T3, responsible for the synthesis of N,N'-diacetyllactosediamine, galNAc beta 1-4GlcNAc. J Biol Chem 278(48):47534–47544

Taniguchi N, Honke K, Fukuda M (2002) Handbook of Glycosyltransferases and Related Genes With contributions by numerous experts Springer, Berlin

β1,3-glycosyltransferase Gene Family and IGnT Gene Family

Akira Togayachi, Takashi Sato, Hisashi Narimatsu

Introduction

This section mainly reviews β1,3-glycosyltransferases (β3GT), which transfer donor substrate sugars via a β1,3-linkage, and IGnTs, which transfer donor substrate sugars via a β1,6-linkage (Hennet 2002; Narimatsu 2004; Togayachi 2006). Information such as accession numbers of these β3GTs and IGnTs are shown in Table 1.

β1,3-glycosyltransferase Family

β1,3-galactosyltransferases (β3Gal-Ts)

These enzymes are involved in the synthesis of the following important carbohydrate chains: type 1 carbohydrate antigens, cancer-associated antigens such as CA19-9 (sialyl Lewis a), fundamental glycan structure such as polylactosamine structure, and glycolipid antigens. β1,3-glycosyltransferases cloned and reported to date form a gene family, which shares a common motif (termed as β3GT motif).

The β3Gal-T1 gene was first isolated using an expression cloning method. With the development of a database, β3Gal-T2 to β3Gal-T4 genes were reported to have high homology to the β3Gal-T1 gene. The activity of β3Gal-T3 was not clarified in the first report, but later it was found to be identical to that of β3GalNAc-T which was cloned as a globoside synthase (newly termed as β3GalNAc-T1). β3Gal-T4 was found to be identical to $GD_{1b}/GM_1/GA_1$ synthase. β3Gal-T5 synthesizes type 1 antigens, which result in the expression of a cancer-associated carbohydrate antigen, CA19-9 (sialyl Lewis a antigen). β3Gal-T5 also has an activity to transfer Gal-residue to core 3 O-glycan as well as stage specific embryonic antigen-3 (SSEA-3; Galβ1-3GalNAcβ1-3Galβ1-4Galβ1-4Glcβ1-1Cer) synthase activity. β3Gal-T6 was reported to be an enzyme which synthesizes the core carbohydrate structure of glycosaminoglycans (Galβ1-3Galβ1-4Xyl).

β1,3-N-acetylglucosaminyltransferases (β3Gn-Ts)

The first β3Gn-T, iGn-T(β3Gn-T1), was isolated by the expression cloning method. The secondary β3Gn-T, β3Gn-T2, was isolated according to structural similarity with β3Gal-T family. Thereafter, two β3Gn-Ts, β3Gn-T3 and -T4, were also subsequently isolated according to structural similarity with the β3Gal-T family. β3Gn-T1 has low sequence homology to other β3Gn-Ts. β3Gn-T2 appeared to have potent in vitro activity on oligosaccharide substrate of polylactosamine structures, indicating that it is a major polylactosamine synthase. β3Gn-T3 is the enzyme which elongates polylactosamine

---

Research Center for Medical Glycoscience (RCMG), National Institute of Advanced Industrial Science and Technology (AIST), OSL Central 2, 1-1-1 Umezono, Tsukuba, Ibaraki 305-8568, Japan
E-mail: a.togayachi@aist.go.jp

Table 1 Official gene symbols and their representative accession number

Enzyme	Official symbols	Previous symbols (Alias)	Representative accession number Nucleotide	Representative accession number Protein	Entrez gene ID	Chromosome	CAZy (http://www.cazy.org)
β1,3-Glucosyltransferase (β3GlcT)							
β1,3-Glucosyltransferase	B3GALTL	Also known as B3GTL; B3Glc-T	NM_194318.2 AB101481	NP_919299.2	145173	13q12.3	GT31
β1,3-N-acetylglucosaminyltransferase (β3GnT)							
β1,3-N-acetylglucosaminyltransferase 1	B3GNT1	iGnT	NM_006876 AF029893	NP_006867	11041	11q13.2	GT49
β1,3-N-acetylglucosaminyltransferase 2	B3GNT2	B3GNT1, B3GNT2	NM_006577 AB049584, AF049584	NP_006568	10678	2p15	GT31
β1,3-N-acetylglucosaminyltransferase 3	B3GNT3		NM_014256 AB049585, AB015630	NP_055071	10331	19p13.1	GT31
β1,3-N-acetylglucosaminyltransferase 4	B3GNT4		NM_030765 AB049586, AB049586	NP_110392	79369	12q24	GT31
β1,3-N-acetylglucosaminyltransferase 5	B3GNT5		NM_032047 AB045278	NP_114436	84002	3q28	GT31
β1,3-N-acetylglucosaminyltransferase 6	B3GNT6		NM_138706 AB073740	NP_619651	192134	11q13.4	GT31
β1,3-N-acetylglucosaminyltransferase 7	B3GNT7		NM_145236 AK000770	NP_660279	93010	2q36.1	GT31
β1,3-N-acetylglucosaminyltransferase 8	B3GNT8		NM_198540 AB175895, AY277592	BAD86525	374907	19q13	GT31
Lunatic fringe	LFNG		NM_001040167.1 NM_001040168.1	NP_001035257.1 NP_001035258.1	3955	7p22.2	GT31

Table 1 Continued

Enzyme	Official symbols	Previous symbols (Alias)	Representative accession number Nucleotide	Representative accession number Protein	Entrez gene ID	Chromosome	CAZy (http://www.cazy.org)
Manic fringe	MFNG		NM_002405.2	NP_002396.2	4242	22q12	GT31
Radical fringe	RFNG		NM_002917.1	NP_002908.1	5986	17q25	GT31
β1,3-Galactosyltransferase (β3GalT) (except core1 β3GalT)							
β1,3-Galactosyltransferase 1	B3GALT1		NM_020981 E07739	NP_066191 NP_003772	8708	2q31.1	GT31
β1,3-Galactosyltransferase 2	B3GALT2		NM_003783 Y15060	NP_003774	8707	1q31	GT31
β1,3-Galactosyltransferase 4	B3GALT4		NM_003782 Y15061	NP_003773	8705	6p21.3	GT31
β1,3-Galactosyltransferase 5	B3GALT5		NM_033171 AB020337	NP_006048	10317	21q22.3	GT31
β1,3-Galactosyltransferase 6	B3GALT6		NM_080605 AY050570	NP_542172	126792	1p36.33	GT31
β1,3-N-acetylgalactosaminyltransferase (β3GalNAcT)							
β1,3-N-acetylgalactosaminyltransferase 1	B3GALNT1	B3GALT3	NM_003781 Y15062	NP_149357.1	8706	3q25	GT31
β1,3-N-acetylgalactosaminyltransferase 2	B3GALNT2		NM_152490 BC029564	NP_689703	148789	1q42.3	GT31
IGnT (β1,6-N-acetylglucosaminyltransferase: β6GnT)							
IGNT1	GCNT2	GCNT2 isoform B	NM_001491 L41605	NP_001482	2651	6p24.2	GT14
IGNT2	GCNT2	GCNT2 isoform A	NM_145649 AB078432	NP_663624	2651	6p24.2	GT14
IGNT3	GCNT2	GCNT2 isoform C	NM_145655 AB078433	NP_663630	2651	6p24.2	GT14

chains on the core 1 O-glycans. β3Gn-T4 also has weak polylactosamine synthase activity. Differential roles of these polylactosamine synthetic enzymes have not been clarified yet. β3Gn-T5, an Lc$_3$Cer synthase, is considered to play a key role in the synthesis of lacto- or neolacto-series carbohydrate chains of glycolipids. β3Gn-T6 is an enzyme which synthesizes the core 3 O-glycans, and it is considered to play an important role in the synthesis and function of mucin-type O-glycans in digestive organs. β3Gn-T7 is known as the enzyme which has the activity to transfer GlcNAc to Galβ1-4(SO$_3^-$-6)GlnNAcβ1-3Galβ1-4(SO$_3^-$-6)GlcNAc (L$_2$L$_2$ oligosaccharide) of keratan sulfate. β3Gn-T8 exhibited activity on tetraantennary N-glycans and elongated polylactosamine chains. It has been reported that the complex of β3Gn-T2 and β3Gn-T8 exhibits potent polylactosamine synthesis activity against tetraantennary N-glycans.

β1,3-N-acetylgalactosaminyltransferases (β3GalNAc-Ts)

As described above, β3GalNAc-T (old β3Gal-T3) had been cloned as globoside (Gb$_4$) synthase by expression cloning. β3GalNAc-T2 synthesizes the GalNAcβ1-3GlcNAc structure. Although the structure formed by this enzyme has not been found in mammals (the structure was found in glycolipids of insects), the discovery of the enzyme suggests the existence of this structure in mammalian.

β1,3-glucosyltransferases (β3Glc-Ts)

Besides protein O-linked fucosylglycan found on the EGF domain in Notch protein, another unique disaccharide structure of O-fucosylglycan, Glcβ1,3Fucα1-Ser/Thr, was detected in human urine and the thrombospondin 1 (TSP1) protein. β1,3-glucosyltransferase (β3Glc-T) transfers glucose to O-linked fucosylglycan on thrombospondin type 1 repeat (TSR) domain of thrombospondin.

IGnT (β1,6-N-Acetylglucosaminyltransferase) Family

The Ii antigens are a kind of blood group antigen (see Chapter by H. Narimatsu, this volume). The determinants were demonstrated to have repeated lactosamine structures (Galβ1-4GlcNAcβ1-3Galβ1-4GlcNAcβ1-) for i antigen and branched structures (Galβ1-4GlcNAcβ1-(Galβ1-4GlcNAcβ1-6)3Galβ1-4GlcNAcβ1-) for I antigen. The gene (IGnT1) encoding the enzyme which catalyzes a β1-6 branched glycan was isolated. Later, additional two isoforms, IGnT2 and IGnT3, were isolated by analyzing a genome database with sequence of IGnT1 gene.

Method for Glycosyltransferase Reaction

Enzyme Preparation (Expression of Recombinant Enzymes)

For the preparation of recombinant enzymes, glycosyltransferase genes are engineered for heterologous expression in mammalian or insect cells as a fusion protein with various protein-tags such as FLAG, c-Myc, and His at the N-terminus of polypeptide (see 2 Chapters by T. Sato and H. Narimatsu, this volume). To obtain active soluble ezymes, the catalytic domain of each glycosyltransferase gene, without the transmembrane domain, is subcloned into expression vectors such as pFLAG-CMV vector (SIGMA) of the mammalian expression vector, pVL1393 (PharMingen) for the baculovirus expression system or Gateway system (Invitrogen). Expression in mammalian cells is a desirable method for obtaining active enzymes. In both expression systems, each enzyme is purified from a culture medium using anti-tag antibodies. Enzymatic reactions of all glycosyltransferases are carried out using the suspension of recombinant proteins as an enzyme source.

Substrates for Glycosyltransferase Assays

1. For donor substrates, UDP-glucose (Glc), UDP-*N*-acetylglucosamine (GlcNAc), UDP-galactose (Gal), UDP-*N*-acetylgalactosamine (GalNAc) (SIGMA (MO, USA), Calbiochem (Merck, La Jolla, CA, USA), GE Healthcare Biosicences (England), American Radiolabeled Chemicals, Inc. (MO, USA), etc.) are utilized.

2. For acceptor substrates, the various acceptor substrates, such as monosaccharides, oligosaccharides, glycolipids, glycopeptides, and glycoproteins, were purchased from Calbiochem (Merck, La Jolla, CA, USA), Toronto Research Chemicals Inc. (Ontario, Canada), Seikagaku Kogyo (Tokyo, Japan), TaKaRa (Okaka, Japan), Glycotech (MD, USA), SIGMA, etc.

Reaction mixture

1. Glycosyltransferase reactions are performed in 20 µl of reaction mixture which contains the following components.
   Soluble (or immobilized to resin) enzyme suspension
   10 nmol of each of acceptor mixture
   2.5~700 µM of various donor substrates (radioactive donor substrates for detection by TLC and scintillation counter)
   10 mM $MnCl_2$
   Reaction is optimized in various buffer systems, i.e., MES, HEPES, Tris-HCl, and Na-cacodylate buffer at different pH values, usually pH 7.4.
2. Incubation at 37°C for various periods
3. Stop the reaction by heating at 100°C for 3 min.
4. Centrifuge at 15,000 rpm for 5 min at 4°C and recover the supernatant.
5. Analyze substrate and product by the methods described below.

General Methods for Detection of Reaction Products

Because the glycosyltransferase activity for acceptor substrates was assayed in various reaction mixtures containing donor substrates as described above, the suitable system should be selected to separate substrates and products. Glycosyltransferase reaction products are identified by the following methods.

1. High-performance liquid chromatography (HPLC): various substrates, such as oligosaccharides, glycopeptides, glycoproteins, and glycolipids
2. Thin layer chromatography (TLC): oligosaccharides, especially glycolipids
3. Mass spectrometry (MS): various substrates, oligosaccharides, glycopeptides, and glycolipids
4. Scintillation counter: various products from radioisotope labeled donor substrates
5. Lectin microarray: oligosaccharides, glycopeptides, and glycoproteins

6. SDS-PAGE: glycopeptides and glycoproteins
7. Sugar-chip(Glycan microarray): use as acceptor substrates.

References

Hennet T (2002) The galactosyltransferase family. Cell Mol Life Sci 59:1081–1095
Narimatsu H (2004) Construction of a human glycogene library and comprehensive functional analysis. Glycoconj J 21:17–24
Togayachi A, Sato T, Narimatsu H (2006) Comprehensive enzymatic characterization of glycosyltransferases with a beta3GT or beta4GT motif. Methods Enzymol 416:91–102

O-Mannosyltransferase and POMGnT1

Tamao Endo, Hiroshi Manya

Introduction

O-Mannosylation is important in muscle and brain development. We previously found that the glycans of α-dystroglycan (α-DG) predominantly include *O*-mannosyl glycans, Siaα2-3Galβ1-4GlcNAcβ1-2Man (Chiba et al. 1997), and then we reported that defects in *O*-mannosyl glycan cause a type of muscular dystrophy. We have identified and characterized glycosyltransferases; protein *O*-mannose β1,2-*N*-acetylglucosaminyltransferase (POMGnT1) (Yoshida et al. 2001) and protein *O*-mannosyltransferase 1 (POMT1) and its homolog POMT2 (Manya et al. 2004) are involved in *O*-mannosyl glycan synthesis. This protocol describes assay methods for the mammalian POMT and POMGnT1.

Methods

The POMT activity is based on the amount of [^{3}H]-mannose transferred from Dol-P-Man to GST-α-DG. The reaction product is purified with a glutathione-Sepharose column and radioactivity of mannosyl GST-α-DG is measured. The POMGnT1 activity is based on the amount of [^{3}H]GlcNAc transferred from UDP-GlcNAc to benzyl-α-mannose (Benzyl-Man) or mannosylpeptide [Ac-Ala-Ala-Pro-Thr(Man)-Pro-Val-Ala-Ala-Pro-NH$_2$]. The reaction product is purified with HPLC and radioactivity is measured. The mannosylpeptide is not commercially available but it is possible to use Benzyl-Man, which is commercially available, as a substitute.

POMGnT1 and POMT activities are detected in various mammalian cells and tissues. Here we describe the methods that use microsomal membrane fraction of rat brain and human embryonic kidney 293T (HEK293T) cells as an enzyme source. Furthermore, to demonstrate that gene products of *POMGnT1*, *POMT1*, and *POMT2* have enzymatic activity, the cells transfected with *POMGnT1* or *POMT1* and *POMT2* are used.

POMT Assay

1. The POMT reaction buffer [10 mM Tris-HCl, pH 8.0, 2 mM 2-mercaptoethanol, 10 mM EDTA, 0.5% *n*-octyl-β-D-thioglucoside (DOJINDO LABORATORIES)] is added to the microsomal membrane fraction at a protein concentration of 4 mg/ml. The fraction

---

Glycobiology Research Group, Tokyo Metropolitan Institute of Gerontology, Foundation for Research on Aging and Promotion of Human Welfare, 35-2 Sakae-cho, Itabashi-ku, Tokyo 173-0015, Japan
Phone: +81-3-3964-3241, Fax: +81-3-3579-4776
E-mail: endo@tmig.or.jp

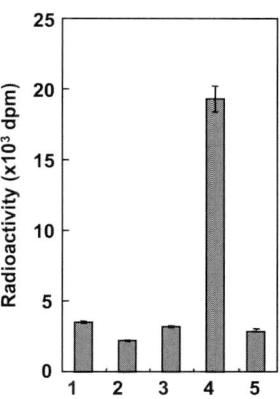

Fig. 1 POMT activity of human POMT1 and POMT2 expressed in HEK293T cells. *Lane 1* Cells transfected with vector alone; *lane 2* cells transfected with human *POMT1*; *lane 3* cells transfected with human *POMT2*; *lane 4* cells co-transfected with *POMT1* and *POMT2*; *lane 5* a mixture of the membrane fractions from the *POMT1*-transfected cells and *POMT2*-transfected cells. Reprinted with permission from Manya et al. (2004). Copyright National Academy of Sciences, USA

is suspended by moderate pipetting and solubilized for 30 min on ice with mild stirring occasionally.

2. Twenty microliter of solubilized fraction and 2 µl of Dol-P-Man solution [Mannosylphosphoryldolichol 95 (Mannose-6-^{3}H) (1.48–2.22 TBq/mmol, American Radiolabeled Chemical, St. Louis, MO, USA). Adjust to 40,000 cpm/µl in 20 mM Tris-HCl (pH 8.0), 0.5% CHAPS] are added to the dried GST-α-DG (10 µg in microcentrifugal tubes), vortexed and spun down gently. Immediately, the reaction mixture is incubated at 25°C for 1 h. The reaction is stopped by adding 200 µl of 1% Triton-PBS (POMT activity is inactivated in the presence of Triton X-100.).

3. The reaction mixture is centrifuged at 10,000g for 10 min. The supernatant is transferred into a screw-cap tube with a packing seal. About 400 µl of 1% Triton-PBS and 40 µl of 25% slurry glutathione-Sepharose beads are mixed with the supernatant, and rotated with rotary mixer at 4°C for 1 h.

4. After centrifugation at 1,000g for 1 min, the supernatant is removed by aspiration, and the beads are washed three times with 0.5% Triton-Tris buffer. 2% SDS is added to the beads and boiled at 100°C for 3 min. The radioactivity adsorbed to the beads is measured (Fig. 1).

POMGnT1 Assay

1. Ten microliter of 1 mM UDP-GlcNAc, 10 µl of UDP-[^{3}H]GlcNAc (100,000 dpm/nmol), and 10 µl of 2 mM mannosylpeptide (or 100 mM Benzyl-Man) are mixed in a microcentrifugal tube and dried up with a centrifugal evaporator.

2. The POMGnT1 reaction buffer [140 mM MES, pH 7.0, 2% Triton X-100, 5 mM AMP, 200 mM GlcNAc, 10% glycerol, 10 mM $MnCl_2$. Store at −20°C without $MnCl_2$. $MnCl_2$ is added just before use.] is added to the microsomal membrane fraction at a protein concentration of 2 mg/ml. The fraction is suspended with a bath-type sonicator on ice and solubilized by moderate pipetting until transparent. After centrifugation at 10,000g for 10 min, 20 µl of the supernatant is added to dried substrate (prepared in step 1), vortexed gently and incubated at 37°C for 2 h. The reaction is stopped by boiling at 100°C for 3 min. Water (180 µl) is added to the reaction mixture and filtered with a centrifugal filter device.

3. The filtrate is analyzed using reversed-phase HPLC under the condition as follows. The gradient solvents are aqueous 0.1% TFA (solvent A) and acetonitrile containing 0.1%

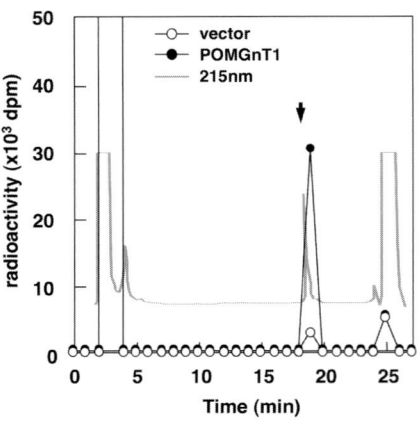

Fig. 2 POMGnT activity of human POMGnT1. UDP-[^{3}H]GlcNAc and mannosylpeptide were reacted with membrane fraction in POMGnT1 reaction buffer and then subjected to reversed-phase HPLC. *Arrow* indicates the elution position of the mannosylpeptide. Vector (*open circle*), cells transfected with vector alone; POMGnT1 (*closed circle*), cells transfected with human *POMGnT1*

TFA (solvent B). The mobile phase consists of (1) 100% A for 5 min, (2) a linear gradient to 75% A, 25% B for 20 min, (3) a linear gradient to 100% B for 1 min, and (4) 100% B for 5 min. The peptide separation is monitored at 214 nm and the radioactivity of each fraction (1 ml) is measured (Fig. 2).

Acknowledgments This study was supported by Research Grants for Nervous and Mental Disorders (17A-10) from the Ministry of Health, Labour, and Welfare of Japan, and for Scientific Research on Priority Area (14082209) from the Ministry of Education, Culture, Sports, Science, and Technology of Japan.

References

Chiba A, Matsumura K, Yamada H, Inazu T, Shimizu T, Kusunoki S, Kanazawa I, Kobata A, Endo T (1997) Structures of sialylated *O*-linked oligosaccharides of bovine peripheral nerve α-dystroglycan: the role of a novel *O*-mannosyl-type oligosaccharide in the binding of α-dystroglycan with laminin. J Biol Chem 272:2156–2162

Manya H, Chiba A, Yoshida A, Wang X, Chiba Y, Jigami Y, Margolis RU, Endo T (2004) Demonstration of mammalian protein *O*-mannosyltransferase activity: coexpression of POMT1 and POMT2 required for enzymatic activity. Proc Natl Acad Sci USA 101:500–505

Yoshida A, Kobayashi K, Manya H, Taniguchi K, Kano H, Mizuno M, Inazu T, Mitsuhashi H, Takahashi S, Takeuchi M, Herrmann R, Straub V, Talim B, Voit T, Topaloglu H, Toda T, Endo T (2001) Muscular dystrophy and neuronal migration disorder caused by mutations in a glycosyltransferase, POMGnT1. Dev Cell 1:717–724

Glycosyltransferases Involved in O-fucose Glycan Synthesis

Tetsuya Okajima, Tsukasa Matsuda

Introduction

Modification with O-fucose glycans is a rare type of post-translational modification found mainly on epidermal growth factor (EGF) domains. Although EGF domains are generally found on a number of secreted or cell surface glycoproteins, only a subset can be O-fucosylated. Founding members of those O-fucosylated proteins include urokinase in human urine and the coagulation/fibrinolytic proteins in blood plasma, such as blood clotting Factors VII, IX, and XII, and tissue plasminogen activator. Recently, a number of transmembrane proteins involved in the Notch signaling pathway have been identified as novel O-fucosylated glycoproteins. The extracellular domain of the Notch receptor is composed largely of a tandem array of EGF repeats (36 in *Drosophila* Notch and mammalian Notch1 and Notch2) and many of them are considered to be O-fucosylated. Similarly, Notch ligands (Delta, Serrate/Jagged) as well as Dlk-1/Pref-1, a negative regulator for Notch signaling, are also EGF repeat-containing proteins that are known to be O-fucosylated. By comparing the sequences surrounding the sites of O-fucose modification of those glycoproteins, the consensus amino acid sequence for O-fucosylation is proposed to be $C^2X_{3-5}S/TC^3$ (where C_2 and C_3 are the second and third conserved cysteine residue, respectively, X_{3-5} are any 3–5 amino acids, and S/T is the fucosylated amino acid) (Haltiwanger and Stanley 2002).

Structurally, O-fucose glycans can have either an O-fucose monosaccharide or a tetrasaccharide form; fucose-α1,3-GlcNAc-β1,4-Gal-α2.3/2.6-Sia (where GlcNAc is *N*-acetylglucosamine, Gal is galactose, and Sia is sialic acid), or a di- or trisaccharide intermediate. For synthesizing the unique structure of O-fucose glycans, two glycosyltransferases are specifically employed. One of them is a polypeptide O-fucosyltransferase1 (OFUT1/O-FucT-1), which acts in the initial step of the glycosylation pathway and transfers fucose onto serine or threonine residues within the consensus sequence. Elongation of O-fucose monosaccharide is achieved by an O-fucose-specific GlcNAc transferase, Fringe; *fringe* was originally identified as a mutant that results in a defect of dorsal-ventral boundary formation during wing development. Three mammalian homologues have been identified, namely, Radical (Rfng), Lunatic (Lfng), and Manic fringe (Mfng). OFUT1 and Fringe have been shown to function at a distinct compartment in the secretion pathway. OFUT1 is a soluble endoplasmic reticulum (ER) protein that carries a KDEL sequence at the carboxyl terminus, which serves as a retrieval signal from the *cis*-Golgi to the ER (Okajima et al. 2003). Although Fringe was originally identified as a secreted molecule, it is reported to localize in the Golgi apparatus. The

---

Department of Applied Molecular Biosciences, Nagoya University Graduate School of Bioagricultural Sciences, Furo-cho, Chikusa-ku, Nagoya 464-8601, Japan
Phone: +81-52-789-4131, Fax: +81-52-789-4128
E-mail: tokajima@agr.nagoya-u.ac.jp

Fig. 1 Assay for O-fucosyltransferase 1 activity. O-fucosyltransferase activity was measured using an EGF domain from Factor VII as an acceptor substrate and GDP-[^{14}C]fucose as a donor substrate. As enzyme sources, wild-type OFUT1 with a V5His tag and its two derivatives carrying either OFUT1^{R245A} or OFUT1^{R245K} mutation were used. Both mutants exhibit negligible enzyme activity. Below is a Western blot probed with anti-V5 antibody (Sigma), which confirms the expression of each OFUT1 construct

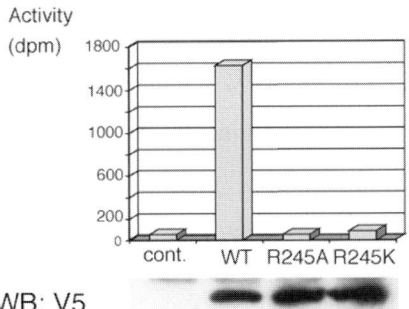

transfer of galactose and sialic acid is mediated by enzymes belonging to the family of β1,4-galactosyltransferases and α2,3 or 2,6-sialyltransferases. These glycosyltransferases are considered to be the same enzymes as those involved in the synthesis of the sialyl lactosamine structure found on N- and O-glycans.

Genetic and biochemical techniques have been emplyed to investigate the biological roles of *Ofut1* and *fringe* for Notch receptors. In both mouse and *Drosophila*, *Ofut1* and its mammalian homolog *Pofut1* have been demonstrated to be essential for Notch signaling (Haltiwanger and Stanley 2002; Okajima and Irvine 2002). On the other hand, Fringe acts as a modulator of Notch–ligand interaction; Fringe facilitates Notch–Delta binding whereas it inhibits Notch–Serrate binding. This effect of Fringe is considered to be the central mechanism for local Notch activation at the Fringe-expression boundary (Haines and Irvine 2003).

Procedure

O-Fucosyltransferase 1 Assay (Fig. 1)

For assay of O-fucosyltransferase 1, folded EGF domians derived from blood clotting Factor VII or Notch have been utilized as acceptor substrates. The enzyme does not exhibit activity toward the denatured EGF domain or synthetic peptide. GDP-[^{14}C]fucose was used as a donor substrate, and the labeled enzyme products were purified using reverse-phase liquid chromatography and subjected to liquid scintillation counting.

1. The purified OFUT1 and 20 µM EGF Factor VII are incubated in 1× assay buffer (100 mM imidazole-HCl, pH 7.0, 50 mM MnCl$_2$) containing 0.1 mM GDP-[^{14}C]fucose (20,000 dpm/nmol), in a total volume of 20 µl.
2. During the incubation, an LC-18 SPE tube (Supelco) is prepared by applying 1 ml of elution solution (80% acetonitrile, 0.052% TFA), then washing with 1 ml of H$_2$O twice.
3. After incubation at 37°C for 2 h, the reaction mixture is applied to the LC-18 SPE tube.
4. The tube is washed twice with 1 ml of H$_2$O, then the labeled substrates are eluted with 1 ml of elution solution.
5. Radioactivity in the eluate is measured using a liquid scintillation counter.

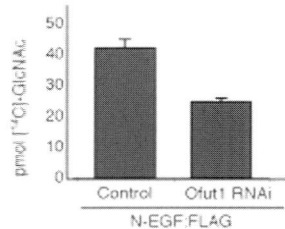

Fig. 2 Assay of Fringe activity. In vitro glycosylation of N-EGF:FLAG by Fringe. N-EGF:FLAG was produced from wild-type S2 cells (control) or those with lower OFUT1 expression (Ofut1 RNAi). These two differentially O-fucosylated N-EGF:FLAGs were used as acceptors. The amount of product formed is indicated by the amount of GlcNAc transferred to equivalent amounts of N-EGF:FLAG (Okajima et al. 2003).

Fringe Assay (Fig. 2)

For Fringe assay, pNp-fucose or the O-fucosylated EGF domain from Factor VII and Notch is typically used as a donor substrate. This assay is also performed to measure the amont of O-fucose on EGF repeat-containing proteins such as Notch or Delta, exploiting the fact that Fringe specifically modifies O-fucose on EGF domains.

1. Notch EGF repeats with a FLAG tag (N-EGF:FLAG) are immobilized on anti-FLAG antibody-conjugated agarose (Sigma).
2. The beads are washed with Hanks' balanced salt solution (HBSS, Invitrogen) three times and then washed once with 1 × Fringe assay buffer (50 mM Hepes, pH 7.0, 150 mM NaCl, 50 mM $MnCl_2$).
3. The purified Fringe:6 × His (Moloney et al. 2000) is incubated with the FLAG beads in 1 × Fringe assay buffer containing 18 µM UDP-[^{14}C]GlcNAc (Amersham Biosciences) in a total volume of 50 µl.
4. After 4 h of incubation at 30°C with constant gentle mixing, the beads are washed four times with HBSS.
5. Radioactivity on the beads is measured using a liquid scintillation counter.

References

Haines N, Irvine KD (2003) Glycosylation regulates Notch signalling. Nat Rev Mol Cell Biol 4:786–797
Haltiwanger RS, Stanley P (2002) Modulation of receptor signaling by glycosylation: fringe is an O-fucose-beta1,3-N-acetylglucosaminyltransferase. Biochim Biophys Acta 1573:328–335
Moloney DJ, Panin VM, Johnston SH, Chen J, Shao L, Wilson R, Wang Y, Stanley P, Irvine KD, Haltiwanger RS, Vogt TF (2000) Fringe is a glycosyltransferase that modifies Notch. Nature 406:369–375
Okajima T, Irvine KD (2002) Regulation of Notch signaling by O-linked fucose. Cell 111:893–904
Okajima T, Xu A, Irvine KD (2003) Modulation of Notch–ligand binding by protein O-fucosyltransferase 1 and fringe. J Biol Chem 278:42340–42345

Fucosyltransferase (α1,2/ α1,3/ α1,4-fucosyltransferases)

Takashi Kudo[1], Hisashi Narimatsu[2]

Many studies using monoclonal antibodies have demonstrated that fucosylated carbohydrate chains are involved in various physiological functions including development, neuronal activity, and cancerous transformation. The genes encoding fucosyltransferases which synthesize fucosylated carbohydrate antigens have been cloned, and their function has been gradually elucidated by analyses of their substrate specificities and biological functions (Taniguchi et al. 2002).

There are two types in mammalian fucosyltransferases, the enzymes which use GDP-fucose as the carbohydrate donor; one is enzymes which transfer fucose to carbohydrate acceptor by α1-2, α1-3, α1-4, or α1-6 linkages, and the other is enzymes which transfer it to serine or threonine residues of proteins by O-linked glycosylation (Table 1). In addition, there are other types of fucosyltransferases that synthesize core fucose structure (α1-3) in plants and insects, which has not been found in mammals. To date, nine genes of α1,2-, α1,3-, α1,3/4-, or α1,6-fucosyltransferases have been isolated and named as FUT1–FUT9. α1,2-fucosyltransferases are classified into two types: FUT1 which synthesizes H (O-type) antigens, the blood group antigens on the erythrocyte surface, and FUT2 which is necessary for the synthesis of ABH antigen that is secreted into mucus including saliva, and Lewis b antigen that is one of Lewis blood group antigens. The blood group on erythrocytes based on Bombay and Lewis blood systems is determined by polymorphism of respective gene (refer to a separate section elsewhere) (Taniguchi et al. 2002). α1,3-fucosyltransferases are classified into six types of enzymes (FUT3, FUT4, FUT5, FUT6, FUT7, and FUT9), and each of them has been proved in vitro to have its peculiar substrate specificity. Especially, FUT3 has α1,4-fucosyltransferase activity in addition to α1,3-fucosyltransferase activity, and genetic polymorphism analysis of the enzyme demonstrated that it was necessary for the synthesis of Lewis system blood group antigens (refer to a separate section elsewhere). CA19-9, a tumor marker frequently used for detecting pancreatic and colon cancer, recognizes sialyl-Lewis a antigen, but individuals having the FUT3 inactive allele in homozygous cannot synthesize this antigen. Therefore, CA19-9 cannot be used as the tumor marker in these individuals (Narimatsu et al. 1998). FUT5 and FUT6 are highly homologous to FUT3, and their genes are arranged within 40 kb in tandem on the chromosome 19. These genes are originated from a single ancestral gene in mammals prior to cattle and exist as a pseudogene in mice. From its substrate specificity and distribution of expression, FUT7 is considered to synthesize sialyl 6-sulfo Lewis X antigen, a ligand for selectin, from its

---

[1] Graduate School of Comprehensive Human Sciences, University of Tsukuba, 1-1-1 Tennodai, Tsukuba, Ibaraki 305-8575, Japan
E-mail: t-kudo@md.tsukuba.ac.jp

[2] Research Center for Medical Glycosciences, National Institute of Advanced Industrial Science and Technology (AIST), 1-1-1 Umezono, Tsukuba, Ibaraki 305-8568, Japan
E-mail: h.narimatsu@aist.go.jp

Table 1 Fucosyltransferase genes cloned to date

Linkage	Human					Mouse			Epitopes synthesized
	Symbol	Other alias	CAZy	Chromosome	AccID (RefSeq)	Symbol	Chromosome	AccID (RefSeq)	
α1,2-	FUT1	H	GT11	19q13.3	NM_000148	Fut1	7	NM_008051	H type 1 and 2
	FUT2	SE	GT11	19q13.3	NM_000511	Fut2	7	NM_018876	
α1,3/4-	FUT3	LE	GT10	19p13.3	NM_000149				Le^a, Le^b, sLe^a, Le^x, Le^y, sLe^x
α1,3-	FUT4	FUC-TIV	GT10	11q21	NM_002033	Fut4	9	NM_010242	Le^x, Le^y, sLe^x
	FUT5	FUC-TV	GT10	19p13.3	NM_002034				Le^x, Le^y, sLe^x
	FUT6		GT10	19p13.3	NM_000150				Le^x, Le^y, sLe^x
	FUT7		GT10	9q34.3	NM_004479	Fut7	2	NM_013524	sLe^x
α1,6-	FUT8		GT23	14q24.3	NM_004480	Fut8	12	NM_016893	
α1,3-	FUT9	Fuc-TIX	GT10	6q16	NM_006581	Fut9	4	NM_010243	Le^x, Le^y
Peptide	POFUT1	O-FucT-1	Not classified	20q11	NM_015352	Pofut1	2	NM_080463	O-fucose on the EGF domain
Peptide	POFUT2		Not classified	21q22.3	NM_133635	Pofut2	10	NM_030262	O-fucose on the TSR domain
Not identified	FUT10		Not classified	8p12	NM_032664	Fut10	8	NM_001012517	Not identified
Not identified	FUT11		Not classified	10q22.2	NM_173540	Fut11	14	NM_028428	Not identified

substrate specificity and expression pattern. Studies in Fut7-knockout mice also revealed that the enzyme was deeply involved in extravascular migration of leukocytes and lymphocyte homing in inflammation. The remaining activity of L-selectin ligand in the lymph node in Fut7-knockout mice was completely disappeared by simultaneous knockout of Fut4, indicating that both FUT4 and FUT7 are involved in the synthesis of the carbohydrate chain of L-selectin ligand (Homeister et al. 2001). Although FUT9 does not react with sialylated carbohydrate acceptors, it synthesizes Lewis X antigens more efficiently than other α1,3-fucosyltransferases do. It has been also demonstrated by an analysis with Fut9-knockout mice that FUT9 synthesizes SSEA-1 (stage-specific embryonic antigen-1) which is considered to be involved in intercellular interaction in the early embryo (Kudo et al. 2004). As Fut9-knockout mice are phenotypically normal but completely lack Lewis X antigens in the brain, stomach, and kidney, further analysis on the knockout mice are expected. In addition, two more genes having the amino acid motif common to existing α1,3-fucosyltransferases have been found in a DNA database and named as FUT10 and FUT11, respectively, but enzymatic activity related to these genes has not been reported yet.

α1,3-Fucosyltransferase Assay (Nishihara et al. 1999)

1. The reaction mixture for the α1,3-fucosyltransferase assay contained the following components in a reaction mixture:
   100 mM Cacodylate buffer (pH 6.8)
   5 mM ATP
   25 mM MnCl2
   10 mM L-Fucose
   25 µM PA-sugar (ex. LNnT-PA for α1,3-Fucosyltransferase assay, LNT-PA for α1,4-fucosyltransferase assay)
   75 µM GDP-fucose.
2. Incubate at 37°C for 2 h.
3. Stop the reaction by heating at 98°C for 3 min.
4. Centrifuge at 15,000g for 5 min at 4°C and recover the supernatant.
5. Apply the supernatant to HPLC on a TSK-gel ODS-80TS column (TOSOH, 4.6 mm × 300 mm) to separate substrate and product.
6. Elute at 35°C with 20 mM ammonium acetate (pH 4.0) (ex. 320 nM, em. 400 nM).
7. Calculate the enzymatic activity.

References

Homeister JW, Thall AD et al (2001) The α(1,3)fucosyltransferases FucT-IV and FucT-VII exert collaborative control over selectin-dependent leukocyte recruitment and lymphocyte homing. Immunity 15:115–126

Kudo T, Kaneko M et al (2004) Normal embryonic and germ cell development in mice lacking α1,3-fucosyltransferase IX (Fut9) which show disappearance of stage-specific embryonic antigen 1. Mol Cell Biol 24:4221–4228

Narimatsu H, Iwasaki H et al (1998) Lewis and secretor gene dosages affect CA19-9 and DU-PAN-2 serum levels in normal individuals and colorectal cancer patients. Cancer Res 58:512–518

Nishihara S, Hiraga T et al (1999) Molecular behavior of mutant Lewis enzymes in vivo. Glycobiology 9:373–382

Taniguchi N, Honke K, Fukuda M (eds) (2002) Handbook of glycosyltransferases and related genes. Springer, Tokyo

FUT8 Assay

Hideyuki Ihara, Hiroaki Korekane, Akio Matsumoto, Naoyuki Taniguchi

Introduction

Mammalian alpha1,6-fucosyltransferase (FUT 8) catalyses the transfer of a fucose residue from a donor substrate guanosine 5′-diphosphate-beta-L-fucose to the reducing terminal N-acetylglucosamine (GlcNAc) of the core structure of an asparagine-linked oligosaccharide. The enzymatic product is called as core-fucosylated N-glycans, which are ubiquitously distributed in many mammalian glycoproteins. Serum levels of fucosylated alpha fetoprotein are very low under normal conditions, but it increases dramatically in malignant diseases and therefore it is one of the best biomarkers for primary hepatomas (see reviews by Miyoshi et al.). Core fucosylation plays a pivotal role in ADCC activity in antibody therapy (see reviews by Sato et al.). KO mice of Fut 8 exhibit severe growth retardation and develop lung emphysema. Crystal structure of this enzyme was reported very recently by our group.

Stock Solution

(A) 500 mM MES solution: Dissolve MES (5.33 g) in 40 ml distilled water and adjust pH to 7.0 by adding 5 N NaOH and make it up to 50 ml and store at $-20°C$.
(B) Triton X-100 solution
  10% (w/v) Triton X-100 solution: dissolve 1 g of Triton X-100 in water and make it up to 10 ml and keep at 4°C.
(C) 5 mM GDP-Fucose: Take GDP-β-L-Fucose bis (triethylammonium) salt in a 1.5-ml sample tube and dissolve in water in a final concentration of 3.959 mg/ml.
(D) 100 µM GnGnbi-Asn-PNSNB (Wako Chemicals) PNSNB; N-(2-(2-pyridylamino) ethyl) succinamic acid 5-norbornene-2,3-dicarobxylmide.

2× Reaction Mixture

Mix with 4 ml of A, 1 ml of B, 850 mg of GlcNAc, 20 mg of BSA, and add water to make it up to 10 ml and store at $-20°C$. The final concentration of each reagent is as follows: 200 mM MES-NaOH (pH 7.0), 1% Triton X-100, 400 mM GlcNAc, and 2 mg/ml bovine serum albumin.

Department of Disease Glycomics Research Institute for Microbial Diseases, Osaka University, 2-1 Yamadaoka, Suita, Osaka 565-0871, Japan
Phone: +81-6-6879-4137, Fax: +81-6-6879-4137
E-mail: tani52@wd5.so-net.ne.jp

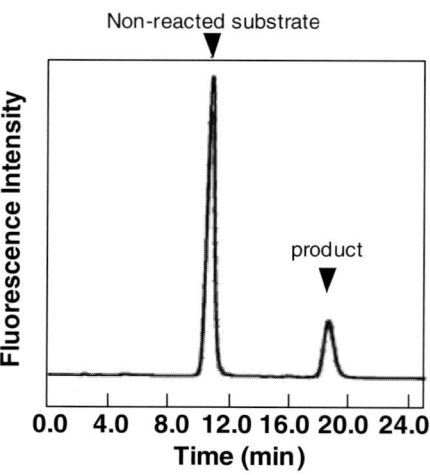

Fig. 1 A typical chromatogram for the enzyme assay is shown when N-[2-(2-Pyridylamino)ethyl]-succinamic acid 5-norbornene-2,3-dicarboxyimide ester (Wako Chemicals) is used for fluorescence labeling. Non-reacted substrate and product were eluted at 11 and 18 min, respectively

5× Stock Buffer Solution for HPLC Analysis

0.1 M Ammonium acetate (pH 4.0)

Acetic acid (for HPLC grade) 5.724 ml (6.01 g)
Ultrapure water obtained by milli Q (MQ) 950 ml
Adjust pH by ammonia solution adjust pH 4.0, and make it up to 10 ml

Assay Method

1. Assay mixture
2× reaction mixture	5 µl
100 mM GnGnbi-Asn-PNSNB	1 µl
5 mM GDP-Fuc	1 µl
Enzyme sample	3 µl
Total 10 µl	
2. Incubate for 1 h or appropriate times at 37 C
3. Stop the reaction by boiling for 5 min
4. Following addition of 40 µl MQ water to the mixture, centrifuge at 15,000× g for 10 min
5. Apply the supernatant (20 µl) onto an HPLC column (TSK-gel, ODS-80 TM (4.6 × 150 mm) (Toso) which had been equilibrated with 20 mM acetate buffer (pH 4.0) containing 0.15% 1-butanol at 55°C.
6. Elute the enzymatic product isocratically at a flow rate of 1.0 ml/min.
7. Detect the fluorescence of the column elute with a fluorescence detector at excitation and emission wavelengths of 315 and 380 nm, respectively
8. Estimate the product amounts from the fluorescence intensity. Specific activity is calculated as pmol product/mg protein/h.

Comment

It is possible to use 4-(2-pyridylamino) butylamine (Uozumi et al. 1996a, b) and N-[2-(2-Pyridylamino)ethyl]-succinamic acid 5-norbornene-2,3-dicarboxyimide ester (Ihara et al. 2006) as reagents for the preparation of a fluorescence-labeled acceptor substrate.

HPLC Analysis

1. Set up an HPLC equipped with a TSK-gel, ODS-80TM column (4.6 × 150 mm) (Tosoh), and with a fluorescence detector (model RF-10AXL, Shimadzu).
2. Equilibrate the column using a 20 mM ammonium acetate buffer (pH 4.0) containing 0.15% 1-butanol at 55°C.
3. Apply 20 μl of the supernatants to HPLC analysis.
4. Elute the enzyme product isocratically at a flow rate of 1.0 ml/min.
5. Detect the fluorescence of the column elute with a fluorescence detector at excitation and emission wavelengths of 315 and 380 nm, respectively.
6. Estimate the product amounts from the fluorescence intensity.

References

Ihara H, Ikeda Y, Taniguchi N (2006) Reaction mechanism and substrate specificity for nucleotide sugar of mammalian α1,6-fucosyltransferase: a large-scale preparation and characterization of recombinant human FUT8. Glycobiology 16:333–342

Miyoshi E, Noda K, Yamaguchi Y, Inoue S, Ikeda Y, Wang W, Ko JH, Uozumi N, Li W, Taniguchi N (1999) The α1-6-fucosyltransferase gene and its biological significance. Biochim Biophys Acta 1473:9–20

Satoh M, Iida S, Shitara K (2006) Non-fucosylated therapeutic antibodies as next-generation therapeutic antibodies. Expert Opin Biol Ther 6:1161–1173

Uozumi N, Yanagidani S, Miyoshi E, Ihara Y, Sakuma T, Gao CX, Teshima T, Fujii S, Shiba T, Taniguchi N (1996a) Purification and cDNA cloning of porcine brain GDP-L-Fuc:N-acetyl-β-D-glucosaminide α1-6fucosyltransferase. J Biol Chem 271:27810–27817

Uozumi N, Teshima T, Yamamoto T, Nishikawa A, Gao YE, Miyoshi E, Gao CX, Noda K, Islam KN, Ihara Y, Fujii S, Shiba T, Taniguchi N (1996b) A fluorescent assay method for GDP-L-Fuc:N-acetyl-β-D-glucosaminide α1–6fucosyltransferase activity, involving high performance liquid chromatography. J Biochem (Tokyo) 120:385–392

Sialyltransferases

Shuichi Tsuji

Introduction

In general, sialic acid is found at the terminal positions of sialylated glycoconjugates where it has roles in cell–cell recognition, cell differentiation, and receptor–ligand binding. Sialyltransferases (EC no. 2.4.99) catalyze the transfer of sialic acid from the nucleotide sugar donor CMP-Sia to acceptor oligosaccharides found on glycoproteins, glycolipids, and polysaccharides. Genomic databases on various organisms revealed the occurrence of many sialyltransferases and their homologs including vertebrates, insects, virus, plants, and so on. Here, I mainly summarized the cloned sialyltransferases that had been characterized their linkage and acceptor specificities in vitro. The abbreviated nomenclature for cloned sialyltransferases follows the system of Tsuji et al. (1996).

List and Comments

Mammalian Sialyltransferases

From mouse and human, 20 sialyltransferase genes and their essential genomic structure have been identified, respectively. Their cDNAs have been cloned and their enzymatic properties have been characterized (Table 1) (Takashima et al. 2002). The mouse and human sialyltransferases have a type II transmembrane topology and contain highly conserved domains called as sialylmotifs L (long), S (short), and VS (very short). These motifs are highly conserved regions comprising about 20% of the total protein sequences. The L- and S-sialylmotifs were shown to bind to the donor CMP-NeuAc and acceptor saccharide substrates. The VS-sialylmotif is considered to be participated in the catalytic center although the strict data have not been published. Each sialyltransferase exhibits strict specificity for acceptor substrates and linkages with which they synthesize (Table 1).

The cloned sialyltransferases can be classified into four families according to the carbohydrate linkages with which they synthesize, i.e., the β-galactoside α2,3-sialyltransferase family (ST3Gal-I–VI), the β-galactoside α2,6-sialyltransferase family (ST6Gal-I,II), the GalNAc α2,6-sialyltransferase family (ST6GalNAc-I–VI), and the α2,8-sialyltransferase family (ST8Sia-I–VI). They can also be classified into subfamilies according to the amino acid sequence similarities, substrate specificity differences, and genomic structures. For example, ST6GalNAc-V and –VI may be classified as a family, because both have the sialic acid transfer activity toward NeuAcα2,3Galβ1,3GalNAc-, and also can transfer sialic acid to GlcNAc in the NeuAcα2,3Galβ1,3GlcNAc-structure (Tsuchida et al. 2003).

Future Science and Thechnology Joint Research Center, Institute of Glycotechnology, Tokai University, 1117 Kitakaname, Hiratsuka, Kanagawa 259-1292, Japan
Phone: +81-463-58-1211, Fax: +81-463-50-2432
E-mail: stsuji@keyaki.cc.u-tokai.ac.jp

Table 1 Acceptor specificities of mouse and human sialyltransferases, and chromosomal localization of human genes

Enzyme	Essential substrates	Chromosomal localization (human)
ST3Gal-I	Galβ1,3GalNAc-(protein)*	8q24.2
ST3Gal-II	Galβ1,3GalNAc-(lipid)*	16q21–22.3
ST3Gal-III	Galβ1,3(4)GlcNAc-	1pter-p32.3
ST3Gal-IV	Galβ1,4(3)GlcNAc-	11q23–q24
ST3Gal-V	Galβ1,4Glcβ1,1Cer (Lac-Cer)	2p24.3–p24.1
ST3Gal-VI	Galβ1,4GlcNAc-	3p21.1–q13.2
ST6Gal-I	Galβ1,4(3)GlcNAc-	3q27–q28
ST6Gal-II	Galβ1,4GlcNAc-	2q11.2–q12.1
ST6GalNAc-I	GalNAcα1,O-Ser/Thr	17q25
ST6GalNAc-II	(Galβ1,3)GalNAcα1,O-Ser/Thr	17q25
ST6GalNAc-III	NeuAcα2,3Galβ1,3GalNAc-(lipid)*	1p31.1–p31.2
ST6GalNAc-IV	NeuAcα2,3Galβ1,3GalNAc-(protein)*	9q34
ST6GalNAc-V	NeuAcα2,3Galβ1,3GalNAc-(GM1b)	1p31.1
ST6GalNAc-VI	NeuAcα2,3Galβ1,3GalNAc-(GM1b, GT1b, GD1a)	9q34
ST8Sia-I	NeuAcα2,3Galβ1,4Glcβ1,1Cer (GM3)	12p12.1–p11.2
ST8Sia-II	N-glycan on NCAM	15q26
ST8Sia-III	NeuAcα2,3Galβ1,4GlcNAc-	18q21
ST8Sia-IV	N-glycan on NCAM	5q21
ST8Sia-V	GM1b, GT1b, GD1a, GD3	18q12.1–q12.3
ST8Sia-VI	NeuAcα2,3(6)Gal-	10p13

Asterisk means preferential but not specific substrate

There are several sets of sialyltransferase genes that encode similar enzymes and have similar genomic structures. Among them, the ST6GalNAc I and II genes, ST6GalNAc III and V genes, and ST6GalNAc IV and VI genes are located close to each other, respectively, suggesting that each gene pair is closely related from an evolutional standpoint. Probably, each gene pair arose from a common ancestral gene through tandem duplication. On the other hand, other pairs of similar sialyltransferase genes are not located on the same chromosome, suggesting that these genes arose from a common ancestral gene through gene duplication and were subsequently dispersed in the genome through translocation.

It is interesting that all the so-far-cloned sialyltransferases each have a counterpart with similar enzymatic properties and genomic structure. The biological significance of these multiple genes is unclear at present. One interpretation is that they may be important for fine control of the expression of sialylglycoconjugates, resulting in a variety of developmental stage-specific and tissue-specific glycosylation patterns. Characterization of each sialyltransferase and analysis of the transcriptional regulation of each gene will help elucidate the biological significance of each sialyltransferase and the sialylglycoconjugates produced by them.

Bacterial Sialyltransferases

Several bacterial sialyltransferases (ST3Gal, ST6Gal, and ST8Sia) have been cloned and characterized (Gilbert et al. 1997; Yamamoto et al. 1998; Bozue et al. 1999 (ST3); McGowen et al. 2001). Interestingly, the sialylmotif has not been found in the amino acid sequences of these bacterial sialyltransferases even though they transfer NeuAc from CMP-NeuAc to oligosaccharides in analogous to the mammalian enzymes. Generally, bacterial enzymes have broader acceptor tolerances compared to mammalian enzymes. For example, *Neisseria meningitidis* ST3Gal can utilize both α- and (3-linked terminal galactoses as acceptors (Gilbert et al. 1997).

The crystallographic analysis of a bacterial sialyltransferase CstII from *Campylobacter jejuni* was performed (Chiu et al. 2004).

Viral Sialyltransferases

A novel myxoma virus early gene, MST3N, is a member of the eukaryotic sialyltransferase gene family (Jackson et at. 1999). Myxoma virus is a member of the poxvirus family of double-stranded DNA viruses, and it is known to infect Old World or European rabbits causing highly lethal myxomatosis. Mammalian cells infected by Myxoma virus produce viral α2,3-sialyltransferase (v-ST3Gal-I) in the cells. v-ST3Gal-I is closely related to mammalian ST3Gal-IV, even though it has a very broad acceptor specificity that is not found among the mammalian or bacterial α2,3-sialyltransferases. Acceptors include not only type I to III disaccharides but also fucosylated Lewis^x and Lewis^a (Sujino et al. 2000).

Plant Sialyltransferases

Sialic acids are widely distributed among living creatures, from bacteria to mammals, but it has been commonly accepted that they do not exist in plants, although the possibility remains that sialylated glycoconjugates may exist in plant cells. Interestingly, however, putative gene homologs for mammalian sialyltransferases and CMP-Sia transporters have been detected in the genome and/or expressed sequence tag (EST) databases of some plants, such as *Arabidopsis thaliana* (thale-cress) and *Oryza sativa* (Japanese rice). This suggests that plants have potential ability of sialylation despite the absence of sialic acids in them. Takashima et al. (2006) cloned three genes from *O. sativa*, each encoding a protein having sialyl motif-like sequences, and analyzed the enzymatic activity of the proteins. One of them, called as OsSTLP1, really transferred sialic acid from CMP-NeuAc to galactose of Galβ1,4GlcNAc through α2,6-linkage.

Sialyltransferase Assays and Product Characterization

Sialyltransferase assay is usually performed as follows.

Reaction Mixture

Enzyme activity is measured in 50 mM MES buffer (pH 6.0), 10 mM MgCl$_2$, 0.5% Triton CF-54, 50~100 μM CMP-[^{14}C]NeuAc, acceptor substrate, and enzyme preparation, in a total volume of 10 μl.

Substrate Concentration

Concentrations of acceptor substrates are 0.5~5 mg/ml of glycoproteins, 0.05~1.0 mM of glycolipids, or 0.05~1.0 mM of oligosaccharides.

Enzyme Reaction

The enzyme reaction is performed at 37°C for 3–20 h.

For glycoproteins, the reaction is terminated by the addition of SDS-PAGE loading buffer, and the reaction mixtures are directly subjected to SDS-PAGE.

For glycolipids, the reaction mixtures are applied to a Sep-Pak Vac C-18 column (100 mg; Waters, Milford, MA, USA), and the purified glycolipids are subjected to high-performance thin-layer chromatography (HPTLC, Silica-Gel 60; Merck, Darmstadt, Germany) with a solvent system of chloroform, methanol, and 0.02% $CaCl_2$ (55:45:10).

For oligosaccharides, the reaction mixtures are directly subjected to HPTLC with a solvent system [for example, 1-propanol/aqueous ammonia/water (6:1:2.5)].

Linkage Analysis of Sialic Acids

For linkage analysis of sialic acids, [^{14}C]NeuAc-incorporated product is digested with a linkage-specific exosialidase: NANase I (specific for α2,3-linked sialic acids units/ml; Glyko), NANase II (specific for α2,3-and α2,6-linked sialic acids 5 units/ml; Glyko), *V. cholerae* sialidase (specific for α2,3-, α2,6-linked, and α2,8-linked sialic acids 1 unit/ml; Boehringer Mannheim), or Newcastle disease virus sialidase (specific for α2,3- and α2,8-linked sialic acids 5 units/ml; Oxford Glycosystems).

After digestion, the following procedures are performed according to the desialylated substrate.

The desialylated glycolipid is purified by C-18 column chromatography, dried, and subjected to HPTLC with solvent systems of chloroform/methanol/0.02% CaC_2 (55:45:10).

The desialylated oligosaccharide is subjected to HPTLC with a solvent system [e.g., 1-propanol/aqueous ammonia/water (6:1:2.5)].

The desialylated glycoprotein is subjected to SDS-polyacrylamide gradient gel (5–20%) electrophoresis. In some cases, oligosaccharide portion of desialylated glycoprotein should be obtained for further analysis.

Observation of the Radioactive Materials

The radioactive materials are visualized and quantified with a Fuji BAS2000 radio image analyzer. The intensity of the radioactivity is converted into moles using the radio activities of various amounts of CMP-[^{14}C]NeuAc (12.0 GBq/mmol, 925 kBq/ml) as standards. Quantification is performed within the linear range of the standard radioactivity.

References

Chiu CPC, Watts AG, Lairson LL, Gilbert M, Lim D, Wakarchuk WW, Withers SG, Strynadka NCJ (2004) Structural analysis of the sialyltransferase CstII from *Campylobacter jejuni* in complex with a substrate analog. Nat Struct Mol Biol 11:163–170

Gilbert M, Cunningham A-M, Watson DC, Martin A, Richards JC, Wakarchuk WW (1997) Characterization of a recombinant *Neisseria meningitidis* α2,3-sialyltransferase and its acceptor specificity. Eur J Biochem 249:187–194

McGowen MM, Vionnet J, Vann WF (2001) Elongation of alternating α2,8/2,9 polysialic acid by the *Escherichia coli* K92 polysialyltransferase. Glycobiology 8:513–620

Sujino K, Jackson RJ, WC Chan N, Tsuji S, Palcic MM (2000) A novel viral α2,3-sialyltransferase (v-ST3Gal I):transfer of sialic acid to fucosylated acceptors. Glycobiology 10:313–320

Takashima S, Tsuji S, Tsujimoto M (2002) Characterization of the second type of human β-galactoside α2,6-sialyltransferase (ST6Gal II) that sialylates Galβ1,4GlcNAc structures on oligosaccharides preferentially. J Biol Chem 277:45719–45728

Takashima S, Abe T, Yoshida S, Kawahigashi H, Saito T, Tsuji S, Tsujimoto M (2006) Analysis of sialyltransferase-like proteins from *Oryza sativa*. J Biochem 139:279–287

Tsuchida A, Okajima T, Furukawa K, Ando T, Ishida H, Yoshida A, Nakamura Y, Kannagi R, Kiso M, Furukawa K (2003) Synthesis of Disialyl Lewis a (Lea) Structure in colon cancer cell lines by a sialyltransferase, ST6GalNAc VI, responsible for the synthesis of α-series Gangliosides. J Biol Chem 278:22787–22794

Tsuji S, Datta AK, Paulson JC (1996) Systematic nomenclature for sialyltransferases. Glycobiology 6(7):v–vii

Yamamoto T, Nakashizuka M, Terada J (1998) Cloning and expression of a marine bacterial β-galactoside α2,6-sialyltransferase gene from *Photobacterium damsela* JT0160. J Biochem 123:94–100

Glucuronyltransferases Involved in the HNK-1 Biosynthesis

Shinako Kakuda[1], Toshisuke Kawasaki[2], Shogo Oka[1]

Introduction

The monoclonal antibody HNK-1 was raised against the membrane fraction of the human HSB-2 T-cell line. The antigen was originally found to be a marker of human natural killer (HNK) cells and is called as CD57 in immunology. After that, it was found that the HNK-1 carbohydrate epitope is highly expressed in the nervous system, especially on a series of cell adhesion molecules, including neural cell adhesion molecule (NCAM), myelin-associated glycoprotein (MAG), L1, P0, telencephalin, and also some glycolipids. The structure of the HNK-1 epitope was demonstrated to comprise the sulfated trisaccharide HSO_3-3GlcAβ1-3Galβ1-4GlcNAc, which is shared by glycolipids and glycoproteins. The expression of the HNK-1 carbohydrate epitope is spatially and temporally regulated during the development of the nervous system, and it functions in cell adhesion, migration, neurite outgrowth and synaptic plasticity (Yamamoto et al. 2002).

Genes for GlcAT-P and GlcAT-S

The biosynthesis of the HNK-1 carbohydrate is mainly regulated by two glucuronyltransferases (GlcAT-P and GlcAT-S) and a sulfotransferase (HNK-1 ST). We purified GlcAT-P to apparent homogeneity from 2-week postnatal rat forebrains. On the basis of the partial amino acid sequence, cDNA encoding the full-length GlcAT-P was cloned. The primary structure deduced from the cDNA sequence predicted a type II transmembrane protein comprising 347 amino residues. Using the cDNA as a probe, both mouse and human GlcAT-P cDNAs were cloned. Alignment of the deduced amino acid sequence of the mouse GlcAT-P with those of the rat and human GlcAT-Ps revealed 99.7 and 98.2% sequence identity, respectively. The mouse and human GlcAT-P genes were mapped to 9A4 and 11q25, respectively.

We used GlcAT-P as a probe and cloned the second glucuronyltransferase (GlcAT-S) cDNA from rat and mouse. The predicted amino acid sequence of mouse GlcAT-S consists of 324 amino acid residues, and is more than 98% similar to that of rat GlcAT-S and about 50% similar to that of mouse GlcAT-P. The mouse GlcAT-S genes are mapped to the A4-5 region of mouse chromosome 1. Human GlcAT-P and GlcAT-S are named as B3GAT1 and 2, respectively.

---

[1] Department of Biological Chemistry, School of Health Sciences, Faculty of Medicine, Kyoto University, Kyoto 606-8507, Japan
[2] Research Center for Glycobiotechnology, Ritsumeikan University, Shiga 525-8577, Japan
Phone: +81-75-751-3959, Fax: +81-75-751-3959
E-mail: shogo@hs.med.kyoto-u.ac.jp

Protocols

Glucuronyltransferase Assay for Glycoprotein Acceptors

An equivalent amount of each enzyme was incubated at 37°C for 1 h in a reaction mixture with a final volume of 50 µl comprising 100 mM MES, pH 6.5, 0. 2% NP-40, 20 mM $MnCl_2$, 20 µg ASOR, 100 µM UDP-[^{14}C]-GlcA (200,000 dpm), and 0.5 mM ATP. After incubation, the assay mixture was spotted onto a 2.5-cm Whatman no. 1 disc. The disc was washed with a 10% (w/v) trichloroacetic acid solution three times, followed by with ethanol/ether (2:1, v/v) and then with ether. The disc was air-dried and then the radioactivity of [^{14}C]GlcA-ASOR on it was counted with a liquid scintillation counter.

Glucuronyltransferase Assay for Glycolipids

An equivalent amount of each enzyme was incubated at 37°C for 1 h in a reaction mixture with a final volume of 50 µl comprising 80 mM sodium cacodylate buffer, pH 6.0, 0.4% NP-40, 10 mM $MnCl_2$, 7.5 µg of paragloboside, 100 µM UDP-[^{14}C]-GlcA (200,000 dpm), and 10 mM ATP. The reaction was terminated by the addition of 1 ml of chloroform/methanol (2:1, v/v). The radioactive reaction products were separated from labeled precursors by passage through a Sephadex G-25 superfine column (1 ml), which had been equilibrated with chloroform/methanol/water (60:30:4.5, v/v/v), and then the radioactivity of [^{14}C]-GlcA-paragloboside, eluted at the void volume of the column, was counted with a liquid scintillation counter.

Glucuronyltransferase Assay for Oligosaccharides

An equivalent amount of each enzyme was incubated at 37°C for 1 h in a reaction mixture with a final volume of 50 µl comprising different concentrations of oligosaccharides, 200 mM MES (pH 6.5), 0.2% NP-40, 20 mM $MnCl_2$, 100 µM UDP-[^{14}C]-GlcA (200,000 dpm), and 0.5 mM ATP. Following incubation, the reaction was terminated by the addition of 1 ml of 5 mM phosphate buffer, pH 6.8. The radioactive reaction products were separated from UDP-[^{14}C]-GlcA by passage through an anion exchange resin AG1-X4 column (0.5 ml), which had been equilibrated with 5 mM phosphate buffer, pH 6.8. The column was washed with 5 ml of 5 mM phosphate buffer, pH 6.8, and then the effluent and washings were collected. The radioactivity was counted with a liquid scintillation counter.

Different Acceptor Specificities of GlcAT-P and GlcAT-S

For comparing their enzymatic properties, we used soluble forms of GlcAT-P and GlcAT-S fused with the IgG-binding domain of protein A (Kakuda et al. 2004). Both GlcAT-P and GlcAT-S transferred glucuronic acid (GlcA) not only to a glycoprotein acceptor, asialo-olosomucoid (ASOR) but also to a glycolipid acceptor, paragloboside. The activity of GlcAT-P toward ASOR was enhanced fivefold in the presence of sphingomyelin (SM), but there was no effect on that of GlcAT-S. The activities of the two enzymes toward paragloboside were only detected in the presence of phospholipids such as phosphatidylinositol (PI). Kinetic analysis revealed that the K_m value of GlcAT-P for ASOR was 10 times lower than that of paragloboside. Furthermore, acceptor specificity analysis involving various oligosaccharides revealed that GlcAT-P specifically recognized

N-acetyllactosamine (Galβ1-4GlcNAc) at the non-reducing terminals of acceptor substrates. In contrast, GlcAT-S recognized not only the terminal Galβ1-4GlcNAc structure but also the Galβ1-3GlcNAc structure, and showed the highest activity toward triantennanry N-linked oligosaccharides. These lines of evidence indicate that these two enzymes have significantly different acceptor specificities, suggesting that they may synthesize functionally and structurally different HNK-1 carbohydrates in the nervous system.

The X-ray crystal structures of recombinant human soluble forms of GlcAT-P and GlcAT-S have been solved (Kakuda et al. 2004; Shiba et al. 2006). The results revealed that GlcAT-P and GlcAT-S have an asymmetric unit which contains two independent molecules. The Val320 and Asn321 residues in GlcAT-P, which are located on the C-terminal long loop of a neighboring molecule, and Phe245 play a key role in establishing the acceptor substrate specificity of Galβ1-4GlcNAc. These amino acids of GlcAT-S are different from those of GlcAT-P (i.e., Phe245 is tryptophan and Val320 is alanine in GlcAT-S). These differences contribute to the distinct specificity of GlcAT-S.

References

Kakuda S, Shiba T, Ishiguro M, Tagawa H, Oka S, Kajihara Y, Kawasaki T, Wakatsuki S, Kato R (2004) Structural basis for acceptor substrate recognition of a human glucuronyltransferase, GlcAT-P, an enzyme critical in the biosynthesis of the carbohydrate epitope HNK-1. J Biol Chem 279: 22693–22703

Kakuda S, Sato Y, Tonoyama Y, Oka S, Kawasaki T (2005) Different acceptor specificities of two glucuronyltransferases involved in the biosynthesis of HNK-1 carbohydrate. Glycobiology 15:203–210

Shiba T, Kakuda S, Ishiguro M, Morita I, Oka S, Kawasaki T, Wakatsuki S, Kato R (2006) Proteins 65:499–508

Yamamoto S, Oka S, Inoue M, Shimuta M, Manabe T, Takahashi H, Miyamoto M, Asano M, Sakagami J, Sudo K, Iwakura Y, Ono K, Kawasaki T (2002) Mice deficient in nervous system-specific carbohydrate epitope HNK-1 exhibit impaired synaptic plasticity and spatial learning. J Biol Chem 277:27227–27231

α1,4-Linkage Glycosyltransferases

Jun Nakayama[1], Koichi Furukawa[2]

Introduction

Glycosyltransferases that transfer GlcNAc and Gal to βGal residues with α1,4-linkages on glycoprotein and glycolipid templates, respectively, are categorized as α1,4-linkage glycosyltransferases. At present, two enzymes, namely, α1,4-N-acetylglucosaminyltransferase (α4GnT) and α1,4-galactosyltransferase (α4GalT, Gb3/CD77 synthase) are assigned to this family. α4GnT is responsible for biosynthesis of GlcNAcα1 → 4Galβ → R attached to O-glycans, whereas α4GalT is a key enzyme in the formation of Gb3/CD77 (Fig. 1). Since cDNAs encoding both enzymes were cloned by expression cloning by the authors (Nakayama et al. 1999; Kojima et al. 2000), we describe here the procedure for expression cloning of α1,4-linkage glycosyltransferases, focusing primarily on α4GnT.

Procedure for Expression Cloning of α4GnT

1. Prepare COS-1 cells (1.2×10^7 cells) cultured in 15-cm dishes as recipient cells because they express core 2 β1,6-N-acetylglucosaminyltransferase-I but not GlcNAcα1 → 4Galβ → R itself.

2. Cotransfect COS-1 cells with 30 μg of a human stomach cDNA library constructed in pcDNAI and the same amount of leukosialin vector, pRcCMV-leu using LipofectAmine (Invitrogen, Carlsbad, CA, USA), as leukosialin has 80 O-glycosylation sites in the extracellular domain.

3. Sixty hours following transfection, collect COS-1 cells expressing GlcNAcα1 → 4Galβ → R by cell sorting using a mixture of HIK1083, PGM36, and PGM37 antibodies, which are specific for terminal α1,4-linked GlcNAc.

4. Rescue plasmid DNA from sorted cells using the Hirt method (Hirt 1967), and in the presence of ampicillin and tetracycline transform host E. coli MC1061/P3 cells with the rescued plasmid by electroporation. Note that bacteria transformed by pcDNAI become resistant to both antibiotics because pcDNAI vector contains the *sup* F gene, which corrects the defect of both ampicillin- and tetracycline-resistant genes in the P3 episome. By contrast, bacteria transformed by the leukosialin vector only are resistant

---

[1] Department of Pathology, Shinshu University School of Medicine, Asahi 3-1-1, Matsumoto 390-8621, Japan
Phone: +81-263-37-3394, Fax: +81-263-37-2581
E-mail: jun@hsp.md.shinshu-u.ac.jp

[2] Department of Biochemistry II, Nagoya University School of Medicine, Tsurumai, Showa-ku, Nagoya 466-0065, Japan
Phone: +81-52-744-2027, Fax: +81-52-744-2069
E-mail: koichi@med.nagoya-u.ac.jp

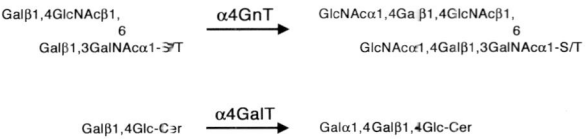

Fig. 1 Enzymatic reactions of two α1,4-linkage glycosyltransferases, α4GnT and α4GalT. GlcNAcα1 → 4Galβ → R on O-glycans is formed by α4GnT, whereas Gb3/CD77 is synthesized by α4GalT

to ampicillin but not to tetracycline. Thus, due to this differential selection, only plasmids derived from the library are amplified.

5. Make a replica of the bacteria plate by placing a nitrocellulose membrane on the aforementioned *E. coli* plate for 4 h, prepare a pool of plasmid DNA from the replica, and cotransfect COS-1 cells cultured in 6-well dishes with 1 μg of the rescued plasmid DNA and the same amount of pRcCMV-leu in each well using LipofectAmine. After 60 h, assay for expression of GlcNAcα1 → 4Galβ → R on transfected COS-1 cells using the antibody mixture specific for terminal α1,4-linked GlcNAc and an FITC-conjugated anti-mouse IgM antibody as secondary antibody under immunofluorescent microscopy.

6. Once positive COS-1 cells are identified, make a replica again as described in step 5, cut the replica into 10 pieces, place them on LB plates containing ampicillin and tetracycline for 4 h to make smaller replicas of each piece, and prepare plasmid DNA from each replica.

7. Cotransfect COS-1 cells cultured in 6-well dishes with 1 μg of each prepared plasmid DNA and the same amount of leukosialin cDNA. After 60 h, check for expression of GlcNAcα1 → 4Galβ → R on transfected COS-1 cells under an immunofluorescent microscope in a manner similar to that described above.

8. Once a section of the replica is identified that enables COS-1 cells to express GlcNAcα1 → 4Galβ → R by transfection, divide bacteria on that piece into smaller pools, prepare plasmid DNA from the each bacteria pool, and repeat step 7.

9. Repeat steps 8 and 7 (so-called sibling selection) until a single plasmid encoding human α4GnT is isolated.

For expression cloning of α4GalT (Kojima et al. 2000), mouse fibroblast L cells were used as recipient cells, because they express a precursor of Gb3, LacCer. The cDNA library was prepared from human melanoma SK-MEL-37 cells constructed with pCDM8. L cells were then cotransfected with that library and *pdl3027* encoding polyoma T antigen. After transfection, L cells expressing Gb3 were enriched by panning with mAb 38.13 specific for Gb3, and then *E. coli* MC1061/P3 cells were transformed with plasmid DNA rescued from the panned cells. Sibling selection was repeated until a single plasmid encoding human α4GalT was isolated.

Results

Both human α4GnT and α4GalT are typical type II membrane-bound proteins of 340 and 353 amino acid residues, respectively (Nakayama et al. 1999; Kojima et al. 2000), showing 35% overall sequence similarity. In normal human tissues, expression of α4GnT is exclusively limited to gland mucous cells of the gastric mucosa, Brunner's gland of the duodenal mucosa, and accessory glands of the pancreatico-biliary tract. Because α4GnT is also expressed in gastric and pancreatic cancer cells, quantitative RT-PCR for α4GnT mRNA expressed in the mononuclear cell fraction of peripheral blood can detect circulating tumor cells in these malignancies (Ishizone et al. 2006). By contrast, α4GalT is strongly expressed in the heart, kidney, spleen, liver, testis, and placenta in humans (Kojima et al. 2000). In tumors, α4GalT is expressed in hematological malignancies including Burkitt's lymphoma and megakaryoblastic leukemia (Furukawa et al. 2002). Recently, it was shown that α4GalT is identical with the blood group P1 synthase (Iwamura et al. 2003). In addition, Gb3/CD77 and its derivatives have been shown to function as a receptor for verotoxins in vivo by analysis of knockout mice lacking α4GalT (Okuda et al. 2006).

References

Furukawa K, Yokoyama K, Sato T, Wiels J, Hirayama Y, Ohta M, Furukawa K (2002) Expression of the Gb3/CD77 synthase gene in megakaryoblastic leukemia cells: implication in the sensitivity to verotoxins. J Biol Chem 277:11247–11254

Hirt B (1967) Selective extraction of polyoma DNA from infected mouse cell cultures. J Mol Biol 26:365–369

Ishizone S, Yamauchi K, Kawa S, Shimizu F, Harada O, Sugiyama A, Miyagawa S, Fukuda M, Nakayama J (2006) Clinical utility of quantitative RT-PCR targeted to α1,4-N-acetylglucosaminyltransferase mRNA for detection of pancreatic cancer. Cancer Sci 97:119–126

Iwamura K, Furukawa K, Uchikawa M, Sojka BN, Kojima Y, Wiels J, Shiku H, Urano T, Furukawa K (2003) The blood group P1 synthase gene is identical to the Gb3/CD77 synthase gene: a clue to the solution of the P1/P2/p puzzle. J Biol Chem 278:44429–44438

Kojima Y, Fukumoto S, Furukawa K, Okajima T, Wiels J, Yokoyama K, Suzuki Y, Urano T, Ohta M, Furukawa K (2000) Molecular cloning of globotriaosylceramide/CD77 synthase, a glycosyltransferase that initiates the synthesis of globo series glycosphingolipids. J Biol Chem 275:15152–15156

Nakayama J, Yeh J-C, Misra AK, Ito S, Katsuyama T, Fukuda M (1999) Expression cloning of a human α1,4-N-acetylglucosaminyltransferase that forms GlcNAcα1 → 4Galβ → R, a glycan specifically expressed in the gastric gland mucous cell-type mucin. Proc Natl Acad Sci USA 96:8991–8996

Okuda T, Tokuda N, Numata S, Ito M, Ohta M, Kawamura K, Wiels J, Urano T, Tajima O, Furukawa K, Furukawa K (2006) Targeted disruption of Gb3/CD77 synthase gene resulted in the complete deletion of globo-series glycosphingolipids and loss of sensitivity to verotoxins. J Biol Chem 281:10230–10235

Glycosyltransferases Involved in the Biosynthesis of Glycolipids

Koichi Furukawa, Akiko Tsuchida, Keiko Furukawa

Introduction

Of the glycosyltransferases involved in the synthesis of glycosphingolipids, those involved in the common steps, those responsible for the synthesis of globo-series glycolipids, and those for the synthesis of lacto/neolacto-series glycolipids were summarized here. The common and fundamental steps consist of reactions catalyzed by Glc-Cer synthase (Ichikawa et al. 1996), lactosylceramide synthase (Nomura et al. 1997), or Gal-Cer synthase (Schulte and Stoffel 1993). Globo-series includes the reactions catalyzed by Gb3/CD77 synthase (Furukawa et al. 2007), Gb4 synthase (Furukawa et al. 2007), Gb5 synthase (Zhou et al. 2000), or sialyl-Gb5 synthase (monosialyl-galactosylgloboside, MSGG) (Saito et al. 2003). Furthermore, DSGG (disialyl-galactosylgloboside) (Furukawa et al. 2007) is synthesized from MSGG by ST6GalNAcVI. Lacto/neolacto-series are generated by various enzymes consisting of amino-CTH synthase (β3GlcNAcTV) (Togayachi et al. 2001), core 1 precursor synthase (β3GalTV) (Isshiki et al. 1999) β4-galactosyltransferase, α2,3/6-sialyltransferase, α1,3/4-fucosyl transferase, ST6GalNAc-VI (α2,6-sialyltransferase) (Furukawa et al. 2007), etc.

List

As shown in Table 1, glycosyltransferase genes involved in the synthesis of common structures were isolated in the 1990s and knockout mice of some genes were also established. Glycosyltransferase genes responsible for the synthesis of globo-series glycolipids were cloned and analyzed after 2000. Knockout mice of some genes were established. As for glycosyltransferase genes involved in the synthesis of lacto/neolacto-series, only a critical gene, β3GlcNAcT-V, and a newly defined sialyltransferase gene responsible for the synthesis of disialyl-Lewis a were listed. This is because β4GalTs and α2,3-SiaTs are redundant.

Protocols

Enzyme activities of these enzymes (Table 1) are measured under optimal conditions for individual enzymes. Therefore, references in which each enzyme was characterized should be used to actually establish enzyme activity-determining system for some glycosyltransferase. Here, enzyme assay protocol for the syntheses of disialyl Lewis a is shown as a representative example.

Department of Biochemistry II, Nagoya University Graduate School of Medicine,
65 Tsurumai, Showa-ku, Nagoya 456-0065, Japan
Phone: +81-52-744-2070, Fax: +81-52-744-2069
E-mail: koichi@med.nagoya-u.ac.jp

Table 1 Glycosyltransferases for the synthesis of glycosphingolipids

Gene name	Synonyms	Enzyme	Accession no.	Map	Acceptor tissue	EC	Reference
Glc-Cer	Glc-Cer synthase	Ceramide: glucosyltransferase					Ichikawa et al. (1996)
Gal-Cer	Gal-Cer synthase	Ceramide: galactosyltransferase					Schulte and Stoffel (1993)
β4GalT6	LacCer synthase	Glc-Cer: β4-galactosyltransferase					Nomura et al. (1997)
α4GalT	Gb3/CD77 synthase	LacCer: α4-galactosyltransferase					Furukawa et al. (2007)
α3GalT	Iso-Gb3 synthase	LacCer: α3galactosyltransferase					Keusch et al. (2000b)
β3GalT3	Gb4 synthase	Gb3: β3GalNAc transferase					Furukawa et al. (2007)
β3GalTV	Gb5 synthase	Gb4: β3galactosyltransferase					Zhou et al. (2000)
ST3GalTII	Sialyl-Gb5 synthase	Gb5: α3galactosyltransferase					Saito et al. (2003)
ST6GalNAcVI	DSGG synthase	MSGG: α6sialyltransferase					Furukawa et al. (2007)
β3GlcNAcTV	Amino-CTH synthase	LacCer: β3GlcNAc transferase					Togayachi et al. (2001)
β3GalTV	Core 1 precursor synthase	AminoCTH: β3-galactosyltransferase					Isshiki et al. (1999)
ST6GalNAcVI	Disialyl-Lewis a synthase	Sialyl-Lc4: α6-sialyltransferase					Furukawa et al. (2007)

Preparation of Membrane Fraction

1. L cells (3×10^6) are plated in 10 cm dishes at least 48 h prior to transfection. Cells are transiently transfected with an expression plasmid by the DEAE-dextran method.
2. After 48 h of culture, the cells are harvested by trypsinization. Cells were lysed in ice-cold PBS containing 1 mM PMSF using a nitrogen cavitation apparatus at 400 p.s.i. for 30 min.
3. After removing nuclei by low speed centrifugation, supernatants are centrifuged at $100,000 \times g$ for 1 h at 4°C.
4. The pellet is resuspended in ice-cold 100 mM sodium cacodylate buffer, pH 7.0, and used as an enzyme source.

Sialyltransferase Assay

1. The sialyltransferase assay is performed in a mixture containing 100 mM sodium cacodylate buffer, pH 6.0, 10 mM $MgCl_2$, 0.3% Triton CF-54, 0.66 mM CMP-NeuAc (Sigma), 6,000 dpm/µl CMP-[^{14}C]NeuAc (Amersham Biosciences), the enzyme solution, and 10 µg of acceptors in a total volume of 50 µl.
2. The reaction mixture is incubated at 37°C for 2 h.
3. The products were isolated using a C_{18} Sep-Pak cartridge (Waters, Milford, MA, USA) and analyzed by TLC.
4. The radioactivity on each plate was visualized with an image analyzer (Fuji Film, Tokyo, Japan).
5. For kinetic analysis, incubation was performed using different concentrations of acceptor substrates, 0–0.2 mM sialyl Lc4.

References

Furukawa K, Tsuchida A, Furukawa K (2007) Biosynthesis of glycolipids. In: Kamerling JP (ed.) Comprehensive glycoscience pp 105–114, Elsevier, Oxford, UK

Ichikawa S, Sakiyama H, Suzuki G et al. (1996) Expression cloning of a cDNA for human ceramide glucosyltransferase that catalyzes the first glycosylation step of glycosphingolipid synthesis. Proc Natl Acad Sci USA 93:4638–4643

Isshiki S, Togayachi A, Kudo T et al. (1999) Cloning, expression, and characterization of a novel UDP-galactose:beta-N-acetylglucosamine beta1,3-galactosyltransferase (beta3Gal-T5) responsible for synthesis of type 1 chain in colorectal and pancreatic epithelia and tumor cells derived therefrom. J Biol Chem 274:12499–12507

Keusch JJ, Manzella SM, Nyame KA et al. (2000b) Expression cloning of a new member of the ABO blood group glycosyltransferases, iGb3 synthase, that directs the synthesis of isoglobo-glycosphingolipids. J Biol Chem 275:25308–25314

Nomura T, Takizawa M, Wakisaka E et al. (1997) Purification, cDNA cloning, and expression of UDP-Gal: glucosylceramide beta-1,4-galactosyltransferase from rat brain. J Biol Chem 273: 13570–13577

Saito S, Aoki H, Ito A et al. (2003) Human alpha2,3-sialyltransferase (ST3Gal II) is a stage-specific embryonic antigen-4 synthase. J Biol Chem 278:26474–26479

Schulte S, Stoffel W (1993) Ceramide UDPgalactosyltransferase from myelinating rat brain: purification, cloning, and expression. Proc Natl Acad Sci USA 90:10265–10269

Togayachi A, Akashima T, Ookubo R et al. (2001) Molecular cloning and characterization of UDP-GlcNAc:lactosylceramide beta 1,3-N-acetylglucosaminyltransferase (beta 3Gn-T5), an essential enzyme for the expression of HNK-1 and Lewis X epitopes on glycolipids. J Biol Chem 276:22032–22040

Zhou D, Henion TR, Jungalwala FB et al. (2000) The beta 1,3-galactosyltransferase beta 3GalT-V is a stage-specific embryonic antigen-3 (SSEA-3) synthase. J Biol Chem 275:22631–22634

Sialyltransferases and Other Enzymes Involved in the Biosynthesis of Gangliosides

Koichi Furukawa, Keiko Furukawa

Introduction

The sialyltransferases involved in the biosynthesis of gangliosides were summarized in Table 1. The simplest sialosylglycolipid is GM3 except for GM4 (sialylgalactosyl ceramide). GM3 is located at the starting point of all ganglio-series glycolipids, extending to a-series, b-series and c-series gangliosides. GM3 synthase is called ST3Gal-V (Tsuji et al. 1996; Furukawa et al. 2007). GD3 synthase is the key enzyme for the synthesis of b-series and c-series gangliosides (GT3 synthase, Nakayama et al. 1996). GD3 synthase is called ST8Sia-I (Tsuji et al. 1996; Furukawa et al. 2007). After extending ganglio-core chain by GM2/GD2/GA2 synthase (Furukawa et al. 2007) and GM1/GD1b/GA1 synthase (Furukawa et al. 2007), GD1a/GT1b/GM1b synthase adds NeuAc to Gal at the non-reducing end (Tsuji et al. 1996). Then GT1a/GQ1b/GD1c synthase catalyzes biosynthesis of GT1a, GQ1b, or GD1c from GD1a, GT1b, or GM1b, respectively (Tsuji et al. 1996).

In addition, α-series gangliosides are also synthesized by ST6GalNAc-III (Sjoberg et al. 1996; Furukawa et al. 2007), ST6GalNAc-V (Furukawa et al. 2007), and ST6GalNAc-VI (Furukawa et al. 2007). ST6GalNAc-V synthesizes GD1α in brain tissues, and ST6GalNAc-VI synthesizes GT1aα and GQ1bα from GM1b or GD1a and GT1b, respectively. This enzyme also produces GD1α.

Besides ganglioside-series gangliosides, sialyl-glycolipids are generated with globo-series and lacto/neolacto-series glycolipids. All these structures are synthesized via common precursors, such as Glc-Cer and lactosylceramide. In globo-series, MSGG (monosialyl-galactosylgloboside) and DSGG (disialyl-galactosylgloboside) are synthesized via Gb3/CD77, Gb4, and Gb5. MSGG is synthesized by ST3Gal-II (Saito et al. 2003), and DSGG is synthesized by ST6GalNAc-VI (Senda et al. 2007). In lacto/neolacto-series, sialyl-paragloboside (Furukawa et al. 2007), sialyl-Lewis a, or sialyl-Lewis x are synthesized via the biosynthesis of amino-CTH synthase, β3/4-galactosyltransferase, α2.3-sialyltransferase, and α1,3/4-fucosyltransferase. Recently identified disialyl-Lewis a is synthesized by ST6GalNAc-VI (α2,6-sialyl-transferase) (Furukawa et al. 2007).

List

In Table 1, sialyltransferases and other glycosyltransferases, responsible for the synthesis of ganglio-series gangliosides, are summarized first. Sialyltransferaess involved in the biosynthesis of gangliosides with globo-core are then listed. As for sialyltransferases

---

Department of Biochemistry II, Nagoya University Graduate School of Medicine, 65 Tsurumai, Showa-ku, Nagoya 466-0065, Japan
Phone: +81-52-744-2070, Fax: +81-52-744-2069
E-mail: koichi@med.nagoya-u.ac.jp

Table 1 Sialyltransferases involved in the biosynthesis of gangliosides

Gene name	Anonyms	Enzyme	Accession No.**	EC	References
GM3 synthase	ST3 Sia-V, SAT-I	alpha2,3Sia-T	NM_003896 NP_003887	2.4.99.9	Furukawa et al. 2007
GD3 synthase	ST8Sia-I, SAT-II	alpha2,8Sia-T	NM_003034 NP_003025	2.4.99.8	Furukawa et al. 2007
GT3 synthase	ST8Sia-I	alpha2,8Sia-T	NM_003034	2.4.99.8	Nakayama et al. 1996
GM2/GD2 synthase	Beta4GalNAcT-I	beta,4-GalNAc-T	NM_001478	2.4.1.92	Furukawa et al. 2007
GM1/GD1b/GA1 synthase	Beta3GalT IV	beta3 Gal T	NM_017420 (M) NP_062293	2.4.1.62	Furukawa et al. 2007
GD1a/GT1b/GM1b synthase	SAT-IV, ST3Gal-II	alpha2,3Sia-T	U63090 Q16842	2.4.99.2	Tsuji 1996
GQ1b/GT1a/GD1c synthase	ST8Sia-V, SAT-V	alpha2,8Sia-T	NM_013305 NP_037437		Tsuju 1996
ST6GalNAcIII	STY, ST6O-II	alpha2,6Sia-T	NM_152996 NP_694541		Sjoberg et al. 1996
ST6GalNAcV	GD1α synthase	alpha2,6Sia-T	NM_030965 Q9BVH7		Furukawa et al. 2007
ST6GAINAcVI	GD1α/GT1aα/GQ1bα synthase	alpha2,6Sia-T	NM_013443 NP_038471		Furukawa et al. 2007
	Disialyl-Lewis a synthase DSGG synthase*				Furukawa et al. 2007 Senda M et al. 2007
Sialy-Gb5 synthase	α3SiaTII, MSGG synthase, SATIV	alpha2,3Sia-T	U63090 Q16842	2.4.99.2	Saito S et al. 2003

* Disialyl-galactosylgloboside synthase
** Unless otherwise mentioned, human gene or proteins are shown

involved in the synthesis of lacto/neolacto-series gangliosides, only a newly defined sialyltransferase responsible for the synthesis of disialyl-Lewis a was listed.

References

Furukawa K, Tsuchida A, Furukawa K (2007) Biosynthesis of glycolipids. In: Kamerling JP (ed.) Comprehensive glycoscience. Elsevier, Oxford, UK, pp 105–114

Nakayama J, Fukuda NM, Hirabayashi Y et al. (1996) Expression cloning of a human GT3 synthase, GD3 AND GT3 are synthesized by a single enzyme. J Biol Chem 271:3684–3691

Saito S, Aoki H, Ito A et al. (2003) Human alpha2,3-sialyltransferase (ST3Gal II) is a stage-specific embryonic antigen-4 synthase. J Biol Chem 278:26474–26479

Senda M, Ito A, Tsuchida A et al. (2007) Identification and expression of a sialyltransferase responsible for the synthesis of disialylgalactosylgloboside in normal and malignant kidney cells: downregulation of ST6GalNAc VI in renal cancers. Biochem J 402:459–470

Sjoberg ER, Kitagawa H, Glushka J et al. (1996) Molecular cloning of a developmentally regulated N-acetylgalactosamine alpha2,6-sialyltransferase specific for sialylated glycoconjugates. J Biol Chem 271:7450–7459

Tsuji S (1996) Molecular cloning and functional analysis of sialyltransferases. J Biochem (Tokyo) 120:1–13

Heparan Sulfate Synthases and Related Genes

Shuji Mizumoto[1], Hiroshi Kitagawa[2]

Introduction

Heparan sulfate (HS) and heparin (HP) are linear polysaccharides composed of repeating disaccharide units, [-4GlcAβ1(IdoAα1)-4GlcNAcα1-]n, where GlcA, IdoA, and GlcNAc represent glucuronic acid, iduronic acid, and N-acetylglucosamine, respectively, which are covalently linked to specific core proteins through the glycosaminoglycan (GAG)-protein linkage region tetrasaccharide, GlcAβ1-3Galβ1-3Galβ1-4Xylβ1-O-Ser (where Gal, Xyl, and Ser stand for galactose, xylose, and serine, respectively) forming proteoglycans (PGs). HS and HP backbones are synthesized by five glycosyltransferases, human exostosin (EXT) family members, which are associated with hereditary multiple exostoses (HME). Subsequently, HS and HP are modified by GlcNAc N-deacetylase/N-sulfotransferase, GlcA C5-epimerase, and O-sulfotransferases, resulting in the formation of functional domains and/or sequences (Mizumoto et al. 2005).

Human GAG-Protein Linkage Region Glycosyltransferases

Heparan sulfate and HP, in addition to chondroitin sulfate (CS) and dermatan sulfate (DS), have the GAG-protein linkage region tetrasaccharide, -GlcA-Gal-Gal-Xyl-, at the reducing side (Fig. 1a). After translation of a core protein, xylosylation of the Ser residue(s) in the core protein is initiated by the xylosyltransferase (XylT), which catalyzes the transfer of a Xyl residue through a β-linkage from uridine diphosphate (UDP)-Xyl to a Ser residue(s). In humans, two different β-XylTs, XylT-1 and -2, have been cloned. Subsequently, the Gal residue is transferred from UDP-Gal to the Xyl residue through a β1-4 linkage by galactosyltransferase-I (GalT-I) encoded by β4GalT7. The second Gal residue is transferred to the Gal residue through a β1-3 linkage by galactosyltransferase-II (GalT-II) encoded by β3GalT6. Finally, the linkage region is completed by the transfer of GlcA through a β1-3 linkage from UDP-GlcA to Galβ1-3Galβ1-4Xylβ1-O-Ser, which is catalyzed by glucuronosyltransferase-I (GlcAT-I) encoded by β3GAT3.

Human EXT Family Members

Chain polymerization of the HS-repeating disaccharide region is initiated by α1-4N-acetylglucosaminyltransferase-I (GlcNAcT-I), which transfers the first GlcNAc residue

---

[1] Laboratory of Proteoglycan Signaling and Therapeutics, Faculty of Advanced Life Sciences, Hokkaido University, Frontier Research Center for Post-Genomic Science and Technology, Kita 21-jo, Nishi 11-chome, Kita-ku, Sapporo 001-0021, Japan

[2] Department of Biochemistry, Kobe Pharmaceutical University, 4-19-1 Motoyamakita-machi, Higashinada-ku, Kobe 658-8558, Japan
Phone: +81-78-441-7570, Fax: +81-78-441-7569
E-mail: kitagawa@kobepharma-u.ac.jp

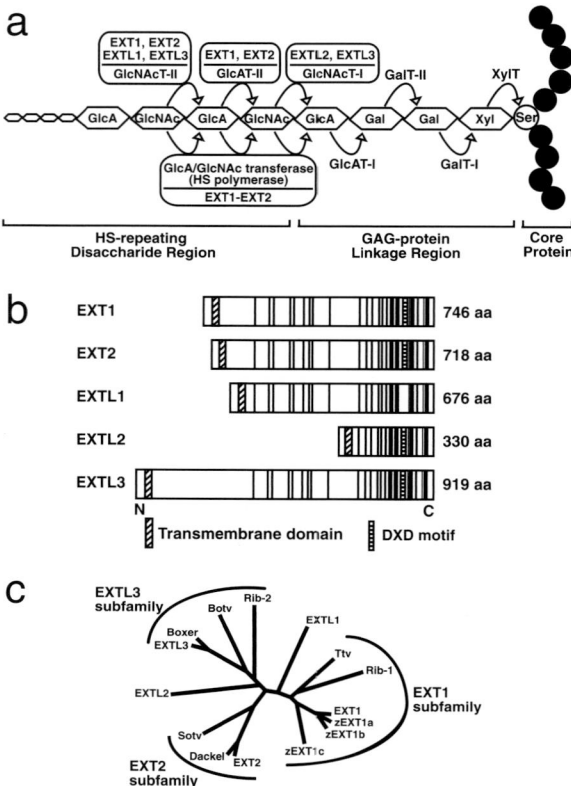

Fig. 1 a Biosynthetic assembly of HS and HP backbones by various glycosyltransferases. HS and HP are synthesized as PGs on specific Ser residues in the so-called GAG-protein linkage region tetrasaccharide, which is common to HS, HP, CS, and DS chains. Synthesis of the linkage region is initiated by XylT. Subsequently, two Gals and GlcA are attached to the resultant sugar chain by GalT-I, GalT-II, and GlcAT-I, respectively. Chain polymerization of the repeating disaccharide region is evoked by transfer of the first GlcNAc, which is mediated by GlcNAcT-I, and then is elongated by the alternating addition of GlcA and GlcNAc residues. **b** Comparison of the five cloned members of the human *EXT* gene family. All members harbor glycosyltransferases involved in HS and/or HP biosynthesis. Each protein shows significant homology at C-terminus. Putative transmembrane domains, D*X*D (D and *X* represent aspartate and any amino acid) motifs for UDP-sugar binding and the conserved region in these proteins are indicated by *oblique lines*, *horizontal lines*, and *black-lacquered boxes*, respectively. **c** Phylogenetic tree of EXT family members among human (EXT1, EXT2, EXTL1, EXTL2, and EXTL3), zebrafish (zEXT1a, zEXT1b, zEXT1c, Dackel, and Boxer), *D. melanogaster* (Ttv, Sotv, and Botv), and *C. elegans* (Rib-1 and Rib-2). These members can be classified into EXT1, EXT2, and EXTL3 subfamilies.

from UDP-GlcNAc to naked GlcA in GAG-protein linkage region tetrasaccharides (Fig. 1a). GlcNAcT-I is encoded by *exostosin like*-genes 2 and 3 (*EXTL2* and *EXTL3*). Three *EXTL* genes are homologous to the tumor suppressor genes, *exostosins* (*EXTs*) (Fig. 1b), which are related to a bone disease, hereditary multiple exostoses (HME). EXTL2 is an *N*-acetylhexosaminyltransferase that can transfer not only GlcNAc but also GalNAc to the linkage region. After transfer of the first GlcNAc residue, the resultant nascent pentasaccharide is elongated by alternate additions of GlcA and GlcNAc from UDP-GlcA

Table 1 Human heparan sulfate/heparin glycosyltransferases and epimerase

Abbreviation	Enzymatic activity	Chromosomal location	Amino acid	mRNA expression	mRNA accession
EXT1	GlcAT-II, GlcNAcT-II	8q24.11–q24.13	746	Ubiquitous	NM_000127
EXT2	GlcAT-II, GlcNAcT-II	11p12–p11	718	Ubiquitous	NM_000401
EXTL1	GlcNAcT-II	1p36.1	676	Skeletal muscle, brain, heart	NM_004455
EXTL2	GlcNAcT-I, α-GalNAcT	1p21	330	Ubiquitous	NM_001439
EXTL3	GlcNAcT-I, GlcNAcT-II	8p21	919	Ubiquitous	NM_001440
GLCE	GlcA C5-epimerase	15q23	617	Ubiquitous	NM_015554

and UDP-GlcNAc, which are catalyzed by β1-4glucuronyltransferase-II (GlcAT-II) and α1-4*N*-acetylglucosaminyltransferase-II (GlcNAcT-II), respectively (Fig. 1a, Table 1). EXTL1 and EXTL3 have GlcNAcT-II activity but not GlcAT-II activity. On the other hand, EXT1 and EXT2 have both GlcAT-II and GlcNAcT-II activities (Table 1). EXT1 and EXT2 form a hetero-dimeric complex in vivo and this protein complex (EXT1–EXT2) shows much higher glycosyltransferase activity than the individual proteins. Furthermore, the recombinant EXT1–EXT2 hetero-dimeric complex can synthesize HS polymer chains on the linkage region tetrasaccharide or HS oligosaccharides (Kim et al. 2003, see "Protocol" section). These observations can explain why mutations in either one cause HME with very similar clinical pathologies despite the presence of the two EXT proteins, and suggest that they cannot compensate for the loss of each other.

Nematode, Fruit Fly, and Zebrafish EXT Family Members

Heparan sulfate is present in model animals such as *Caenorhabditis elegans*, *Drosophila melanogaster*, and zebrafish (*Danio rerio*) (Mizumoto et al. 2005). In addition, EXT genes are also conserved between humans and these animals (Fig. 1c, Table 2). In *C. elegans*, *RIB-1* and *RIB-2* exhibit the highest homology to human EXT1 and EXTL3, respectively (Fig. 1c). RIB-2 harbors GlcNAcT-I and GlcNAcT-II activities but not GlcAT-II activity. Although RIB-1 shows little individual glycosyltransferase activity, the polymerization of HS chains was observed when RIB-1 was coexpressed with RIB-2, indicating that the mechanism of HS biosynthesis in *C. elegans* is similar to that of humans (Kitagawa et al. 2007). Both *rib-1* and *rib-2* mutant worms exhibit reduced synthesis of HS and embryonic lethality.

Drosophila homologs of *EXT* genes, *tout-velu* (*ttv*), *sister of tout-velu* (*sotv*), and *brother of tout-velu* (*botv*) that correspond to human *EXT1*, *EXT2*, and *EXTL3*, respectively, are involved in Hedgehog (Hh), Wingless (Wg), and Decapentaplegic (Dpp) signaling (Fig. 1c, Table 2). TTV and SOTV have both GlcNAcT-II and GlcAT-II activities, and no GlcNAcT-I activity (Izumikawa et al. 2006). In addition, the coexpression of TTV with SOTV markedly augmented both GlcNAcT-II and GlcAT-II activities when compared with the expression of TTV or SOTV alone (Izumikawa et al. 2006). On the other hand, BOTV has GlcNAcT-I and GlcNAcT-II activities, and no GlcAT-II activity. TTV-SOTV hetero-dimeric complex can polymerize HS chains on the linkage region pentasac-

Table 2 Model organisms with the loss- or gain-of-functions caused by mutations of the genes encoding *EXT* family members and epimerase

Affected genes	Encoded enzymes	Phenotypes
C. elegans		
rib-1	EXT1	Developmental abnormalities at embryonic stage
rib-2	EXTL3	Developmental abnormalities at embryonic stage
hse-5	C5-epimerase	Abnormal axon branching
D. melanogaster		
ttv	EXT1	Defects in Hh, Wg, Dpp signaling
sotv	EXT2	Defects in Hh, Wg, Dpp signaling
botv	EXTL3	Defects in Hh, Wg, Dpp signaling
Zebrafish (*Danio rerio*)		
dackel	EXT2	Defects in optic tract sorting and pectoral fins
boxer	EXTL3	Defects in optic tract sorting and pectoral fins
Mouse		
EXT1$^{-/-}$	EXT1	Disruption of gastrulation
EXT1$^{-/-}$ (specific for brain)	EXT1	Defects in the midbrain-hindbrain region, disturbed Wnt-1 distribution
EXT2$^{-/-}$	EXT2	Disruption of gastrulation
EXT2$^{+/-}$	EXT2	Exostoses
EXT2 transgenic mice	EXT2	Upregulation of the formation of trabeculae
Hsepi$^{-/-}$	C5-epimerase	Neonatal lethality with renal agenesis, lung defects, skeletal malformation
Human		
Hereditary multiple exostoses	EXT1, EXT2	An autosomal dominant disorder characterized by the formation of cartilage-capped tumors (exostoses) that develop from the growth plate of endochondral bones, especially of long bones

charide, GlcNAc-GlcA-Gal-Gal-Xyl-, but not on the linkage region tetrasaccharide, GlcA-Gal-Gal-Xyl- (Izumikawa et al. 2006). These findings suggest that in contrast to humans, BOTV, which possesses GlcNAcT-I activity required for the initiation of HS, is indispensable for the biosynthesis of HS chains in *Drosophila*.

In zebrafish, *dackel* and *boxer* mutants that are defective in *ext2* and *extl3*, respectively, show inappropriate projection of some dorsal retinal ganglion cell axons into the optic tract and developmental abnormality of the limb buds caused by the reduction of HS; however, DACKEL and BOXER have not been demonstrated to have glycosyltransferase activities.

Protocol

In vitro Polymerization

1. Construct each expression vector of EXT1 and EXT2 with certain features, e.g., insulin signal, His6 tag, etc. as desired.
2. Co-transfect with the expression vectors of both EXT1 and EXT2 into appropriate mammalian cells.
3. Purify recombinant EXT1–EXT2 hetero-oligomeric protein using appropriate resin beads corresponding to the tag used.

4. Prepare the reaction mixture for polymerization assays (total volume of 20 µl), which contains 10 µl of the purified enzyme, 100 mM MES–NaOH, pH 6.5, 10 mM $MnCl_2$, 1 mM ATP, 250 µM UDP-[^{3}H]GlcNAc (about 5.5×10^5 dpm), 250 µM UDP-GlcA, and a synthetic GAG-protein linkage region analog such as GlcAβ1-3Galβ1-O-C_2H_4NH-benzyloxycarbonyl and GlcAβ1-3Galβ1-O-naphthalenemethanol as an acceptor substrate.
5. Incubate at 37°C for 12 h.
6. Apply the reaction products to a Superdex 200 column (GE Healthcare) and measure the radioactivity of the collected fractions by liquid scintillation counting.
7. Digest the reaction products with 6 mIU of heparitinase I (Seikagaku Corp.) in a total volume of 100 µl of 20 mM sodium acetate buffer, pH 7.0, containing 2 mM calcium acetate at 37°C overnight and then analyze the digest on a Superdex Peptide column (GE Healthcare) to confirm the reaction products.

Comment

It is suggested that in vitro HS polymerization requires the enzyme complex of EXT1–EXT2.

References

Izumikawa T, Egusa N, Taniguchi F, Sugahara, K, Kitagawa H (2006) Heparan sulfate polymerization in *Drosophila*. J Biol Chem 281:1929–1934

Kitagawa H, Izumikawa T, Mizuguchi S, Dejima K, Nomura KH, Egusa N, Taniguchi F, Tamura J, Gengyo-Ando K, Mitani S, Nomura K, Sugahara K (2007) Expression of *rib-1*, a *Caenorhabditis elegans* homolog of the human tumor suppressor *EXT* genes is indispensable for heparan sulfate synthesis and embryonic morphogenesis. J Biol Chem 282:8533–8544

Kim BT, Kitagawa H, Tanaka J, Tamura J, Sugahara K (2003) In vitro heparan sulfate polymerization: crucial roles of core protein moieties of primer substrates in addition to the EXT1–EXT2 interaction. J Biol Chem 278:41618–41623

Mizumoto S, Kitagawa H, Sugahara K (2005) Biosynthesis of heparin and heparan sulfate. In: Garg HG, Linhardt RJ, Hales CA (eds) Chemistry and biology of heparin and heparan sulfate. Elsevier, Oxford, pp 203–243

Chondroitin Sulfate Biosynthesis and Related Genes

Hideto Watanabe, Koji Kimata

Introduction

Chondroitin sulfate (CS) comprises repeating disaccharide units of *N*-acetylgalactosamine (GalNAc) and glucuronic acid (GlcA) residues with sulfate residues at different positions. CS biosynthesis (Fig. 1) is initiated by the transfer of a GalNAc residue to the linkage region of a GlcA-β1, 3-galactose (Gal)-β1, 3-Gal-β1,4-Xyl tetrasaccharide primer that is attached to a serine residue of the core protein. Then, chain elongation occurs by the alternate addition of GalNAc and GlcA residues. Enzyme activities that catalyze the initiation and elongation processes are termed as glycosyltransferase-I and II activities, respectively. To date, six glycosyltransferases involved in CS synthesis have been identified (Fig. 2): chondroitin sulfate synthase-1 (CSS-1)/chondroitin synthase (CSy) (Kitagawa et al. 2001), chondroitin sulfate synthase-2 (CSS-2)/chondroitin polymerizing factor (ChPF) (Yada et al. 2003), chondroitin sulfate synthase-3 (CSS-3), chondroitin sulfate glucuronyltransferase (CSGlcAT) (Gotoh et al. 2002a), chondroitin sulfate *N*-acetylgalactosaminyltransferase-1 (Gotoh et al. 2002b), and 2 (Sato et al. 2003) (CSGalNAcT-1, 2). All these enzymes have an N-terminal transmembrane domain and are localized to the Golgi apparatus where CS biosynthesis takes place. CSS-1, CSS-2, CSS-3, and CSGlcAT form a family of glycosyltransferase enzymes. CSS-1, CSS-2, and CSS-3 contain two glycosyltransferase domains and exhibit both *N*-acetylgalactosaminyltransferase (GalNAcT) and glucuronyltransferase (GlcAT) activities in chain elongation. Thus, they have glycosyltransferase-II (both GalNAcT-II and GlcAT-II) activity. CSGlcAT has an inactive GalNAcT domain and exhibits only GlcAT-II activity. CSGalNAcT-1 and 2 have one glycosyltransferase domain and exhibit GalNAcT activity in both the initiation and elongation processes, indicating that CSGalNAcT-1 and 2 have both GalNAcT-I and II activities. Thus, CS chain elongation may involve all six enzymes, whereas CS chain initiation likely involves only CSGalNAcT-1 and 2.

Recent studies have revealed that CSGalNAcT-1 plays a critical role in CS biosynthesis in cartilage, which contains a large amount of CS. The initiation rather than the elongation is likely the limiting step of CS biosynthesis.

Glycosyltransferase Assay

A. GalNAcT-II Assay Using CS Polymer as an Acceptor:

1. Add 3.0 μl 0.5 M MES, pH 6.2, and 0.1 M $MnCl_2$ to a microtube.
2. Add 10 μl sample solution.

Institute for Molecular Science of Medicine, Aichi Medical University, Nagakute, Aichi 480-1195, Japan
Phone: +81-561-62-3311, Fax: +81-561-63-3532
E-mail: wannabee@aichi-med-u.ac.jp

Chondroitin Sulfate Biosynthesis and Related Genes 65

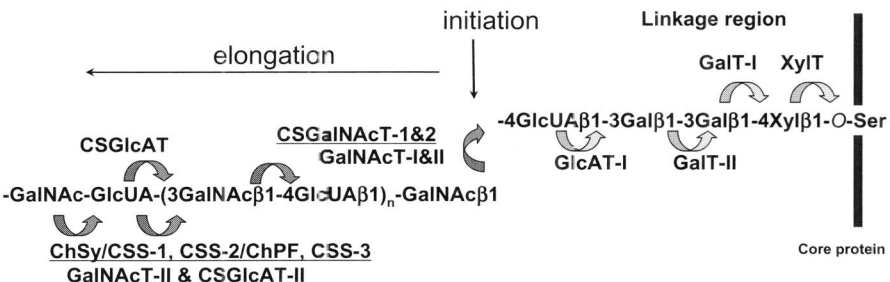

Fig. 1 Catalytic activities of glycosyltransferases in CS biosynthesis

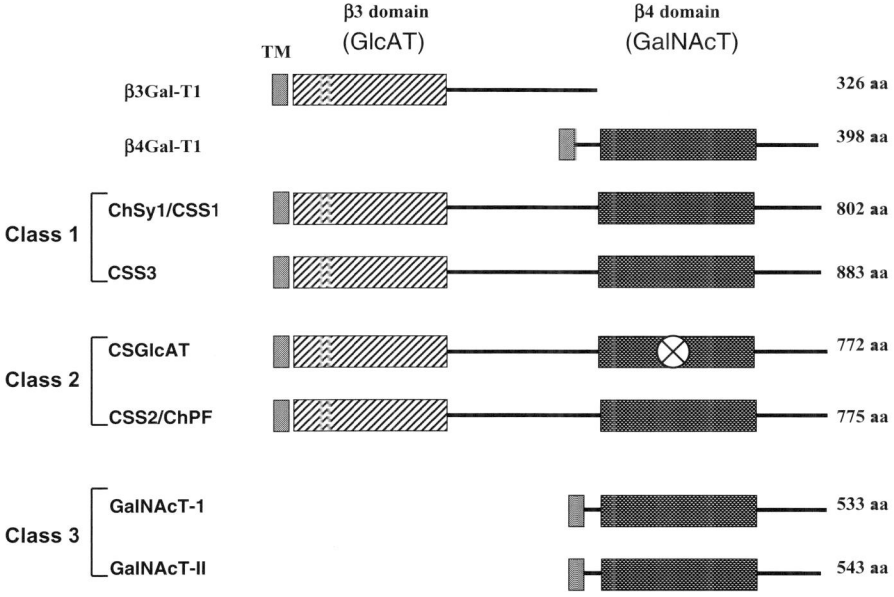

Fig. 2 Structural comparison of the six glycosyltransferases involved in CS biosynthesis

3. Add 1-100 μg CS.
4. Add 0.32 nmol of UDP-[^{3}H]GalNAc (6.66×10^5 dpm).
5. Add dH$_2$O to final volume of 30 μl.
6. Incubate at 37°C for 1 h or longer.
7. Stop the reaction by boiling the sample for 5 min.
8. Apply the reaction mixture to Superdex® peptide HR10/30 column (10 × 300 mm) with 0.2 M NaCl as an eluant and remove free UDP-[^{3}H]GalNAc.
9. Measure the radioactivity of the sample recovered by liquid scintillation counting.

B. GlcAT-II Assay Using CS Polymer as an Acceptor:

1. Add 3.0 µl 0.5 M MES, pH 6.2, and 0.1 M $MnCl_2$ to a microtube.
2. Add 10 µl sample solution.
3. Add 1-100 µg CS.
4. Add 0.3073nmol of UDP-[^{12}C]GlcA (2.22 × 10^5 dpm).
5. Add dH_2O to final volume of 30 µl.
6. Incubate at 37°C for 1 h or longer.
7. Stop the reaction by boiling the sample for 5 min.
8. Apply the reaction mixture to Superdex® peptide HR10/30 column (10 × 300 mm) with 0.2 M NaCl as an eluant and remove free UDP-[^{12}C]GlcA.
9. Measure the radioactivity of the sample recovered by liquid scintillation counting.

Comments

Recent studies have suggested that a heterodimer of CS synthetic enzymes exhibits CS polymerizing activity. For assay of the polymerizing activity, cold UDP-GlcA in addition to UDP-[^{3}H]GalNAc or cold UDP-galNAc to UDP-[^{14}C]glcA could be added in the reaction.

References

Kitagawa H, Uyama T, Sugahara K (2001) Molecular cloning and expression of human chondroitin synthase. J Biol Chem 276:38721–38726

Gotoh M, Yada T, Sato T, Akashima T, Iwasaki H, Mochizuki H, Inaba N, Togayachi A, Kudo T, Watanabe H, Kimata K, Narimatsu H (2002a) Molecular cloning and characterization of a novel chondroitin sulfate glucuronyltransferase that transfers glucuronic acid to N-acetylgalactosamine. J Biol Chem 277:38179–38188

Gotoh M, Sato T, Akashima T, Iwasaki H, Kameyama A, Mochizuki H, Yada T, Inaba N, Zhang Y, Kikuchi N, Kwon YD, Togayachi A, Kudo T, Nishihara S, Watanabe H, Kimata K, Narimatsu H (2002b) Enzymatic synthesis of chondroitin with a novel chondroitin sulfate N-acetylgalactosaminyltransferase that transfers N-acetylgalactosamine to glucuronic acid in initiation and elongation of chondroitin sulfate synthesis. J Biol Chem 277:38189–38196

Sato T, Gotoh M, Kiyohara K, Akashima T, Iwasaki H, Kameyama A, Mochizuki H, Yada T, Inaba N, Togayachi A, Kudo T, Asada M, Watanabe H, Imamura T, Kimata K, Narimatsu H (2003) Differential roles of two N-acetylgalactosaminyltransferases, CSGalNAcT-1, and a novel enzyme, CSGalNAcT-2. Initiation and elongation in synthesis of chondroitin sulfate. J Biol Chem 278: 3063–3071

Yada T, Gotoh M, Sato T, Shionyu M, Go M, Kaseyama H, Iwasaki H, Kikuchi N, Kwon Y D, Togayachi A, Kudo T, Watanabe H, Narimatsu H, Kimata K (2003) Chondroitin sulfate synthase-2. Molecular cloning and characterization of a novel human glycosyltransferase homologous to chondroitin sulfate glucuronyltransferase, which has dual enzymatic activities. J Biol Chem 278:30235–30247

Biosynthesis of Keratan Sulfate

Akira Seko, Katsuko Yamashita

Introduction

Keratan sulfate (KS) is a poly-sulfated glycosaminoglycan with a poly-N-acetyllactosamine linear chain structure. KS is a rich component in cornea, and is also present in cartilage and brain. In cornea, KS plays important roles in maintaining water content and developing collagen matrix. Recent studies have demonstrated its association with Alzheimer's disease and arthritis.

Keratan sulfate is biosynthesized by sequential steps of galactosylation, N-acetylglucosaminylation, and sulfation, which are catalyzed by each specific glycosyl- or sulfotransferase. Fukuta et al. (1997) primarily described Gal 6-O-sulfotransferase (Gal6ST). Next, Akama et al. (2000) and Hayashida et al. (2006) showed that 6-O-sulfation of GlcNAc residues is catalyzed by GlcNAc6ST-5 in cornea, whereas Zhang et al. (2006) showed that GlcNAc6ST-1 is responsible for KS sulfation in brain. Seko et al. (2003, 2004) showed that β4GalT-IV and β3Gn-T7 have efficient enzymatic activities for elongating KS-related oligosaccharides. Recently, Kitayama et al. (2007) showed that β4GalT-IV and β3Gn-T7 are involved in the biosynthesis of keratan sulfate in vivo. From a compilation of these studies, a biosynthetic pathway for KS is proposed as shown in Fig. 1.

Perspectives

Hyaluronic acids and chondroitin sulfate are widely utilized in medical treatment and in the cosmetic industry. KS is expected to serve as a material for supporting eye function. Since enzymes involved in its biosynthesis are being identified, it is theoretically possible to supply KS in various sizes and degree of sulfation, some of which may have biological activities.

Protocol

Assay of β3Gn-T7

Reaction mixtures (20 µl), containing 50 mM HEPES–NaOH (pH 7.2), 10 mM MnCl$_2$, 0.5%(w/v) Triton X-100, 1 mM L2L2[Galβ1 → 4(SO$_3^-$ → 6)GlcNAcβ1 → 3Galβ1 → 4(SO$_3^-$ → 6)GlcNAc] (a gift from Seikagaku, Tokyo, Japan), 2.5 µM UDP-[^{3}H]GlcNAc (6.7 × 10^6 dpm), 30 µM UDP-GlcNAc, 1 mM AMP, and enzyme fractions, were incubated at 37°C for 1 h. The ^{3}H-labeled products were purified by paper electrophoresis (pyridine:acetic acid:water, 3:1:337, pH 5.4). A peak with the R_f value ~0.9 (bromophenol blue as standard) was [^{3}H]GlcNAcβ1 → 3L2L2, and its radioactivity was counted.

Innovative Research Initiatives, Tokyo Institute of Technology, Nagatsuta 4259, Midori-ku, Yokohama 226-8503, Japan
Phone: +81-45-921-4308, Fax: +81-45-921-4308
E-mail: kyamashi@bio.titech.ac.jp

Fig. 1 Biosynthesis of keratan sulfate. First, non-reducing terminal GlcNAc residues in N-linked/O-linked glycan chains are 6-O-sulfated by GlcNAc6ST-1 (brain) or -5 (cornea) (step 1). The 6-O-sulfated GlcNAc residues are next galactosylated at the 4-OH position by β4GalT-IV (step 2). Further elongation of backbone poly-N-acetyllactosamine chains occurs by β3Gn-T7 (step 3). Sequential reaction of steps 1, 2, and 3 allows the production of long 6-O-sulfated poly-N-acetyllactosamine chains, and then Gal6ST acts on part of internal and non-reducing Gal residues to sulfate at the 6-OH position

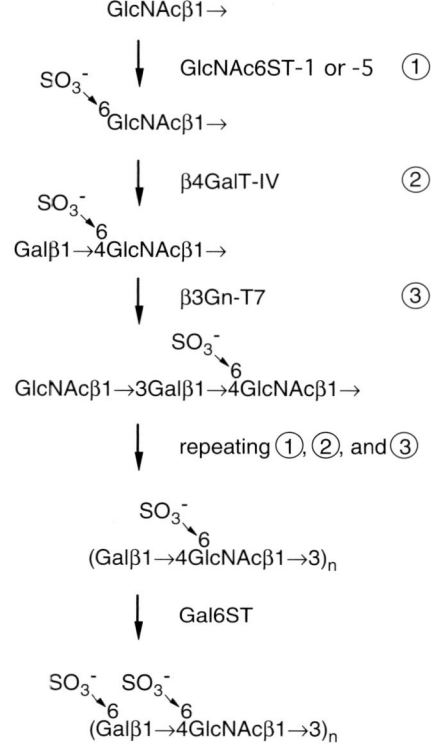

Proposed scheme of the biosynthesis of keratan sulfate

Assay of β4GalT-IV

Reaction mixtures (20 μl), containing 50 mM HEPES–NaOH (pH 7.2), 10 mM MnCl$_2$, 0.5%(w/v) Triton X-100, 1 mM agL2L2[SO$_3^-$ → 6GlcNAcβ1 → 3Galβ1 → 4(SO$_3^-$ → 6)GlcNAc], 0.3 μM UDP-[^{3}H]Gal (4.9 × 10^5 dpm), 250 μM UDP-Gal, 0.5 mM spermine, and enzyme fractions, were incubated at 37°C for 1 h. AgL2L2 was prepared from L2L2 by *Streptococcus* 6646K β-galactosidase (Seikagaku, Tokyo, Japan). The reaction mixtures were added to 500 μl of 0.01 N HCl and heated at 100°C for 10 min to destroy residual UDP-[^{3}H]Gal. The ^{3}H-labeled products were purified by paper electrophoresis (pyridine:acetic acid:water, 3:1:387, pH 5.4). A peak with the R_f value ~1.0 (bromophenol blue as standard) was [^{3}H]L2L2, and its radioactivity was counted.

References

Akama TO, Nishida K, Nakayama J, Watanabe H, Ozaki K, Nakamura T, Dota A, Kawasaki S, Inoue Y, Maeda N, Yamamoto S, Fujiwara T, Thonar EJMA, Shimomura Y, Kinoshita S, Tanigami A, Fukuda MN (2000) Macular corneal dystrophy type I and type II are caused by distinct mutations in a new sulphotransferase gene. Nat Genet 26:237–241

Fukuta M, Inazawa J, Torii T, Tsuzuki K, Shimada E, Habuchi O (1997) Molecular cloning and characterization of human keratan sulfate Gal-6-sulfotransferase. J Biol Chem 272:32321–32328

Hayashida Y, Akama TO, Beecher N, Lewis P, Young RD, Meek KM, Kerr B, Hughes CE, Caterson B, Tanigami A, Nakayama J, Fukuda MN, Tano Y, Nishida K, Quantock AJ (2006) Matrix morphogenesis in cornea is mediated by the modification of keratan sulfate by GlcNAc 6-O-sulfotransferase. Proc Natl Acad Sci USA 103:13333–13338

Kitayama K, Hayashida Y, Nishida K, Akama TO (2007) Enzymes responsible for synthesis of corneal keratan sulfate glycosaminoglycans. J Biol Chem 282:30085–96

Seko A, Dohmae N, Takio K, Yamashita K (2003) Beta 1,4-galactosyltransferase (beta4GalT)-IV is specific for GlcNAc 6-O-sulfate. Beta 4GalT-IV acts on keratan sulfate-related glycans and a precursor glycan of 6-sulfosialyl-Lewis X. J Biol Chem 278:9150–9158

Seko A, Yamashita K (2004) beta 1,3-N-Acetylglucosaminyltransferase-7 (beta3Gn-T7) acts efficiently on keratan sulfate-related glycans. FEBS Lett 556:216–220

Zhang H, Muramatsu T, Murase A, Yuasa S, Uchimura K, Kadomatsu K (2006) N-Acetylglucosamine 6-O-sulfotransferase-1 is required for brain keratan sulfate biosynthesis and glial scar formation after brain injury. Glycobiology 16:702–710

Hyaluronan Synthase Assay

Naoki Itano

Introduction

Hyaluronan (HA) is a polysaccharide composed of repeating GlcNAcβ(1 → 4)-GlcUAβ(1 → 3) disaccharide units. This polysaccharide has a molecular mass ranging from 10^3 to 10^7 Da, depending on the tissue source and physiological conditions. Three members of the HA synthase family, HAS1, HAS2 and HAS3, have been thus far identified in a wide variety of vertebrates. Sequence analysis has shown that all HAS proteins are composed of multiple membrane-spanning regions and have large cytoplasmic loops. The HAS proteins also possess a DXD motif that is common in the glycosyltransferases of many organisms, and a QXXRW motif conserved in cellulose and chitin synthases. Each HAS isoform differs in enzymatic properties such as stability, substrate kinetics, and rate of sugar chain elongation, and may thus synthesize distinct forms of HA. The HAS1 and HAS3 synthesize HA with an estimated molecular mass of 2×10^5 to 2×10^6 Da, whereas very large molecules of more than 2×10^6 Da are synthesized by HAS2. The HAS genes also show temporally and spatially distinct expression patterns.

Enzyme Reaction

The HAS activity is monitored in a cell-free HA synthesis system using UDP-[^{14}C]GlcUA and UDP-GlcNAc as donors and a membrane-rich fraction of HAS-expressing cells as an enzyme source (Itano et al. 1999).

1. Prepare the reaction buffer in a final volume of 0.2 mL of 25 mM Hepes–NaOH buffer, pH 7.1, 5 mM DTT, 15 mM $MgCl_2$, 0.1 mM UDP-GlcNAc, 2 μM UDP-GlcUA, and 2 μCi of UDP-[^{14}C]GlcUA.
2. Harvest HAS-expressing cells and disrupt by sonication in lysis buffer (10 mM Hepes–NaOH buffer, pH 7.1, 0.5 mM dithiothreitol, 0.25 M sucrose) on ice.
3. Ultracentrifuge the cell lysate at $105,000 \times g$ at 4°C for 1 h to give high-speed pellets.
4. Suspend the crude membrane fractions (2–20 μg protein) with reaction buffer.
5. Incubated at 37°C for 1 h with or without 1 TRU of *Streptomyces* hyaluronidase.
6. Stop the reaction by the addition of SDS to a final concentration 2% (w/v).
7. Spot the mixtures onto Whatman no. 3 MM paper.
8. Transfer the paper to a paper chromatography chamber.
9. Perform descending paper chromatography in the solvent containing 1 M ammonium acetate (pH 5.5) and ethanol (65:35 v/v) for 3 days.

Department of Molecular Oncology, Division of Molecular and Cellular Biology, Institute on Aging and Adaptation, Shinshu University Graduate School of Medicine, Matsumoto, Nagano 390-8621, Japan
Phone: +81-263-37-2722, Fax: +81-263-37-2724
E-mail: itano@sch.md.shinshu-u.ac.jp

10. Cut out the origin of the paper strip where the synthesized polymers are retained.
11. Determine the amount of radioactivity in the hyaluronidase-sensitive HA by liquid scintillation counting.

Determination of HA Size by Agarose Gel Electrophoresis

The size distribution of radiolabeled HA synthesized in the cell-free reaction is analyzed by agarose gel electrophoresis (Lee et al. 1994).

1. Incubate the synthesized HA at 37°C for 1 h with or without 1 TRU of *Streptomyces* hyaluronidase.
2. Mix 14 µl of the radiolabeled HA sample or HA standard with 2 µl of TAE buffer (40 mM Tris–HCl buffer, pH 7.9, 5 mM sodium acetate, 0.8 mM sodium EDTA) containing 2 M socrose.
3. Prepare a 0.5% (w/v) agarose gel (20 × 20 cm^2) by melting 0.6 g agarose in 108 ml of TAE buffer.
4. Transfer the gel plate to the horizontal electrophoresis apparatus.
5. Apply the sample to each lane of the agarose gel.
6. Electrophorese at 2 V/cm for 10 h in TAE buffer at room temperature.
7. Stain the gel for 4 h under light-protective cover at room temperature in a solution containing 0.005% Stains-All in 50% ethanol.
8. Destain the gel in water and then dry.
9. Photograph and detect the radioactive HA on X-ray film.

Comment

HA samples with average molecular mass ranging from 10^5 to 2×10^6 Da are used for standards.

References

Itano N, Sawai T, Yoshida M, Lenas P, Yamada Y, Imagawa M, Shinomura T, Hamaguchi M, Yoshida Y, Ohnuki Y, Miyauchi S, Spicer AP, McDonald JA, and Kimata K (1999) Three isoforms of mammalian hyaluronan synthases have distinct enzymatic properties. J Biol Chem 274:25085–25092

Lee HG, Cowman MK (1994) An agarose gel electrophoretic method for analysis of hyaluronan molecular weight distribution. Anal Biochem 219:278–287

Glycosylphosphatidylinositol-Anchored Protein Biosynthesis and Related Genes in Mammalian Cells

Yusuke Maeda, Taroh Kinoshita

Introduction

In the late 1970s, Hiroh Ikezawa and Martin Low found that phosphatidylinositol (PI)-specific phospholipase C (PI-PLC) released membrane-bound hydrolases from the mammalian cell surface and predicted the presence of enzymes, which were anchored to the plasma membrane via PI. A number of researchers who studied *Trypanosome brucei*, which is a parasite causing a sleeping sickness, noticed that variant surface glycoproteins (VSG) were anchored to the surface membrane via lipids. These findings led to the discovery of glycosylphosphatidylinositol (GPI)-anchored proteins (GPI-APs). In 1988, the first complete GPI structures for VSG and rat Thy1 were solved. Studies of *T. brucei* in which the activity for GPI biosynthesis is relatively high have described the biosynthesis pathway. Nevertheless, identification of enzymes and genes involved in the GPI biosynthesis pathway had been difficult and unsuccessful.

Characterization of Mammalian Genes Involved in the Biosynthesis of GPI-APs

Many T cell lymphoma cell lines lacking surface Thy1, as well as other cell lines, are deficient in various steps of GPI biosynthesis. Kinoshita et al. considered that the expression cloning of genes responsible for these cells should be a good approach for revealing the biosynthesis of GPI-APs. They established a method in which they transfected a cDNA library into GPI-AP-negative cells, collected cells with restored surface expression of GPI-APs by a cell sorter, and rescued cDNA clones. Thus, the first gene involved in the biosynthesis of GPI-APs was identified in 1993 (Miyata et al. 1993). This gene encoded the catalytic component of an enzyme complex, GPI-N-acetylglucosamine transferase (GPI-GnT), which mediates the first step in GPI biosynthesis. It was named PIG-A (Phosphatidyl Inositol Glycan-class A) because the mutant cells used in the expression cloning belonged to complementation class A. They also found that a defect of the PIG-A gene causes paroxysmal nocturnal hemoglobinuria (PNH), a GPI-deficient disease (Takeda et al. 1993) (refer to Chapter by Y. Murakami and T. Kinoshita, this volume).

Kinoshita's laboratory then aimed to identify all the genes involved in this pathway. They not only looked for new classes of GPI mutant cells from various cell lines but also established GPI mutant cell lines from Chinese hamster ovary (CHO) cells, and

---

Research Institute for Microbial Diseases, Osaka University, 3-1 Yamada-oka, Suita, Osaka 565-0871, Japan
Phone: +81-6-6879-8328, Fax: +81-6-6875-5233
E-mail: tkinoshi@biken.osaka-u.ac.jp

identified several new genes such as PIG-B, PIG-F, PIG-L, PIG-M, PIG-V, PIG-W and PIG-X by expression cloning. Tetsu Kamitani cloned PIG-H in a similar way.

In budding yeast, the biosynthesis of GPI is essential for survival. Research groups in USA and Switzerland established temperature-sensitive mutant strains and cloned new genes involved in the GPI pathway by monitoring the restoration of growth at non-permissive temperatures. Based on sequence homology with yeast genes identified using these temperature sensitive mutant strains, Kinoshita's laboratory found mammalian homologue candidates from the genome database. The involvement of these genes in GPI biosynthesis was verified by examining the capability to complement GPI-defective mutant cells in which the gene responsible had not been cloned; PIG-C, GPI7, GPI8 (PIG-K) and DPM1 were identified in this way. In the case that the genes did not complement any mutant cell lines, the genes were knocked out in F9 cells, and GPI1 (PIG-Q), GAA1, PIG-N and PIG-O were identified.

As the knowledge about biosynthetic enzymes increased, it appeared that many enzymes are protein complexes. We purified protein complexes from cells over-expressing a tagged known component to identify new components. We expressed tandem-tagged GPI8 in GPI8-defective cells, affinity-purified the GPI-transamidase protein complex from the lysate, and identified PIG-S, PIG-T and PIG-U as new components of GPI-transamidase. Using similar methods, we identified PIG-P as a component of GPI-GnT involved in the first step of the biosynthetic pathway, and identified DPM2 and DPM3 as components of dolichol-phosphate mannose synthase.

Twenty-six human genes involved in GPI biosynthesis have been identified by these methods (Fig. 1). Thus, more than 90% of our aim of revealing all genes involved in the biosynthesis has been achieved (Maeda et al. 2006).

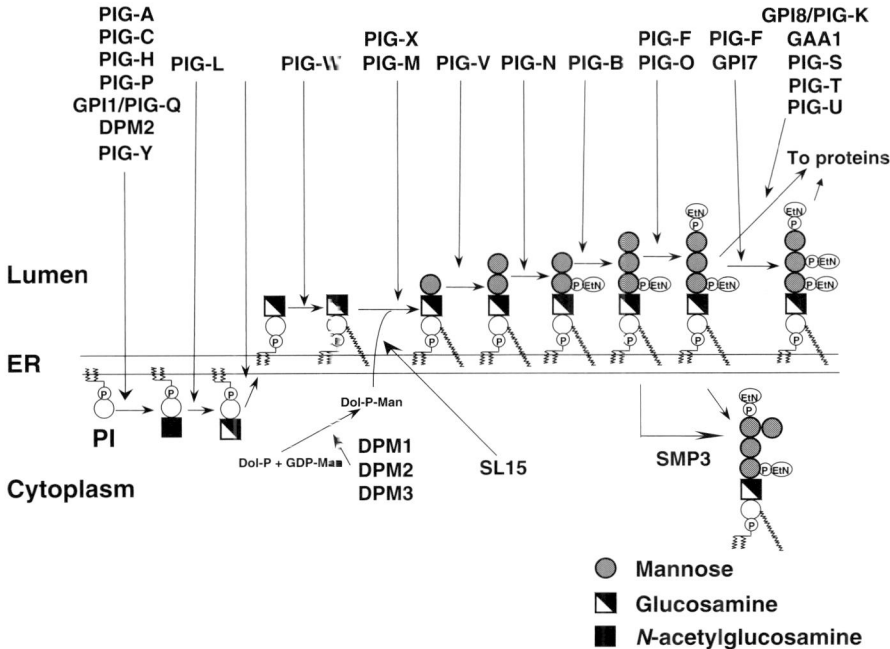

Fig. 1 Biosynthesis of GPI in mammalian cells

As a result of many studies, it was found that (1) the reactions of early steps take place in the cytosolic side of the endoplasmic reticulum (ER), and the second intermediate flip-flops and translocates into the lumenal side, on which the later steps of biosynthesis and the transamidation to proteins are executed, (2) four mannosyltransferases, PIG-M, PIG-V, PIG-B and SMP3, belong to a super-family including many glycosyltransferases, which utilize dolichol-phosphate mono-sugar as a substrate in the ER, (3) among them, PIG-B and SMP3 form a family with Alg9 and Alg12, indicating that the biosynthesis of GPI is evolutionally related with that of N-glycosylation, and (4) many glycosyltransferases involved in GPI biosynthesis are composed of several protein components suggesting the presence of elaborate regulation. Thus, there has been a lot of clarification about genes involved in GPI biosynthesis and the addition to precursor proteins. In the future, a higher level of studies such as analysis of structure and reaction mechanism of each enzyme and how the total biosynthetic pathway is spatially and temporally orchestrated in the ER are required.

Many genes involved in GPI biosynthesis in parasites such as *Trypanosoma* and *Plasmodium* are also being cloned based on the human and yeast sequence information. They are promising targets for the development of new drugs against global parasitic diseases. The GPI biosynthetic pathway is also considered to be a good target for fungicides, as GPI is essential for their survival.

GPI-APs and Lipid Rafts

GPI-APs are transported along the secretary pathway via the Golgi to the plasma membrane and concentrated in microdomains, so called lipid rafts, which are enriched in glycosphingolipids, cholesterol and non-receptor type tyrosine kinases. Whereas most of the cellular PIs, from which GPI is biosynthesized, contain unsaturated fatty chains, GPI-APs usually have two saturated chains in the PI moiety that are considered to be compatible with the liquid-ordered raft membrane. Very recently we reported that GPI-APs with two saturated fatty chains are generated from those bearing an unsaturated chain by fatty acid remodeling that most likely occurs in the Golgi and requires PGAP2 and PGAP3 (Maeda et al. 2007). We have also shown that the remodeling requires the preceding PGAP1-mediated modification of GPI-APs, namely deacylation of palmitate from inositol, which is critical to render GPI-APs competent for association with lipid rafts. We are now focusing on the identification of PGAP (Post GPI Attachment to Proteins) genes involved in lipid remodeling and other events occurring after attachment of GPI to precursor proteins. In polarized epithelial cells, GPI-APs are sorted to the apical membrane, presumably using rafts as carriers. Upon ligand binding, many receptors are sorted to lipid rafts and form platforms to start intracellular signaling. Such lipid rafts are also utilized by many microbes for invasion. Further studies of GPI-AP biosynthesis are required to elucidate the functions of lipid rafts and the mechanisms of raft formation.

References

Maeda Y, Ashida H, Kinoshita T (2006) CHO glycosylation mutants: GPI anchor. Meth Enzymol 416:182–205

Maeda Y, Tashima Y, Houjou T, Fujita M, Yoko-o T, Jigami Y, Taguchi R, Kinoshita T (2007) Fatty acid remodeling of GPI-anchored proteins is required for their raft association. Mol Biol Cell 18:1497–1506

Miyata T, Takeda J, Iida Y, Yamada N, Inoue N, Takahashi M, Maeda K, Kitani T, Kinoshita T (1993) Cloning of PIG-A, a component in the early step of GPI-anchor biosynthesis. Science 259:1318–1320

Takeda J, Miyata T, Kawagoe K, Iida Y, Endo Y, Fujita T, Takahashi M, Kitani T, Kinoshita T (1993) Deficiency of the GPI anchor caused by a somatic mutation of the PIG-A gene in paroxysmal nocturnal hemoglobinuria. Cell 73:703–711

Cloning of Genes for N-linked Oligosaccharide Synthesis in Endoplasmic Reticulum

Ken-ichi Nakayama[1], Yoshifumi Jigami[2]

Introduction

A congenital disorder of glycosylation syndrome (CDGS) is an autosomal recessive disorder on protein N-glycosylation. The mutated genes causing CDGS are found in N-linked oligosaccharide synthesis gene in the endoplasmic reticulum (ER) (Chantret et al. 2003; Frank et al. 2004). The N-glycosylation pathway in the ER is almost the same in any eukaryotic cells from yeast to human. Since most of the N-glycosylation genes and corresponding mutant strains have been isolated from yeast, we cloned and characterized human N-glycosylation genes in the ER by the phenotypic complementation of yeast N-glycosylation mutant cells.

Procedure

Cloning of Human ALG11 Gene

For the cloning of human *ALG11* gene, *Shizosaccharomyces pombe gmd3* mutant was used (Umeda et al. 2000). Since the *S. pombe gmd3+* gene is a homolog of *Saccharomyces cerevisiae ALG11* gene and shows temperature sensitivity (*ts*, growth defect at 37°C) and protein glycosylation deficiency at 37°C, we cloned human *ALG11* gene by the complementation of both *ts* phenotype and glycosylation defect of *S. pombe gmd3* mutant.

1. Human *ALG11* candidate gene FLJ21803 ORF was amplified by PCR (94°C: 15 s, 49°C: 30 s, 72°C: 3 min, 30 cycles) using human cDNA library as a template.
2. FLJ21803 ORF was introduced into the *S. pombe* expression vector pREP1.
3. *S. pombe gmd3* mutant was transformed with pREP-FLJ21803. Isolated transformant cells were cultured at 37°C to confirm the complementation of *ts* phenotypes.
4. Human *ALG11* transformed *S. pombe gmd3* cells (*gmd3*/FLJ21803), *gmd3* mutant *S. pombe* cells, yeast *ALG11* transformed *S. pombe* cells (*gmd3*/*ALG11*) and wild type

---

[1] Health Technology Research Center, National Institute of Advanced Industrial Science and Technology (AIST), 2217-14 Hayashi, Takamatsu, Kagawa 761-0395, Japan
Phone: +81-29-861-6769, Fax: +81-87-869-3593
E-mail: k-nakayama@aist.go.jp

[2] Research Institute for Cell Engineering, National Institute of Advanced Industrial Science and Technology (AIST), 1-1-1 Higashi, Tsukuba, Ibaraki 305-8566, Japan
Phone: +81-29-861-6160, Fax: +81-29-861-6161
E-mail: jigami.yoshi@aist.go.jp

Fig. 1 Human *ALG11* gene (FLJ221803) complements the phenotypic defect of *S. pombe gmd3* N-glycosylation. The cell homogenates from individual cell were subjected to electrophoresis on 6% native polyacrylamide gel. The band of acid phosphatase was detected by its activity

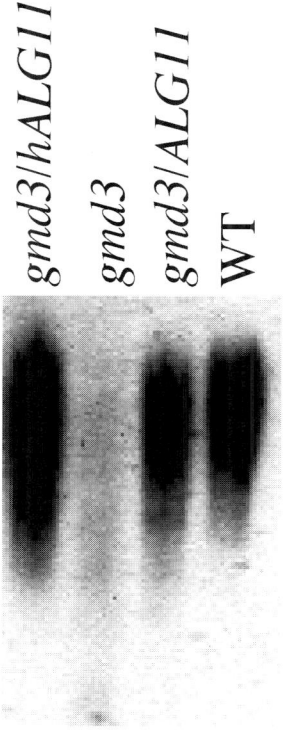

S. pombe cells (WT, JY746) were cultured in a low-phosphate medium and homogenized by glass beads. Supernatants of individual homogenate were subjected to the native polyacrylamide gel electrophoresis. The band of acid phosphatase was detected by its activity on the gel and the complementation of glycosylation defect was confirmed by the recovery of the mobility of acid phosphatase in mutant cells to that in transformed cells and wild type cells (Fig. 1).

Cloning of Human ALG8, ALG9 and ALG10 Genes

For the cloning of human *ALG8*, *ALG9* and *ALG10* genes, *S. cerevisiae alg8 wbp1*, *alg9 wbp1*, and *alg10 wbp1* double mutants were used. Since single mutations of *alg8*, *alg9*, and *alg10* were reported not to show any *ts* phenotypes, it is difficult to clone human corresponding genes by the complementation of *ts* phenotypes of each *alg* mutants of *S. cereviaies*. However, since the double mutants including *wbp1* mutation, which is defective in *WBP1* gene encoding one of the components of oligosaccharyltransferase complex, show the *ts* phenotypes, it is possible to clone the human *ALG* genes by the complementation of *ts* phenotypes of the double mutant cells (*alg8 wbp1*, *alg9 wbp1*, *alg10 wbp1*). Human *ALG8*, *ALG9* and *ALG10* genes were cloned by this method and further confirmed by the complementation of glycosylation defect in mutant cells.

    1. Human *ALG8* candidate gene MGC2840 was amplified by PCR (94°C: 30 s, 50°C: 30 s, 72°C: 2 min, 30 cycles) using human cDNA library as a template.

    2 MGC2840 ORF was inserted into the *S. cerevisiae* expression vector YEp352GAO.

Fig. 2 Human *ALG8* gene (MGC2840) complements the phenotypic defect of *S. cerevisiae alg8 wbp1* N-glycosylation. The cell homogenates from each cell were subjected to Western blot analysis and the band of CPY was detected by anti-CPY monoclonal antibody

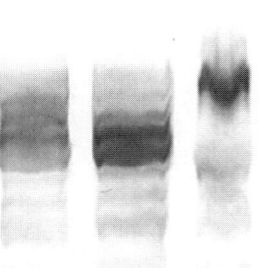

3. The *alg8 wbp1* mutant of *S. cerevisiae* was transformed with YEp-MGC2840. Isolated transformant cells were cultured at 37°C to confirm the complementation of *ts* phenotype.

4. Human *ALG8* transformed *S. cerevisiae alg8 wbp1* cells (*alg8 wbp1*/MGC2840), *alg8 wbp1* mutant *S. cerevisiae* cells and *S. cerevisiae* wild type cells (WT, W303-1A) were cultured and homogenized by glass beads. Supernatants of individual homogenate were subjected to the SDS-PAGE. Carboxypeptidase Y (CPY), one of the typical vacuolar glycoproteins in *S. cerevisiae*, was detected by Western blot analysis and confirmed the complementation of glycosylation defect in mutants (Fig. 2).

5. Human *ALG9* and *ALG10* were also cloned by the same method as described above.

References

Chantret I, Dancourt J, Dupre T, Delenda C, Bucher S, Vuillaumier-Barrot S, Ogier de Baulny H, Peletan C, Danos O, Seta N, Durand G, Oriol R, Codogno P, Moore SE (2003) A deficiency in dolichyl-P-glucose:Glc1Man9GlcNAc2-PP-dolichyl alpha3-glucosyltransferase defines a new subtype of congenital disorders of glycosylation. J Biol Chem 278:9962–9971

Frank CG, Grubenmann CE, Eyaid W, Berger EG, Aebi M, Hennet T (2004) Identification and functional analysis of a defect in the human *ALG9* gene: definition of congenital disorder of glycosylation type IL. Am J Hum Genet 75:146–150

Umeda K, Yoko-o T, Nakayama K, Suzuki T, Jigami Y (2000) *Schizosaccharomyces pombe gmd3*(+)/*alg11*(+) is a functional homologue of *Saccharomyces cerevisiae ALG11* which is involved in *N*-linked oligosaccharide synthesis. Yeast 16:1261–1271

Section II
Sugar-Modified Enzymes

N-acetylglucosamine-6-*O*-sulfotransferases

Kenji Uchimura,[1] Takashi Muramatsu[2]

Introduction

N-acetylglucosamine-6-*O*-sulfotransferase (GlcNAc6ST) transfers a sulfate group from PAPS to an *N*-acetylglucosamine residue, which is usually located at the non-reducing end of glycoconjugates. It is important to note that sulfation proceeds the elongation of the glycan chain. This enzyme is involved in the synthesis of keratan sulfate and sialyl 6-sulfo Lewis X, which is present on the luminal surface of the high endothelial venule (HEV) of lymph nodes. So far, five isozymes of GlcNAc6ST have been identified in humans (Table 1). They share about 30% sequence identity, and have the configuration of type II transmembrane proteins, a feature typical of Golgi-located enzymes.

The sulfate group is transferred only to C-6 of *N*-acetylglucosamine by GlcNAc6ST-1, the first cloned enzyme, which was shown by a biochemical analysis of the product (Uchimura et al. 1998). Convenient substrates to assay GlcNAc6STs are oligosaccharides with an *N*-acetylglucosamine terminus, and radioactively labeled PAPS as the sulfate donor. Usually, the product of the enzymatic reaction is separated from PAPS by high-performance liquid chromatography or thin-layer chromatography. The latter procedure is useful for assaying a large number of samples (Uchimura et al. 2002). Substrate specificities of GlcNAc6STs assayed in vitro are not dramatically different from each other. However, among GlcNAc6ST-1, -2, and -3, only GlcNAc6ST-2 can act on core 3 structure (Galβ1-3GalNAc) (Uchimura et al. 2002). As reported by de Graffenried and Bertozzi, Golgi localization of GlcNAc6STs is different: GlcNA6ST-1 is confined to the trans-Golgi network, GlcNAc6ST-3 is confined to the early secretory pathway, and GlcNAc6ST-2 is distributed throughout the Golgi. This difference in localization influences enzymatic activities observed in intact cells.

Comments

Important roles played by GlcNAc6STs have been revealed especially by employing genetical approaches. Sialyl 6-sulfo Lewis X is a ligand structure of L-selectin, which

---

[1] Section of Pathophysiology and Neurobiology, Department of Alzheimer's Disease Research, National Institute for Longevity Sciences (NILS), 35-3 Gengo, Morioka, Obu, Aichi 474-8522, Japan
Phone: +81-562-46-2311, Fax: +81-562-46-3157
E-mail: arumihcu@nils.go.jp

[2] Department of Health Science, Faculty of Psychological and Physical Science, Aichi Gakuin University, 12 Araike, Iwasaki-cho, Nisshin, Aichi 470-0195, Japan
Phone: +81-561-73-1111, Fax: +81-561-73-1142
E-mail: tmurama@dpc.aichi-gakuin.ac.jp

Table 1 N-acetylglucosamine-6-sulfotransferases

Name	Other abbreviated names	Gene name	Map	Accession No.	Major sites of expression	PubMed ID No. for key references
GlcNAc6ST-1		CHST2	3q24	NM-004267	Multiple tissues	9712885, 10200296, 11726653, 12855678, 15175329, 15220337, 15728736, 1622785, 16227986, 16624895, 16772045
GlcNAc6ST-2	LSST HEC-GLCNAcST	CHST4	16q22.3	NM-005769	HEV	10330415, 10435581, 11520459, 11726653, 12107080, 12855678, 14597732, 16227985, 16227986, 16772045, 16897186
GlcNAc6ST-3	I-GlcNAcST	CHST5	16q22	NM-024533	Intestine, cornea	10491328, 11278593, 11726653, 12855678, 16938851
GlcNAc6ST-4	C6ST-2	CHST7	Xp11.3	NM-019886	Multiple tissues	10781596, 10913333, 10956661
GlcNAc6ST-5	C-GlcNAcST	CHST6	16q22	NM-021615	Cornea	11017086, 11278593, 15013869

Gene name, chromosomal location, and accession No. are those of human enzymes

regulates the initial step of lymphocyte homing to lymph nodes by inducing the rolling of lymphocytes on the luminal surface of HEV. GlcNAc6ST-2 is preferentially expressed in HEV, and GlcNAc6ST-1 is expressed in many tissues including HEV. In GlcNAc6ST-2-deficient mice, lymphocyte homing to peripheral lymph nodes is reduced to 50–60% of that observed in wild-type mice. However, in double knockout (DKO) mice, deficient in both GlcNAc6ST-1 and -2, the value is reduced to 25%, indicating that both enzymes are involved in the synthesis of L-selectin ligand (Kawashima et al. 2005; Uchimura et al. 2005). The complete loss of sialyl 6-sulfo Lewis X structure from HEV of DKO mice is indicated by biochemical and immunohistochemical analyses. In HEV of DKO mice, the rolling of lymphocytes is still observed, but the rolling velocity is greatly increased. Thus, sialyl Lewis X structure without GlcNAc6-sulfation is sufficient to induce a minimum level of rolling, and its GlcNAc-6-sulfation normalizes rolling by decreasing the rolling velocity.

GlcNA6ST-5 is expressed in the cornea and is involved in the synthesis of keratan sulfate. Null mutation of GlcNAc6ST-5 in human leads to macular corneal dystrophy type I and II (Akama et al. 2000). GlcNAc6ST-3 is closely related to GlcNAc6ST-5. Hayashida et al. have revealed that knockout mice deficient in GlcNAc6ST-3 lack highly sulfated keratan sulfate in the cornea and have thinner corneas compared to wild-type mice. The collagen fibrillar architecture is markedly altered in the cornea of the mutant mice. Zhang et al. have shown that GlcNAc6ST-1 is involved in the synthesis of highly sulfated keratan sulfate in the brain. Synthesis of highly sulfated keratan sulfate is induced upon brain injury. In GlcNAc6ST-1 knockout mice, this induction is impaired, and glial scar formation, which inhibits repair of the brain, is suppressed.

Upregulation of GlcNAc6STs under pathological conditions is a subject of significant interest. The GlcNAc6ST-2 level is increased in mucinous adenocarcinoma of the colon, and in chemotherapy-resistant ovarian adenocarcinoma. Expression of GlcNAc6ST-1 and -2 is increased in the synovium of mice upon collagen-induced arthritis, a model of rheumatoid arthritis.

Protocol: GlcNAc-6-O-sulfotransferase Assay

Microsomal fractions in mammalian cell lines that are transfected with expression plasmids encoding GlcNAc-6-O-sulfotransferases are used as enzymes. Various oligosaccharides derived from N-linked or O-linked glycans and keratan sulfate are applicable to the assay described here. Oligosaccharide substrates tested are GlcNAcβ1-3Galβ1-4GlcNAc, GlcNAcβ1-6Man-O-methyl, GlcNAcβ1-2Man, GlcNAcβ-6[Galβ1-3]GalNAc-p-nitrophenyl and GlcNAcβ1-3GalNAc-p-nitrophenyl. Thin-layer chromatography (TLC) is employed to handle a large number of samples.

1. The standard reaction mixture in a 1.5 mL tube contains 1 μmol Tris–HCl, pH 7.5, 0.2 μmol MuCl$_2$, 0.04 μmol adenosine 5′-monophosphate, 2 μmol sodium fluoride, 20 μmol oligosaccharides, 150 pmol [^{35}S]-3′phosphoadenosine 5′-phosphosulfate (PAPS) (1.5×10^6 cpm), 0.05% of Triton-X and 5 μL of enzymes in a final volume of 20 μL.
2. Incubate the tube containing the reaction mixture at 30°C for 1 h.
3. Aliquots of 2 μL of the reaction mixture are applied to TLC plates. The plates are precoated with 0.1-mm thick cellulose.
4. Develop the plates with a developing solvent, ethanol/pyridine/n-butyl alcohol/water/acetic acid (100:10:10:30:3, by volume).
5. End the development when the solvent front reaches to the top of the plates. The ^{35}S-labeled oligosaccharides migrate faster than [^{35}S]-PAPS.

The radioactivity of the ^{35}S-labeled oligosaccharides is visualized and measured using a BAS2000 bioimaging analyzer.

References

Akama TO, Nishida K, Nakayama J, Watanabe H, Ozaki K, Nakamura T, Dota A, Kawasaki S, Inoue Y, Maeda N, Yamamoto S, Fujiwara T, Thonar EJ, Shimomura Y, Kinoshita S, Tanigami A, Fukuda MN (2000) Macular corneal dystrophy type I and type II are caused by distinct mutations in a new sulphotransferase gene. Nat Genet 26:237–241

Kawashima H, Petryniak B, Hiraoka N, Mitoma J, Huckaby V, Nakayama J, Uchimura K, Kadomatsu K, Muramatsu T, Lowe JB, Fukuda M (2005) N-acetylglucosamine-6-O-sulfotransferases 1 and 2 cooperatively control lymphocyte homing through L-selectin ligand biosynthesis in high endothelial venules. Nat Immunol 6:1096–1104

Uchimura K, Muramatsu H, Kadomatsu K, Fan QW, Kurosawa N, Mitsuoka C, Kannagi R, Habuchi O, Muramatsu T (1998) Molecular cloning and characterization of an N-acetylglucosamine-6-O-sulfotransferase. J Biol Chem 273:22577–22583

Uchimura K, El-Fasakhany FM, Hori M, Hemmerich S, Blink SE, Kansas GS, Kanamori A, Kumamoto K, Kannagi R, Muramatsu T (2002) Specificities of N-acetylglucosamine-6-O-sulfotransferases in relation to L-selectin ligand synthesis and tumor-associated enzyme expression. J Biol Chem 277:3979–3984

Uchimura K, Gauguet JM, Singer MS, Tsay D, Kannagi R, Muramatsu T, von Andrian UH, Rosen SD (2005) A major class of L-selectin ligands is eliminated in mice deficient in two sulfotransferases expressed in high endothelial venules. Nat Immunol 11:1105–1113

Sulfotransferases Involved in Sulfation of Glycosaminoglycans

Osami Habuchi

Introduction

Sulfated glycosaminoglycans are linear polysaccharides composed of various types of repeating disaccharide units. Combination of different types of disaccharide units generates huge structural diversity in each glycosaminoglycan chain. Glycosaminoglycans have been reported to be involved in various biological phenomena through specific interactions with various proteins. These interactions are mediated by specific structural domain of glycosaminoglycans (Bulow and Hobert 2006; Kimata et al. 2007). Each sulfotransferase involved in the synthesis of glycosaminoglycans has strict acceptor specificity, and plays crucial roles in the formation of the specific structural domain (Habuchi 2000; Kusche-Gullberg and Kjellen 2003; Taniguchi et al. (eds) 2001). Almost all glycosaminoglycan sulfotransferases have been cloned so far, and the substrate specificity of each sulfotransferase has been studied. In Table 1, glycosaminoglycan sulfotransferases so far cloned and characterized are shown.

Molecular Cloning and Characterization of Sulfotransferases

Chondroitin 6-sulfotransferase (C6ST)

C6ST-1, which sulfates position 6 of GalNAc residue of CS, has been cloned on the basis of amino acid sequence of the enzyme purified from the culture medium of chick chondrocytes. Human chondrodysplasia Omani type was reported to be caused by the mutation in C6ST-1 gene through altered sulfation of CS chains.

Chondroitin 4-sulfotransferase (C4ST)

C4ST-1, which sulfates position 4 of GalNAc residue of CS, has been cloned on the basis of amino acid sequence of the protein purified from the culture medium of rat chondrosarcoma cells. C4ST-2, C4ST-3 and D4ST have been cloned as the homologous proteins to C4ST-1. Mice deficient in C4st1 gene were born at Mendelian ratio, but died within 6 h of birth. Multiple skeletal abnormalities were observed.

GalNAc 4-sulfate 6-O-sulfotransferase (GalNAc4S-6ST)

GalNAc4S-6ST, which sulfates position 6 of GalNAc(4S) residue of CS, was purified from the squid cartilage. On the basis of amino acid sequence of the purified enzyme, partial cDNA of squid GalNAc4S-6ST was cloned. From the amino acid sequence

---

Department of Chemistry, Aichi University of Education, Igaya-cho, Kariya 448-8542, Japan
Phone: 81-566-26-2642, Fax: 81-566-26-2649
E-mail: ohabuchi@auecc.aichi-edu.ac.jp

Table 1 Glycosaminoglycan sulfotransferases and their substrate specificities

Name	Abbreviation and isoform[a]	Species[b]	Amino acid	Accession No.	Acceptors[c]	Products[c]
Chondroitin 6-sulfotransferase	C6ST-1	Chick	458	D49915	CS	-GlcA-GalNAc-
					CS/DS	-GlcA-GalNAc-IdoA-GalNAc-GlcA-
					DS	-IdoA-GalNAc-
		Human	479	AB012192	KS	-GlcNAc-Gal-
						-GlcNAc(6S)-Gal-
	C6ST-2	Human	486	AB037187 AF280089	CS	-GlcA-GalNAc-
Chondroitin 4-sulfotransferase	C4ST-1	Mouse Human	352 352	AB030378 AB042326	CS	-GlcA-GalNAc-
				AF239820	CS/DS	-GlcA-GalNAc-IdoA-GalNAc-GlcA-
	C4ST-2	Human	414	AF239822	CS	-GlcA-GalNAc-
	C4ST-3	Human	341	AY120869	CS	-GlcA-GalNAc-
Dermatan 4-sulfotransferase	D4ST	Human	376	AF401222	DS	-IdoA-GalNAc-
GalNAc 4-sulfate 6-O-sulfotransferase	GalNAc4S-6ST	Squid	425	AB292855	CS	-GlcA-GalNAc(4S)-GalNAc(4S)-GlcA(2S)-GalNAc(6S)-
						-GlcA-GalNAc(4S)-GlcA(2S)-GalNAc(6S)-
					DS	-IdoA-GalNAc(4S)-
		Human	561	AB062423	CS	-GlcA-GalNAc(4S)-GalNAc(4S)-GlcA(2S)-GalNAc(6S)-
					DS	-IdoA-GalNAc(4S)-

Products[c]:
- -GlcA-GalNAc(**6S**)-
- -GlcA-GalNAc-IdoA-GalNAc(**6S**)-GlcA-
- -IdoA-GalNAc(**6S**)-
- -GlcNAc-Gal(**6S**)-
- -GlcNAc(6S)-Gal(**6S**)-
- -GlcA-GalNAc(**6S**)-
- -GlcA-GalNAc(**4S**)-
- -GlcA-GalNAc(**4S**)-IdoA-GalNAc-GlcA-
- -GlcA-GalNAc(**4S**)-
- -GlcA-GalNAc(**4S**)-
- -IdoA-GalNAc(**4S**)-
- -GlcA-GalNAc(4S, **6S**)-GalNAc(4S, **6S**)-GlcA(2S)-GalNAc(6S)-
- -GlcA-GalNAc(4S, **6S**)-GlcA(2S)-GalNAc(6S)-
- -IdoA-GalNAc(4S, **6S**)-
- -GlcA-GalNAc(4S, **6S**)-GalNAc(4S, **6S**)-GlcA(2S)-GalNAc(6S)-
- -IdoA-GalNAc(4S, **6S**)-

Enzyme	Abbreviation	Species	aa	Accession	GAG	Substrate	Product
Chondroitin/dermatan sulfate uronyl 2-O-sulfotransferase	CS/DS2ST	Human	406	AB020316	DS	-IdoA-GalNAc(4S)--IdoA-GalNAc-	-IdoA(2S)-GalNAc(4S)--IdoA(2S)-GalNAc-
					CS	-GalNAc(4S)-GlcA-GalNAc(6S)-	-GalNAc(4S)-GlcA(2S)-GalNAc(6S)-
Keratan sulfate Gal-6-sulfotransferase	KSGal6ST	Human	411	AB003791	KS	-GlcNAc-Gal--GlcNAc(6S)-Gal-	-GlcNAc-Gal(6S)--GlcNAc(6S)-Gal(6S)-
GlcNAc 6-O-sulfotransferase	GlcNAc6ST-1	Mouse	483	AB011451	KS	GlcNAc-Gal-GlcNAc	GlcNAc(6S)-Gal-GlcNAc
		Human	484	AB014679			
	C-GlcNAc6ST	Human	395	AF219990		GlcNAc-Gal-GlcNAc(6S)-	GlcNAc(6S)-Gal-GlcNAc(6S)-
Heparan sulfate-N-deacetylase/N-sulfotransferase	NDST-1	Rat	882	M92042	HS	-GlcA-GlcNAc-	-GlcA-GlcN-
		Human	882	U18918 U36600 U17970			-GlcA-GlcNS-
	NDST-2	Mouse	883	U02304 X75885			
		Human	883	U36601			
	NDST-3	Human	873	AF074924			
	NDST-4	Human	872	AB036429			
Heparan sulfate 2 sulfotransferase	HS2ST	Hamster	356	E17300	HS	-IdoA-GlcNS- >> -GlcA-GlcNS-	-IdoA(2S)-GlcNS- >> -GlcA(2S)-GlcNS-
		Human	356	AB024568			
Heparan sulfate 6-sulfotransferase	HS6ST-1	Hamster		AB006180	HS	-IdoA GlcNS- > -IdoA(2S)-GlcNS- >> -GlcA-GlcNS-	-IdoA GlcNS(6S)- > -IdoA(2S)-GlcNS(6S)- >> -GlcA-GlcNS(6S)-
		Human	410	AB006179			
	HS6ST-2	Mouse	506	AB024565		-GlcA-GlcNS-, -IdoA(2S)-GlcNS- > -IdoA-GlcNS-	-GlcA-GlcNS(6S)-, -IdoA(2S)-GlcNS(6S)- > -IdoA-GlcNS(6S)-
		Human	499	AB067776			
	HS6ST-2S	Human	459	AB067777			
	HS6ST-3	Mouse	470	AB024567		-IdoA-GlcNS	-IdoA-GlcNS(6S)-
		Human	471	NM_153456 AF539426		-IdoA(2S)-GlcNS-GlcA-GlcNS-	-IdoA(2S)-GlcNS(6S)-GlcA-GlcNS-

Table 1 Continued

Name	Abbreviation and isoform[a]	Species[b]	Amino acid	Accession No.	Acceptors[c]	Products[c]
Heparan sulfate 3-*O*-sulfotransferase	HS3ST-1	**Mouse**	311	AF019385	HS	-GlcA-GlcNS(**3S**)-
		Human	307	AF019386	-GlcA-GlcNS-	-GlcA-GlcNS(**3S**,6S)-
					-GlcA-GlcNS(6S)-	
	HS3ST-2	Human	367	NM_006043	-GlcA(2S)-GlcNS-	-GlcA(2S)-GlcNS(**3S**)-
					-IdoA(2S)-GlcNS-	-IdoA(2S)-GlcNS(**3S**)-
	HS3ST-3A	Human	406	AF105376	-IdoA(2S)-GlcN-	-IdoA(2S)-GlcN(**3S**)-
	HS3ST-3B	Human	390	AF105377		
	HS3ST-4	Human	456	AF105378	?	?
	HS3ST-5	Human	346	AF503292	-GlcA-GlcNS(6S)-	-GlcA-GlcNS(**3S**,6S)-
					-IdoA(2S)-GlcNS-	-IdoA(2S)-GlcNS(**3S**)-
					-IdoA(2S)-4GlcNS(6S)-	-IdoA(2S)-GlcNS(**3S**,6S)-
	HS3ST-6	Human	342	AY574375	-IdoA(2S)-GlcNS-	-IdoA(2S)-GlcNS(**3S**)-
					-IdoA(2S)-GlcNS(6S)-	-IdoA(2S)-GlcNS(**3S**,6S)-

[a] Some sulfotransferases (C6ST-1, C6ST-2, KSGal6ST, GlcNAc6ST, and C-GlcNAc6ST) are able to sulfate sugar chains of glycoproteins, but these activities are not shown in this table
[b] If prototype of the sulfotransferase was not obtained from human, the species from which the prototype was obtained are listed. Bold letters indicate the species from which the sulfotransferase was purified to the homogeneity and cloned on the basis of the amino acid sequences of the purified protein
[c] Abbreviations for glycosaminoglycans and back bone structures of glycosaminoglycans: CS, chondroitin sulfate; DS, dermatan sulfate; KS, keratan sulfate; HS, heparan sulfate; -*GlcA-GalNAc-* (*CS*), -GlcAβ1–3GalNAcβ1–4; -*IdoA-GalNAc-* (*DS*), -IdoAα1–3GalNAcβ1–4; -*GlcNAc-Gal-* (*KS*), -GlcNAcβ1–3Galβ1–4; -*IdoA-GlcNS-* (*HS*), -IdoAα1–4GlcNSα1–4; -*GlcA-GlcNS-* (*HS*), -GlcAβ1–4GlcNSα1–4

deduced from the cDNA, human GalNAc4S-6ST cDNA was identified. GalNAc4S-6ST of human and mouse was expressed in immunologically relevant tissues, and appears to be involved in the synthesis of CS-E in the bone marrow-derived mast cells. Recently, the full-length cDNA of squid GalNAc4S-6ST has been cloned. The recombinant squid GalNAc4S-6ST was found to synthesize a unique chondroitin sulfate containing E-D hybrid tetrasaccharide structure.

Chondroitin/Dermatan Sulfate Uronyl 2-O-sulfotransferase (CS/DS2ST)

CS/DS2ST transfers sulfate to position 2 of GlcA of CS or IdoA of DS, and cloned as an enzyme having homology with HS2ST. Sequences in CS and DS that are recognized by CS/DS2ST are different (see Table 1).

Keratan Sulfate Gal-6-sulfotransferase (KSGal6ST)

KSGal6ST transfers sulfate to position 6 of Gal residue of KS, and cloned as an enzyme having homology with C6ST-1.

GlcNAc 6-O-sulfotransferase (GlcNAc6ST)

GlcNAc6ST transfers sulfate to position 6 of non-reducing terminal GlcNAc residue. Among five isoforms, GlcNAc6ST-1 and C-GlcNAc6ST are thought to be involved in the synthesis of KS. C-GlcNAc6ST has been identified as the gene responsible for macular corneal dystrophy.

Heparan Sulfate N-deacetylase/N-sulfotransferase (NDST)

NDST-1 and NDST-2 have been cloned on the basis of amino acid sequences of the purified enzymes obtained from rat liver and mast mastocytoma, respectively. NDST catalyzes initial modification reaction of HS. N-deacetylation of GlcNAc residue and N-sulfation of the resulting GlcN. Four isoforms of NDST have been cloned. From the studies of the systemic null mice, NDST-1 and NDST-2 are thought to be involved mainly in the synthesis of HS in all tissues and the synthesis of mast cell heparin, respectively.

Heparan Sulfate 2-sulfotransferase (HS2ST)

HS2ST, which sulfates position 2 of IdoA residue of HS and forms FGF2 binding domain of HS, has been cloned on the basis of amino acid sequence of the enzyme purified from the CHO cells. HS2ST is essential for the normal development of mice, because HS2ST-deficient mice are either stillborn or die within 24 h of birth and exhibit bilateral renal agenesis.

Heparan Sulfate 6-sulfotransferase (HS6ST)

HS6ST-1, which sulfates position 6 of GlcN residue of HS, has been cloned on the basis of amino acid sequence of the enzyme purified from the culture medium of CHO cells. Three isoforms and one alternatively spliced form (HS6ST-2S) have been cloned; the individual isoforms exhibit characteristic preferences in their substrate specificities for the uronic acid residue neighboring the N-sulfoglucosamine. Most

HS6ST-1 null mice die during E15.5 and the perinatal stage and are smaller than wild-type mice.

Heparan Sulfate 3-sulfotransferase (HS3ST)

HS3ST-1, which sulfates position 3 of GlcNS residue of HS and forms anticoagulant HS, has been cloned on the basis of amino acid sequence of the enzyme purified from the culture medium of L-33 cells. Six isoforms have been so far cloned. HS3ST-2, -3, and -6 generate an entry receptor for herpes simplex virus 1. HS3ST-5 produces both anticoagulant HS and an entry receptor for herpes simplex virus 1.

Recognition of Sequences of the Acceptor Glycosaminoglycans by Sulfotransferases

It has been clarified that glycosaminoglycan sulfotransferases recognize not only the position of the sugar residues to which sulfate is transferred but also the structure of neighboring sugar residues or sulfation pattern around the targeting sugar residue. In some case, individual isoform of the same sulfotransferase shows different recognition from the other isoforms. In Table 1, acceptor substrate specificities of individual isoforms of the glycosaminoglycan sulfotransferases are indicated.

Glycosaminoglycan Sulfotransferase Assay

Glycosaminoglycan sulfotransferase activities are determined by the radioactivity of $^{35}SO_4$ transferred to acceptor glycosaminoglycans from the universal sulfate donor, [^{35}S]PAPS. The ^{35}S-labeled glycosaminoglycans are then degraded with appropriate lyases to unsaturated disaccharides or monosaccharides, and the degradation products are separated with HPLC. Positions to which sulfate is transferred are determined by identifying the unsaturated disaccharides or monosaccharides having radioactivity. As a representative, assay of GalNAc4S-6ST is shown below.

    1. The standard reaction mixture contains 2.5 µmol of imidazole–HCl, pH 6.8, 0.5 µmol $CaCl_2$, 1 µmol reduced glutathione, 25 nmol (as galactosamine) CS-A, 50 pmol [^{35}S]PAPS (about 5.0×10^5 cpm), and enzyme in a final volume of 50 µl.

    2. The reaction mixtures are incubated at 37°C for 20 min and the reaction is stopped by heating at 100°C for 1 min.

    3. The reaction product is precipitated with three volumes of ethanol containing 1.3% potassium acetate in the presence of 50 nmol CS-A as a carrier.

    4. Solution of the ^{35}S-labeled product is injected to a Sephadex G-25 superfine column (10 mm × 10 cm) equilibrated with 0.1 M NH_4HCO_3. The column is run at a flow rate of 2 ml/min. The ^{35}S-labeled glycosaminoglycan is recovered in the void volume and separated from $^{35}SO_4^{2-}$ and [^{35}S]PAPS. Radioactivity of the ^{35}S-labeled glycosaminoglycans is measured.

    5. The ^{35}S-labeled glycosaminoglycan is then digested with chondroitinase ACII or chondroitinase ABC and the digests are injected to a Partisil-10 SAX column equilibrated with 10 mM KH_2PO_4. The column is developed with 10 mM KH_2PO_4 for 10 min followed by a 45-min linear gradient from 10 to 500 mM KH_2PO_4, and then washed with 500 mM KH_2PO_4 for 10 min. Column temperature is maintained at 40°C. Fractions

(0.5 ml) are collected at a flow rate of 1 ml/min. Radioactivity of each fraction is measured. The radioactive peaks are identified by the retention time of standard disaccharides and monosaccharides.

References

Bulow HE, Hobert O (2006) The molecular diversity of glycosaminoglycans shapes animal development. Annu Rev Cell Dev Biol 22:375–407

Habuchi O (2000) Diversity and functions of glycosaminoglycan sulfotransferases. Biochim Biophys Acta 1474:115–127

Kimata K, Habuchi O, Habuchi H, Watanabe H (2007) Knockout mice and proteoglycans. In: Kamerling JP et al (eds) Comprehensive glycoscience, vol 3, Chap 4.10. Elsevier, Amsterdam, pp 159–191

Kusche-Gullberg M, Kjellen L (2003) Sulfotransferases in glycosaminoglycan biosynthesis. Curr Opin Struct Biol 13:605–611

Taniguchi N, Honke K, Fukuda M (eds) (2001) Sulfotransferases. In: Handbook of glycosyltransferases and related genes. Springer, Tokyo, pp 413–508

Sulfatide Synthase

Koichi Honke[1,2]

Introduction

Two major sulfoglycolipids exist in the mammal: one being the sulfatide, which is a sphingolipid and the other being the seminolipid, which is an ether glycerolipid. Sulfatide is a major lipid component of the myelin sheath and is synthesized in the myelin-generating cells, oligodendrocytes in the central nervous system and Schwann cells in the peripheral nervous system. Seminolipid is synthesized in spermatocytes and maintained in the subsequent germ-cell stages. The carbohydrate moiety of sulfatide and seminolipid has the same structure and are biosynthesized via sequential reactions catalyzed by common enzymes: ceramide galactosyltransferase (CGT, EC 2.4.1.45) and cerebroside sulfotransferase (CST, EC 2.8.2.11) (Fig. 1). This sulfotransferase is also known as sulfatide synthase. CST catalyzes the transfer of sulfonate group from 3′-phosphoadenosine 5′-phosphosulfate (PAPS) to the C3 position of the non-reducing terminal galactose of glycolipid oligosaccharides. We purified CST homogeneously from human renal cancer cells (Honke et al. 1996) and subsequently isolated a cDNA clone of *CST* from a cDNA library of human renal cancer cells on the basis of the partial amino acid sequences of the purified enzyme (Honke et al. 1997). Furthermore, we cloned human genomic DNA (Tsuda et al. 2000) and mouse cDNA and genomic DNA (Hirahara et al. 2000). CST includes PAPS recognizing motifs as the other sulfotransferases do. CST possesses two *N*-glycans, and the C-terminal one is essential for its enzymatic activity. CST is expressed in a tissue-specific manner, and is highly expressed in brain, kidney, testis and alimentary system. This tissue-specific expression of the CST gene is regulated by the alternative usage of multiple promoters (Hirahara et al. 2000). CST is responsible for the biosynthesis of sulfatide, which is rich in myelin sheath, and seminolipid, which is unique in spermatogenic cells. *CST*-null mice generated by gene targeting manifest some neurological disorders due to myelin dysfunction and an arrest of spermatogenesis, indicating that CST and its products, or sulfoglycolipids are essential for organisms (Honke et al. 2002). CST is the first member of βGal 3-*O*-sulfotransferases. The other three members, Gal3ST-2–4, have been identified in the EST database by making use of homology to the *CST* gene (Honke et al. 2002). Gal3ST-2–4 act on glycoproteins.

Enzyme Assay and Substrate Specificity

A convenient assay for CST activity was developed using anion-exchange chromatography (Kawano et al. 1989). The reaction mixture contains 5 nmol of GalCer, 0.5 μmol

---

[1] Department of Biochemistry, Kochi University Medical School, Kohasu, Oko-cho, Nankoku, Kochi 783-8505, Japan
Phone: +81-88-880-2313, Fax: +81-88-880-2314
E-mail: khonke@kochi-u.ac.jp

[2] CREST, Japan Science and Technology Agency, Japan

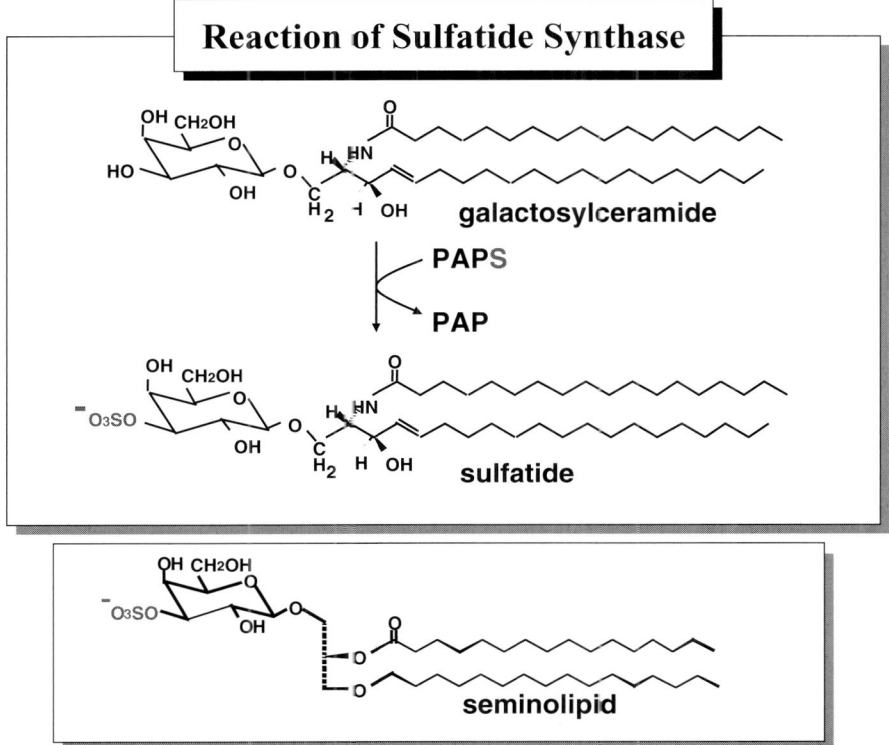

Fig. 1 Enzyme reaction of sulfatide synthase. Sulfatide synthase synthesizes not only sulfatide but also seminolipid

of MnCl$_2$, 1 nmol of [^{35}S]PAPS (100 cpm/pmol), 0.5 mg of Lubrol PX, 12.5 nmol of dithiothreitol, 0.25 μmol of NaF, 0.1 μmol of ATP, 20 mg of BSA and enzyme protein in 25 mM Na cacodylate–HCl, pH 6.5, in a total volume of 50 μl. After incubating at 37°C for 1 h, the reaction is terminated with 1 ml of chloroform/methanol/water (30:60:8). The reaction product is isolated on a DEAE-Sephadex A-25 column and assayed for radioactivity using a liquid scintillation counter. The values are corrected for a blank value, which is obtained by using a reaction mixture devoid of the acceptor. The substrate specificity of CST purified from human renal cancer cells is investigated (Honke et al. 1996). GalCer is the best acceptor and LacCer, galactosyl 1-alkyl-2acyl-sn-glycerol (the precursor for seminolipid), and galactosyl diacylglycerol are also good acceptors. GlcCer, Gg3Cer, Gg4Cer, Gb4Cer, and nLc4Cer serve as acceptors although the relative activities are low. On the other hand, the enzyme cannot act on Gb3Cer, which possesses α-galactoside at the non-reducing terminus. Neither galactose nor lactose serves as an acceptor. These observations suggest that CST prefers β-galactoside at the non-reducing termini of sugar chains attached to a lipid moiety.

Enzyme Assay Protocol

1. Prepare the reaction mixture in a total volume of 50 μl as follows:

			Final concentration
250 mM	Na cacodylate, pH 6.5	5 μl	25 mM
0.5 mM	GalCer in 5% TRX-100	5 μl	50 μM
0.1 M	MnCl2	5 μl	10 mM
10%	Lubrol PX	5 μl	1%
12.5 mM	DTT	1 μl	0.25 mM
0.25 M	NaF	1 μl	5 mM
0.1 M	ATP	1 μl	2 mM
0.5 M	NaCl	5 μl	50 mM
1 mM	[^{35}S]PAPS (ca. 100 cpm/pmol)	2 μl	40 μM
	Subtotal	30 μl	
	Enzyme	20 μl	
	Total	50 μl	

2. Incubate at 37°C for 1–2 h.
3. Terminate the reaction with 1 ml of chloroform/methanol/water (30:60:8, v/v).
4. Apply the whole reaction product onto a mini column packed with 1 ml of DEAE-Sephadex A-25 resin.
5. Wash with 3 ml of chloroform/methanol/water (30:60:8, v/v).
6. Wash with 6 ml of methanol.
7. Elute with 5 ml of 90 mM ammonium acetate in methanol. Eluate is directly collected into a scintillation vial.
8. Put a scintillation cocktail into the eluate and count the radioactivity with a liquid scintillation counter. The values are corrected for a blank value, which is obtained using a reaction mixture devoid of GalCer.

References

Hirahara Y, Tsuda M, Wada Y, Honke K (2000) cDNA cloning, genomic cloning, and tissue-specific regulation of mouse cerebroside sulfotransferase. Eur J Biochem 267:1909–1917

Honke K, Taniguchi N (2002) Sulfotransferases and sulfated oligosaccharides. Med Res Rev 22:637–654 (review)

Honke K, Yamane M, Ishii A, Kobayashi T, Makita A (1996) Purification and characterization of 3′-phosphoadenosine-5′-phosphosulfate:GalCer sulfotransferase from human renal cancer cells. J Biochem (Tokyo) 119:421–427

Honke K, Tsuda M, Hirahara Y, Ishii A, Makita A, Wada Y (1997) Molecular cloning and expression of cDNA encoding human 3′-phosphoadenylylsulfate:galactosylceramide 3′-sulfotransferase. J Biol Chem 272:4864–4868

Honke K, Hirahara Y, Dupree J, Suzuki K, Popko B, Fukushima K, Fukushima J, Nagasawa T, Yoshida N, Wada Y, Taniguchi N (2002) Paranodal junction formation and spermatogenesis require sulfoglycolipids. Proc Natl Acad Sci USA 99:4227–4232

Kawano M, Honke K, Tachi M, Gasa S, Makita A (1989) An assay method for ganglioside synthase using anion-exchange chromatography. Anal Biochem 182:9–15

Tsuda M, Egashira M, Niikawa N, Wada Y, Honke K (2000) Cancer-associated alternative usage of multiple promoters of human GalCer sulfotransferase gene. Eur J Biochem 267:2672–2679

CMP-*N*-acetylneuraminic Acid Hydroxylase

Akemi Suzuki

Concept

CMP-*N*-acetylneuraminic acid (CMP-NeuAc) hydroxylase catalyzes the hydroxylation of CMP-NeuAc to CMP-*N*-glycolylneuraminic acid (CMP-NeuGc) in the presence of hydrogen donor (NADH), cytochrome b5, and NADH-cytochrome b5 reductase (Kozutsumi et al. 1990; Kawano et al. 1994). This biosynthetic step is the rate limiting reaction for the expression of NeuGc, and only one gene encodes the hydroxylase in humans, chimpanzee, and mouse. In human genome, the deletion of 92 bp fragment which corresponds to exon 6 of mouse hydroxylase gene takes place. Humans are defective in the production of hydroxylase activity (Irie et al. 1998; Chou et al. 1998) and, therefore, lack the expression of NeuGc in normal tissues as a major component seen in other mammals.

In humans, it is well documented that adult and normal tissues do not contain NeuGc as the major component of glycans in the glycoproteins and glycolipids, but small amount of NeuGc was detected in various normal tissues, colon cancers and meconium. At this stage, we still do not understand the molecular mechanism for this NeuGc expression but it was proposed that incorporation from digested food into our body as NeuGc or NeuGc glycans and into cells by endocytosis is responsible for this distribution. Fetal calf serum (FCS) is widely used for cell culture, and NeuGc and NeuGc glycoconjugates in FCS are incorporated into the cultured cells and reused for the building of new glycoconjugates containing NeuGc. If human ES cells or various types of progenitor cells are cultured in FCS containing media and used for human treatment, the presence of NeuGc would be a matter of consideration (Martin et al. 2005). On the other hand, if we can induce the expression of NeuGc in cancer cells, drugs using NeuGc as the target would be candidates for cancer drugs.

Knock out mice of CMP-NeuAc hydroxylase gene were generated and their phenotypes are reported. Hyper-activation of B cells is observed and more interestingly it is revealed that the change of expression from NeuGc- to NeuAc-containing glycans happens during the activation of B cells in the germinal center of mouse spleen (Naito et al. 2007).

Answering the question, what is lacking for the NeuGc expression in human species in terms of evolution?, is an attractive subject and further studies are required. Here, a protocol for measuring CMP-NeuAc hydroxylase is summarized.

Institute of Glycotechnology, Future Science and Technology Joint Research Center, Tokai University, 1117 Kitakaname, Hiratsuka, Kanagawa 259-1292, Japan
Phone: +81-463-58-1211 ext. 4643, 4656, Fax: +81-463-50-2432
E-mail: akmszk@keyaki.cc.u-tokai.ac.jp

Measuring CMP-hydroxylase Activity

Preparations for the Enzyme Source

Mouse liver (about 1 g) is homogenized in four volumes of 10 mM Tris–HCl buffer, pH 7.5, containing 0.25 M sucrose, 1 mM EDTA, and 2 mg/ml pepstatin A. The homogenate is centrifuged at 10,000×g for 20 min and resulting supernatant is centrifuged at 140,000×g for 3 h. The supernatant, cytosol, is dialyzed against 10 mM Tris–HCl buffer, pH 7.5, containing 0.15 M NaCl, 1 mM EDTA, and 0.1 mM DTT and then 10 mM Tris–HCl buffer, pH 7.5, containing 0.1 mM DTT, and finally used as the enzyme source. This fraction contains necessary components, probably soluble types of cytochrome b5 and NADH-cytochrome b5 reductase (Kozutsumi et al. 1990).

Incubation

Incubation mixture is composed of 34 nmol of NADH, 2 nmol of CMP-NeuAc, 1 mM DTT in 10 mM Tris–HCl buffer, pH 7.5, and the above enzyme source in a total volume of 50 μl. After 1 or 2 h incubation, 0.3 ml of ice-cold ethanol was added and mixed well. After keeping on ice for 15 min, the mixtures are centrifuged at 10,000 rpm for 5 min. The supernatant is analyzed by HPLC.

HPLC Determination of CMP-NeuAc and -NeuGc

Aliquots of the incubation mixture are injected to HPLC with a C18 column (4.5 mm i.d. × 250 mm, STK-gel, ODS-80TM, Tosoh, Tokyo) and 50 mM $NH_4H_2PO_4$ as the elution solvent. Peaks are monitored by a UV detector at 271 nm. CMP-NeuGc is eluted before CMP-NeuAc. The synthesized amount of CMP-NeuGc is determined by calibrating the curves obtained with standard CMP-NeuAc and CMP-NeuGc and the peak area of the incubation mixtures.

Comments

1. Salts inhibit the assay of the activity.
2. If you purify the hydroxylase, the addition of essentially enough amounts of cytochrome b5 and NADH-cytochrome b5 reductase is required in the incubation mixtures.

References

Chou HH, Takematsu H, Diaz S, Iber J, Nickerson E, Wright KL, Muchmore EA, Nelson DL, Warren ST, Varki A (1998) A mutation in human CMP-sialic acid hydroxylase occurred after the Homo-Pan divergence. Proc Natl Acad Sci USA 95:11751–11756

Irie A, Koyama S, Kozutsumi Y, Kawasaki T, Suzuki A (1998) The molecular basis for the absence of N-glycolylneuraminic acid in humans. J Biol Chem 273:15866–15871

Kawano T, Kozutsumi Y, Kawasaki T, Suzuki A (1994) Biosynthesis of N-glycolylneuraminic acid-containing glycoconjugates. Purification and characterization of the key enzyme of the cytidine monophospho-N-acetylneuraminic acid hydroxylation system. J Biol Chem 269:9024–9029

Kozutsumi Y, Kawano T, Yamakawa T, Suzuki A (1990) Participation of cytochrome b5 in CMP-N-acetylneuraminic acid hydroxylation in mouse liver cytosol. J Biochem 108:704–706

Martin MJ, Muotri A, Gage F, Varki A (2005) Human embryonic stem cells express an immunogenic nonhuman sialic acid. Nat Med 11:228–232

Naito Y, Takematsu H, Koyama S, Miyake S, Yamamoto H, Fujinawa R, Sugai M, Okuno Y, Tsujimoto G, Yamaji T, Hashimoto Y, Itohara S, Kawasaki T, Suzuki A Kozutsumi Y (2007) Germinal center marker GL7 probes activation-dependent repression of N-glycolylneuraminic acid, a sialic acid species involved in the negative modulation of B-cell activation. Mol Cell Biol 27:3008–3022

Section III
Transporters

Nucleotide Sugar Transporter Genes and Their Functional Analysis

Shoko Nishihara

Introduction

Carbohydrate structures on glycoproteins and glycolipids play important roles in various biological processes, such as morphogenesis/organ development, viral and bacterial infections/immune response, and cancer invasion. Nucleotide sugar transporters (NSTs) are crucial components in the synthesis of glycoconjugates (Fig. 1) (Caffaro and Hirschberg 2006). Glycosylation can be performed by various types of glycosyltransferase in the lumens of the endoplasmic reticulum and the Golgi apparatus. All glycosyltransferases require donor sugars activated by the addition of a nucleoside mono- or diphosphate (UDP, GDP, or CMP), that is, nucleotide sugars. Nucleotide sugars are synthesized in the cytosol (or in the nucleus in the case of CMP-sialic acid). The translocation of nucleotide sugars from the cytosol into the lumen compartment is mediated by specific NSTs. NSTs are multiple-membrane-spanning proteins that transport nucleotide sugars in coupling with the antiport of nucleoside monophosphate (NMP), which is produced as the result of a glycosyltransferase reaction and a subsequent luminal nucleoside diphosphatase (NDPase) reaction. Recently, NSTs have been suggested to be possible crucial players in the synthesis of glycoconjugates (Suda et al. 2004; Caffaro and Hirschberg 2006).

The nucleotide sulfate, 3'-phosphoadenosine 5'-phosphosulfate (PAPS), is a universal sulfuryl donor for sulfation. Sulfate is transferred from PAPS to a defined position on the sugar residue by sulfotransferases. In an analogous fashion to nucleotide sugars, PAPS is synthesized in the cytosol by a PAPS synthetase and is translocated from the cytosol into the Golgi lumen through a PAPS transporter (Kamiyama et al. 2003, 2006; Goda et al. 2006). Therefore, PAPS transporters belong to the nucleotide sugar transporter family (Fig. 2).

Nine human NST genes, including PAPS transporters, have now been cloned and their protein activities identified (Fig. 2). Among these, four transporters have multi-substrate specificities (Suda et al. 2004). Three others transport single-specific nucleotide sugars, and a further two act similarly for PAPS (Kamiyama et al. 2003, 2006; Goda et al. 2006). The inactivation of the GDP-Fuc transporter causes congenital disorders of glycosylation (CGD)-IIc/leukocyte adhesion deficiency type II (Caffaro and Hirschberg 2006). More recently, mutation of the CMP-Sia transporter was found to cause a new type of CGD-II.

Laboratory of Cell Biology, Department of Bioinformatics, Faculty of Engineering, Soka University, 1-236 Tangi-cho, Hachioji, Tokyo 192-8577, Japan
Phone: +81-426-91-8140, Fax: +81-426-91-8140
E-mail: shoko@scc1.t.soka.ac.jp

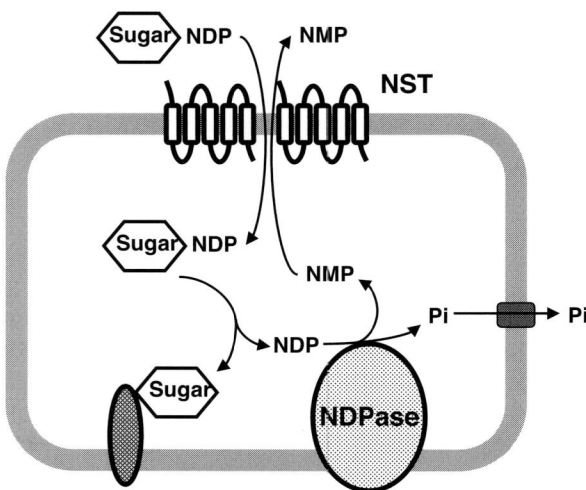

Fig. 1 Nucleotide sugar transporter controls glycosylation. Nucleotide sugar transporters (*NSTs*) transport nucleotide sugars from the cytosol to ER/Golgi lumens. Glycosyltransferases transfer sugars to acceptor substrates from donor nucleotide sugar substrates. Released nucleoside diphosphates (*NDPs*) are then hydrolyzed by nucleoside diphosphatase (*NDPase*) to form nucleoside monophosphates (*NMPs*) and inorganic phosphates (*Pis*). NMPs are exported in antiport with incoming nucleotide sugars. Pis may exit via a phosphate transporter

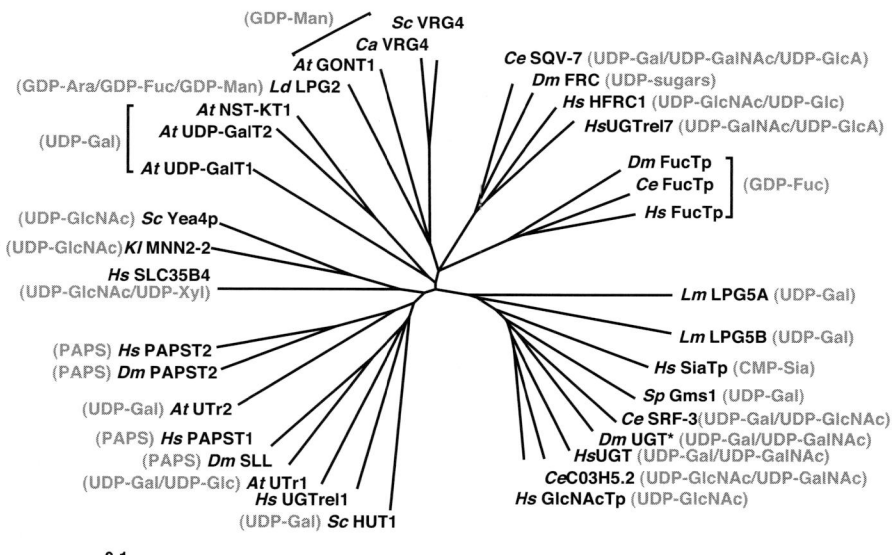

Fig. 2 Phylogenetic tree of nucleotide sugar transporters. The phylogenetic tree is based on amino acid sequences and was constructed using the ClustalX program. Branch lengths indicate evolutionary distances between members. The *scale at the bottom* represents the evolutionary distance. The nucleotide sugars that each transporter transfers from the cytosol to the ER or Golgi lumen are shown in the *parentheses*. The *Drosophila CG2675* gene product is identified by an *asterisk*. At, *Arabidopsis thaliana*; Ca, *Candida albicans*; Ce, *Caenorhabditis elegans*; Dm, *Drosophila melanogaster*; Hs, *Homo sapiens*; Kl, *Kluyveromyces lactis*; Ld, *Leishmania donovani*; Lm, *Leishmania major*; Sc, *Saccharomyces cerevisiae*; Sp, *Schizosaccharomyces pombe*

In model organisms, such as *C. elegans*, *Drosophila*, *Leishmania*, and yeast, mutations of NST genes have been identified and used for analysis of their biological functions. Functional analysis of NSTs will help to clarify the physiological functions of sugar chains.

Procedure

Assays for Nucleotide Sugar Transporter Activity

There are two types of assay for nucleotide sugar transporter (NST) activity: the 'heterologous expression system' (Kamiyama et al. 2003; Suda et al. 2004) and the 'proteoliposome system' (Caffaro and Hirschberg 2006). This latter system is not considered further in this chapter. The former method makes use of expression of human NST in a yeast expression system. This is feasible as the yeast microsome normally shows only low endogenous NST activity, with the exception of GDP-Man transporter activity.

Protocol for Subcellular Fractionation of Yeast (*Saccharomyces cerevisiae*) (Kamiyama et al. 2003, 2006; Suda et al. 2004; Goda et al. 2006)

1. Insert an NST coding region into the yeast expression vector YEp352GAP-II with three copies of HA epitope tags (YPYDVPDYA) at the position corresponding to the C terminus[1] of the NST to be expressed.
2. Transform the yeast strain W303-1a (*MATa, ade2-1, ura3-1, his3-11,15, trp1-1, leu2-3,112,* and *can1-100*) by the lithium acetate procedure using the yeast expression vector YEp352GAP-II.
3. Grow the transformed yeast cells at 30°C in a synthetic defined medium that lacks uracil in order to select for transformants.
4. Harvest the cells and wash with ice-cold 10 mM NaN_3.
5. Convert the cells into spheroplasts by incubation at 37°C for 30 min in spheroplast buffer [1.4 M sorbitol, 50 mM potassium phosphate (pH 7.5)], 10 mM NaN_3, 40 mM 2-mercaptoethanol, and 1 mg/g of cells of zymolyase 100T (Seikagaku Kogyo Co. Ltd., Tokyo, Japan).
6. Pellet the spheroplasts using a refrigerated centrifuge and wash twice with 1.0 M ice-cold sorbitol to remove traces of zymolyase.
7. Suspend the cells in ice-cold lysis buffer [0.8 M sorbitol in 10 mM triethanolamine (pH 7.2), 5 μg/ml pepstatin A, and 1 mM phenylmethylsulfonyl fluoride], and homogenize using a Dounce homogenizer.
8. Centrifuge the lysate at 1,000×g for 10 min to remove unlysed cells and cell wall debris.
9. Centrifuge the supernatant at 10,000×g for 15 min at 4°C to obtain the P10 membrane fraction pellet.
10. Centrifuge the supernatant further at 100,000×g to obtain the P100 membrane fraction pellet.
11. Use each pellet in a transporter activity assay.

---

[1] An HA tag could be inserted at the position corresponding to the N terminus if the NST has an ER retention signal, a dilysine motif, at its C terminus.

The expression level of HA-tagged NST in each membrane fraction is checked by Western blot analysis using an anti-HA mouse mAb (Santa Cruz Biotechnology, Inc., Santa Cruz, CA, USA).

Transporter Activity Assay (Kamiyama et al. 2003, 2006; Suda et al. 2004; Goda et al. 2006)

1. Incubate each of the pellets described above (200 μg protein) at 30°C for 5 min in 100 μl of reaction buffer [20 mM Tris–HCl (pH 7.5), 0.25 M sucrose, 5.0 mM $MgCl_2$, 1.0 mM $MnCl_2$, and 10 mM 2-mercaptoethanol] that contains 1 μM radiolabeled nucleotide sugars or PAPS substrate.
2. Stop the reaction by adding 1 ml of stop buffer [20 mM Tris–HCl (pH 7.5), 0.25 M sucrose, 5.0 mM $MgCl_2$].
3. Trap the radioactivity incorporated in the microsomes using a 0.45 μm nitrocellulose filter. Wash the filter with 10 ml of stop buffer, and then measure radioactivity using a liquid scintillation counter. The amount of incorporated radioactivity is calculated as the difference from a background value obtained using the same assay at 30°C for 0 min for each sample.

Functional Analysis of Nucleotide Sugar Transporters by RNA Interference

Recently, RNA interference (RNAi) has been used for functional analysis of a wide range of biologically important molecules. The introduction and expression of double-stranded RNAs in cultured cells or model animals (see Chapter by S. Nishihara, this volume) are comparatively straightforward and results in the knockdown of expression of the target gene (Kamiyama et al. 2003; Goda et al. 2006). Nucleotide sugar transporter (NST) genes are also no exception. In mammalian cell lines, 21–25 nt synthetic small interfering RNAs (siRNAs) are more useful than short hairpin RNAs (shRNAs) expressed using shRNA expression vectors (Kamiyama et al. 2006). In order to eliminate off-target effects, at least two different siRNA sequences should be used. siRNA-induced knockdown of NSTs had a delayed effect on glycosylation compared to knockdown of genes coding for core-proteins and glycosyltransferases. For detailed descriptions of experimental conditions, refer to Kamiyama et al. 2003, 2006; Goda et al. 2006.

Lists and Comments

A phylogenetic tree of the currently identified nucleotide sugar transporters (NSTs) and their transporter activities is shown in Fig. 2. In humans, more than one NST transport the same nucleotide sugar, e.g., UDP-GlcNAc, UDP-GalNAc (Suda et al. 2004), and PAPS (Kamiyama et al. 2003, 2006). This apparent redundancy raises the important question of whether or not these transporters work in different metabolic pathways. Considering the localization of NSTs and their interactions with glycosyltransferases, this issue has considerable importance for our understanding of the basic mechanisms of the carbohydrate modification system. In addition to the RNA interference approach, mutation of NST genes in model organisms will also provide us with excellent tools for the functional analysis of NSTs.

References

Caffaro CE, Hirschberg CB (2006) Nucleotide sugar transporters of the Golgi apparatus: from basic science to diseases. Acc Chem Res 39:805–812

Goda E, Kamiyama S, Uno T, Yoshida H, Ueyama M, Kinoshita-Toyoda A, Toyoda H, Ueda R, Nishihara S (2006) Identification and characterization of a novel *Drosophila* 3′-phosphoadenosine 5′-phosphosulfate transporter. J Biol Chem 281:28508–28517

Kamiyama S, Suda T, Ueda R, Suzuki M, Okubo R, Kikuchi N, Chiba Y, Goto S, Toyoda H, Saigo K, Watanabe M, Narimatsu H, Jigami Y, Nishihara S (2003) Molecular cloning and identification of 3′-phosphoadenosine 5′-phosphosulfate transporter. J Biol Chem 278:25958–25963

Kamiyama S, Sasaki N, Goda E, Ui-Tei K, Saigo K, Narimatsu H, Jigami Y, Kannagi R, Irimura T, Nishihara S (2006) Molecular cloning and characterization of a novel 3′-phosphoadenosine 5′-phosphosulfate transporter, PAPST2. J Biol Chem 281:10945–10953

Suda T, Kamiyama S, Suzuki M, Kikuchi N, Nakayama K, Narimatsu H, Jigami Y, Aoki T, Nishihara S (2004) Molecular cloning and characterization of a human multisubstrate specific nucleotide-sugar transporter homologous to *Drosophila* fringe connection. J Biol Chem 279:26469–26474

Section IV
Glycosidases

Processing Enzymes Involved in *N*-glycan Biosynthesis and Related Genes: the Golgi *N*-glycan Processing α-mannosidase II and α-mannosidase IIx

Tomoya O. Akama[1], Michiko N. Fukuda[2]

Introduction

N-linked carbohydrate modification is found in many secretary and cell surface proteins. This post-translational modification is thought to be important for biological function of proteins in vivo. Recent gene knockout studies demonstrated biological function of glycoproteins, and revealed as yet unknown biosynthetic pathway of *N*-glycans. We have analyzed the functions of α-mannosidase II (MII) and α-mannosidase IIx (MX) in the mouse by targeted disruption of each gene, and demonstrated that enzymatic activity of either MII or MX enzymes is essential for *N*-glycan biosynthesis in vivo. Mutant mice lacking both enzymatic activities die shortly after birth owing to respiration failure, thus suggesting the essential role of complex-type *N*-glycan in the lung of neonates.

Concept

MII is one of the enzymes responsible for conversion of the hybrid-type *N*-glycan to the complex-type *N*-glycan (Fig. 1). Hybrid-type *N*-glycan $GlcNAc_1Man_5GlcNAc_2$, which is produced by β1,2-*N*-acetylglucosaminyltrasnferase-I (GnTI), is processed by MII and the product, $GlcNAc_1Man_3GlcNAc_2$, is further modified by β1,2-*N*-acetylglucosaminyl-transferase-II (GnTII) for formation of the complex-type *N*-glycan core structure. Mutant mice lacking GnTI show several developmental abnormalities in neural tube formation and vascularization, and die at embryonic days 9–10 (Ioffe and Stanley 1994; Metzler et al. 1994). Mutant mice lacking GnTII survive at neonatal stage, but exhibit severe phenotypes, such as gastrointestinal and hematologic abnormalities, and most of the mutant mice die by 4 weeks (Wang et al. 2001). Contrary to these severe phenotypes, MII gene knockout mutant mice are fertile and live out a natural life span, which are of much milder phenotype than that of GnTI and GnTII knockout mice. MII-deficient mice show anemic phenotype, which resembles to CDA type II or HEMPAS in human, and also develop autoimmune phenotype in kidney (Chui et al. 1997, 2001). Carbohydrate

---

Glycobiology Program, Cancer Research Center, Burnham Institute for Medical Research, 10901 North Torrey Pines Road, La Jolla, San Diego, CA 92037, USA
Phone: +1-858-646-3680
E-mail: [1]takama@burnham.org, [2]michiko@burnham.org

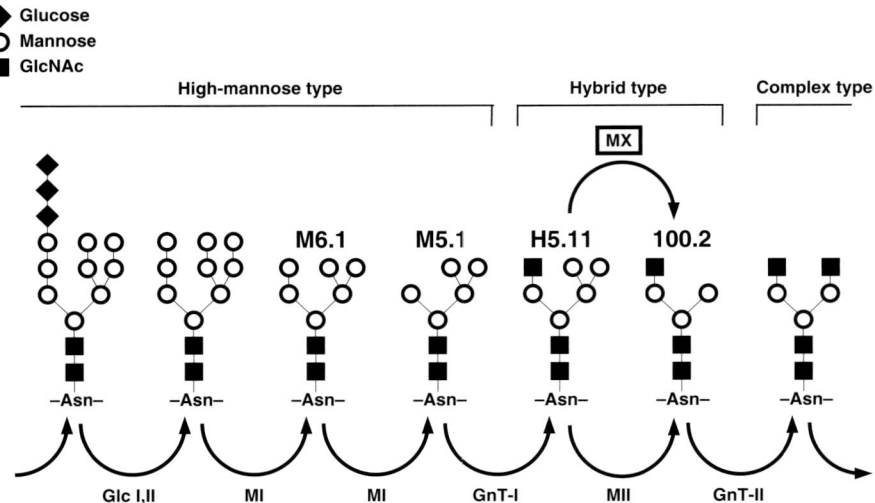

Fig. 1 Biosynthetic pathway of N-glycan. High-mannose type N-glycan is processed by GnTI and converted to hybrid type N-glycan, GlcNAc$_1$Man$_5$GlcNAc$_2$ (*H5.11*). This structure is next processed by either MII or MX to form GlcNAc$_1$Man$_3$GlcNAc$_2$ (*100.2*), and further converted to form complex type core structure GlcNAc$_2$Man$_3$GlcNAc$_2$ by GnTII. GnTI-deficient mice cannot produce hybrid type N-glycan and GnTII-deficient as well as MII/MX double-deficient mice have no complex type N-glycan in the tissues

structural analysis revealed that MII-deficient mice have reduced but significant amount of complex-type N-glycan structures, whereas both GnTI and GnTII knockout mice show no complex-type N-glycans, indicating an alternative MII-independent pathway for N-glycan processing.

MX was identified as an MII-like mannosidase because of its homology to MII (Misago et al. 1995). MX enzyme hydrolyzed synthetic α-mannosidase substrate, but specific substrate of MX has not been determined due to its weak enzymatic activity in vitro. MX-deficient mice showed no apparent phenotype other than male infertility (Akama et al. 2001). Carbohydrate structural analysis showed no significant change of N-glycan structure in MX null tissues, except testis, in which MX may have a major role on N-glycan processing. Double mutant mice lacking both MII and MX enzymes exhibit more severe phenotype than each single gene knockout mouse does. Some of double mutant mice die around embryonic days 15–18 and rest of them die at neonatal stage (within 2 days after birth) owing to respiration failure (Akama et al. 2006). Carbohydrate structural analysis demonstrated that the double mutant mice have no complex-type N-glycans at embryonic day 15 and in neonatal stage. Enzymatic analysis of recombinant MX demonstrated that the MX hydrolyses GlcNAc$_1$Man$_5$GlcNAc$_2$ structure, which is identical to the structure of specific substrate for MII enzyme. These findings clearly demonstrated that MX is involved in N-glycan processing and either MII or MX is necessary for proper N-glycan biosynthesis. The results also suggest the importance of the complex type N-glycans for the lung development (Fig. 2).

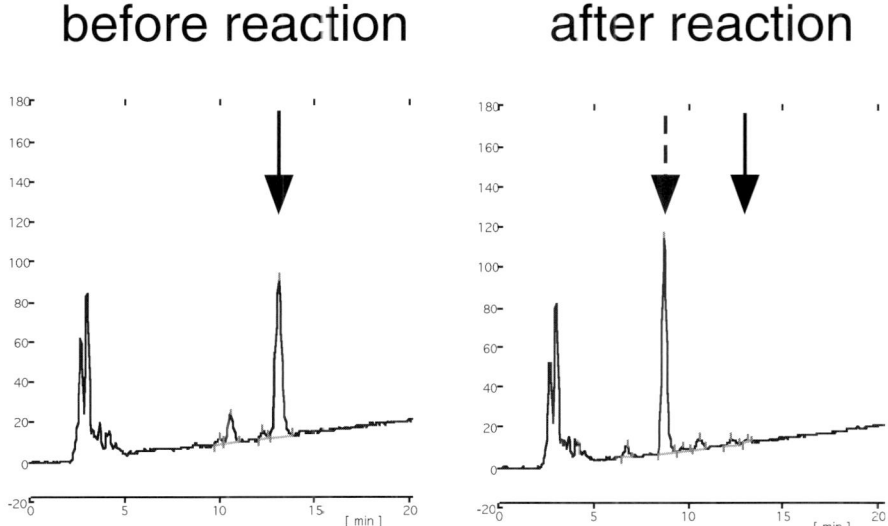

Fig. 2 Mannosidase activity analysis. Recombinant MX enzyme was incubated with pyridylaminated GlcNAc$_1$Man$_5$GlcNAc$_2$ for overnight and the reaction mixture was analyzed by amide column HPLC. *Solid* and *dashed arrows* indicate elution position of GlcNAc$_1$Man$_5$GlcNAc$_2$ (original substrate) and GlcNAc$_1$Man$_3$GlcNAc$_2$ (hydrolyzed product), respectively

Procedure

Activity Assay for MII and MX

1. Prepare enzyme source. Recombinant soluble enzyme can be used as an enzyme source of MII and MX (Akama et al. 2006). Immunoprecipitated sample can also be used as an enzyme source of MII (Chui et al. 1997).
2. Prepare reaction mixture in 50 µl of 0.1 M NaOAc pH 5.8 containing 15 pmol of pyridylaminated substrate and enzyme source of interest.
3. Incubate the mixture at 37°C for 1 h or an appropriate time.
4. Stop enzymatic reaction by placing the sample tube in boiled water for 5 min.
5. Analyze the product by Amide column HPLC. Each sample applied to Amide-80 column (Tosoh Bioscience LLC, Montgomeryville, PA) is separated by a linear gradient elution from 65% acetonitrile–175 mM triethylamine-AcOH, pH 7.3 (initial solvent) to 56% acetonitrile–220 mM triethylamine-AcOH, pH 7.3 (final solvent) over 30 min at 1 ml/min flow rate. Eluted carbohydrates are detected by a fluorescence detector (320 nm excitation and 400 nm emission)

Comments

GlcNAc$_1$Man$_5$GlcNAc$_2$ structure can be a substrate of both MII and MX (Akama et al. 2006). However, this substrate can also be hydrolyzed by other α-mannosidases such as lysosomal mannosidase. Therefore isolation of the target α-mannosidase is essential to determine enzymatic activity of MII and MX. Immunoprecipitation is effective for MII enzyme, but specific antibody for MX is not currently available.

References

Akama TO, Nakagawa H, Sugihara K, Narisawa S, Ohyama C, Nishimura S, O'Brien DA, Moremen KW, Millan JL, Fukuda MN (2001) Germ cell survival through carbohydrate-mediated interaction with Sertoli cells. Science 295:124–127

Akama TO, Nakagawa H, Wong NK, Sutton-Smith M, Dell A, Morris HR, Nakayama J, Nishimura S, Pai A, Moremen KW, Marth JD, Fukuda MN (2006) Essential and mutually compensatory roles of alpha-mannosidase II and alpha-mannosidase IIx in N-glycan processing in vivo in mice. Proc Natl Acad Sci USA 103:8983–8988

Chui D, Oh-Eda M, Liao YF, Panneerselvam K, Lal A, Marek KW, Freeze HH, Moremen KW, Fukuda MN, Marth JD (1997) Alpha-mannosidase-II deficiency results in dyserythropoiesis and unveils an alternate pathway in oligosaccharide biosynthesis. Cell 90:157–167

Chui D, Sellakumar G, Green RS, Sutton-Smith M, McQuistan T, Marek KW, Morris HR, Dell A, Marth JD (2001) Genetic remodeling of protein glycosylation in vivo induces autoimmune disease. Proc Natl Acad Sci USA 98:1142–1147

Ioffe E, Stanley P (1994) Mice lacking N-acetylglucosaminyltransferase I activity die at mid-gestation, revealing an essential role for complex or hybrid N-linked carbohydrates. Proc Natl Acad Sci USA 91:728–732

Metzler M, Gertz A, Sarkar M, Schachter H, Schrader JW, Marth JD (1994) Complex asparagine-linked oligosaccharides are required for morphogenic events during post-implantation development. EMBO J 13:2056–2065

Misago M, Liao YF, Kudo S, Eto S, Mattei MG, Moremen KW, Fukuda MN (1995) Molecular cloning and expression of cDNAs encoding human alpha-mannosidase II and a previously unrecognized alpha-mannosidase IIx isozyme. Proc Natl Acad Sci USA 92:11766–11770

Wang Y, Tan J, Sutton-Smith M, Ditto D, Panico M, Campbell RM, Varki NM, Long JM, Jaeken J, Levinson SR, Wynshaw-Boris A, Morris HR, Le D, Dell A, Schachter H, Marth JD (2001) Modeling human congenital disorder of glycosylation type IIa in the mouse: conservation of asparagine-linked glycan-dependent functions in mammalian physiology and insights into disease pathogenesis. Glycobiology 11:1051–1070

Sialidase Genes

Taeko Miyagi

Introduction

Sialidases catalyze the removal of α glycosidically linked sialic acid residues from glycoproteins, glycolipids, and oligosaccharides, which is an initial step in the degradation of these glycoconjugates. They demonstrate great variation in their expression during cell differentiation, cell growth, and malignant transformation, and have been shown to be intimately involved in many biological processes (Miyagi et al. 2004). Herein, four types of mammalian sialidase have been cloned and identified so far, and summarized briefly in terms of their properties and possible functions (see Table 1). Although the sequences of Neu1, Neu2, Neu3, and Neu4 are not particularly similar to those of bacterial and parasite sialidases, sequence alignment has revealed that they all contain several Asp boxes (-Ser-X-Asp-X-Gly-X-Thr-Trp-), and an Arg-Ileu-Pro sequence, conserved sequences found in sialidases from microorganisms.

Procedures

Sialidasea Assay with Ganglioside Substrates

1. Place the reaction mixture in a test tube
    (a) Sodium acetate buffer (pH 4.5)[1]   5 µmol
    (b) Gangliosides (releasable sialic acid sites)   10 nmol
    (c) TritonX-100 or sodium cholate   50 µg
    (d) Bovine serum albumin   50 µg
    (e) Enzyme (~0.5 mg protein) in 50 µl
2. Incubate for 0.5~2 h at 37°C
3. Determine released sialic acids by TBA or HPLC

Determination of Released Sialic Acids[2]

TBA Methods (Warren 1963)

1. Terminate the reaction (50 µl) by adding 25 µl metaperiodate solution[3]
2. Stand for 20 min at room temperature
3. Add 250 µl arsenite solution[4] and voltex the tube until a yellow-brown color disappears
4. Add 750 µl TBA solution[5] and voltex and heat in a boiling water bath for 15 min

---

Division of Biochemistry, Miyagi Cancer Center Research Institute, Natori, Miyagi 981-1293, Japan

---

[1] Optimal pH in activity assay: 4.5 for Neu1, Neu3 and Neu4, and 5.5~6.0 for Neu2.
[2] TBA method has been generally used but the color reaction interferes with other sugar compounds and is much less sensitive than HPLC methods with fluorogenic reagents.
[3] 0.2 M sodium metaperiodate in 9 M phosphoric acid.
[4] 10% Sodium arsenite in 0.1 N sulfuric acid and 0.5 M sodium sulfate.
[5] 0.6% Thiobarbituric acid in 0.5 M sodium sulfate.

Table 1 Comparison of the four types of mammalian sialidase

	Neu1	Neu2	Neu3	Neu4
Major location	Lysosomes	Cytosol	Plasma membrane	Lysosomes*
				Mitochondria**
				Intracellular membranes**
Major substrate	Oligosaccharides	Oligosaccharides	Gangliosides	Oligosaccharides
	4MU-NeuAc	4MU-NeuAc		4MU-NeuAc
		Glycoproteins		Glycoproteins
		Gangliosides		Gangliosides
Optimal pH	4.4–4.6	6.0–6.5	4.6–4.8	4.4–4.5
Total amino acids (human)	415	380	428	496 (484)
Chromosome location (human)	6p 21.3	2q 37	11q13.5	2q37.3
Possible function	Degradation in lysosomes	Myoblast differentiation	Neural differentiation	Apoptosis
	Immune function		Apoptosis	
			Cell signaling	

*Seyrantepe et al. (2004)
**Yamaguchi et al. (2005)

5. Place in cold water for 5 min
6. Add 750 µl cyclohexanone and voltex and centrifuge at 3000 rpm for 10 min
7. Measure the upper phase at 549 nm

HPLC Analysis with Fluorimetric Detection[6] (Li 1992):

1. Add 0.8% malononitrile 50 µl into the reaction mixture (25 µl)
2. Add 0.15 M tetraborate (pH 9.5) 355 µl
3. Place for 20 min at 80°C
   Apply 10~20 µl to a reversed-phase column (C_{18}) and separate with a methanol and ammonium acetate buffer (15:85 v/v, pH 5.5)
4. Measure fluorometrically (emission 434 nm, excitation 357 nm)

Sialidase Assay with 4MU-NeuAc Substrate

1. Prepare the reaction mixture
   Sodium phosphate buffer (pH 4.5)   10 µmol
   4MU-NeuAc   30 nmol
   Bovine serum albumin   0.1 mg
   Enzyme in 0.1 ml
2. Incubate for 20–60 min at 37°C

---

[6] DDB (1, 2-diamino-4,5 methylene dioxybenzene) can also be used as a fluorogenic reagent for sialic acid determination (Hara et al. 1986), but the reaction with DDB needs longer time (2.5 h at 50°C in the dark).

3. Terminate the reaction by addition of 1.25 ml 0.25 M glycine–NaOH (pH 10.4)
4. Measure fluorometrically 4-methylumbelliferylone(4-MU) released (emission 448 nm, excitation 365 nm)

Comments

The lysosomal sialidase, Neu1, is a target gene for a sialidase deficiency, sialidosis, and is known to be associated with a protective protein (carboxypeptidase A) and β-galactosidase as a complex in lysosomes, dissociation of the complex leading to sialidase inactivation. In 1996–1998, the *Neu1* gene was shown by three groups in humans and in mice to encode a major histocompatibility complex (MHC)-related sialidase. With an optimum pH of about 4.5, the expressed sialidase desialylates mainly glycopeptides and oligosaccharides, as well as 4-methylumbelliferyl-neuraminic acid (4MU-NeuAc), a synthetic substrate. The human form features a lysosomal C-terminal targeting motif, and evidence has been generated for the existence of a protective protein responsible for transport to lysosomes. However, recent observations revealed an intracellular distribution of sialidase from the human homolog NEU1 gene in plasma membranes as well as within lysosomes under conditions of cell stimulation. Examination of sialidosis patients has revealed various mutations in genomic DNA including frameshift insertions and missense changes.

Cytosolic sialidase, Neu2, provided the first example of a mammalian sialidase for which cDNA cloning was achieved. In addition to location in cytosol, the enzyme has also been found in nucleoplasm of muscle fibers by immuno-histochemical analysis on electron microscopy, probably due to the presence of a nuclear localization signal near the N-terminus. The expressed sialidase shows an optimum pH of about 6.5 and can desialylate fetuin and gangliosides as well as synthesized 4MU-NeuAc. Homologs cloned from cDNA libraries of CHO, mouse brain, and thymus and from a genomic library of human skeletal muscle have all shown high amino acid identity (98–70%) with the rat gene. Recent studies on the crystal structure of the human recombinant enzyme have furthermore provided evidence for a canonical six-blade beta-propeller (Chavas et al. 2005). This has similarly been observed for viral and bacterial sialidases, with the active site in a shallow crevice. However, the presence of residues recognizing the *N*-acetyl and glycerol moieties of DANA is not shared by bacterial and viral sialidases.

The plasma membrane-associated sialidase, Neu3, was first cloned from a bovine brain library, and then from a cDNA library of human brain and later from the human genome database. In COS-7 cells transiently expressing the bovine sialidase, hydrolysis was found to be essentially specific for gangliosides other than GM1 and GM2, in the presence of Triton X-100. The major subcellular localization was established to be plasma membranes by Percoll density gradient centrifugation of cell homogenates and by immunofluorescence staining of cells. Analysis of the membrane topology of protease protection further suggested that this sialidase has a type II membrane orientation, with its carboxy-terminus facing the extracytoplasmic side. The primary sequences covering the entire coding region of the corresponding human, mouse, and rat genes display an 83, 79, and 78% overall identity with the bovine gene, respectively. Unlike the bovine and mouse Neu3 sialidases, the human ortholog NEU3 is not always detected on cell surfaces but can be moved to and concentrated at leading edges in response to growth stimuli. NEU3 expression is markedly increased in human cancers and is closely associated with neurite formation.

The fourth sialidase, Neu4, was only recently identified, based on cDNA sequences in public databases. Examples of murine and human genes have been characterized in transfected cells, and the murine gene contains four exons and an open reading frame of 501 amino acids, demonstrating greatest similarity to Neu3 (42%) among the known murine sialidases. With regard to the subcellular localization of the human ortholog, NEU4, two different sites have been reported on the basis of gene transfection studies: one the lysosomal lumen (Seyrantepe et al. 2004), and the other the mitochondria and intracellular membranes (Yamaguchi et al. 2005). NEU4 seems to consist of iso-forms differing in the presence and absence of 12 N-terminal amino acid residues which act on mitochondrial targeting so that this might explain the differences in localization. The isoforms are also differentially expressed in a tissue-specific manner, brain, muscle and kidney containing both, and the liver and colon possessing predominantly the short form, as assessed by RT-PCR. Unlike other human sialidases, the enzyme possesses broad substrate specificity, including sensitivity to mucin.

References

Chavas LM, Tringali C, Fusi P, Venerando B, Tettamanti G, Kato R, Monti E, Wakatsuki S (2005) Crystal structure of the human cytosolic sialidase Neu2. Evidence for the dynamic nature of substrate recognition. J Biol Chem 280:469–475

Hara S, Yamaguchi M, Takemori Y, Nakamura M, Ohkura Y (1986) Fluorometric high-performance liquid chromatography of N-acetyl- and N-glycolylneuraminic acids and its application to their microdetermination in human and animal sera, glycoproteins, and glycolipids. J Chromatogr 377:111–119

Li K (1992) Determination of sialic acids in human serum by reversed-phase liquid chromatography with fluorimetric detection. J Chromatogr 579:209–213

Miyagi T, Kato K, Ueno S, Wada T (2004) Aberrant expression of sialidase in cancer. Trends Glycosci Glycotech 92:371–381

Seyrantepe V, Landry K, Trudel S, Hassan JA, Morales CR, Pshezhetsky AV (2004) Neu4, a novel human lysosomal lumen sialidase, confers normal phenotype to sialidosis and galactosialidosis cells. J Biol Chem 279:37021–37029

Warren L (1963) Thiobarbituric acid assay of sialic acids. In: Colowick SP, Kaplan NO (eds) Methods in enzymology. Academic Press, New York, vol 6, p 463

Yamaguchi K, Hata K, Koseki K, Shiozaki K, Akita H, Wada T, Moriya S, Miyagi T (2005) Evidence for mitochondrial localization of a novel human sialidase (NEU4). Biochem J 390:85–93

Release of Sugar Chains from Glycosphingolipids

Makoto Ito

Introduction

Sugar chains of glycosphingolipids (GSLs) can be released by ozonolysis or periodate oxidation; however, these methods are not suitable for alkaline-sensitive GSLs because they are performed under alkaline conditions. Furthermore, these chemical procedures are time-consuming, somewhat troublesome, and of low yield; and thus a simple and reproducible method for obtaining intact sugar chains from GSLs is required. This section deals with the release of sugar chains from GSLs by endoglycoceramidase (EGCase, EC3.2.1.123).

The EGCase is an enzyme that specifically hydrolyzes the glycosidic linkage between oligosaccharides and ceramides of various GSLs (Ito and Yamagata 1989). Thus, intact sugar chains as well as ceramides can be obtained simultaneously using the enzyme with high yield. Three molecular species of EGCase (EGCase I, II, and III) have been found in the supernatant of *Rhodococcus* sp. M-750. EGCase II, the gene of which was the first to be cloned, is able to hydrolyze lacto-/neolacto-, and ganglio-series GSLs very efficiently (Izu et al. 1997). A recombinant EGCase III (EGALC) specifically hydrolyzes 6-gala-series GSLs which possess the common structure, R-Galβ1-6Galβ1-1′Cer. However, EGALC cannot hydrolyze GSLs, sugar chains of which are linked to a ceramide moiety via a β1-1′glucosidic linkage (Ishibashi et al. 2007). Very recently, the gene of EGCase I was cloned, and a recombinant form of the protein was successfully expressed in Rhodococcal cells. The substrate specificity of the recombinant EGCase I is relatively broad compared with that of EGCase II, i.e., EGCase I can hydrolyze globo-series GSLs much faster than EGCase II. The mode of action of EGCases against GSLs is shown in Fig. 1, and the specificity of the three molecular species of EGCase is summarized in Table 1. None of the species described here is able to degrade glucosylceramide (GlcCer) and sulfatide, while galactosylceramide (GalCer) is hydrolyzed, but with low reaction velocity, by EGALC.

Protocol for Obtaining Intact Sugar Chains of GSLs with EGCase

Hydrolysis of GSLs by EGCase[1]

1. Dissolve GSLs at a concentration of 1 mM in 10 mM sodium acetate buffer, pH 5.0~5.5, containing 0.4% (w/v) Triton X-100.
2. Subject to sonic-treatment for 1 min. When difficult to dissolve, the sample should be heated.

Department of Bioscience and Biotechnology, Graduate School Kyushu University, 6-10-1 Hakozaki, Higashi-ku, Fukuoka 812-8581, Japan
Phone: +81-92-641-2898, Fax: +81-92-641-2907
E-mail: makotoi@agr.kyushu-u.ac.jp

[1] The recombinant EGCase II is commercially available from Takara Bio Co., Otsu, Japan.

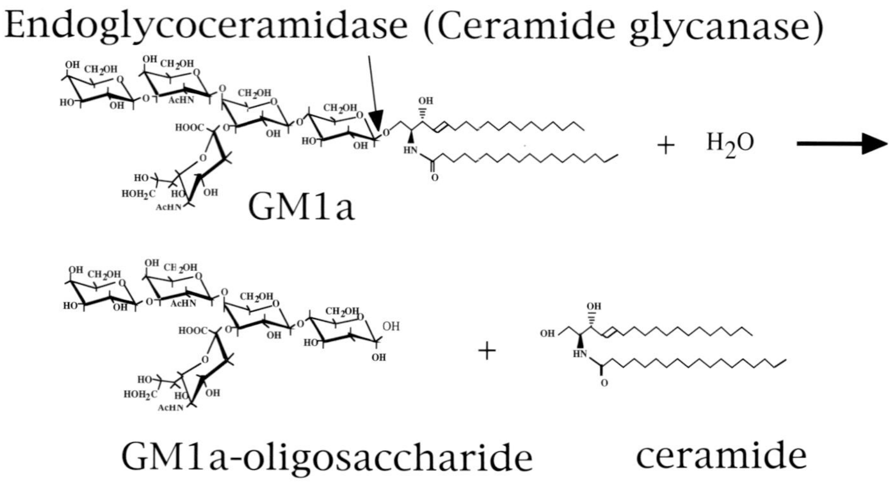

Endoglycoceramidase (Ceramide glycanase)

GM1a + H$_2$O →

GM1a-oligosaccharide + ceramide

Fig. 1 Action mode of endoglycoceramidase on a glycosphingolipid, GM1a.

Table 1 Specificity of three molecular species of EGCase

Substrate	EGCase I	EGCase II	EGCase III EGALC
GlcCer	0	0	0
LacCer	100	100	0
GM3	100	100	0
GM2	100	100	0
GM1a	100	100	0
GD1a	100	100	0
Gb3Cer	80	5	0
Gb4Cer	80	5	0
GalCer	0	0	5
6-Gal2Cer	0	0	80
6-Gal3Cer	0	0	80
Sulfatide	0	0	0
SM	0	0	0

Reaction condition, 10 mU of EGCase is incubated with 10 nmol of GSLs in 20 μl of 20 mM sodium acetate buffer, pH 5.5, containing 0.2% Triton X-100 at 37°C overnight.

GlcCer, Glcβ1-1Cer; LacCer, Galβ1-4Glcβ1-1Cer; GM3, NeuAcα2-3 Galβ1-4Glcβ1-1Cer; GM2, GalNAcβ1-4(NeuAcα2-3)Galβ1-4Glcβ1-1Cer; GM1a, Galβ1-3GalNAcβ1-4(NeuAcα2-3) Galβ1-4Glcβ1-1Cer; GD1a, NeuAcα2-3Galβ1-3GalNAcβ1-4(NeuAcα2-3) Galβ1-4Glcβ1-1Cer; Gb3Cer, Galα1-4Galβ1-4Glcβ1-1Cer; Gb4Cer, GalNAcβ1-3Galα1-4Galβ1-4Glcβ1-1Cer; GalCer, Galβ1-1Cer; 6-Gal2Cer, Galβ1-6Galβ1-1Cer; 6-Gal3Cer, Galβ1-6Galβ1-6Galβ1-1Cer; sulfatide, HO$_3$S-3 Galβ1-1Cer; SM (sphingomyelin), phosphochorineCer; Cer, Ceramide

3. Incubate a reaction mixture at 37°C overnight which contains 50 nmol of various GSLs and 1.5–10 mU of EGCase in 50 µl of 20 mM sodium acetate buffer, pH 5.0~5.5, containing 0.2% Triton X-100. The amount of EGCase depends on the species of GSLs (see Table 1).

Examination of Extent of Hydrolysis of GSLs

1. Dry a reaction mixture corresponding to 10 nmol of each GSL by Speed Vac concentrator.
2. Dissolve in 10 µl of 50% methanol and apply onto a TLC plate (precoated Silica Gel TLC plates, Merck, Germany).
3. Develop the plate with n-butanol/acetic acid/water (2/1/1, v/v) as the developing solvent.
4. Visualize the oligosaccharide which migrates slower than the original GSL by using orcinol-H_2SO_4.
5. Scan the TLC plate with a Shimadzu CS-9000 TLC chromatoscanner with the reflectance mode set at 540 nm.
6. Calculate the extent of hydrolysis of GSLs as follows:
Hydrolysis (%) = DR for oligosaccharide released/DR for remaining GSL + DR for oligosaccharide released × 100, where DR = densitometric response at 540 nm.

Isolation of Intact Sugar Chains from Reaction Mixtures

Method 1

1. Partition the reaction mixture with chloroform/methanol (2/1, v/v).
2. Take the upper layer as the fraction containing sugar chains and dry under a stream of N_2-gas.

Method 2

1. Apply the reaction mixture onto an SEP-PAK cartridge previously equilibrated with methanol/water (1/1, v/v).
2. Take the pass fraction as the fraction containing the sugar chains. Ceramides and GSLs undigested by EGCase are adsorbed onto an SEP-PAK cartridge.
3. Dry the pass fraction under a stream of N_2-gas.

In the case of small quantity samples (Method 2),

1. Take some of the SEP-PAK from the cartridge and place it in a Millipore filtration tube (Ultra free C3LGC, Millipore Co.).
2. Mix the reaction mixture and keep at room temperature for 30 min with shaking.
3. Insert the filtration tube into an Eppendorf centrifugation tube (1 ml).
4. Centrifuge at 7,000 rpm for 20 min.
5. Dry the filtrate under a stream of N_2-gas.

Further Comments

All EGCases require detergents such as Triton X-100 for hydrolysis of GSLs in vitro. Interestingly, *Rhodococcus* sp. M-750 produces EGCases as well as their activator proteins (activator I and II) in the culture supernatant. Activator protein II was found to stimulate the hydrolysis of GSLs by EGCase II in the absence of detergents (Ito et al. 1991). In contrast to Triton X-100, the purified activator II is not toxic to intact cells. Therefore, hydrolysis of the cell-surface GSLs of living cells can be performed in the presence of an activator protein instead of detergents.[2] Very recently, the gene for an activator protein was cloned, and a recombinant form of the protein was expressed in Rhodococcal cells as a host.

References

Ishibashi Y, Nakasone T, Kiyohara M, Horibata Y, Sakaguchi K, Hijikata A, Ichinose S, Omori A, Yasui Y, Imamura A, Ishida H, Kiso M, Ito M (2007) A novel endoglycoceramidase hydrolyzes oligogalactosylceramides to produce galactooligosaccharides and ceramides. J Biol Chem 282: 11386–11396

Ito M, Yamagata T (1989) Purification and characterization of glycosphingolipid-specific endoglycosidaseses (endoglycoceramidases) from a mutant strain of *Rhodococcus* sp. J Biol Chem 264:9510–9519

Ito M, Ikegami Y, Yamagata T (1991) Activator proteins for glycosphingolipid hydrolysis by endoglycoceramidase. J Biol Chem 266:7919–7926

Izu H, Izumi Y, Kurome Y, Sano M, Kondo A, Kato I, Ito M (1997) molecular cloning, expression, and sequence analysis of the endoglycoceramidase II gene from *Rhodococcus* sp. strain M-777. J Biol Chem 272:19846–19850

---

[2]EGCase II with activator II (EGCase-ACT) can be obtained from Takara Bio Co., Japan.

Heparan Sulfate Endosulfatase Assay

Kazuko Keino-Masu, Masayuki Masu

Introduction

Heparan sulfate (HS) shows enormous structural heterogeneity generated by the combined modification of epimerization, N-sulfation, and O-sulfation in the polysaccharide chain. A novel family of extracellular sulfatases (SulfFP1/sulf-1 and SulfFP2/sulf-2) identified in quail, rats, humans, and mice (Dhoot et al. 2001; Morimoto-Tomita 2002; Ohto et al. 2002) act as endosulfatases, which remove 6-O sulfate from N-acetylglucosamine residues in intact heparin/HS. Previous studies have revealed that these endosulfatases modulated the signaling of many heparin-binding factors, suggesting that they play a role in the regulation of cellular signaling in vivo. Here, we describe a method to measure the endosulfatase activity.

Procedure

Enzyme Reaction

1. Prepare a conditioned medium from the 293 cells transfected with SulfFP1 or SulfFP2 together with sulfatase modifying factor 1 (SUMF1).
2. Concentrate the conditioned medium using a Microcon YM-30 (Millipore).
3. Incubate 10 μg heparin or HS with the conditioned medium (5 μl) in 5 mM Tris–HCl (pH 7.5), 150 mM NaCl and 10 mM $MgCl_2$ in a total volume of 10 μl at 37°C for 24 h.
4. Stop the reaction by heating at 95°C for 2 min.
5. Add 10 μl of 60 mM sodium acetate buffer (pH 7.0) containing 6 mM calcium acetate and a mixture of 0.5 U heparinase I (Sigma-Aldrich), 1 mIU of heparitinase I (Seikagaku), and 1 mIU of heparitinase II (Seikagaku), and incubate at 37°C for 24 h.
6. After stopping the reaction by heating at 95°C for 2 min, clean the mixture using an Ultrafree-MC Biomax-5 filter (Millipore).

HPLC Analysis (Reversed-phase Ion-pair Chromatography Described by Toyoda et al. 2000)

1. Set up an HPLC equipped with a Senshu Pak Docosil column (4.6 × 150 mm, particle size 5 μm, Senshu Scientific) and a spectrophotometer.
2. Equilibrate the column using a buffer containing 2 mM NaCl, 1.2 mM tetra-n-butylammonium hydrogen sulfate, and 8.5% acetonitrile at 55°C.

Department of Molecular Neurobiology, Institute of Basic Medical Sciences, Graduate School of Comprehensive Human Sciences, University of Tsukuba, 1-1-1 Tennoudai, Tsukuba, Ibaraki 305-8577, Japan
Phone: +81-29-853-3249, Fax: +81-29-853-3498
E-mail: kazumasu@md.tsukuba.ac.jp

Fig. 1 A typical chromatogram for the endosulfatase assay. The retention time of 8 standard unsaturated disaccharides is shown on the *top*. Incubation of heparin with the conditioned medium of the 293 cells transfected with SulfFP1 and SUMF1, or SulfFP2 and SUMF1, resulted in decrease of a trisulfated disaccharide unit ΔUA2S-GlcNS6S and concomitant increase of a disulfated unit ΔUA2S-GlcNS, compared with the control using the cells transfected with SUMF1 alone

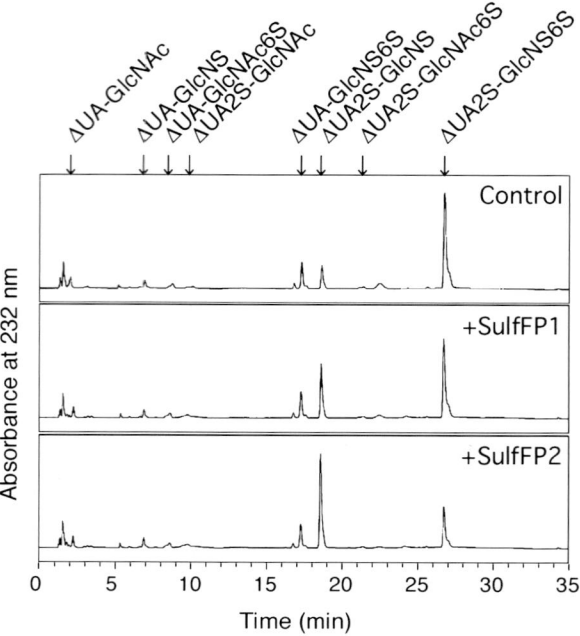

3. Apply the digested product to HPLC analysis.
4. Elute the product by increasing gradient of NaCl (from 2 to 106 mM) at a flow rate of 1.1 ml/min at 55°C.
5. Monitor the absorbance at 232 nm.
6. Estimate the product amounts from the absorbance at 232 nm (Fig. 1).

References

Dhoot GK et al (2001) Regulation of Wnt signaling and embryo patterning by an extracellular sulfatase. Science 293:1663–1666
Morimoto-Tomita M, Uchimura K, Werb Z, Hemmerich S, Rosen SD (2002) Cloning and characterization of two extracellular heparin-degrading endosulfatases in mice and humans. J Biol Chem 277:49175–49185
Ohto T et al (2002) Identification of a novel nonlysosomal sulphatase expressed in the floor plate, choroid plexus and cartilage. Genes Cells 7:173–185
Toyoda H, Kinoshita-Toyoda A, Selleck SB (2000) Structural analysis of glycosaminoglycans in Drosophila and Caenorhabditis elegans and demonstration that tout-velu, a Drosophila gene related to EXT tumor suppressors, affects heparan sulfate in vivo. J Biol Chem 275:2269–2275

Sphingolipid Activator Proteins

Junko Matsuda

Introduction

Glycosphingolipids (GSLs) are amphiphilic constituents of the outer leaflet of eukaryotic plasma membranes. In vivo degradation of GSLs takes place predominantly in the lysosome by the stepwise release of monosaccharide units from the nonreducing end of the oligosaccharide chain via specific exohydrolases (Fig. 1). Several of these enzymes need assistance of small glycoprotein cofactors, the so-called 'sphingolipid activator proteins'. These are low molecular weight glycoproteins, which by themselves are catalytically inactive but are required as cofactors to facilitate interactions between membrane-bound hydrophobic sphingolipids and water-soluble exohydrolases in the lysosome, either by direct activation of their respective enzymes or as biological detergents that lift substrates out of the membrane in which sphingolipids are embedded. Five sphingolipid activation proteins (GM2 activator protein, and four saposins) encoded by two distinct genes are known. One gene encodes the GM2-activator protein, the other encodes prosaposin, the precursor protein of four saposins (Sandhoff et al. 2001; Kolter and Sandhoff 2005). This review limits its scope primarily to the activator function of the sphingolipid activator proteins but these proteins have other functions, such as lipid-binding and lipid-transfer proteins.

GM2 Activator Protein

The GM2-activator protein is encoded in the gene (*GM2A*) on chromosome 5 (5q31.3–q33.1) in humans. It is a glycoprotein which acts as a cofactor essential for in vivo degradation of ganglioside GM2 by β-hexosaminidase A (Fig. 1). The inherited deficiency of the GM2-activator protein leads to the AB variant of GM2-gangliosidosis. An X-ray crystallographic structure of the nonglycosylated GM2-activator protein shows a hydrophobic cavity that harbors the ceramide moiety of ganglioside GM2, suggesting that the GM2-activator protein may bind to the ganglioside GM2-containing intra-lysosomal lipid vesicles and lift the ganglioside GM2 out of the membrane and form water-soluble glycolipid–protein complexes, and then present ganglioside GM2 to the active site of β-hexosaminidase A.

Institute of Glycotechnology, Future Science and Technology Joint Research Center, Tokai University, 1117 Kita-kaname, Hiratsuka, Kanagawa 259-1292, Japan
Phone: +81-463-58-1211, Fax: +81-463-50-2432
E-mail: matsujun@keyaki.cc.u-tokai.ac.jp

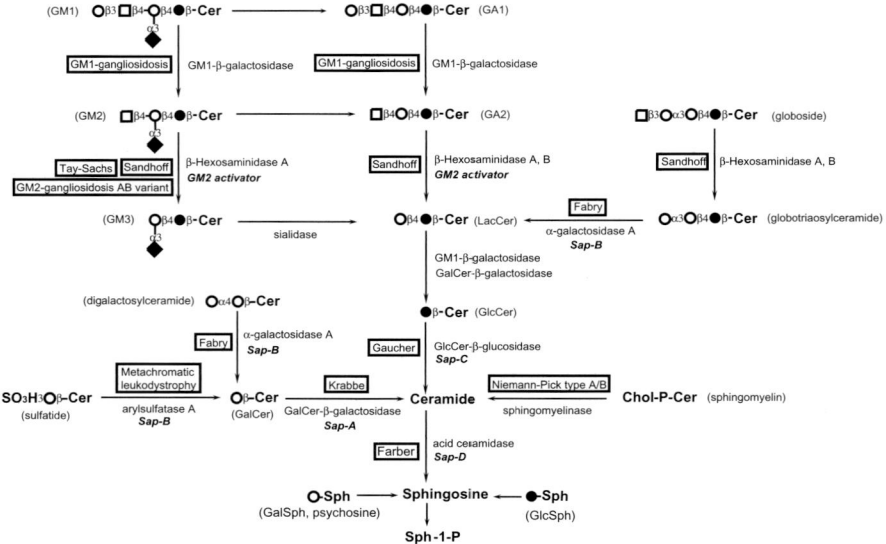

Fig. 1 Degradation pathway of major sphingolipids in mammals. The respective exohydrolases and activator proteins whose *in vivo* functions have been established either by human disorders or mouse models are indicated for each degradation step. The lysosomal sphingolipid storage disorders resulting from either enzyme or activator protein deficiency are shown in *frames*. The symbol nomenclatures of glycan structure are follows: *open circle* galactose, *filled circle* glucose, *open square* N-acetyl-galactosamine and *filled diamond* NeuAc. The linkage information is represented in *text* next to the symbols

Saposins

The four saposins (saposins A, B, C, and D) are required for lysosomal degradation of various GSLs by exohydrolases. They are derived from a single precursor protein, prosaposin. In humans, prosaposin is encoded by the gene (*PSAP*) located on chromosome 10 (10q21-q22), and the four saposins are encoded in tandem within the prosaposin sequence in the order of A, B, C, and D. The prosaposin is processed to individual saposins by protease in the lysosome. The four saposins are homologous, and each consists of about 80 amino acids containing one *N*-glycosylation site, with the exception of human saposin A carrying two *N*-glycosylation sites, and six highly conserved cysteine residues that can form three intramolecular disulfide bridges in the processed saposins. Despite these structural similarities, they activate degradation of sphingolipids by lysosomal enzymes with relatively high substrate/enzyme specificity. Activated degradation of a relatively large number of sphingolipids has been reported by tests in the test tube. For example, at least three saposins were said to be activators of sphingomyelin degradation. The in vivo functions of each saposin have been clarified most definitively by the observation of human disorders of specific saposin deficiencies and of the specific knockout mouse models. No abnormal accumulation of sphingomyelin in the complete deficiency of all saposins, both in the human patients and in the knockout mouse, indicates the need for serious caution in interpreting the results of the test tube.

Saposin A

Saposin A is required for in vivo degradation of galactosylceramide by galactosylceramide-β-galactosidase which is the enzyme genetically deficient in human Krabbe disease (Fig. 1). This function was convincingly demonstrated by the phenotype of saposin A-specific knockout mouse (Matsuda et al. 2001). These animals presented the phenotype of a late-onset form of Krabbe disease with accumulation of galactosylceramide in the kidney, psychosine in the brain and the spinal cord, and of the seminolipid precursor in the testis. To date, one human patient with an in-frame 3-bp deletion in the saposin A domain showing the symptom of infantile Krabbe disease has been reported, and the second unrelated patient of the Portuguese origin with the same mutation is known (M. T. Vanier, personal communication). Recent evidence indicates that saposin A has lipid binding and lipid mobilization capacity under acidic pH conditions.

Saposin B

Historically, saposin B was the first activator protein to be identified in the mid-1960s. It mediates the in vivo degradation of sulfatide by arylsulfatase A and of globotriaosylceramide and digalactosylceramide by α-galactosidase A (Fig. 1). Inherited defects of saposin B lead to an atypical form of metachromatic leukodystrophy, with late infantile or juvenile onset. The disease can be distinguished from the arylsulfatase A deficiency by excess urinary excretion of globotriaosylceramide in addition to sulfatide. Similar to the GM2-activator protein, the crystal structure of saposin B shows a shell-like homodimer that encloses a large hydrophobic cavity.

Saposin C

Saposin C exists as a homodimer and was initially isolated from the spleens of patients with Gaucher disease in the late 1960s. It is required for the lysosomal degradation of glucosylceramide by glucosylceramide-β-glucosidase (Fig 1). Saposin C deficiency leads to a juvenile form of Gaucher disease with an accumulation of glucosylceramide. The solution structure of saposin C consists of five tightly packed α-helices that form half of a sphere. In contrast to the mode of action of the GM2-activator or saposin B, saposin C can directly activate glucosylceramide-β-glucosidase in an allosteric manner. Saposin C also supports interaction of the enzyme with the substrate embedded in vesicles containing anionic phospholipids by destabilizing these vesicles. Recently saposin C and other saposins are said to be required in the lipid antigen presentation by CD1 immunoreceptors of T cells (Kolter et al. 2005).

Saposin D

Saposin D stimulates lysosomal ceramide degradation by acid ceramidase in vitro and in cultured cells (Fig. 1). Saposin D-specific knockout mouse showed an accumulation of α-hydroxyl fatty acid-containing ceramide in the kidney and the cerebellum (Matsuda et al. 2004). No human patient with saposin D deficiency has been reported so far. The detailed physiological function and mode of action of saposin D is unclear. It can bind to vesicles containing negatively charged lipids and to solubilize them at an appropriate pH.

Prosaposin

Prosaposin itself has been suggested to be involved in a variety of biological processes. Secreted prosaposin is rich in milk, cerebrospinal fluid, and semen. Many functions have been attributed to the secreted form. In the nervous system, the function as a neurotrophic factor is suggested. In the reproductive organs, prosaposin is also the major sulfated glycoprotein secreted by the Sertoli cells in the testis. Four families of the prosaposin deficiency, with three different mutations causing total loss of prosaposin, and all four saposins have been reported so far. In all cases so far studied in detail, patients showed a rapidly progressive, severe neurological disease with accumulation of multiple sphingolipids in the brain and other organs leading to early death. A mouse model of prosaposin deficiency showed clinical, pathological, and biochemical abnormalities similar to human patients.

Determination of Relative Tissue Level of Prosaposin in the Mouse

Relative tissue level of prosaposin in the mouse is analyzed by Western blot analysis using the polyclonal antibody against mouse prosaposin (anti-PSAP) generated by immunizing rabbits with three synthetic oligopeptides corresponding to the mouse *Psap* sequences that do not encode any saposins.

1. Mouse tissue is dissected freshly and is homogenized in an ice-cold 10 mM Tris–HCl (pH 7.0) containing 0.25 M sucrose, 1 mM EDTA and 1.5% protease inhibitor cocktail (STE buffer) to 25% homogenates. The 10 µL of 25% homogenate is used for protein assay.

2. The homogenate is first centrifuged by microcentrifuge at 800 g for 15 min at 4°C, and the supernatant is additionally centrifuged at 100,000 g for 1 h at 4°C. The separated pellet (microsome fraction) is suspended by STE buffer in the same volume of supernatant (cytosole fraction). Each fraction is proceeded to sodium dodecil sulfate polyacrylamide gel electrophoresis (SDS-PAGE).

3. The solubilized proteins (50 µg) are separated by SDS-PAGE on 10 or 15% polyacrylamide gels and transferred to polyvinylidene difluoride membranes.

4. The membranes are incubated with 1% blocking solution (BM Chemiluminescence Western Blotting Kit [Mouse/Rabbit], Roche Molecular Biochemicals) in TBS buffer (20 mM Tris–HCl, pH 7.4, 0.15 M NaCl) for 16 h at 4°C, then incubated with rabbit polyclonal anti-mouse prosaposin antibody (3.74 µg/mL) for 2 h at room temperature.

5. The membrane is washed with TBS-T (20 mM Tris–HCl, pH 7.4, 0.15 M NaCl, 1% Tween-20) buffer and then incubated with peroxidase-labeled anti-rabbit IgG (40 mU/ml; Roche Molecular Biochemicals) for 1 h at room temperature, followed by the treatment with the detection solution of a BM chemiluminescence kit (Roche Molecular Biochemicals). The immunoreactive protein bands are visualized with an LAS-1000 Plus luminescence image analyzer (Fujifilm).

Comments

Anti-PSAP antibody recognizes a major band with molecular weight about 65 kDa corresponding to the size of prosaposin, but no band is detected at around 13 kDa corresponding to the size of mature saposins. The majority of 65 kDa band is glycopeptidase F sensitive and shifted to 58 kDa band.

References

Kolter T, Sandhoff K (2005) Principal of lysosomal membrane digestion–stimulation of sphingolipid degradation by sphingolipid activator proteins and anionic lysosomal lipids. Annu Rev Cell Dev Biol 21:81–103

Kolter T, Winau F, Schaible UE, Leippe M, Sandhoff K (2005) Lipid-binding proteins in membrane digestion, antigen presentation, and antimicrobial defense. J Biol Chem 280:41125–41128

Matsuda J, Vanier MT, Saito Y, Tohyama J, Suzuki K, Suzuki K (2001) A mutation in the saposin A domain of the sphingolipid activator protein (prosaposin) gene results in a late-onset, chronic form of globoid cell leukodystrophy in the mouse. Hum Mol Genet 10:1191–1199

Matsuda J, Kido M, Tadano-Aritomi K, Ishizuka I, Tominaga K, Toida K, Takeda E, Suzuki K, Kuroda Y (2004) Mutation in saposin D domain of sphingolipid activator protein gene causes urinary system defects and cerebellar Purkinje cell degeneration with accumulation of hydroxy fatty acid-containing ceramide in the mouse. Hum Mol Genet 13:2709–2723

Sandhoff K, Kolter T, Harzer K (2001) Sphingolipid activator proteins. In: Scriver CR, Beaudet AL, Sly WS, Valle D, Childs B, Vogelstein B (eds) The metabolic and molecular bases of inherited disease. McGraw-Hill, New York, pp 3371–3388

Section V
Animal Lectins

Affinity Purification of Recombinant Galectins

Jun Iwaki, Jun Hirabayashi

Introduction

Galectins are multifunctional animal lectins defined as β-galactoside-binding proteins with conserved carbohydrate-recognition domains (CRDs). They are largely classified into three structural types: proto, chimera, and tandem-repeat types (Kasai and Hirabayashi 1996). The proto- and chimera-type galectins are composed of a single CRD and form noncovalent multimers, typically dimer of prototype ones. On the other hand, the tandem-repeat type galectins consist of two distinct CRDs connected by a linker polypeptide. Most galectins exhibit hemagglutinin activity inhibited by β-galactosides, such as lactose.

In general, galectins have no signal sequence and post-translational modifications, typical to secreted proteins, such as glycosylation and disulfide bonding. Therefore, it is relatively convenient to produce recombinant galectins in *Escherichia coli* for various biological studies. Here, we describe the production of recombinant galectins in their intact forms and their purification by affinity chromatography.

Procedures

Production of Galectins in Bacteria

Reagents

1. *Escherichia coli* BL21 (DE3) strain
2. Luria Bertani (LB) medium
3. Ampicillin (MP Biomedicals)
4. Kanamycin (Wako)
5. Isopropylthio-β-D-galactoside (IPTG, Fermentas)
6. Phosphate buffered sarine (PBS) [20 mM sodium phosphate, pH 7.4, 150 mM NaCl]
7. MEPBS [PBS containing 4 mM β-mercaptoethanol and 2 mM ethylenediaminetetraacetic acid (EDTA)] (β-mercaptoethanol is essential for galectin-1 activity but it is not substantially important for preservation of other galectins. EDTA is assumed to inhibit metal-dependent protease activity).

Recombinant galectins are produced in *E. coli* BL21 (DE3) strain under the control of T7 promoter. The genes encoding galectins are subcloned into an appropriate expression vector of pET Systems (Novagen), and the resulting plasmids are used for transformation of *E. coli* BL21 (DE3) strain. The transformants are cultured in an appropriate volume (e.g., 250 ml) of LB medium containing 50 µg ml^{-1} ampicillin or 20 µg ml^{-1} kanamycin

---

Lectin Application and Analysis Team, Research Center for Medical Glycoscience, National Institute of Advanced Industrial Science and Technology, AIST Tsukuba Central 2, 1-1-1, Umezono, Tsukuba, Ibaraki 305-8568, Japan
E-mail: jun-hirabayashi@aist.go.jp

Fig. 1 SDS-PAGE of purified recombinant human galectins. Molecular mass markers are indicated on the *left* (kDa)

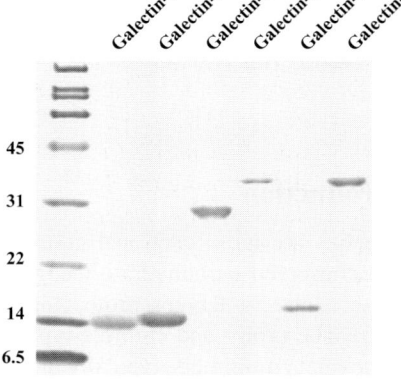

at 37°C to an optical density ($OD_{600} = 0.6$). After the induction of galectin expression by the addition of IPTG to a final concentration of 1 mM, cells are cultivated under optimal conditions (i.e., 20–37°C, 2–16 h) for each galectin. Cells are harvested by centrifugation (6,000 g for 20 min) and suspended in 8 ml of cold MEPBS.

Affinity Purification

Instrument and Reagents

1. Sonifire (Branson) (note that continuous use of this equipment should generate heat, which may inactivate recombinant proteins)
2. 0.45 µm Millex filter (Millipore)
3. Lactose–agarose* (Seikagaku Co.)
4. Lactose (Sigma)
5. Amicon Ultra, MWCO 5,000 (Millipore)

All purification steps are performed at 4°C. *E. coli* cells are disrupted by sonication, and the derived homogenate is centrifuged at 8,000 g for 30 min. If necessary, supernatant is filtrated with a 0.45 µm Millex filter. The clarified fraction is applied to lactose–agarose (e.g., 1 ml of bed volume) equilibrated with PBS. After extensive washing (e.g., 10 ml) of the column with MEPBS, a desorbed galectin is eluted with MEPBS containing 200 mM lactose. Fractions containing galectin are checked by protein assay and sodium dodecyl sulfate polyacrylamide gel electrophoresis (SDS-PAGE). They are combined and dialyzed extensively against MEPBS (e.g., 2 l, twice) to remove lactose. If necessary, the dialyzate is concentrated with Amicon Ultra MWCO 5,000 to an appropriate concentration (e.g., ~1 mg/ml). The purity is checked again by SDS-PAGE, and the final concentration is defined by Bradford or bicinchoninic acid (BCA) protein assay. This purification strategy can be used for most of the other recombinant galectins including human galectin-1~8 (Fig. 1).

Comments

Lactose–Sepharose and asialofetuin–Sepharose can also be used for affinity purification of recombinant galectins. Lactose–Sepharose is prepared as described (Levi and Teichberg 1981). Briefly, 10 ml of Sepharose 4B beads (GE Healthcare) are extensively washed with 0.5 M NaCO$_3$, pH 11.0. One milliliter of divinylsulfone (Sigma) is added, and the suspension is stirred gently at 23°C for 70 min. Thus activated Sepharose 4B beads are washed again with 0.5 M NaCO$_3$, pH 11.0, and are resuspended in 10 ml of 0.5 M NaCO$_3$ containing 10% lactose. The coupling reaction is allowed to proceed by gentle stirring at 23°C for 15 h. The resultant beads are washed sequentially with 0.5 M NaCO$_3$, distilled water, and PBS. The amount of lactose coupled to the activated beads is determined by using the 3,5-dinitrosalicylic acid (DNS) method (Bailey et al. 1992). Asialofetuin–Sepharose is prepared by conjugation of asialofetuin (Sigma) to NHS-activated Sepharose 4FF (GE Healthcare) according to the manufacturer's instruction.

Acknowledgments We thank Dr. Nishi for providing the expression plasmids for human galectins-2, 4, 7, and 8.

References

Bailey MJ, Biely P, Poutanen K (1992) Interlaboratory testing of methods for assay of xylanase activity. J Biotech 23:257–270

Kasai K, Hirabayashi J (1996) Galectins: a family of animal lectins that decipher glycocodes. J Biochem (Tokyo) 119:1–8 (Review)

Levi G, Teichberg VI (1981) Isolation and physicochemical characterization of electrolectin, a β-D-galactoside binding lectin from electric organ of *Electrophorus electricus*. J Biol Chem 256:5735–5740

Siglec Family

Takashi Angata

Introduction

Siglec family is the largest-known group of vertebrate lectins that recognize sialylated glycans (Fig. 1A; Varki and Angata 2006). Humans have more than 10 Siglecs, most of which are expressed on the cells in the immune system. Most Siglecs also interact with proteins involved in intracellular signaling. Different species have different set of Siglecs (Angata 2006), and the glycan binding specificity of each Siglec differs markedly (Blixt et al. 2003). Glycan recognition is essential for Siglec function (Poe et al. 2004). Therefore, it is essential to know the glycan-binding specificity of each Siglec to understand its higher-order function.

Recombinant proteins fused with Fc portion of IgG have been used extensively in biological research. Here, a protocol for Siglec–Fc fusion protein production and glycan binding analysis using the fusion protein is explained.

Procedure

Production of Siglec–Fc Fusion Protein

Materials

    293T cells (available from ATCC), maintained in DMEM + 10% FCS
    Siglec–Fc expression constructs (available from the author upon request)
    Spin-X 0.22 μm microfilter (Corning)
    LipofectAMINE 2000 (Invitrogen)
    Opti-MEM (Invitrogen)
    Low IgG-fetal calf serum (HyClone)
    1 M Tris–HCl buffer, pH 8.0
    Tris-buffered saline (10 mM Tris–HCl, pH 8.0, 150 mM NaCl)
    0.1 M citrate–NaOH buffer, pH 5.8
    0.1 M glycine–HCl buffer, pH 3.0
    Protein A-Sepharose (GE Healthcare)
    Disposable plastic column (e.g., Polyprep column, Bio-Rad)
    20 mM HEPES–NaOH buffer, pH 7.0
    Arthrobacter ureafaciens sialidase (Sigma or Nacalai)
    Disposable ultrafiltration device (e.g., Amicon Ultra, cut off = 30 kDa, Millipore)

Methods (for a T75 Flask; This Protocol is Scalable)

1. Seed 293T cells to a T75 flask at ~50% confluence a day before the transfection. The flask should be near confluent the next day.

Research Center for Medical Glycoscience, National Institute of Advanced Industrial Science and Technology (AIST), 1-1-1 Umezono, Tsukuba, Ibaraki 305-8568, Japan
E-mail: takashi-angata@aist.go.jp

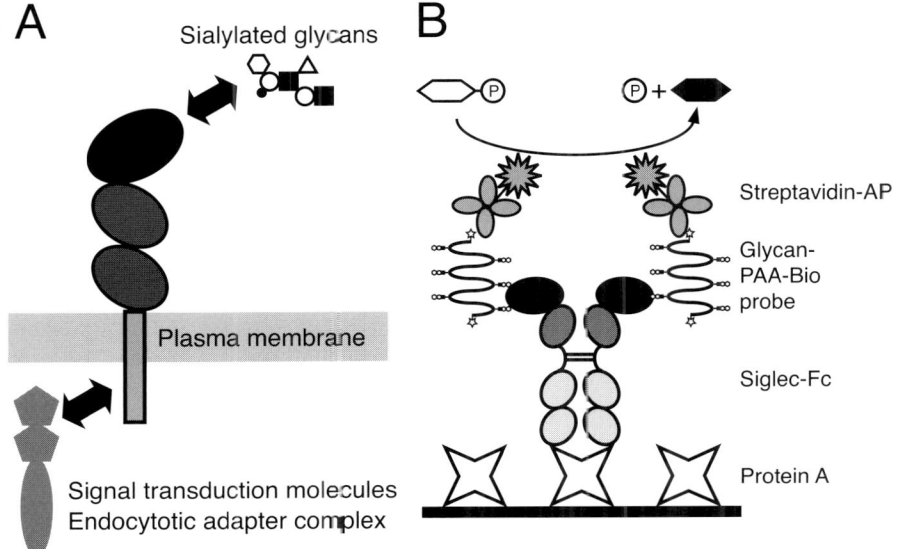

Fig. 1 Siglec and glycan binding assay. **A** Siglec in its native form. Siglec is a type I transmembrane protein, and its extracellular domain recognizes glycans, while its intracellular (or transmembrane) domain interacts with some signal transduction molecules and other proteins. **B** ELISA-like glycan binding assay using Siglec–Fc chimera. Siglec–Fc is captured onto protein A immobilized on the surface of multi-well plate. Its binding to biotinylated and multiple glycosylated polyacrylamide probe (Glycan-PAA-Bio) is colorimetrically analyzed using streptavidin-alkaline phosphatase (Streptavidin-AP) and p-nitrophenyl phosphate

2. Filter-sterilize the plasmid (30 μg) solution using Spin-X (12,000 g, 5 min). Mix 3 ml Opti-MEM and 75 μl LipofectAMINE 2000 in a 15 ml tube and leave for 5 min at room temperature. Add the filter-sterilized plasmid into the tube and mix well. Leave for 20 min at room temperature, then add evenly to the semi-confluent flask of 293T cells. Culture overnight in CO_2 incubator.
3. Remove medium, gently wash once with 5 ml Opti-MEM or PBS, and add 15 ml of Opti-MEM + 2% low IgG–FCS. Culture for 3 days in CO_2 incubator.
4. Transfer culture sup from the flask (by decantation) to 50 ml tube. Add 15 ml fresh Opti-MEM + 2% low IgG–FCS to the flask and put them back into CO_2 incubator, and culture for 3 days. Meanwhile, centrifuge the collected culture sup (at 1,000~1,500g, 5 min) to remove debris. Store at 4°C.
5. Collect culture sup from the flask into 50 ml tube. Centrifuge the collected culture sup (1,000~1,500g, 5 min) to remove debris. Culture sup harvested on day 3 and day 6 may be combined.
6. Add 1/50 vol. of 1 M Tris–HCl buffer to the clarified culture sup. Add 0.1 ml (as packed gel) protein A-Sepharose, prepared as instructed by the manufacturer. Incubate overnight at 4°C with gentle agitation.
7. Prepare disposable column assembly. Pour the culture sup + protein A-Sepharose into the column. Wait until all culture sup has passed through the column.
8. Add 2 ml Tris-buffered saline to the column and drain (= 'wash the column'). Wash the column once again with 2 ml Tris-buffered saline.

9. Wash the column with 1 ml of 20 mM HEPES–NaOH buffer. Cap the column tip and add the same buffer (final volume should be ~500 µl). Add 10 mU of sialidase and incubate at 37°C for 1 h. Remove the bottom cap and wash the column 5 times with 1 ml Tris-buffered saline to remove sialidase.
10. Wash the column twice with 1 ml of 0.1 M citrate–NaOH buffer, pH 5.8.
11. Elute Siglec–Fc protein from the column twice with 1 ml 0.1 M glycine–HCl buffer, pH 3.0, and collect the eluate. Immediately add 0.2 ml of 1 M Tris–HCl buffer to neutralize. The protein A-Sepharose column can be regenerated by washing with 1 ml of 6 M guanidine HCl, followed by a wash with Tris-buffered saline.
12. Transfer the neutralized eluate to ultrafiltration device and concentrate. Discard flow-through. Replace to a desired buffer by repeated concentration and buffer addition, if necessary. Recover the Siglec–Fc fusion protein and analyze quantity (by protein assay) and quality (by SDS-PAGE, etc.). Store the protein at −20°C.

Glycan Binding Analysis (Fig. 1B)

Materials

    96-well plate (NUNC #269620)
    Immobilization buffer (50 mM sodium bicarbonate buffer, pH 9.5)
    ELISA buffer (20 mM HEPES–NaOH buffer, pH 7.45, 125 mM NaCl, 1% BSA, 0.02% sodium azide)
    Protein A (Sigma #P6031), dissolved at 1 mg/ml in PBS
    Biotinylated and multiply glycosylated polyacrylamide probes ('glycan-PAA-Bio', Glycotech; also available from the Consortium for Functional Glycomics)
    Streptavidin-conjugated alkaline phosphatase (Invitrogen or Jackson ImmunoResearch Labs)
    Alkaline phosphatase substrate (100 mM Na_2CO_3, 10 mM *p*-nitrophenyl phosphate, 1 mM $MgCl_2$)

Methods

1. Prepare 5 µg/ml solution of protein A in immobilization buffer and add 100 µl/well to wells of 96-well plate. Each Siglec–probe combination should be assayed in triplicate wells. Incubate overnight at 4°C.
2. Remove protein A solution. Pat the plate lightly against paper towel to remove excess solution lingering to the wells. Add 150 µl ELISA buffer to each well, then revert to discard the solution, and pat lightly against paper towel (= 'wash the plate'). Wash the plate once again.
3. Add 150 µl ELISA buffer to each well and incubate at room temperature for 1 h. Meanwhile, prepare 1~5 µg/ml solution of Siglec–Fc or human IgG in ELISA buffer.
4. Remove the ELISA buffer and lightly pat the plate against paper towel. Add 100 µl Siglec–Fc (or human IgG) solution in each well. Incubate at room temperature for 2 h.
5. Remove Siglec–Fc solution by aspiration, and wash the plate 3 times with 150 µl/well ELISA buffer. Add 100 µl/well of 1~10 µg/ml glycan-PAA-Bio probes, diluted in ELISA buffer. Incubate at room temperature for 2 h.

6. Remove probe solution by aspiration, and wash the plate 3 times with 150 μl/well ELISA buffer. Dilute streptavidin-alkaline phosphatase at 1:1,000 in ELISA buffer. Add 100 μl/well. Incubate at room temperature for 1 h.
7. Remove streptavidin-alkaline phosphatase solution, and wash the plate 3 times with 150 μl/well ELISA buffer.
8. Add 100 μl/well alkaline phosphatase substrate and incubate at room temperature for 5 min to overnight. Measure absorbance at 405 nm using a plate reader. Check the plate at 5, 15, 30 min, 1 and 2 h after addition of the substrate.

Expected Results

Production of Recombinant Siglec–Fc Fusion Protein

Yield differs markedly for each Siglec. You should expect 0~1,000 μg of Siglec–Fc per T75 flask. Protein purity generally exceeds >90% but depends on the protein yield. A major contaminant (if any) is bovine IgG.

Glycan Binding Analysis

Some Siglecs (e.g., CD22/Siglec-2 and myelin-associated glycoprotein/Siglec-4) show strong binding signals to the preferred probes within a few minutes, whereas others show weak binding signals even after extended incubation. There is no strong correlation between the Siglec–Fc protein yield and the probe binding signal, although a Siglec–Fc that produce extremely poorly tends to bind glycan probes very poorly, a fact that suggests that protein instability is the key factor in such cases.

Troubleshooting

Production of Recombinant Siglec–Fc Fusion Protein

If the yield of Siglec–Fc is not satisfactory, it is recommended to use CHO cells. If both 293T and CHO cell lines fail to produce Siglec–Fc of your interest, then consider using S2 insect (*Drosophila*) cell line. Choice of culture media is critically important if you use S2 cell line, and a serum-free medium (HyQ SFX-Insect from HyClone) gave the best result in my experience. A construct for production of recombinant Siglec–Fc in S2 cells is available from the author upon request.

Glycan Binding Analysis

Sialidase treatment of Siglec–Fc is essential to expose glycan binding ability of Siglec–Fc. The choice of 96-well plate also affects assay outcome, and in my experience NUNC's plate (catalog #269620) has worked consistently well.

References

Angata T (2006) Molecular diversity and evolution of the Siglec family of cell-surface lectins. Mol Divers 10:555–566

Blixt O, Collins BE, van den Nieuwenhof IM, Crocker PR, Paulson JC (2003) Sialoside specificity of the siglec family assessed using novel multivalent probes: identification of potent inhibitors of myelin-associated glycoprotein. J Biol Chem 278:31007–31019

Poe JC, Fujimoto Y, Hasegawa M, Haas KM, Miller AS, Sanford IG, Bock CB, Fujimoto M, Tedder TF (2004) CD22 regulates B lymphocyte function in vivo through both ligand-dependent and ligand-independent mechanisms. Nat Immunol 5:1078–1087

Varki A, Angata T (2006) Siglecs—the major subfamily of I-type lectins. Glycobiology 16:1R–27R

Part 2
Functional Analyses of Sugar Chains

Section VI
Immunology

NKT Cells and Their Recognition of Glycolipids

Ken-ichiro Seino[1], Masaru Taniguchi[2]

Introduction

CD1d-restricted Natural Killer T (NKT) cells expressing a single invariant T cell receptor alpha chain recognize glycolipid antigens presented by a CD1d molecule (Taniguchi et al. 2003; Kronenberg 2005)[2]. Studies have indicated that NKT cells recognize various kinds of glycolipids derived from not only mammals but also from bacteria or marine organisms. NKT cells can produce various biological proteins such as cytokines or cytotoxic molecules upon activation. Through their production, NKT cells influence other immune cells and involve various kinds of immune responses, including autoimmunity, microbe infection, transplant immunity, tumor immunity, and allergic inflammation (Seino and Taniguchi 2004). It is expected to synthesize some glycolipids which could manipulate NKT cell function and the immune system, especially intending to treat some diseases better.

Principle

CD1d-restricted NKT cells constitute a distinct lymphocyte subpopulation with unique characteristics. In the peripheral tissues such as spleen or liver, NKT cells co-express an invariant T cell receptor (TCR) alpha chain and NK cell markers, such as NK1.1 in mice. The alpha chain of the TCR expressed on mouse NKT cells is encoded by invariant Vα14-Jα281 gene segment and has a human homolog, Vα24-JαQ. Both receptors recognize glycolipid antigens presented by a monomorphic antigen-presenting molecule, CD1d. Activated NKT cells produce various cytokines such as IFN-γ, IL-4, IL-10, or IL-13, and cytotoxic molecules such as Fas ligand or perforin. Then they can exert substantial immune-modulating functions in various immune responses such as infection, autoimmunity, transplantation, or allergy (Seino and Taniguchi 2004; Bendelac et al. 2007).

α-Galactosylceramide and Its Analogs

CD1d is highly conserved among mammalian species and has an MHC-like fold with two large hydrophobic binding grooves that are adapted to present nonpeptidic antigens. In 1997, one of the glycolipid ligands for NKT cells presented by CD1d was found (Taniguchi et al. 2003). That is α-galactosylceramide (α-GalCer), a glycolipid composed of a hydrophilic carbohydrate moiety with an α-linkage to the hydrophobic ceramide portion, whose original form was derived from marine sponge (Taniguchi et al. 2003). Obviously, α-GalCer does not seem to be an endogenous ligand for NKT cells, because

---

[1] Laboratory for Molecular Therapy, Institute of Medical Science, St. Marianna University, 2-16 Sugao, Miyamae-ku, Kawasaki, Kanagawa 216-8511, Japan

[2] Laboratory for Immune Regulation, RIKEN Research Center for Allergy and Immunology, Suehiro-cho 1-7-22, Tsurumi, Yokohama, Kanagawa 230-0045, Japan

α-glycosphingolipids could not be detected in mammalians. Nevertheless, α-GalCer has been extensively used to study the in vivo functions of NKT cells in various immune responses (Seino and Taniguchi 2004; Bendelac et al. 2007).

Synthetic α-GalCer analogs have been also reported, such as OCH or α-C-GalCer (Tsuji 2006). Although α-GalCer stimulation of NKT cells induces rapid production of both IFN-γ and IL-4, these α-GalCer analogs show biased cytokine production pattern. OCH, which is truncated at the sphingosine chain of α-GalCer, induces the production of IL-4-biased cytokines from NKT cells. On the other hand, α-C-GalCer, a C-glycoside analog of α-GalCer, mainly induces IFN-γ-biased cytokine production. Besides OCH or α-C-GalCer, numerous other glycolipids were also synthesized and tested for their ability to stimulate NKT cells (Tsuji 2006), and a variant of self-glycolipid, sulfatide was found to have the ability to activate NKT cells in CD1d-dependent fashion. From a pharmaceutical point of view, the synthesis of new glycolipids, which can better manipulate NKT cells, is expected and actually performed in various laboratories in the world.

Glycolipid Antigens Other Than α-GalCer

The NKT cells can recognize glycolipid antigens derived from Gram-negative, LPS-negative bacteria such as *Escherichia* or *Sphingomonas*, belonging to the same class of alpha-Proteobacteria (Kronenberg 2005). Compounds containing glycosphingolipids derived from these bacteria, and also synthetic glycosphingolipids designed to copy those in the bacterial cell wall, strongly activate NKT cells in CD1d-dependent fashion (Kronenberg 2005; Bendelac et al. 2007). Injection of *Sphingomonas* into immune-competent mice caused inflammatory septic shock and lethality, whereas in NKT cell-deficient mice, the *Sphingomonas*-induced lethality decreased. Therefore, NKT cells can be directly activated by some of the bacteria by recognizing the glycolipids on the cell wall and modulating systemic immune responses.

In 2004, Bendelac's group reported that they identified isoglobotrihexosylceramide (iGb3) as the endogenous ligand for NKT cells in mammals (Bendelac 2007). They showed that mice deficient in the enzymes β-hexosaminidase A and B, which degrade iGb4 to generate iGb3 in lysosomes, show defective NKT cell development in the thymus. Synthetic iGb3 or natural iGb3 derived from cat intestine could stimulate mature NKT cells, inducing robust cytokine production comparably to α-GalCer. However, quite recently in 2007, it was reported that NKT cells are normally developed and function in mice with iGb3-deficient mice (Porubsky et al. 2007). It was also shown that iGb3 could not be detected in any human tissues including thymus and dendritic cells by a highly sensitive HPLC assay (Speak et al. 2007). Therefore, iGb3 does not seem to be the real endogenous NKT cell ligand in mammals, although it may have a capacity to stimulate NKT cells. The search for the true endogenous ligand for NKT cells became an issue of burning concern again.

Structure of CD1d/Glycolipid Complex

In 2005, the crystal structures of docking of CD1d and α-GalCer or its analogs were reported (Kronenberg 2005; Bendelac et al. 2007). In these crystal structures, the lipid part of α-GalCer tightly fitted in the human CD1d binding groove, with the sphingosine chain bound in the C' pocket and the longer acyl chain anchored in the A' pocket, thereby providing a structural basis for the fact that glycolipid antigens are presented by CD1d. In the analysis of mouse CD1d, a short-chain variant of α-GalCer having eight-carbon acyl chain loaded more efficiently with mouse CD1d, suggesting the presence of an endogenous 'filler' lipid in the CD1d A' pocket. The proposed filler lipid in the CD1d A' pocket may stabilize this pocket and would have to be displaced by the longer 26-carbon acyl chain of α-GalCer. These findings are consistent with the previous observations that some α-GalCer analogs with various lengths of the sphingosine or acyl chain can bind to CD1d with distinct affinity and induce different patterns of cytokine production (Tsuji 2005).

Analysis of NKT Cell Functions with the Glycolipids

For analyzing biological functions of NKT cells, α-GalCer has been widely used. For example, nanogram-order of α-GalCer per milliliter of culture medium could fully activate NKT cells to secrete cytokines in the presence of CD1d-expressing antigen-presenting cells. The concentration of cytokines can be assessed with enzyme-linked immunosorbent assay (ELISA). When α-GalCer is intravenously injected into mice, 1–2 µg of the compound can induce vigorous cytokine production in the serum. However, it is also known that nanogram-order of α-GalCer can substantially influence other immune cells, such as dendritic cells or NK cells, although the serum cytokine production is not prominent.

In flow cytometer, NKT cells are recognized as TCR- and NK marker-positive cells. However, this population includes other T cells than NKT cells. To analyze NKT cells more strictly, a fluorescence-conjugated fusion protein, CD1d-tetramer or dimer is used. To stain NKT cells, the fluorescence-conjugated CD1d-tetramer or dimer should be loaded with α-GalCer first. Then cell suspensions including NKT cells are incubated with α-GalCer-loaded CD1d-tetramer or dimer. After incubation, the cells are washed and analyzed on flow cytometer. At present, the precise definition of NKT cells on flow cytometric analysis is TCR-positive and CD1d-tetramer- (or dimer-) reactive cells.

Perspective

The NKT cell/CD1d system is now well established as a new immunological system based on glycolipid-recognition and with the capability to regulate the immune responses. From a therapeutic point of view, the manipulation of NKT cell functions with α-GalCer or new glycolipids is of particular interest. Further studies investigating the molecular mechanisms governing the NKT cells' recognition of glycolipids will contribute to better understanding the entire spectrum of immune system.

References

Bendelac A, Savage PB, Teyton L (2007) The biology of NKT cells. Annu Rev Immunol 25:297–336
Kronenberg M (2005) Toward an understanding of NKT cell biology: progress and paradoxes. Annu Rev Immunol 23:877–900
Porubsky S, Speak AO, Luckow B, Cerundolo V, Platt FM, Gröne HJ (2007) Proc Natl Acad Sci USA. 104:5977–5982
Seino K, Taniguchi M (2004) Functional roles of NKT cell in the immune system. Front Biosci 9:2577–2587
Speak AO, Salio M, Neville DC, Fontaine J, Priestman DA, Platt N, Heare T, Butters TD, Dwek RA, Trottein F, Exley MA, Cerundolo V, Platt FM (2007) Proc Natl Acad Sci USA 104:5971–5976
Taniguchi M, Harada M, Kojo S, Nakayama T, Wakao H (2003) The regulatory role of Valpha14 NKT cells in innate and acquired immune response. Annu Rev Immunol 21:483–513
Tsuji M (2006) Glycolipids and phospholipids as natural CD1d-binding NKT cell ligands. Cell Mol Life Sci 63:1889–1898

Carbohydrate Ligands for Selectins in Immune Cell Trafficking

Reiji Kannagi[1], Katsuyuki Ohmori[2]

Introduction

The first step of leukocyte extravasation is adhesion of leukocytes to endothelial cells mediated by selectins and their carbohydrate ligands (Fig. 1). This cell adhesion system tethers leukocytes in the bloodstream to endothelial walls, and this in turn triggers the action of chemokines that are expressed on the endothelial surface. Signal transduction through leukocyte chemokine receptors activates leukocyte integrins, which lead to extravasation of leukocytes.

Principle

Carbohydrate Ligand Expression Determines which Leukocyte Subset is Recruited in Inflammatory Response

A massive leukocyte extravasation occurs specifically at the inflammatory lesion. This spatial specificity of leukocyte extravasation for inflammatory lesion is supported by the induction of E- and P-selectins' expression on endothelial cells by inflammatory cytokines. Expression of these selectins is suppressed on resting endothelial cells.

On the other hand, the cell lineage specificity of leukocyte extravasation is determined by carbohydrate ligands for selectins. Granulocytes and monocytes constitutively express the sialyl Lewis x determinant, a potent ligand for E- and P-selectins, and these cells are thus ready to undergo extravasation where the vascular bed expresses these selectins. This is due to constitutive transcription of fucosyltransferase genes, which encode the rate-limiting step enzymes for sialyl Lewis x synthesis.

Most resting lymphocytes, in contrast, lack sialyl Lewis x expression except NK cells and a small subset of T-lymphocytes (Ohmori et al. 1989, 1993). Therefore, most lymphocytes remain in the general circulation even in severe inflammatory responses. Lymphocytes, however, acquire sialyl Lewis x expression when activated by antigenic stimulus. Such activated lymphocytes are selectively recruited to the inflammatory lesion. This selective recruitment is conferred by expression of carbohydrate ligands for selectins. Transcriptional induction of fucosyltransferase genes by inflammatory stimulus induces expression of sialyl Lewis x determinant specifically in activated lymphocytes.

---

[1] Department of Molecular Pathology, Aichi Cancer Center, Nagoya, Japan
Phone: +81-52-762-6111, Fax: +81-52-764-2973
E-mail: rkannagi@aichi-cc.jp

[2] Department of Clinical Pathology, Kyoto University School of Medicine, Kyoto, Japan
Phone: +81-75-751-3599, Fax: +81-75-751-3758
E-mail: ohmori@kuhp.kyoto-u.ac.jp

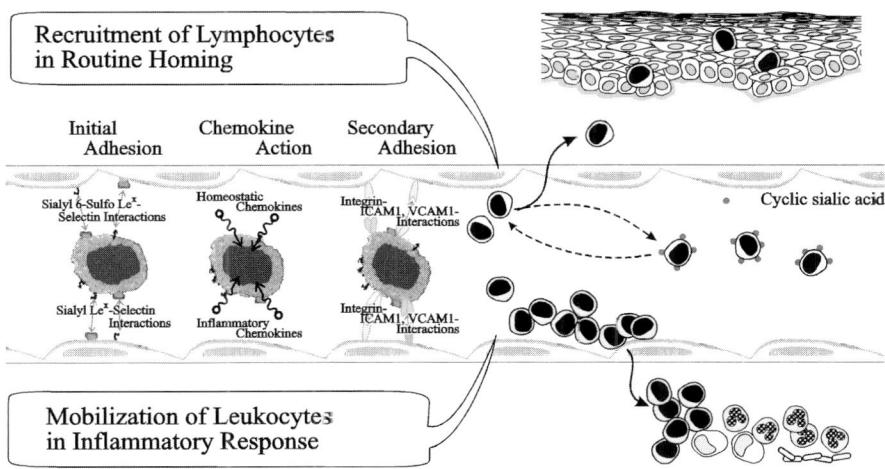

Fig. 1 Schematic illustration of adhesion of leukocytes to endothelial cells (Kannagi 2002). The cell adhesion is initiated by the binding of selectins on endothelial cells and their carbohydrate ligands on leukocytes. This triggers chemokine action, which in turn induces activation of leukocyte integrins. Two ligands for selectins are noted. Sialyl 6-sulfo Lewis x determinant is involved in the routine homing of lymphocytes in healthy individuals, whereas non-sulfated conventional sialyl Lewis x determinant is mainly involved in the large-scale mobilization of leukocytes in inflammatory response

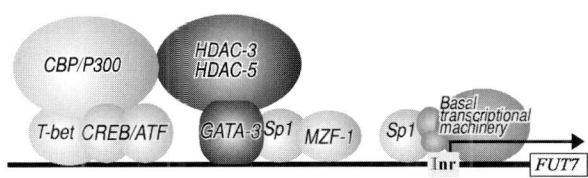

Fig. 2 Transcription factors involved in the regulation of fucosyltransferase VII (*Fuc-T VII*) gene expression (Chen et al. 2006; Kannagi 2002). The factors indicated in *dark gray* (GATA-3 and HDACs) exert suppressing effects, whereas others in *light gray* promote Fuc-T VII transcription

The fucosyltransferases involved in the synthesis of selectin ligands in leukocytes are known to be Fuc-T VII and IV. Transcription of both genes is very actively regulated by a variety of transcription factors (Withers and Hakomori 2000; Chen et al. 2006). The promoter region of the gene encoding Fuc-T VII is equipped with binding sites for important transcription factors including CREB/ATF, T-*bet*, HIF-1, GATA-3, MZF-1, and Sp1, and these factors regulate transcription of the gene (Fig. 2) (Chen et al. 2006; Kannagi 2001, 2002). MZF-1 is related to the constitutive expression of the gene in granulocytes and monocytes. CRE and GC box are involved in activation-induced transcription of the gene. T-*bet* and GATA-3 regulate differential transcription of the gene in Th1- and Th2-lymphocytes (Chen et al. 2006). It is important in developing anti-inflammatory remedies to know the exact molecular biological mechanisms involved in the transcription of Fuc-T VII and IV.

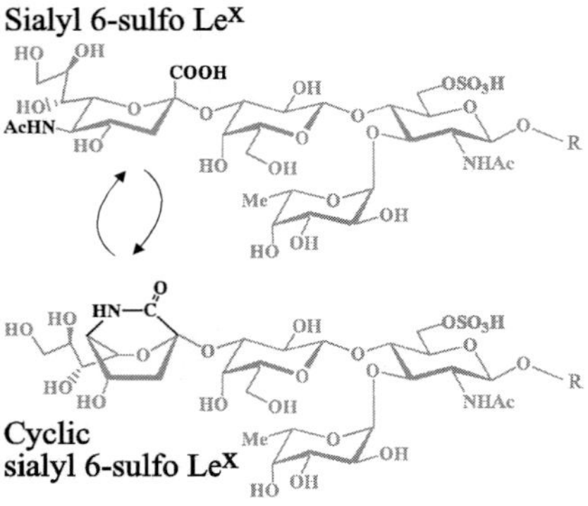

Fig. 3 Regulation of selectin-binding activity of sialyl 6-sulfo Lewis x through intramolecular cyclization of sialic acid moiety (Mitsuoka et al. 1999). De-*N*-acetylation followed by lactam ring formation in sialic acid moiety completely abrogates selectin-binding activity of the carbohydrate determinant

Carbohydrate Ligand Expression Determines also Routine Homing of Leukocytes in Healthy Individuals

Cell adhesion mediated by selectins and their carbohydrate ligands is involved also in the routine homing of lymphocytes. The most well-known lymphocyte homing is the recruitment of naïve T-lymphocytes to the secondary lymphoid tissues such as peripheral lymph nodes. High endothelial venules in the secondary lymphoid tissues serve as an entrance for naive T-lymphocyte homing. Naïve T-lymphocytes express L-selectin, and therefore the high endothelial venules had been assumed to express specific L-selectin ligand. Some time ago we identified the L-selectin ligand on the high endothelial venules as the sialyl 6-sulfo Lewis x determinant (Mitsuoka et al. 1998). At the initial stage of study, although some researchers proposed that some other determinants were the L-selectin ligand on the high endothelial venules, most researchers later concluded it was the sialyl 6-sulfo Lewis x determinant. Two GlcNAcβ:6-*O*-sulfotransferases are involved in the synthesis of the sialyl 6-sulfo Lewis x determinant in lymph nodes. Introduction of either gene can reconstitute L-selectin ligand on the transfected cells, whereas disruption of both genes was required to obtain the mice lacking the sialyl 6-sulfo Lewis x determinant in lymph nodes.

The sialyl 6-sulfo Lewis x determinant is involved not only in the homing of naïve T-lymphocytes, but also in the homing of helper memory T-lymphocytes, which undergo skin- and gut-homing (Ohmori et al. 2006; Kannagi 2002). In contrast to the non-sulfated conventional sialyl Lewis x determinant, which is involved mainly in the inflammatory mobilization of leukocytes, the sialyl 6-sulfo Lewis x determinant preferentially engages in the routine trafficking of lymphocytes, which occurs daily in the body of every healthy individual, and maintains homeostasis of lymphocyte distribution.

Another interesting aspect of the sialyl 6-sulfo Lewis x determinant is that its binding activity is easily lost during the course of cell–cell interactions, due to some modification occurring at its sialic acid moiety (Fig. 3) (Mitsuoka et al. 1999; Kannagi 2002). We assume this modification partly involves cyclization of sialic acid, which is catalyzed by a calcium-dependent enzyme, tentatively called sialic acid cyclase. Mobilization of intracellular calcium within a few minutes causes rapid loss of the selectin binding activity of sialyl 6-sulfo Lewis x. It is surprising that cell surface carbohydrate determinants undergo such a rapid structural modification process, which has a biologically significant resultant loss of selectin binding activity. This may well support the concept that carbohydrate determinants are biologically relevant molecules, which figure heavily in regulation of important physiological functions. This modification of sialic acid moiety occurs specifically in the sialyl 6-sulfo Lewis x determinant, and no such modification is detected in the non-sulfated conventional sialyl Lewis x determinant.

The routine lymphocyte homing proceeds slowly and steadily with a small cell population (Fig. 1). The negative-feedback system of cell adhesion, built into the structure of the sialyl 6-sulfo Lewis x determinant, seems to play a significant role in maintaining this slow and steady pace of routine lymphocyte homing. In contrast, this negative-feedback system is not involved in the mobilization of leukocytes in an inflammatory response, which rapidly occurs in large-scale leukocyte populations. Carbohydrate determinants not only decide the lineage of leukocytes to be recruited, but also control the pace of leukocyte mobilization.

Immune Cell Trafficking and Cell-Adhesive Carbohydrates in Mucous Membranes

Skin- or gut-homing lymphocytes, after entering the mucous membranes of various organs, cruise along the epithelial cell layer on the lookout for possible pathogens invading the mucous membranes. Constant immune surveillance by these cells is routinely performed in healthy individuals without any sign of inflammatory disease.

During the course of the immune surveillance, lymphocytes and other immune cells interact with each other, and also with epithelial cells. This interaction is also mediated by a set of carbohydrate determinants and carbohydrate-recognition proteins. CD44 and hyaluronic acid are involved in the interaction of skin-homing lymphocytes with squamous epithelial cells. GlcNAc 6-O-sulfation of carbohydrate determinants on CD44 is known to be important in the binding to hyaluronic acid. A similar 6-O-sulfated determinant is also known to be specifically expressed on a subset of dendritic cells that are particularly active in cell trafficking. Fully differentiated epithelial cells in the intestine are known to carry a specific carbohydrate ligand for Siglec-7 and to interact with a certain population of tissue macrophages (Fig. 4). Study of the carbohydrate-mediated interaction of immune cells in the mucous membrane has just started only recently, and

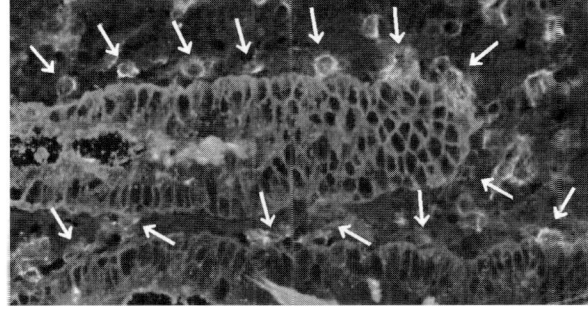

Fig. 4 Adhesion of tissue macrophages to epithelial cells in normal colonic mucosal membranes. Confocal microscopic observation of normal colonic section where siglec-7 expressed on tissue macrophages is stained in *green*, and a specific carbohydrate ligand for siglec-7, the disialyl f a determinant, is stained in *red* (Miyazaki et al. 2004)

promises to shed much light on the diagnosis and therapy of diseases in the skin and intestine in the near future.

Protocols

Protocols for multi-color analyses of the T-lymphocyte subset expressing sialyl 6-sulfo Lewis x are as follows (Ohmori et al. 2006). Note that the G152 determinants on T cells are unstable and that the direct staining of whole blood samples is necessary for maximum detection. Purification of lymphocytes using Ficoll-Hypaque solution and involving multiple centrifugation processes is not recommended.

1. Prepare a tube containing 10 µl of anti-sialyl 6-sulfo Lewis x antibody G152.
2. To this, 100 µl of freshly obtained whole human blood sample (anti-coagulated using EDTA) is added, and the tube is incubated at 4°C for 30 min.
3. Add 4 ml of PBS and wash the cells by centrifuging at 500×g for 5 min. Repeat the washing procedure twice.
4. Add 50 µl of FITC-labeled anti-murine IgM antibody (100× diluted) and incubate at 4°C for 30 min.
5. Add 4 ml of PBS and wash the cells by centrifuging at 500×g for 5 min. Repeat the washing procedure twice.
6. Add 5 µl each of APC-labeled anti-CD3 antibody SK7 (Leu4), PC5-labeled anti-CD4 antibody 13B8.2, PE-labeled anti-CD45RO antibody UCHL-1 (Leu45RO), and incubate at 4°C for 30 min.
7. Add 2 ml of FACS Lysing Solution (Becton Dickinson, diluted 10× with distilled water) and allow to stand at room temperature for 10 min to lyse red blood cells.
8. Centrifuge the cells at 500×g for 5 min and discard the supernatant.
9. Add 4 ml of PBS and wash the cells once by centrifugation at 500×g for 5 min.
10. Re-suspend the cells in 0.5 ml PBS. The stained cells are then subjected to four-color flow-cytometric analyses.

References

Chen G-Y, Osada H, Santamaria-Babi LF, Kannagi R (2006) Interaction of GATA-3/T-bet transcription factors regulates expression of sialyl Lewis X homing receptors on Th1/Th2 lymphocytes. Proc Natl Acad Sci USA 103:16894–16899

Kannagi R (2001) Transcriptional regulation of expression of carbohydrate ligands for cell adhesion molecules in the selectin family. In: Wu AM (ed) Advances in experimental medicine and biology, vol 491. Molecular Immunology of Complex Carbohydrates II. Plenum, New York, pp 267–278

Kannagi R (2002) Regulatory roles of carbohydrate ligands for selectins in homing of lymphocytes. Curr Opin Struct Biol 12:599–608

Mitsuoka C, Sawada-Kasugai M, Ando-Furui K, Izawa M, Nakanishi H, Nakamura S, Ishida H, Kiso M, Kannagi R (1998) Identification of a major carbohydrate capping group of the L-selectin ligand on high endothelial venules in human lymph nodes as 6-sulfo sialyl Lewis x. J Biol Chem 273:11225–11233

Mitsuoka C, Ohmori K, Kimura N, Kanamori A, Komba S, Ishida H, Kiso M, Kannagi R (1999) Regulation of selectin binding activity by cyclization of sialic acid moiety of carbohydrate ligands on human leukocytes. Proc Natl Acad Sci USA 96:1597–1602

Ohmori K, Yoneda T, Shigeta K, Hirashima K, Kanai M, Itai S, Sasaoki T, Arii S, Arita H, Kannagi R (1989) Sialyl SSEA-1 antigen as a carbohydrate marker of human natural killer cells and immature lymphoid cells. Blood 74:255–261

Ohmori K, Takada A, Ohwaki I, Takahashi N, Furukawa Y, Maeda M, Kiso M, Hasegawa A, Kannagi M, Kannagi R (1993) A distinct type of sialyl Lewis X antigen defined by a novel monoclonal antibody is selectively expressed on helper memory T cells. Blood 82:2797–2805

Ohmori K, Fukui F, Kiso M, Imai T, Yoshie O, Hasegawa H, Matsushima K, Kannagi R (2006) Identification of cutaneous lymphocyte-associated antigen as sialyl 6-sulfo Lewis x, a selectin ligand expressed on a subset of skin-homing helper memory T cells. Blood 107:3197–3204

Withers DA, Hakomori SI (2000) Human α(1,3)-fucosyltransferase IV (FUTIV) gene expression is regulated by elk-1 in the U937 cell line. J Biol Chem 275:40538–40593

Carbohydrate Recognition by Cytokines and Its Relevance to Their Physiological Activities

Keiko Fukushima, Katsuko Yamashita

Introduction

Cytokines regulate the immune response, inflammation, cell proliferation and differentiation as mediators of the regulatory network among lymphoid cells, hematopoietic cells, and endothelial cells. Many signal transduction mechanisms are evoked after cytokines bind to their respective receptors. Most cytokine receptors consist of several specific receptor subunits. These subunits form high-affinity complexes with their respective cytokines, which then induce various physiological effects. The formation of cytokine–receptor complexes cannot be explained by the simple interaction between a cytokine and its receptor. Lectin-like characteristics of cytokines may provide an insight into their multiple functions, although their physiological significance has not generally been elucidated (Yamashita and Fukushima 2007).

Glycosaminoglycan-Binding Cytokines

A large number of cytokines have been described as glycosaminoglycan (GAG)-binding proteins. Most GAG-binding cytokines interact with heparin/heparan sulfate (HS), while a few are known to interact with chondroitin sulfate or dermatan sulfate. Early studies focused on FGF-1 and FGF-2, but systematic analyses of the GAG-structure involved in the interaction with GAG-binding cytokines are still limited. In the case of FGF-2, a biochemical approach used to define the minimal HS structure required for binding of FGF-2 implicated a pentasaccharide sequence with three hexuronic acid units, including two N-sulfated glucosamine residues and 2-O-sulfated IdoUA at the reducing terminus, HexUA-GlcNSO$_3$-HexUA-GlcNSO$_3$-IdoUA(2-OSO$_3$). A pentasaccharide containing IdoUA(2-OSO$_3$) was later deduced from crystallographic analyses of an FGF-2 and fully sulfated heparin hexasaccharide complex and an FGF-2/FGF-R1/heparin 10-mer complex. To elucidate direct roles of HS in FGF function, Rapraeger et al. investigated the effects of hepalitinase or sodium chlorate on Swiss 3T3 fibroblast proliferation and MM14-myoblast differentiation. Both direct digestion of HS by hepalitinase and reduced sulfation by sodium chlorate blocked the FGF-2 induced mitogenic response of 3T3 cells. The addition of soluble HS recovered FGF-mediated signaling in hepalitinase-treated or sodium chlorate-treated 3T3 cells. Yayon et al. also demonstrated a requirement of HS for creating high-affinity binding of FGF and FGFR, by using HS-deficient CHO cells. HS-deficient CHO cells, which express abundant chondroitin sulfate proteoglycans, do not bind FGF-2, and free HS promotes binding of FGF-2 to its high-affinity receptor.

Innovative Research Initiatives, Tokyo Institute of Technology, 4259 Nagatsuta-cho, Midori-ku, Yokohama 226-8503, Japan
Phone: +81-45-921-4308, Fax: +81-45-921-4308
E-mail: kyamashi@bio.titech.ac.jp

These studies indicate that HS molecules are required for high-affinity complex formations of the FGF/FGFR and related signal transduction (Fig. 1A).

High-Mannose Type Glycan Binding IL-2

Interleukin-2 (IL-2) is a cytokine synthesized by activated T cells. IL-2 mediates its physiological functions through interaction with the IL-2R complex, which consists of three receptor subunits, α, β, and γ (IL-2Rα, β, and γ) (Fig. 1A). The mechanism underlying the formation of a high-affinity IL-2-IL-2Rα, β, γ, complex has remained unidentified. However, the observation that IL-2 has carbohydrate recognition properties provided an opportunity to resolve this mechanism. Inhibition assay showed that the lectin activity of IL-2 is indispensable for induction of IL-2-dependent CTLL-2 cell proliferation. If a lectin-like interaction between an IL-2 and a specific glycoprotein triggers the formation of the high-affinity receptor complex, it follows that a specific glycoprotein having $Man_{5-6}GlcNAc_2$ should be coimmunoprecipitated with the IL-2 receptor complex in the lysates of IL-2-stimulated CTLL-2 cells using antibodies against the IL-2Rα, β, or γ subunits. Among the components of the IL-2 receptor complex in CTLL-2 cells, only the IL-2Rα subunit has the high-mannose type glycan with $Man_5GlcNAc_2$; IL-2 bi-functionally binds to a high-mannose type glycan and a specific peptide sequence of IL-2Rα (Fukushima et al. 2001), although all of the subunits of IL-2R have several potential N-glycosylation sites (Fig. 1B).

GPI-Anchor Glycans Binding Cytokines

IL-1β, TNF-α, or IL-18 binds to the GPI-anchor glycans of GPI-anchored glycoproteins on the cell surface. To determine the precise carbohydrate binding specificities of IL-1β, TNF-α, or IL-18, binding activity of recombinant human IL-1β, TNF-α, or IL-18 to various glycoproteins was investigated using enzyme-linked immunosorbent assay (ELISA) plates. These studies indicated that a mannose-6-phosphate diester in the GPI-anchor glycan is necessary for the carbohydrate binding of IL-1β, TNF-α, or IL-18.

In the case of TNF-α, human lymphoma U937 cells were used to investigate whether the recognition of GPI anchor glycans on the cell surface is necessary for TNF-α-stimulated apoptosis. Notably, TNF-α-induced apoptosis of these cells was inhibited in the presence of mannose-6-phosphate. A similar result was obtained by observation of DNA ladders, which were detected by agarose gel electrophoresis. These results suggest that the binding of TNF-α to the GlcNAcβ1-phosphate-6mannose residue in a GPI anchor glycan triggers TNF-α-induced intracellular signaling in U937 cells that lead to their apoptosis.

Interleukin-18 (IL-18) is a cytokine that induces T and natural killer cells to produce interferon-γ (IFN-γ). It is known that human leukemia KG-1 cells produce IFN-γ when they are stimulated with IL-18. To determine whether the recognition of the GPI-anchor glycan by IL-18 is important for its physiological functions, it was assessed whether mannose 6-phosphate or PI-PLC treatment blocks the stimulatory effects of IL-18 on IFN-γ production by KG-1 cells. The results indicated that IL-18 binding of the third mannose 6-phosphate diester in the GPI-anchor glycan is required to enhance IL-18-dependent intracellular signal transduction and IFN-γ production. Furthermore, CD48 was specifically immunoprecipitated with IL-18Rα, although many GPI anchored glycoproteins exist on the cell surface. IL-18Rα bound to CD48-coated plates in a

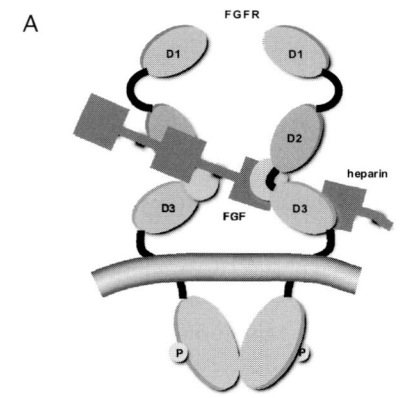
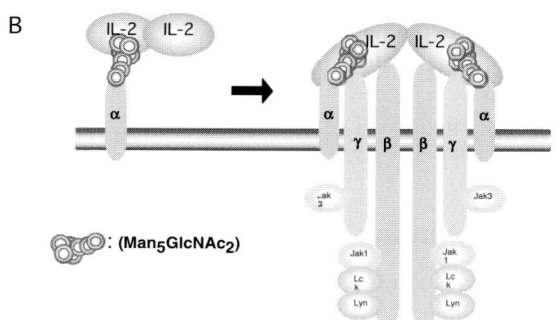
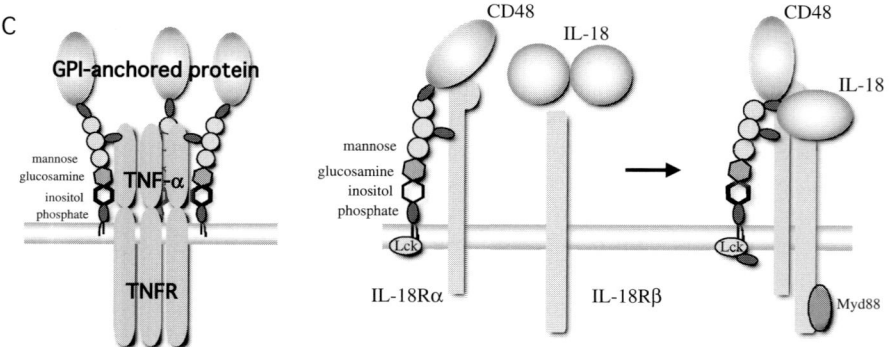

Fig. 1 Tentative schematic models of the formation of high-affinity complex between cytokine and cell surface glycoprotein. **A** $(EGF)_2/Heparin/(EGFR)_2$. **B** $(IL-2)_2/(IL-2R\alpha)_2/\beta_2/\gamma_2$, **C** $(INF-\alpha)_3/(TNFR)_3/$ (GPI-anchored protein)$_3$ (*left*), $(IL-18)_2/(IL-18R\alpha)_2/(CD48)_2(IL-18R\beta)_2$ (*right*)

dose-dependent manner, and this binding was inhibited by the addition of anti-CD48. On the contrary, IL-18Rβ did not bind to CD48. These results suggest that only CD48 is associated with IL-18Rα, because CD48 can bind to IL-18Rα via both GPI-glycan and the protein portion of CD48. Since it is known that the IL-18/IL-18Rα complex binds to IL-18Rβ, it is likely that after IL-18Rα, IL-18, and CD48 form a complex on the cell surface; this complex immediately binds to IL-18Rβ (Fukushima et al. 2005), as presented in the model shown (Fig. 1C).

Protocol

Inhibition Assay of IL-18 Dependent IFN-γ Production by Mannose 6-Phosphate or PI-PLC Treatment

Inhibitory effects of IL-18 dependent IFN-γ production in KG-1 cells are assayed in the presence of mannose 6-phosphate or PI-PLC treatment.

1. Plate 1×10^5 of KG-1 cells (RCB1166) in microtiter plates at 100 µl/well.
2. Add 100 µl of rhIL-18 at various concentrations diluted in complete RPMI 1640 medium.
3. Incubate overnight at 37°C in 0.5% CO_2 atmosphere.
4. ELISA assay by using a human IFN-γ enzyme immunoassay kit.
5. To investigate the effect of mannose 6-phosphate, mix with rhIL-18 and mannose 6-phosphate at various concentrations and preincubate at 37°C for 30 min before being added to cells.
6. To investigate the effect of PI-PLC, KG-1 cells (1×10^6 cells/100 µl PBS) are incubated with 5 mU/ml PI-PLC in PBS at 37° for 60 min before IL-18 stimulation.

Comment

IL-18, whose relative activity is checked before assay, should be used.

References

Fukushima K, Hara-Kuge S, Ideo H, Yamashita K (2001) Carbohydrate recognition site of interleukin-2 in relation to cell proliferation. J Biol Chem 276:31202–31208

Fukushima K, Ikehara Y, Yamashita K (2005) Functional role played by the GPI-anchor glycan of CD48 in IL-18-induced IFN-gamma production. J Biol Chem 280:18056–18062

Yamashita K, Fukushima K (2007) In: Kamerling JP (ed) Comprehensive glycoscience. Elsevier, Amsterdam, pp 337–363

Detection of Mouse MGL1/CD301a by a Specific Monoclonal Antibody

Kaori Denda-Nagai[1], Kiwamu Suzuki[1], Mitsuhiro Goda[1], Akihiko Kudo[2], Hayato Kawakami[2], Tatsuro Irimura[1,3]

Introduction

Macrophage galactose-type and calcium-type lectins (MGL/CD301) are type 2 transmembrane proteins expressed by macrophages and dendritic cells. The bindings of these lectins are effectively blocked by galactose and *N*-acetylgalactosamine as monosaccharides, although some oligosaccharides exhibit higher affinity than these monosaccharides. Mice have MGL1 (CD301a) and 2 (CD301b) genes, whereas humans have a single MGL gene (Tsuiji et al. 2002). Recent studies by the use of *Mgl1-/-* mice revealed that these lectins are involved in the recognition and removal of apoptotic cells generated X-irradiation during the development of embryos and in the inflammatory tissue formation induced by chemically modified proteins (Yuita et al. 2005; Sato et al. 2005). The expression of MGL1 can be assessed at mRNA levels and at protein levels by specific monoclonal antibodies (mAbs). Some mouse MGL-specific mAbs used in early studies such as LOM-14, LOM-11, ER-MP-23 are cross-reactive between MGL1 and 2 (Mizuochi et al. 1997). Although expression of mouse MGL on dendritic cells and its involvement in the uptake of glycoconjugates using dendritic cells induced in vitro, further characterizations of cells expressing these lectins in situ are necessary (Denda-Nagai et al. 2002). The present manual is to detect mouse MGL1 in tissues and on cell surfaces by mAb LOM-8.7.

Procedure

Immunohistochemistry

1. Cells expressing MGL1 can be immunohistochemically detected in acetone-fixed frozen sections of tissues (10 μm thick) by mAb LOM-8.7 (rat IgG2a).
2. The sections were blocked against non-specific bindings by PBS containing 3% BSA, 2% normal goat serum, and 2% normal mouse serum (Solution A) for 30 min. In some cases, further blocking by the use of Avidin-Biotin Blocking kit (Victor) was necessary.
3. Incubations with mAb LOM-8.7 (5~10 μg/ml, purified IgG) diluted in Solution A are performed for 1 h.

---

[1] Laboratory of Cancer Biology and Molecular Immunology, Graduate School of Pharmaceutical Sciences, The University of Tokyo, Tokyo, Japan

[2] Second Department of Anatomy, Kyorin University School of Medicine, Tokyo, Japan

[3] E-mail: irimura@mol.f.u.-tokyo.uc.jp

4. After repeated washing with PBS, the sections are incubated with biotinylated goat anti-rat IgG (H + L) for 30 min.
5. The mAb binding is detected by an alkaline phosphatase-conjugated streptavidin (2 μg/ml). Visualization may be enzymatic/light microscopy. As a negative control, the sections are incubated with rat IgG2a in place of mAb LOM-8.7. The stained sections may also be counter-stained with hematoxylin.

Flow Cytometric Analysis

1. Although MGL1-positive macrophages and dendritic cells can generally be obtained by collagenase digestion of minced tissues, the condition depends on the tissue. Therefore, it is not simple to describe the universal method to obtain MGL1-positive cells.
2. Once single-cell suspensions are obtained, $1–10 \times 10^5$ cells are pre-incubated with 15–50 μl of FCM solution (PBS containing 0.01% BSA, 0.01% NaN_3) containing 1–5% normal mouse serum and 1–5% normal rat serum or rat anti-mouse CD16/32 (2.4G2, 5–10 μg/ml) for 5–10 min on ice.
3. Then, cells are reacted with antibodies diluted in FCM solution for 30 min on ice. As an example, a mixture (15–50 μl) of PE-conjugated mAb LOM-8.7 (1 μg/ml) and biotinylated F4/80 (2 μg/ml) may be used.
4. The cell suspensions are diluted with FCM solution and the cells are precipitated by centrifugation.
5. The cells are incubated with 30–100 μl of APC-conjugated streptavidin (0.1–0.2 μg/ml) for 30 min on ice, washed by centrifugation, and diluted into FCM solution containing 7-amino-actinomycin D (7-AAD, 1 μg/ml).
6. The cells are analyzed on a flow cytometer with an argon laser (488 nm) and a HeNe laser (633 nm).

Results

Immunohistochemistry: Figure 1 shows the distribution of mAb LOM-8.7 binding sites in the ovary. This is an organ in which MGL1/CD301a-positive cells distribute predominantly in the connective tissue, as seen with many other organs. These putative macrophages, dendritic cells, or both are observed around the follicles. However, MGL1-positive cells can also be seen within some follicles (Fig. 1a, c). MGL1-positive cells are observed around the periphery of oviducts (Fig. 1e, g).

Flow cytometric analysis: Figure 2 shows a cell surface profile of a suspension obtained from a mouse tissue. MGL1-positive cells obtained from different organs appear to represent macrophages, dendritic cells, or both, dependent on the organ judging from the expressions of other cell surface markers. In the case of the Mgl1-positive cells shown in Fig. 2, they also express high levels of F4/80, a putative macrophage marker.

Comments

Protocols for immunohistochemistry and flow cytometric analysis may need modifications depending on the tissues and cells to be examined.

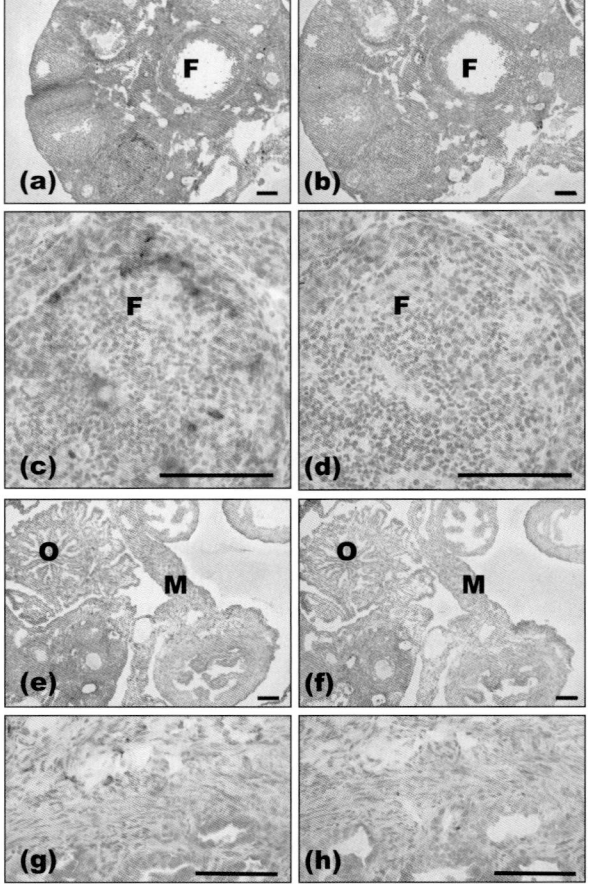

Fig. 1 Mouse ovary and oviducts stained with mAb LOM-8.7 (anti-MGL1) (**a, c, e, g**). Panels **b, d, f,** and **h** represent the results of control staining (rat IgG2a) of the corresponding areas from serial sections. Counterstaining with Hematoxilin is performed. MGL1-positive cells distributed mostly in extrafollicular connective tissues (**a, c**). MGL1-positive cells were also observed along oviducts in the surrounding connective tissues (**e, g**). *F* follicle, *O* oviduct, *M* intraoviductal membrane. *Scale bars* indicate 100 μm

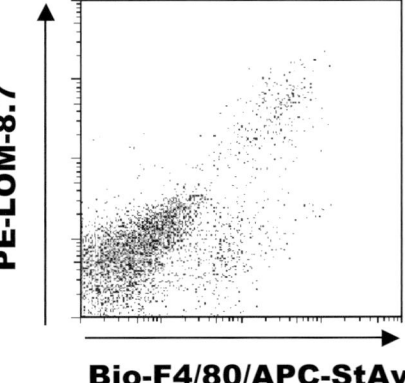

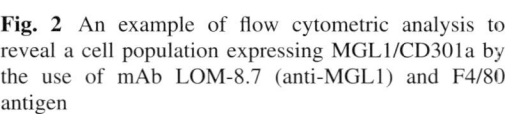

Fig. 2 An example of flow cytometric analysis to reveal a cell population expressing MGL1/CD301a by the use of mAb LOM-8.7 (anti-MGL1) and F4/80 antigen

References

Denda-Nagai K, Kubota N, Tsuiji M, Kamata M, Irimura T (2002) Macrophage C-type lectin (MGL) on bone marrow-derived immature dendritic cells is involved in the recognition and internalization of glycosylated antigens. Glycobiology 12:443–450

Mizuochi S, Akimoto,Y, Imai Y, Hirano H, Irimura T (1997) Unique tissue distribution of a mouse macrophage C-type lectin. Glycobiology 7:137–146

Sato K, Imai Y, Higashi N, Kumamoto Y, Onami TM, Hedrick SM, Irimura T (2005) Lack of antigen-specific tissue remodeling in mice deficient with the macrophage galactose-type calcium-type lectin 1/CD301a. Blood 106:207–215

Tsuiji M, Fujimori M, Ohashi Y, Higashi N, Onami TM, Hedrick SM, Irimura T (2002) Molecular cloning and characterization of a novel mouse macrophage C-type lectin, mMGL2, which has a distinct carbohydrate specificity from mMGL1. J Biol Chem 277:28892–28901

Yuita H, Tsuiji M, Tajika Y, Matsumoto Y, Hirano K, Suzuki N, Irimura T (2005) Retardation of removal of radiation-induced apoptotic cells in developing neural tubes in macrophage galactose-type C-type lectin-1-deficient mouse embryos. Glycobiology 15:1368–1375

Roles of Serum Lectins in Host Defense

Nobuko Kawasaki[1], Bruce Yong Ma[2], Toshisuke Kawasaki[3]

Introduction

Mannan-binding protein (MBP), also called mannose-binding protein (MBP) or mannan-binding lectin (MBL), is a Ca^{2+}-dependent (C-type) mammalian lectin specific for mannose, N-acetylglucosamine (GlcNAc), and fucose, and is an important serum component associated with innate immunity. Human MBP is a homooligomer of an approximately 31 kDa subunit, each subunit containing a carbohydrate recognition domain (CRD) followed by a short neck region on the COOH terminal side and a collagen-like domain followed by a short cysteine-rich region on the NH_2 terminal side. Three subunits form a structural unit, and MBP normally consists of 3–6 structural units joined through disulfide bonds at the NH_2 termini, the whole molecular mass being approximately 300–600 kDa. MBP is a pattern-recognition molecule of the innate immune system that promotes phagocytosis of microorganisms through "the lectin pathway" of complement activation when it binds to ligand sugars such as mannose and GlcNAc on microbes (Kawasaki 1999) (see Fig. 1). More recently, MBP was found to have potent growth inhibitory activity to human colorectal carcinoma cell line, SW1116, in vivo via a complement-independent mechanism (Ma et al. 1999). This was proposed to term MBP-dependent cell-mediated cytotoxicity (MDCC). The MBP-ligand oligosaccharides expressed on the surface of SW1116 cells, which were assumed to be associated with MDCC, were isolated and characterized (Terada et al. 2005). Endogenous ligands for MBP were shown to be expressed highly in the brush border epithelial cells of kidney proximal tubules, and both meprin α and β have been identified as novel endogenous MBP ligand proteins. Interestingly, the interaction of MBP with meprins resulted in significant decreases in the proteolytic activity and matrix-degrading ability of meprins (Hirano et al. 2005). GlcNAc-binding lectins called ficolins in human serum, which have a fibrinogen-like domain but no CRD of C-type lectins, have been shown to activate complement (Matsushita et al. 2002).

---

[1] Research Center for Glycobiotechnology, Ritsumeikan University, 1-1-1 Nojihigashi, Kusatsu, Shiga 525-8577, Japan
Phone: +81-77-561-3452, Fax: +81-77-561-3452
E-mail: nobukokw@se.ritsumei.ac.jp

[2] Phone: +81-77-561-3450, Fax: +81-77-561-3451
E-mail: byma@fc.ritsumei.ac.jp

[3] Phone: +81-77-561-3440, Fax: +81-77-561-3496
E-mail: tkawasak@fc.ritsumei.ac.jp

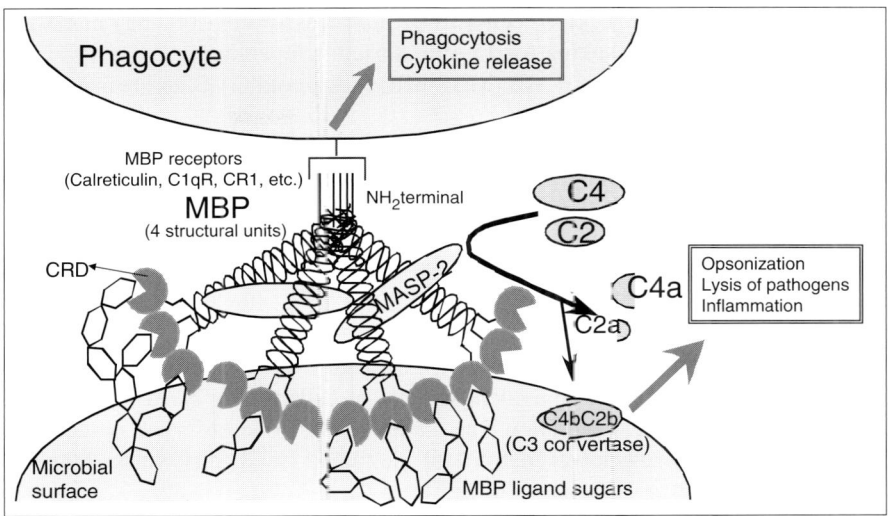

Fig. 1 Complement activation through the lectin (MBP) pathway. Once MBP–MASP-2 complexes bind to carbohydrates on the surfaces of microbes, the activated MASP-2 cleaves C4 and C2 to form C3 convertase (C4bC2b), which cleaves C3, followed by the lytic pathway yielding the membrane-attack complex (C5–C9). Ficolins act in combination with the MASPs. *MBP* mannan-binding protein, *MASP-2* MBP-associated serine protease-2 (Kawasaki et al. 2005)

Procedures and Results

1. The Lectin Pathway of Complement Activation: the Third Complement Activation Pathway

The MBP-mediated complement activation represents the third pathway distinct from both the classical and alternative pathways.

Complement Activation Assay by Passive Hemolysis

Complement activation by MBP was shown by means of the passive hemolysis assay using sheep erythrocytes, which were coated with yeast mannan by the chromium chloride method and sensitized with MBP.

1. Wash the sheep erythrocytes in saline by centrifugation (2,000 × g for 2 min) three times and suspend them to 5% packed cell volumes.

2. Mix 0.5 ml aliquots of mannan solution (5–150 μg) with an equal volume of $CrCl_3$ solution (0.5 mg/ml).

3. Add an equal volume of the erythrocyte suspension (1×10^9 cells) to the above solution.

4. Incubate the mixture with occasional mixing for 5 min at 25°C.

5. Stop the reaction by adding 1.5 ml of ice-cold GVB (gelatin-Veronal-buffered saline; 5 mM Veronal buffer, pH 7.4, containing 0.145 M NaCl, 0.1% gelatin, 2 mM $CaCl_2$ and 0.5 mM $MgCl_2$).

6. Wash the erythrocytes coated with mannan (ME) three times by centrifugation with GVB, and resuspend them to a final concentration of 1×10^9 cells/ml in GVB.

7. Add 0.1 ml of ME (1×10^8 cells) to 0.4 ml of MBP (1–1,000 ng) in GVB and incubate them with gentle shaking for 15 min at room temperature.

8. Wash the MBP-sensitized ME (MBP-ME) in ice-cold GVB and resuspend them to 1×10^9 cells/ml.

9. Use guinea pig serum as a complement source. Prepare MBP-depleted complement by absorbing the serum by passage at 4°C through Sepharose 4B-mannan column to remove endogenous MBP. Three milliliters of serum is absorbed per ml column bed volume.

10. Incubate MBP-ME (1×10^8 cells) with MBP-depleted complement diluted with GVB (0.25–2 CH_{50}) in a final volume of 1.5 ml at 37°C for 1 h and centrifuge the mixture.

11. Determine $A541_{nm}$ of the supernatant. The degree of specific lysis was calculated based on the absorption of an equivalent volume of cells totally lysed in water and is expressed as a percentage. Correction was made for the value observed in the absence of MBP, which was always lower than 5% of the totally lysed cells' value.

The lectin pathway of complement activation was also assayed by MBP-dependent complement component 4 (C4) deposition assay (Gadjeva et al. 2003). The *Escherichia coli* K12 and B strains, which have exposed GlcNAc and L-glycero-D-mannoheptose, respectively, are killed by MBP complement-dependently. In addition, dMM (1-deoxymannojirimycin, an α-mannosidase inhibitor)-treated BHK (baby hamster kidney) cells, which have high-mannose type oligosaccharides exposed on their surfaces, were killed by MBP with the aid of complement (Kawasaki 1999).

2. MBP-dependent Cell-mediated Cytotoxicity (MDCC): In Vivo Assay of Antitumor Activity of MBP

The recombinant vaccinia virus carrying human MBP gene was demonstrated to possess potent growth inhibiting activity against human colorectal carcinoma cells, SW1116, transplanted in KSN nude mice when administered by intratumoral or subcutaneous injection. Moreover, significant prolongation of life span of tumor-bearing mice resulted from the treatment. This effect appears to be a consequence of local production of MBP (Ma et al. 1999).

3. Endogenous MBP-ligands on the Surface of Colorectal Carcinoma Cells: Tumor Markers?

In order to isolate MBP ligands on the surface of colorectal carcinoma cells, pronase glycopeptides were prepared from whole cell lysates of SW1116 cells. Oligosaccharides were liberated by hydrazinolysis followed by being tagged by pyridylamination. The PA-MBP-ligand oligosaccharides were isolated carrying an MBP-affinity column, and then their sequences were determined by MS and MS/MS after permethylation, in combination with endo-β-galactosidase digestion, chemical defucosylation and lectin-HPLC analysis. The MBP ligands were shown to be large, multiantennary *N*-glycans with a highly fucosylated polylactosamine-type structure. At the nonreducing termini, Le^b–Le^a or tandem repeats of the Le^a structure prevail, a substantial proportion of which are attached via internal Le^x or *N*-acetyllactosamine units to the trimannosyl fucosylated core. It is concluded that MBP requires clusters of tandem repeats of the Le^a epitope for

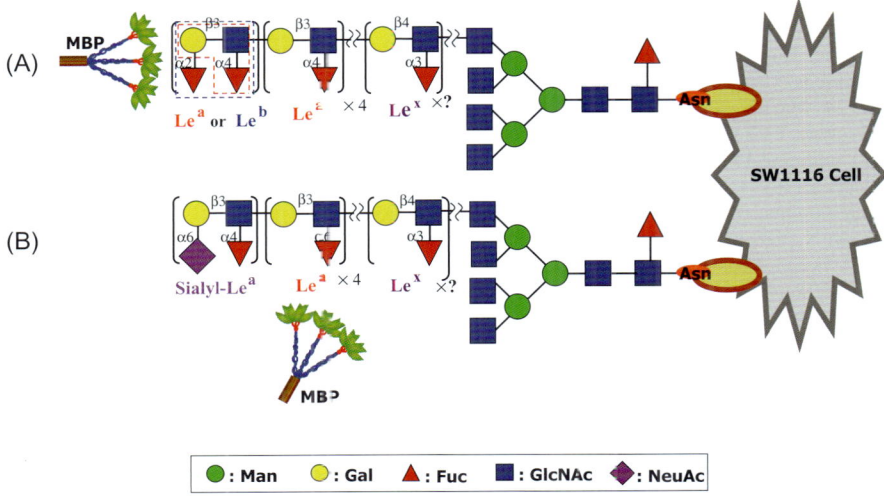

Fig. 2 Structures of MBP-ligand oligosaccharides expressed on human colorectal carcinoma cells, SW1116. **A** MBP-ligand oligosaccharide: neutral fraction; **B** MBP-ligand oligosaccharide: acidic fraction. The MBP ligands on SW1116 cells are mostly composed of tetraantennary core-fucosylated N-glycans with highly fucosylated polylactosamine-type structures. The nonreducing terminal units are mostly Le$^{b/a}$, a substantial portion of which is carried on extended type 1 chain as multimeric Le^a units (the maximum number of repeating Le^a units detected is 4). The inner units are likely to be dominated by type 2 chains and not fully fucosylated. The overall carbohydrate structures of the neutral and acidic MBP ligands are very similar except that the acidic ligands are capped with sialic acid at the nonreducing termini, which yields the sialyl-Le^a structure. MBP does not bind to sialyl-Le^a, but binds to the inner Le^a epitope

recognition (see Fig. 2). This novel structure may be a new type of tumor-associated carbohydrate antigen, which might be used as a tumor marker (Terada et al. 2005).

4. Endogenous MBP Ligands from Mouse Kidney: Antiprotease Activity of MBP

The microsomes prepared from a Balb/c mouse kidney homogenate were solubilized with lysis buffer containing 1% Triton X-100 and then centrifuged at 150,000 × g for 60 min. The supernatant was applied to an MBP-affinity column in the presence of 20 mM CaCl$_2$, and the proteins bound to the column were eluted with TBS buffer (pH 7.5) containing 4 mM EDTA and 0.1% Triton X-100. The eluted proteins were resolved on SDS-PAGE and stained with colloidal Coomassie blue. Two Bands (83 kDa and 91 kDa) were excised from the gel and subjected to in-gel digestion by trypsin. The peptides released from the gel were subjected to LC/MS/MS analysis. Both meprin α and β were identified as novel endogenous MBP ligands, for the 83 kDa and 91 kDa band, respectively. Meprins are extracellular matrix-degrading metalloprotease. Deglycosylation experiments indicated that the MBP ligands on meprins are high-mannose or complex-type N-glycans. Interestingly, the interaction of MBP with meprins resulted in significant decreases in the proteolytic activity and matrix-degrading ability of meprins, suggesting that N-linked oligosaccharides on meprins are associated with the optimal enzymatic activity, and that MBP is an important regulator for modulation of the localized meprin proteolytic activity via N-glycan-binding.

References

Gadjeva M, Thiel S, Jensenius JC (2003) Assays for the mannan-binding lectin pathway. In Coligan JF, et al. (eds) Current protocols in immunology vol.3, supplement 58, Wiley Interscience

Hirano M, Ma BY, Kawasaki N, Okimura K, Baba M, Nakagawa T, Miwa K, Kawasaki N, Oka S, Kawasaki T (2005) Mannan-binding protein blocks the activation of metalloproteases meprin α and β. J Immunol 175:3177–3185

Kawasaki T (1999) Structure and biology of mannan-binding protein, MBP, an important component of innate immunity. Biochim Biophys Acta 1473:186–195

Kawasaki N, Nakagawa T, Kawasaki T (2005) In: Taniguchi N (ed) Book of gene and medicine No. 3, pp 227–233. MEDICAL DO CO, Osaka (in Japanese)

Ma BY, Uemura K, Oka S, Kozutsumi Y, Kawasaki N, Kawasaki T (1999) Antitumor activity of mannan-binding protein in vivo as revealed by a virus expression system: Mannan-binding protein-dependent cell-mediated cytotoxicity. Proc Natl Acad Sci USA 96:371–375

Matsushita M, Endo Y, Fujita T (2002) Complement-activating complex of ficolin and mannose-binding lectin-associated serine protease. J Immunol 164:2281–2284

Terada M, Khoo K-H, Inoue R, Chen C, Yamada K, Sakaguchi H, Kadowaki N, Ma BY, Oka S, Kawasaki T, Kawasaki N (2005) Characterization of oligosaccharide ligands expressed on SW1116 cells recognized by mannan-binding protein. A highly fucosylated polylactosamine type N-glycan. J Biol Chem 280:10897–10913

Siglec-2 Is a Key Molecule for Immune Response

Takeshi Tsubata

Introduction

Siglecs are sialic acid-binding membrane-bound lectins expressed on various hematopoietic cells (Varki and Angata 2006). Siglec-2 (also called as CD22) specifically recognizes α2.6 sialic acid-containing glycans and is expressed on B cells, the precursors of antibody secreting cells, plasma cells (Nitschke and Tsubata 2004). Both biochemical analysis on various B cells and studies using Siglec-2-deficient mice have demonstrated that Siglec-2 negatively regulates signaling through the B cell antigen receptor (BCR), which plays a crucial role in response of B cells to antigen stimulation. B cells express membrane-bound immunoglobulin (Ig) as BCR. Ligation of BCR by antigens activates BCR-associated kinases, which in turn phosphorylate various substrates, leading to activation of various signaling cascades including Ca^{2+} signaling. Siglec-2 negatively regulates BCR signaling by recruitment and activation of SH2-containing phosphatase 1 (SHP-1), which counteracts the phospharylation-mediated activation of signaling molecules by dephosphorylation, and reduces Ca^{2+} signaling by regulating the activity of the Ca^{2+} pump PMCA. In the past few years, we have assessed the immunological function of Siglec-2 and found that it functions as a molecular switch that determines the fate of antigen-interacted B cells whether they undergo apoptosis or activation. Siglec-2 may therefore play a key role in the establishment and maintenance of normal humoral immunity, which produces antibodies to pathogens but not self-antigens, and, after vaccination, respond to pathogens rapidly.

Siglec-2 as a Molecular Switch Determining the Fate of BCR-ligated B Cells

When antigens ligate BCR and generate BCR signaling, B cells undergo apoptosis but not activation (Garside et al. 1998; Hokazono et al. 2003). Antigen-stimulated B cells are activated when these B cells interact with additional signals such as CD40L derived from activated T lymphocytes (T cells). In normal individuals, T cells are activated only when antigen is presented to T cells by pathogen-stimulated antigen-presenting cells such as dendritic cells. Therefore, the requirement of CD40L for B cell activation may avoid B cell response to non-pathogens. i.e., self-antigens and environmental antigens such as pollens, thereby preventing autoimmunity and allergy.

Siglec-2 and CD72, a C-type lectin domain-containing membrane molecule functionally redundant to Siglec-2 in BCR signal regulation, are involved in the fate determination of BCR-ligated B cells whether they undergo apoptosis or activation (Nitschke and Tsubata 2004). When B cells are stimulated with antigens, both Siglec-2 and CD72 markedly down-modulate BCR signaling. In contrast, BCR signaling is only weakly

---

Laboratory of Immunology, School of Biomedical Science, Tokyo Medical and Dental University; CREST, JST, 1-5-45 Yushima, Bunkyo-ku, Tokyo 113-8510, Japan

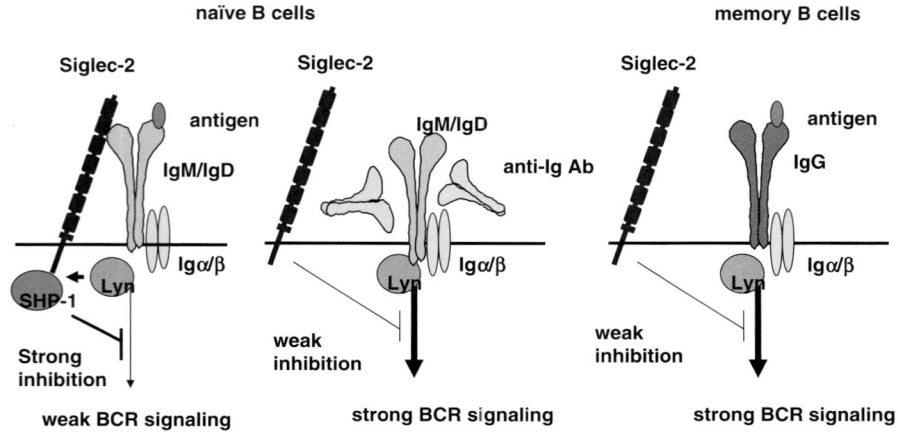

Fig. 1 BCR signal regulation by Siglec-2. The extent of BCR signal regulation by Siglec-2 appears to depend on Ig isotypes and stimuli that ligate BCR. See the text for details

regulated by Siglec2 or CD72 when BCR is ligated by anti-Ig antibodies, agonistic antibodies to BCR (Hokazono et al. 2003). Because anti-Ig antibodies normally bind to the distinct region of BCR from the antigen-binding site, binding of anti-Ig antibodies to BCR may disrupt interaction between BCR and membrane-bound lectins, and prevent signal regulation by these molecules (Fig. 1). The BCR ligation by anti-Ig antibodies induces activation of B cells, a distinct fate from that induced by antigen stimulation. When efficient BCR signal inhibition by Siglec-2 and CD72 is restored by co-ligation of BCR with these membrane-bound lectins, B cells are no longer activated but undergo apoptosis. Therefore, lack of efficient BCR signal regulation by Siglec-2 and CD72 is required for B cell activation induced by anti-Ig antibody-induced BCR ligation. Further, presence of efficient signal regulation by these lectins converts the fate of anti-Ig antibody-treated B cells from activation to apoptosis. Thus, both Siglec-2 and CD72 constitute a molecular switch that determines whether BCR-ligated B cells undergo apoptosis or activation, and install the requirement of anti-apoptotic signaling generated by T cell-derived CD40L for B cell activation.

Ig Isotype-Specific BCR Regulation by Siglec-2

Vaccination and previous infection often protect hosts from infection, a phenomenon recognized as "immunity". The "immunity" is mediated by memory lymphocytes that are generated from naïve lymphocytes after activation by vaccination or previous infection. Memory lymphocytes stay in the body for a long period and, upon encountering the same pathogen, generate immune responses much more rapidly and vigorously than naïve lymphocytes—lymphocytes that have not yet encountered antigens. Quick response enables the host to remove pathogens before they induce symptoms, thereby preventing infectious diseases. Thus, rapid activation of both memory B and T cells is crucial for host protection against infectious diseases by "immunity" generated by vaccination or previous infection.

Siglec-2 efficiently down-regulates signaling through the classes of BCR expressed on naïve B cells but not those expressed on memory B cells (Nitschke and Tsubata 2004;

Sato et al. 2007; Wakabayashi et al. 2002), resulting in more efficient BCR signaling in memory B cells than that in naïve B cells. There are five classes of immunoglobulin molecules, i.e, IgM, IgD, IgG, IgE, and IgA. These different classes of immunoglobulins are not only secreted as antibodies but also expressed on the cell surface as BCR. During immune responses, naïve B cells expressing both IgM and IgD are activated by encounters with antigens, and then undergo class switching to produce IgG, IgE or IgA. Since memory B cells are generated from antigen-activated B cells, most memory B cells express Ig of switched classes, especially IgG. IgG-expressing B cells show stronger BCR signaling and respond to antigens more efficiently than those expressing IgM and IgD. When IgM-containing BCR (IgM-BCR) and IgD-BCR are ligated by antigens, Siglec-2 efficiently recruits SHP-1 and negatively regulates BCR signaling. In contrast, IgG-BCR ligation induces only weak signal regulation mediated by Siglec-2, although Siglec-2 is expressed on memory B cells as well as on naïve B cells. Poor Siglec-2-mediated signal regulation appears to cause strong BCR signaling through IgG-BCR, thereby inducing rapid activation of memory B cells, crucial for "immunity" generated by vaccination or previous infection (Fig. 1). Thus, class-specific signal regulation by Siglec-2 may play a role in differentiating rapid memory responses by memory B cells from slow primary responses by naïve B cells.

Protocol for Assessing Phosphorylation of Siglec-2 and Recruitment of SHP-1 After Ligation of the B Cell Antigen Receptor

Immunoprecipitation of Siglec-2

1. Collect B cells (1×10^7 cells/sample) in a 1.5 ml tube and incubate them in 500 µl medium (RPMI 1640 medium supplemented with 10% FCS, 1 mM Glutamine, and 50 µM β-Mercaptoethanol) at 37°C for 10–15 min.
2. Add antigen or anti-Ig antibody and incubate for various time periods.
3. Spin down and suspend the cells with 500 µl lysis buffer (1% Triton X-100, 10% Glycerol, 150 mM NaCl, 2 mM EDTA, 0.02% NaN_3, 10 µg/ml PMSF, and 1 mM Na_3VO_4)
4. Incubate on ice for 20 min.
5. Centrifuge and transfer the supernatant to a new 1.5 ml tube.
6. Wash Protein G beads, suspend the beads with lysis buffer, and add 20 µl of the beads to the lysate together with 2 µl anti-Siglec-2 antibody.
7. Shake using an orbital shaker at 4°C for 20 min.
8. Wash the beads three times with lysis buffer.
9. Suspend the beads with 65 ml 2× sample buffer (0.1 M Tris–HCl pH 6.8, 4% SDS, 10% β-Mercaptoethanol, 20% Glycerol BPB)
10. Boil the beads for 5 min, centrifuge, and collect the supernatant.

Western Blotting for Phosphotyrosine and SHP-1

1. Separate the proteins by SDS-PAGE.
2. Transfer the proteins to nylon membrane.
3. Wash the membrane with TBST, and incubate the membrane with PBS containing 3% BSA at room temperature for 1 h.
4. Incubate the membrane with anti-phosphotyrosine (4G10) antibody or anti-SHP-1 antibody at room temperature for 1 h.

5. Wash the membrane with TBST (TBS pH7.4, 1% Tween 20) and incubate the membrane with horseradish peroxidase (HRP)-labeled goat anti-mouse IgG or anti-rabbit IgG antibody.
6. Visualize the membrane using a detection kit.

Conclusion

Siglec-2 and CD72 regulate BCR signaling, thereby determining the fate of BCR-ligated B cells whether they undergo apoptosis or activation. In antigen-stimulated naïve B cells, BCR regulation by these membrane-bound lectins appears to induce apoptosis and install requirement of T cell-derived CD40 signaling for activation. In contrast, Siglec-2 does not efficiently regulate BCR signaling through IgG, the major Ig isotype in memory B cells. Lack of Siglec-2-mediated efficient signal regulation may be involved in strong BCR signaling through IgG and contribute to rapid and vigorous activation of memory B cells crucial for host defense induced by vaccination.

References

Garside P, Ingulli E, Merica RR, Johnson JG, Noelle RJ, Jenkins MK (1998) Visualization of specific B and T lymphocyte interactions in the lymph node. Science 281:96–99

Hokazono Y, Adachi T, Wabl M, Tada N, Amagasa T, Tsubata T (2003) Inhibitory coreceptors activated by antigens but not by anti-Ig heavy chain antibodies install requirement of costimulation through CD40 for survival and proliferation of B cells. J Immunol 171:1835–1843

Nitschke L, Tsubata T (2004) Molecular interactions regulate BCR signal inhibition by CD22 and CD72. Trends Immunol 25:543–550

Sato M, Adachi T, Tsubata T (2007) Augmentation of signaling through BCR containing IgE but not that containing IgA due to lack of CD22-mediated signal regulation. J Immunol 178:2901–2907

Varki A, Angata T (2006) Siglecs—the major subfamily of I-type lectins. Glycobiology 16:1R–27R

Wakabayashi C, Adachi T, Wienands J, Tsubata T (2002) A distinct signaling pathway used by the IgG-containing B cell antigen receptor. Science 298:2392–2395

Galectin and Immune System: Toward Clinical Applications

Mitsuomi Hirashima

Introduction

We first cloned galectin-9 (Gal-9) as a T cell-derived eosinophil chemoattractant. Gal-9 plays a role in not only accumulation but also activation of eosinophils in experimental allergic models and human allergic patients, because Gal-9 induces eosinophil chemoattraction in vitro and in vivo and activates eosinophils in many aspects (Matsumoto et al. 1998). However, Gal-9 also has other functions, such as cell differentiation, aggregation, adhesion, and cell death, than eosinophil chemoattraction and activation (Kashio et al. 2003; Dai et al. 2005). For example, in patients with malignant tumor, such as malignant melanoma and breast cancer, high Gal-9 expression in tumor cells links to better clinical features in tumor-bearing patients (Kageshita et al. 2002; Irie et al. 2005). From the above findings, we have established Galpharma, a Bio-Venture company to develop the method for predicting tumor metastasis, especially for breast cancer.

However, we have also found that Gal-9 exhibits regulating activity for a variety of inflammation, such as autoimmune diseases and allergic inflammation. Thus, we are now going to apply Gal-9 as a novel candidate of therapeutics for the regulation of inflammation.

Gal-9 can be Used for Prediction of Metastasis

High Gal-9 expression is linked to better prognosis in breast cancer patients: even in breast cancer patients with metastasis to axillary lymphnode. a few of them have remote metastasis if Gal-9 is positive in breast cancer cells. These results suggest the possible application of Gal-9 in anti-cancer and/or anti-metastatic therapy.

Gal-9 has Immunomodulating Activity

We have described that Gal-9 induces apoptosis of human-activated T cells via Ca^{2+}–calpain–caspase1 pathway. However, Gal-9 is so susceptible to proteolysis that it is easily degradaed resulting in the loss of Gal-9 biological activity. Therefore, we have established the stable form of Gal-9 (sGal-9) that exhibits the same biological activity as natural Gal-9 by truncating its ligand (Nishi et al. 2005). Recently, it has been shown that Gal-9 is a ligand of TIM-3 expressed on the surface of Th1 and Th17 cells, and it induces apoptosis of TIM-3 expressing Th1 cells (Zhu et al. 2005). Application of sGal-9 in rodent models of rheumatoid arthritis has revealed that Gal-9 has not only preventive but also therapeutic effects on them. Moreover, we have found that sGal-9 preferentially induces apoptosis of synovial cells from patients with rheumatoid arthritis. These results

---

Department of Immunology and Immunopathology, Faculty of Medicine, Kagawa University, Kagawa, Japan

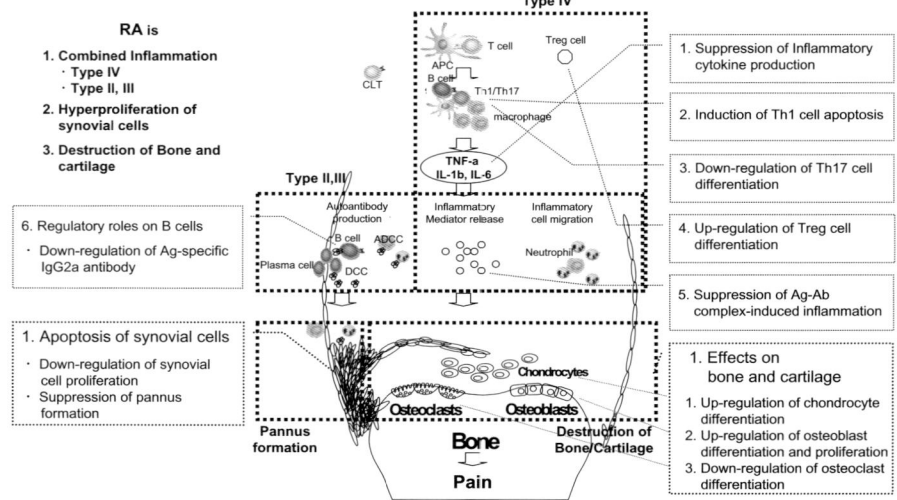

Fig. 1 Therapeutic activities of sGal-9 in immune disorders

suggest that sGal-9 can be a novel therapeutic target for inflammatory diseases including rheumatoid arthritis as multiple targeting therapeutics (Fig. 1).

We have recently found that sGal-9 regulates innate immunity (Tsuboi et al. 2007) and allergic inflammation (Katoh et al. 2007), suggesting the usefulness in those inflammations in addition to autoimmune diseases.

References

Dai S-Y, Nakagawa R, Itoh A, Murakami H, Kashio Y, Abe H, Katoh S, Kontani K, Kihara M, Zhang SL, Hata T, Nakamura T, Yamauchi A, Hirashima M (2005) Galectin-9 induces maturation of human monocyte-derived dendritic cells. J Immunol 175:2974–2981

Irie A, Yamauchi A, Kontani K, Kihara M, Liu D, Shirato Y, Seki M, Nishi N, Nakamura T, Yokomise H, Hirashima M (2005) Galectin-9 as a prognostic factor with antimetastatic potential in breast cancer. Clin Cancer Res 11:2962–2968

Kageshita T, Kashio Y, Yamauchi A, Seki M, Abedin MJ, Nishi N, Shoji H, Nakamura T, Ono T, Hirashima M (2002) Possible role of galectin-9 in cell aggregation and apoptosis of human melanoma cell lines and its clinical significance. Int J Cancer 99(6):809–816

Kashio Y, Nakamura K, Abedin MJ, Seki M, Nishi N, Yoshida N, Nakamura T, Hirashima M (2003) Galectin-9 induces apoptosis through the calcium–calpain–caspase-1 pathway. J Immunol 170:3631–3636

Katoh S, Nobumoto A, Ishii N, Takeshita K, Dai S-Y, Shinonaga R, Niki T, Nishi N, Tominaga A, Yamauchi A, Hirashima M (2007) Galectin-9 inhibits CD44-hyaluronan interaction and suppresses a murine model of allergic asthma. AJRCCM (in press)

Matsumoto R, Matsumoto H, Seki M, Hata M, Asano Y, Kanegasaki S, Stevens RL, Hirashima M (1998) Human ecalectin, a variant of human galectin-9, is a novel eosinophil chemoattractant produced by T lymphocytes. J Biol Chem 273:16976–16984

Nishi N, Itoh A, Fujiyama A, Yoshida N, Araya S, Hirashima M, Shoji H, Nakamura T (2005) Development of highly stable galectins: truncation of the linker peptide confers protease-resistance on tandem-repeat type galectins. FEBS Lett 579:2058–2064

Tsuboi Y, Abe H, Nakagawa R, Oomizu S, Watanabe K, Nishi N, Nakamura T, Yamauchi A, Hirashima M (2007) Galectin-9 protects mice from Shwartzman reacting prostaglandin E2-producing polymorphonuclear leukocytes. Clin Immunol (in press)

Zhu C, Anderson AC, Schubart A, Xiong H, Imitola J, Khoury SJ. Zheng XX, Storm TB, Kuchroo VK (2005) The Tim-3 ligand galectin-9 negatively regulates T helper type 1 immunity. Nat Immunol 6(12):1245–1252

Section VII
Brains and Nerves

Novel Glycolipids in Developing Brain: Determination by Nano-LC/MS/MS

Yoshio Hirabayashi[1,3], Yasuko Nagatsuka[1], Shinya Ito[2]

Introduction

Recent studies have shown that membrane lipids such as glycosphingolipids form distinct and functional microdomains and/or nanodomains. Although extensive studies have characterized membrane lipids, the precise composition of these domains is still not fully elucidated. In 2003, we identified a novel glycolipid, phosphatidylglucoside (PtdGlc), and categorized it as a new member of the lipid microdomain (Nagatsuka et al. 2003). PtdGlc, which is enriched in a detergent-insoluble membrane fraction, appears to be involved in cellular differentiation. Its chemical structure, however, was not fully studied. Recently, the complete chemical structure of PtdGlc isolated from rat embryonic brain tissues has been determined. Surprisingly PtdGlc is exclusively composed of saturated fatty acyl chains (C18:0/C20:0) in its diacylglycerol moiety (Nagatsuka et al. 2006a). The saturated fatty acid composition with single molecular species is barely detectable in known mammalian lipids. Very recently, we found that PtdGlc is a novel cell surface marker for developing radial/astro glia cells, and lyso-derivative secreted from the glia cells has potent biological activities to surrounding neurons. Here, we present an improved method to detect PtdGlc in brain tissues by MS spectrometry (Nagatsuka et al. 2006b).

Procedure

Nano-LC/MS/MS

Glycolipids were isolated from cultured glial cells as described (Nagatsuka et al. 2006a). For LC-MS/MS, we used a Hitachi High-Technologies NanoFrontier L (Hitachi High-Technologies, Tokyo, Japan). The LC-MS/MS consisted of a nano flow HPLC and a linear ion trap time-of-flight mass spectrometer with an electrospray ion source. The sample injector was an Upchurch M-435 microinjection valve (Upchurch Scientific, Oak Harbor, WA, USA). The HPLC column was an Inertsil SIL-100A (3 µm, 0.075 mm i.d., and 150 mm length) Si column from GL Science (Tokyo, Japan).

The eluents were chloroform:methanol (2:1, v/v) in 0.07% aqueous ammonium hydroxide (solvent A), chloroform (solvent B), and methanol (solvent C). The gradient pump delivered solvents at a flow rate of 100 µL/min. The gradient profile used was as follows; solvent A/B/C = 0/75/25 (0 min) → 2/75/23 (5 min) → 5/75/20 (30 min) →

---

[1] Hirabayashi Research Unit, Brain Science Institute, RIKEN, 2-1 Hirosawa, Wako, Saitama, Japan

[2] Hitachi High-technologies Corp., Tokyo, Japan

[3] CREST
Phone: +81-48-467-6372, Fax: +81-48-467-6372
E-mail: hirabaya@riken.jp

10/50/40 (40 min) →10/50/40 (60 min). A flow rate of the nano flow pump was 200 nL/min. ESI spray potential was −4500 V in negative ion mode, curtain gas flow was 1.0 L/min, and AP1 temperature were 140°C. Scan mass range was m/z 200 to 2,000. In MS/MS experiment, data-dependent MS/MS scan mode was used. CID gain energy was determined automatically depending on the target m/z.

Results

A minor component of PtdGlc was detected in total lipid extract from cultured glia cells (Fig. 1). By using normal phase silica gel-based column, PtdGlc could be separated from other lipids such as phosphatidylglycerol and phosphatidylinositol. PtdGlc was detected as a single peak with m/z = 893.6 (negative mode), indicating that PtdGlc is composed of saturated fatty acids, C18:0 and C20:0 (Fig. 2). This fatty acid composition was identical to that of the rat developing brain. Further confirmation of the structure was achieved from MS/MS analysis. A MS^2 spectrum of PtdGlc gave a characteristic peak of m/z = 311i3 derived from C20:0 fatty acid.

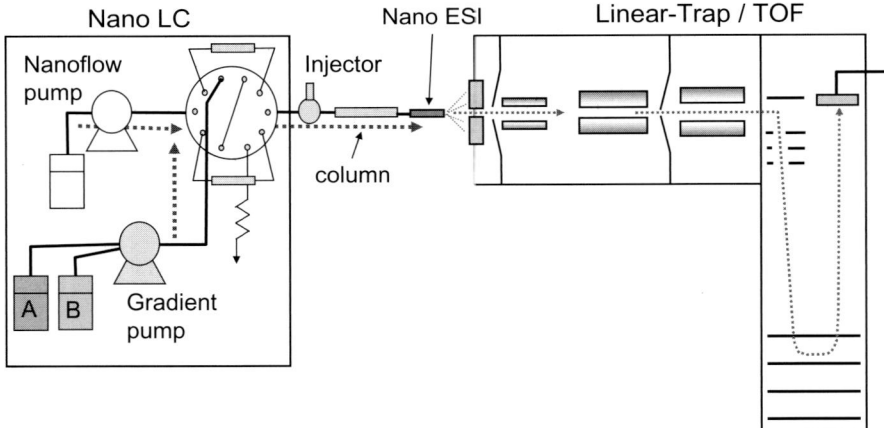

Fig. 1 Schematic diagram of nanoflow liquid chromatography-linear trap time of flight mass spectrometry

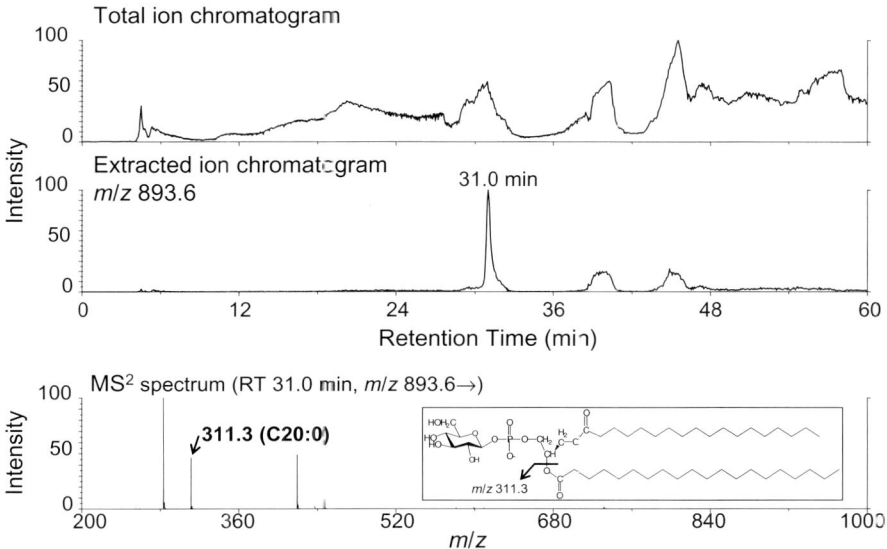

Fig. 2 Detection of PtdGlc from chick cultured glia cells by normal phase LC/MS/MS

References

Nagatsuka Y, Hara-Yokoyama M, Kasama T, Takekoshi M, Maeda F, Ihara S, Fujiwara S, Ohshima E, Ishii K, Kobayashi T, Shimizu K, Hirabayashi Y (2003) Carbohydrate-dependent signaling from the phosphatidylglucoside-based microdomain induces granulocytic differentiation of HL60 cells. Proc Natl Acad Sci USA 100:7454–7459

Nagatsuka Y, Horibata Y, Yamazaki Y, Kinoshita M, Shinoda Y, Hashikawa T, Koshino H, Nakamura T, Hirabayashi Y (2006a) Phosphatidylglucoside exists as a single molecular species with saturated fatty acyl chains in developing astroglial membranes. Biochemistry 45:8742–8750

Nagatsuka Y, Tojo H, Hirabayashi Y (2006b) Identification and analysis of novel glycolipids in vertebrate brains by HPLC/mass spectrometry. Methods Enzymol 417:155–167

Functional Roles of the HNK-1 Carbohydrate and Polysialic Acid in the Nervous System

Yasuhiko Kizuka[1], Shogo Oka[2]

Introduction

In the nervous system, many types of cells recognize each other to form a precise neural network, whereas these cells are morphologically and functionally different. The development and maintenance of this complicated but accurate network involves many molecules. Among them, cell adhesion molecules (CAMs) are considered to play important roles in these processes. On these CAMs in the nervous system, unique carbohydrate epitopes are known to be expressed such as HNK-1 (*H*uman *N*atural *K*iller-1, also called CD57) and PSA (*p*oly*s*ialic *a*cid), which are rarely found in other organs or tissues. Recently, it was revealed that these unique carbohydrates are involved in the formation and maintenance of neural network.

Results

HNK-1 Carbohydrate

The HNK-1 carbohydrate has a very unique structural feature comprising a sulfated trisaccharide (HSO_3-3GlcAβ1-3Galβ1-4GlcNAc-), whose biosynthesis is regulated by two glucuronyltranferases (GlcAT-P and GlcAT-S). Sulfation of glucuronic acid (GlcA) is catalyzed by a specific sulfotransferase called HNK-1ST. Even before elucidation of this biosynthetic pathway, this carbohydrate antigen had been thought to be involved in neural network formation because of its characteristic and well-regulated expression pattern in the nervous system. In addition to its conserved expression in various species from vertebrates to invertebrates, the HNK-1 carbohydrate is observed in migrating neural crest cells, rhombomeres, and the cerebellum during brain development. In more detail, it was revealed that this carbohydrate epitope is actually associated with neural crest cell migration, neuron to glial cell adhesion, outgrowth of astrocytic processes and migration of the cell bodies, as well as the preferential outgrowth of neurites from motor neurons. In order to elucidate the roles of the HNK-1 carbohydrate more clearly, we generated and analyzed GlcAT-P gene-deficient mice. Unexpectedly, the mutant mice seemed to be apparently normal and showed no aberration of brain development, although expression of the HNK-1 carbohydrate was almost completely abolished in the entire brain. However, GlcAT-P deficient mice exhibited reduced synaptic plasticity in hippocampal neurons

---

[1] Department of Biological Chemistry, Graduate School of Pharmaceutical Sciences, Kyoto University, Kyoto 606-8501, Japan

[2] Department of Biological Chemistry, School of Health Sciences, Faculty of Medicine, Kyoto University, Kyoto 606-8507, Japan
Phone: +81-75-751-3959, Fax: +81-75-751-3959
E-mail: shogo@hs.med.kyoto-u.ac.jp

resulting in impairment of learning and memory (Yamamoto et al., 2002). These results indicate that the HNK-1 epitope plays important roles in neural high order function in vivo. Moreover, another group reported that HNK-1ST null mice showed a similar defect to GlcAT-P null mice such as impairment of synaptic plasticity, learning, and memory (Senn et al. 2002). These lines of evidence strongly suggest the necessity of the HNK-1 carbohydrate for maintenance of precise neural function.

PSA (Polysialic Acid)

Polysialic acid is a highly negative charged and a spatially large modification expressed, almost specifically on neural cell adhesion molecule (NCAM). PSA consists of polymerized alpha2-8 linked sialic acids (from several to more than a hundred residues) at the non-reducing terminal of *N*-glycans. It is known that PSA regulates both homophilic and heterephilic interactions of NCAM through its anionic properties and steric hindrance. Expression of PSA is also well-regulated accompanying brain development, that is, the expression level is high during the embryonic period, whereas that in adulthood is decreased, its expression being limited to areas of continued neurogenesis such as the subventricular zone or hippocampus. These results suggest that PSA is involved in neural network formation. Biosynthesis of PSA is regulated by two polysialyltransferases, ST8SiaII (STX) and ST8siaIV (PST). ST8SiaIV is predominant in the adult brain, while the expression of ST8SiaII is higher in the embryonic brain; therefore, these two polysialyltransferases function differently in the nervous system. In fact, ST8SiaIV-deficient mice showed reduced synaptic plasticity in the hippocampus, while ST8SiaII-deficient mice did not. Instead, in the ST8SiaII null mouse hippocampus, misguidance of mossy fibers and ectopic synapses was seen, and the mice showed highly exploratory behavior and reduced responses to fearful situations. These lines of evidence indicate that PSA is related to various functions of the brain. Moreover, in 2005, it was revealed that simultaneous targeting of both the ST8siaII and ST8siaIV genes caused a severer phenotype, such as hydrocephalus, growth retardation, and death at an early age (Weinhold et al. 2005), strongly indicating that PSA is necessary for precise neural network formation.

References

Senn C, Kutsche M, Saghatelyan A, Bosl MR, Lohler J, Bartsch U, Morellini F, Schachner M (2002) Mice deficient for the HNK-1 sulfotransferase show alterations in synaptic efficacy and spatial learning and memory. Mol Cell Neurosci 20:712–729

Weinhold B, Seidenfaden R, Rockle I, Muhlenhoff M, Schertzinger F, Conzelmann S, Marth JD, Gerardy-Schahn R, Hildebrandt H (2005) Genetic ablation of polysialic acid causes severe neurodevelopmental defects rescued by deletion of the neural cell adhesion molecule. J Biol Chem 280:42971–42977

Yamamoto S, Oka S, Inoue M, Shimuta M, Manabe T, Takahashi H, Miyamoto M, Asano M, Sakagami J, Sudo K, Iwakura Y, Ono K, Kawasaki T (2002) Mice deficient in nervous system-specific carbohydrate epitope HNK-1 exhibit impaired synaptic plasticity and spatial learning. J Biol Chem 277:27227–27231

Neuronal Function of Sulfatide

Koichi Honke[1,2]

Introduction

Mammals have acquired property known as myelin during the evolution so as to enhance the conductivity of the neuronal impulse. Oligodendrocytes produce vast amounts of myelin, a unique and lipid-rich biomembranes with a relatively simple array of myelin-specific proteins in the central nervous system. This membrane, an extension of the oligodendrocyte plasma membrane, forms multilamellar and spirally wrapped sheaths around neuronal axons. The gaps between adjacent myelin sheaths are referred to as nodes of Ranvier, and myelin forms lateral loops there. These myelin loops terminate at the paranode region and engage in the formation of a septate-like adhesive junction with the axon membrane, axolemma. This specialized axo-glial junction acts as an electronical and biochemical barrier between nodal and internodal membrane compartments. Voltage-gated sodium channels concentrate in the nodal axolemma, whereas shaker-type K^+ channels, Kv1.1 and Kv1.2, localize within the juxtaparanodal axolemma. Saltatory conduction of the action potential is attributed to this organization. The adhesion of myelin to the axolemma plays a critical role in this clustering of ion channels. Thus myelin serves not only as a simple insulator but also as a functional platform of the neuron–glia interaction (Fig. 1).

Myelin membrane is rich in lipids and contains unique galactolipids. GalCer and sulfatide comprise 23 and 4% of the total lipid content in the myelin sheath, respectively. These galactolipids are biosynthesized via sequential reactions catalyzed by ceramide galactosyltransferase (CGT, EC 2.4.1.45) and cerebroside sulfotransferase (CST, EC 2.8.2.11).

Results

To elucidate the physiological function of sulfatide, Cst-knockout mice were created by means of gene targeting (Honke et al. 2002). CST-deficient mice show a complete loss of sulfatide in brain. CST-null mice were born healthy, but began to display hindlimb weakness by 6 weeks of age and subsequently showed a pronounced tremor and progressive ataxia. Histological analysis revealed that axons were well myelinated in CST-null mice. Electron microscopic analysis of myelinated nerve fibers, however, revealed disorganized termination of the lateral loops at the node of Ranvier, similar to that reported in CGT-deficient mice (Dupree et al. 1998). Furthermore, clustering of Na^+ and K^+ channels at the node is also deteriorated in CST-null mice (Ishibashi et al. 2002) as observed in CGT-null mice (Dupree et al. 1999). These findings strongly suggest that sulfatide is

---

[1] Department of Biochemistry, Kochi University Medical School Kohasu, Oko-cho, Nankoku, Kochi 783-8505, Japan
Phone: +81-88-880-2313, Fax: +81-88-880-2314
E-mail: khonke@kochi-u.ac.jp

[2] CREST, Japan Science and Technology Agency, Japan

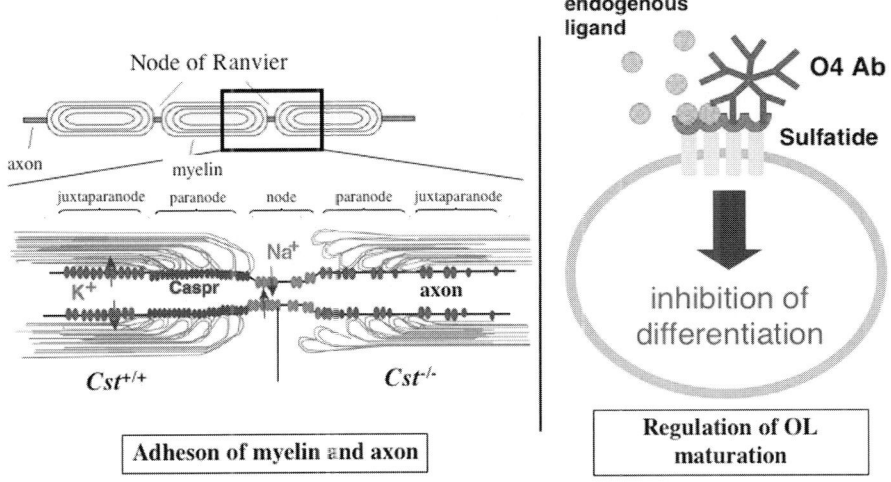

Fig. 1 Neuronal function of sulfatide. Sulfatide participates in adhesion of myelin sheath and axolemma at the paranodal junction (*left*) and negative regulation of oligodendrocyte maturation (*right*)

an indispensable molecule for the adhesive junction of myelin loop and axolemma at the paranode region (Fig. 1). It has been reported that three proteins: Caspr/paranodin, contactin, which are on the axolemma, and the 155 kDa splice isoform of neurofascin (NF155), which is on the myelin membrane, form the axo-glial adhesion apparatus at the paranodal region. It is an important problem in the future on how a lipid component of myelin, sulfatide, interacts with these protein complexes and maintains the adhesive junction.

Despite the significant neurological disorders, CST-null mice are able to survive for more than 1 year. The phenotype of the CST-deficient mice was milder than that of the CGT-deficient mice in terms of the age of onset, life span, and the severity of symptoms and pathological findings (Marcus et al. 2006). This discrepancy suggests that sub-localization on myelin sheath is different between GalCer and sulfatide, and that GalCer acts not only as a precursor for sulfatide synthesis, but also as a functional molecule.

Since sulfatide emerges when oligodendrocyte progenitors cease dividing and commence differentiating, it is supposed to be involved in the regulation of terminal differentiation. In fact, this terminal differentiation is enhanced in CGT-KO mice (Bansal et al. 1999). Furthermore, oligodendrocyte differentiation in wild-type mice was found to be blocked by anti-sulfatide antibody but not by anti-GalCer antibody (Bansal et al. 1999). These findings strongly suggest that sulfatide is a key negative regulator of the oligodendrocyte differentiation. This hypothesis was certified by the fact that terminal differentiation of oligodendrocytes is enhanced in CST-KO mice (Hirahara et al. 2004). Hence, sulfatide plays a critical role in the regulation of oligodendrocyte terminal differentiation, in addition to their eventual roles as structural components of mature myelin (Fig. 1).

References

Bansal R, Winkler S, Bheddah S (1999) Negative regulation of oligodendrocyte differentiation by galactosphingolipids. J Neurosci 19:7913–7924

Dupree JL, Coetzee T, Blight A, Suzuki K, Popko B (1998) Myelin galactolipids are essential for proper node of Ranvier formation in the CNS. J Neurosci 18:1642–1649

Dupree JL, Girault JA, Popko B (1999) Axo-glial interactions regulate the localization of axonal paranodal proteins. J Cell Biol 147:1145–1152

Hirahara Y, Bansal R, Honke K, Ikenaka K, Wada Y (2004) Sulfatide is a negative regulator of oligodendrocyte differentiation: development in sulfatide-null mice. Glia 45:269–277

Honke K, Hirahara Y, Dupree J, Suzuki K, Popko B, Fukushima K, Fukushima J, Nagasawa T, Yoshida N, Wada Y, Taniguchi N (2002) Paranodal junction formation and spermatogenesis require sulfoglycolipids. Proc Natl Acad Sci USA 99:4227–4232

Ishibashi T, Dupree JL, Ikenaka K, Hirahara Y, Honke K, Peles E, Popko B, Suzuki K, Nishino H, Baba H (2002) A myelin galactolipid, sulfatide, is essential for maintenance of ion channels on myelinated axon but not essential for initial cluster formation. J Neurosci 22:6507–6514

Marcus J, Honigbaum S, Shroff S, Honke K, Rosenbluth J, Dupree JL (2006) Sulfatide is essential for the maintenance of CNS myelin and axon structure. Glia 53:372–381

Accumulation of Alzheimer β-amyloid via Ganglioside Clusters

Katsumi Matsuzaki

Introduction

Alzheimer's disease (AD) is the most common form of dementia. The pathological hallmarks of AD brains are extracellular senile plaques and intracellular neurofibrillary tangles. It is widely accepted that the amyloid β-peptide (Aβ), which exists in fibrillar forms as a major component of senile plaques, is central to the development of AD. Aβ is a peptide composed of 40–42 amino acids (Fig. 1) and generated from proteolytic cleavage of amyloid precursor protein by β- and γ-secretases. Aβ is present at a very low concentration ($<10^{-8}$ M) in biological fluids, and its physiological role is unknown. According to the so-called Aβ hypothesis, the conversion of soluble, nontoxic Aβ to aggregated toxic Aβ rich in β-sheet structures ignites the neurotoxic cascade(s) of Aβ. However, it is not clear how soluble Aβ forms aggregates.

Concept

We propose based on liposomal (Kakio et al. 2001, 2002) and cellular (Wakabayashi et al. 2005; Wakabayashi and Matsuzaki 2007) studies that ganglioside clusters, which are formed in a cholesterol-dependent manner, mediate the aggregation of Aβ (Fig. 1). A specific form of Aβ bound to monosialoganglioside GM1 was found in brains exhibiting early pathological changes of AD. Gangliosides are considered to form lipid raft microdomains with other sphingolipids and cholesterol. Aβ selectively binds to raft membranes containing ganglioside clusters, undergoing drastic conformational changes depending on the peptide-to-ganglioside ratio. Aβ assumes an α-helix-rich structure at lower ratios (≤ 0.025), whereas the peptide changed its conformation to a β-sheet at higher ratios (≥ 0.05). The latter form acts as a seed for the formation of amyloid fibrils. This process can be inhibited by small compounds, e.g., nordihydroguaiaretic acid (Matsuzaki et al. 2007).

Our model agrees well with various previous in vitro and in vivo observations. A significant increase in GM1 was reported in Aβ-positive nerve terminals from the AD cortex. Lipid rafts from the frontal cortex and the temporal cortex of AD brains were also found to contain a higher concentration of GM1 compared to an age-matched control. A link between cholesterol, Aβ, and AD has been reported. The amount of cholesterol in the exofacial leaflets of the synaptic plasma membrane increases in aged as well as in apolipoprotein E4-knock-in mice. Both the aging and the apolipoprotein E4

Graduate School of Pharmaceutical Sciences, Kyoto University
46-29 Yoshida-Shimoadachi-cho, Sakyo-ku, Kyoto 606-8501, Japan
Phone: +81-75-753-4521, Fax: +81-75-753-4578
E-mail: katsumim@pharm.kyoto-u.ac.jp

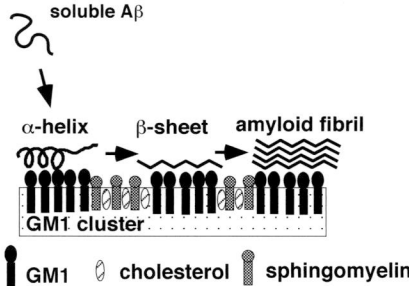

Fig. 1 The amino acid sequence of Aβ-(1–42) and a hypothetical model of the aggregation of Aβ induced by ganglioside clusters in lipid rafts. Aβ is soluble and takes an unordered structure in solution. Once ganglioside clusters are generated in a cholesterol-dependent manner, Aβ binds to the clusters, forming an α-helix-rich structure at lower peptide-to-ganglioside ratios, whereas the peptide changes its conformation to a β-sheet at higher ratios. The β-sheet form facilitates the fibrillization of Aβ, leading to cytotoxicity.

allele are strong risk factors for developing AD. For more information, refer to our recent review (Matsuzaki 2007).

Protocol

Determination of Membrane-Induced Amyloid Formation by Amyloid β-Protein (Aβ)

Preparation of Aβ Solution

1. Dissolve Aβ powder into 0.02% ammonia on ice.
2. Remove large aggregates by ultracentrifugation (540,000 g, 3 h, 4°C).
3. Determine the protein concentration of the supernatant.
4. Store the protein solution at −80°C until use.

Preparation of Small Unilamellar Vesicles

1. Dissolve GM1, cholesterol, and sphingomyelin into methanol/chloroform (1:2, v/v), ethanol, and chloroform, respectively.
2. Mix the solutions at a molar ratio of GM1:cholesterol:sphingomyelin = 4:3:3.
3. Remove the solvent by a rotary evaporator.
4. Redissolve the residual film into ethanol.
5. Completely remove the solvent by a rotary evaporator and *in vacuo* overnight.
6. Hydrate the lipid film with a buffer (10 mM Tris/150 mM NaCl/1 mM EDTA, pH 7.4), producing multilamellar vesicles by vortexing at 60°C.
7. Sonicate the vesicle suspension with a probe-type sonicator (5 min × 3) under a stream of nitrogen, producing small unilamellar vesicles.

Determination of Amyloid Formation by the Thioflavin-T Assay

1. Dilute the Aβ solution with an equal volume of a double concentrated buffer (20 mM Tris/300 mM NaCl/2 mM EDTA, pH 7.4).
2. Mix the above Aβ solution with the small unilamellar vesicles and incubate the mixture at 37°C.
3. Add aliquots of the mixture into a thioflavin-T solution (5 µM thioflavin-T/50 mM glycine, pH 8.5) at a final Aβ concentration of 0.5 µM.
4. Measure fluorescence intensity at an emission wavelength 490 nm (excitation at 446 nm).

References

Kakio A, Nishimoto S, Yanagisawa K, Kozutsumi Y, Matsuzaki K (2001) Cholesterol-dependent formation of GM1 ganglioside-bound amyloid β-protein, an endogenous seed for Alzheimer amyloid. J Biol Chem 276:24985–24990

Kakio A, Nishimoto S, Yanagisawa K, Kozutsumi Y, Matsuzaki K (2002) Interactions of amyloid β-protein with various gangliosides in raft-like membranes: importance of GM1 ganglioside-bound form as an endogenous seed for Alzheimer amyloid. Biochemistry 41:7385–7390

Matsuzaki K (2007) Physicochemical interactions of amyloid β-peptide with lipid bilayers. Biochim Biophys Acta 1768:1935–1942

Matsuzaki K, Noguch T, Wakabayashi M, Ikeda K, Okada T, Ohashi Y, Hoshino M, Naiki H (2007) Inhibitors of amyloid β-protein aggregation mediated by GM1-containing raft-like membranes. Biochim Biophys Acta 1768:122–130

Wakabayashi M, Okada T, Kozutsumi Y, Matsuzaki K (2005) GM1 ganglioside-mediated accumulation of amyloid β-protein on cell membranes. Biochem Biophys Res Commun 328:1019–1023

Wakabayashi M, Matsuzaki K (2007) Formation of amyloids by Aβ-(1-42) on NGF-differentiated PC12 cells: roles of gangliosides and cholesterol. J Mol Biol 371:924–933

Chondroitin Sulfate Proteoglycans and Neuronal Network Formation

Nobuaki Maeda

Introduction

Chondroitin sulfate proteoglycans play important roles in the development of the nervous system including neuronal migration, neurite extension, and synapse formation (Maeda 2007). The physiological functions of chondroitin sulfate proteoglycans in the developing nervous system can be investigated using various in vitro culture systems as described below.

Preparation of Dissociated Neurons

1. Dissect cerebra from embryonic day-17 Sprague–Dawley rats (~6 tissues), and remove meninges in L15 medium on ice. For the preparation of cerebellar neurons, ~5 cerebella from postnatal day-7 rats should be used.
2. Wash the tissues with 1 ml of Ca^{2+}- and Mg^{2+}-free Hanks' balanced salt solution (CMF-HBSS) two times.
3. Incubate the tissues in 1 ml of 0.1% trypsin/CMF-HBSS for 15 min at 37°C.
4. Wash the tissues with 1 ml of CMF-HBSS three times.
5. Triturate the tissues with fire-polished Pasteur pipettes in 2 ml of CMF-HBSS containing 0.025% DNAase I, 0.4 mg/ml soybean trypsin inhibitor, 3 mg/ml BSA, and 12 mM $MgSO_4$.
6. Centrifuge the cell suspension at 160×g for 5 min at 4°C, and then wash the cells with 5 ml of CMF-HBSS.
7. Resuspend the cells in culture medium consisting of a 1:1 mixture of Dulbecco's modified Eagle's medium and F12 medium containing 2% B-27 supplement (Gibco) (DF/B-27 medium).

Comment

The cells obtained from postnatal day-7 rat cerebella are mainly granule cells.

Cell Migration Assay on Glass Fibers

1. Autoclave Whatman GF/A glass fiber filters, and then break them down by vortexing vigorously in distilled water.
2. Coat the glass fibers with 7 μg/ml poly-L-lysine (Mr > 300 × 10^3) for 1 h at room temperature.

Department of Developmental Neuroscience, Tokyo Metropolitan Institute for Neuroscience, 2-6 Musashidai, Fuchu, Tokyo 183-8526, Japan
Phone: +81-42-325-3881, Fax: +81-42-321-8678
E-mail: maedan@tmin.ac.jp

3. Wash the glass fibers three times with distilled water.
4. Coat the glass fibers with 30 µg/ml laminin or pleiotrophin diluted in 5 mM Tris-HCl, pH 8.0 for 2 h at room temperature.
5. Wash the glass fibers with phosphate-buffered saline (PBS) three times and then with DF/B-27 medium once.
6. Add the glass fibers in 200 µl of DF/B27 medium to the wells of a 48-well plate, pre-coated with 20 µg/ml poly-L-lysine.
7. Add the neurons (50,000 cells in 20 µl of DF/B-27 medium) to each well.
8. Incubate the plates for ~15 h at 37°C under 5% CO_2.
9. Monitor cell migration by time-lapse recording. Fields containing neurons bound to the glass fibers should be randomly selected, and their images recorded at 5-min intervals for 2 h (Maeda and Noda 1998).

Comment

The function of chondroitin sulfate proteoglycans may be investigated by adding chondroitinase ABC (Seikagaku Corp.), various chondroitin sulfate preparations and function-blocking antibodies against proteoglycans to the culture medium.

Boyden Chamber Cell Migration Assay

1. Spot 12 µl of 30 µg/ml laminin or pleiotrophin diluted in 5 mM Tris-HCl, pH 8.0 on parafilm.
2. Coat the undersurface of the polycarbonate membrane of the Transwells (Costar; tissue culture treated, 6.5 mm diameter, 3-µm pores) with the protein solution on parafilm for 2 h at room temperature. Coating of the membrane should be done in a moisture chamber.
3. Wash both sides of the membrane of the Transwells three times with PBS.
4. Place the membrane in the wells of a 24-well plate containing 0.5 ml of DF/B-27 medium.
5. Add the neurons (100,000 cells in 0.2 ml DF/B-27 medium) to the upper chamber and incubate for 20 h at 37°C under 5% CO_2.
6. Fix the cells with 4% paraformaldehyde/0.1 M sodium phosphate, pH 7.4 for 10 min at room temperature.
7. Wipe the upper surface of the membrane with a cotton-tip applicator to remove non-migratory cells.
8. Stain the migrated cells on the undersurface of the membrane with 1% crystal violet, 5% ethanol, 0.1 M borate, pH 9.0 for 30 min.
9. Wash the membrane with PBS three times.
10. Cut the membrane and mount it between two glass slides with 50% glycerol.
11. Count the number of nuclei of migrated cells per microscopic field.

Comment

The cells that translocated the nuclei from the upper side to the undersurface of the membrane are regarded as migrated cells. Cell processes without nuclei should not be counted as migrated cells (Maeda and Noda 1998; Qi et al. 2001). Various reagents such as chondroitinase ABC may be added to the culture medium in both or either of the chambers.

Neurite Extension Assay

1. Incubate glass coverslips (13 mm in diameter) in solution containing 0.002% poly-L-lysine ($M_r > 300 \times 10^3$) for 1 h at room temperature.
2. Wash the coverslips with distilled water five times and air-dry.
3. Add neurons (4,000 cells in 40 µl DF/B-27 medium) to the cover slips in a 24-well culture plate, and incubate the plate for 1 h at 37°C under 5% CO_2.
4. Add 0.5 ml of DF/B-27 medium per well, and incubate the plate for 20–50 h at 37°C under 5% CO_2.
5. Wash the coverslips with PBS once, and then fix the cells with 4% paraformaldehyde/0.1 M sodium phosphate, pH 7.4 for 15 min.
6. Wash the coverslips with Tris-buffered saline three times, and then permeabilize the cells with 0.2% Triton X-100/PBS for 30 min.
7. Block the cells with 2% BSA/4% goat serum in PBS for 30 min.
8. Incubate the coverslips in 0.05% BSA/PBS solution containing rabbit anti-MAP2 antiserum (Chemicon, 1:1,000) and mouse anti-phosphorylated neurofilament SMI 312 (Covance, 1:500) for 2 h at room temperature.
9. Wash the coverslips with PBS three times, and then incubate them in the 0.05% BSA/PBS solution containing Alexa Fluor 488 goat anti-rabbit IgG (Molecular Probes, 1:100) and Alexa Fluor 594 goat anti-mouse IgG (Molecular Probes, 1:100) for 1 h at room temperature.
10. Wash the coverslips with PBS three times, and then mount them in Vectashield (Vector).
11. Observe the cells with a fluorescence microscope.

Comment

Anti-MAP2 and SMI 312 stain dendrites and axons, respectively; however, at the initial stage, all neurites are stained by both antibodies (Maeda et al. 1996). Various reagents such as chondroitin sulfate and purified proteoglycans may be coated on the poly-L-lysine-treated coverslips before culture.

Organotypic Slice Culture of Cerebellum

1. Dissect cerebella from 9-day-old Wistar rats, and cut the vermal region parasagittally into ~600-µm-thick slices in CMF-PBS.
2. Mount the slices on a collagen-coated porous Nuclepore polycarbonate membrane (Whatman, 2-µm pores) floated at the interface between the air and culture medium in a culture dish (Fig. 1). The culture medium consists of 15% heat-inactivated horse serum, 25% Earle's balanced salt solution, 60% Eagle's basal medium, 5.6 g/l glucose, 3 mM L-glutamine, 5 µg/ml bovine insulin, 5 µg/ml human transferrin, 30 nM sodium selenite, 20 nM progesterone, 1 mM sodium pyruvate, 50 U/ml penicillin, and 100 µg/ml streptomycin.
3. Incubate the culture dish for 6 days at 33°C under 5% CO_2.
4. Fix the slices with 4% paraformaldehyde/0.1 M sodium phosphate, pH 7.4 for 20 min.
5. Freeze the slices in liquid nitrogen, and then section them at 12 µm on a cryostat.
6. Treat the sections for immunohistochemistry as described above.
7. Obtain fluorescent images using a confocal laser scanning microscope.

Fig. 1 Organotypic slice culture system of cerebellum

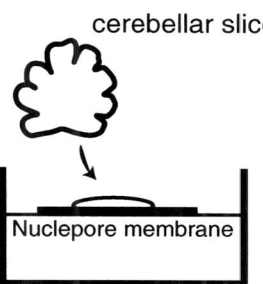

Comment

Since many neurons degenerate in the bottom part of the slices, sections from the top half (air side) should be analyzed (Tanaka et al. 2003). Reagents such as chondroitinase ABC and chondroitin sulfate may be added to the culture medium after overnight culture.

References

Maeda N (2007) A chondroitin sulfate proteoglycan, PTPζ/phosphacan, and neuronal network formation. In: Weiss ML (ed) Neuronal network research horizons. Nova Science Pub, New York, pp 179–203

Maeda N, Noda M (1996) 6B4 proteoglycan/phosphacan is a repulsive substratum but promotes morphological differentiation of cortical neurons. Development 122:647–658

Maeda N, Noda M (1998) Involvement of receptor-like protein tyrosine phosphatase ζ/RPTPβ and its ligand pleiotrophin/heparin-binding growth-associated molecule (HB-GAM) in neuronal migration. J Cell Biol 142:203–216

Qi M, Ikematsu S, Maeda N, Ichihara-Tanaka K, Sakuma S, Noda M, Muramatsu T, Kadomatsu K (2001) Haptotactic migration induced by midkine. - Involvement of protein-tyrosine phosphatase ζ, mitogen-activated protein kinase, and phosphatidylinositol 3-kinase. J Biol Chem 276:15868–15875

Tanaka M, Maeda N, Noda M, Marunouchi T (2003) A chondroitin sulfate proteoglycan PTPζ/RPTPβ regulates the morphogenesis of Purkinje cell dendrites in the developing cerebellum. J Neurosci 23:2804–2814

Processing of Glycosyltransferases by Alzheimer's β-secretase (BACE1)

Shinobu Kitazume[1,2], Shou Takashima[1,2], Yasuhiro Hashimoto[1,2,3]

Introduction

Majority of glycosyltransferases for oligosaccharide biosynthesis are retained in endoplasmic reticulum or the Golgi apparatus. Some of the glycosyltransferases are subsequently cleaved by proteases, and then secreted out of the cell. Indeed, many glycosyltransferases exist as soluble forms in bodily fluids. We previously reported that Alzheimer's β-secretase (BACE1) is involved in the cleavage and secretion of ST6Gal I, representing the first identification of a protease that plays a role in the secretion of glycosyltransferases (Kitazume et al. 2001). With the expectation that other sialyltransferases and glycosyltransferases might also be BACE1 substrates, we started on screening a series of sialyltransferases for possible BACE1 substrates. BACE1 was overexpressed in cultured cells and its effect on secretion of sialyltransferases was examined (Kitazume et al. 2006).

Method

1. COS-7 cells were maintained in DMEM containing 10% fetal bovine serum. The cells were plated on 150-mm tissue culture dishes and grown in a 5% CO_2 incubator at 37°C until they reached 50–70% confluence. An expression vector pSVL that carried a FLAG-tagged sialyltransferase (ST FLAG-pSVL) was co-transfected with either human BACE1-pcDNA3.1 cDNA or control pcDNA vector alone by the Lipofectin or Fugene method using Opti-MEM I. Following transfection, the cells were further cultured for 48 h.

2. Soluble secreted ST-FLAG was pulled down from the media with M2-agarose (Sigma). The soluble ST-FLAG and a membrane-bound form of ST-FLAG in cell lysates (40 μg of proteins) were subjected to 4–20% gradient SDS-PAGE, and then transferred to a nitrocellulose membrane. The membrane was sequentially incubated with an anti-FLAG polyclonal antibody (1:1,000) and a horseradish peroxidase (HRP)-goat anti-rabbit IgG as a secondary antibody (MP Biomedicals, Irvine, CA, USA). FLAG-tagged proteins were visualized with a chemiluminescent substrate (Pierce). The signals were quantified using a Luminoimage Analyzer LAS-1000 PLUS (Fuji, Tokyo, Japan) (Kitazume et al. 2005).

---

[1] Glyco-chain Functions Laboratory, Supra-biomolecular System Group, Frontier Research System, RIKEN, 2-1 Hirosawa, Wako, Saitama 351-0198, Japan

[2] CREST, Japan Science and Technology Agency, Kawaguchi, Saitama 560-0082, Japan

[3] Department of Biochemistry, School of Medicine, Fukushima Medical University, Fukushima 960-12, Japan
Phone: +81-48-467-9613, Fax: +81-48-462-4690
E-mail: yasua@riken.jp

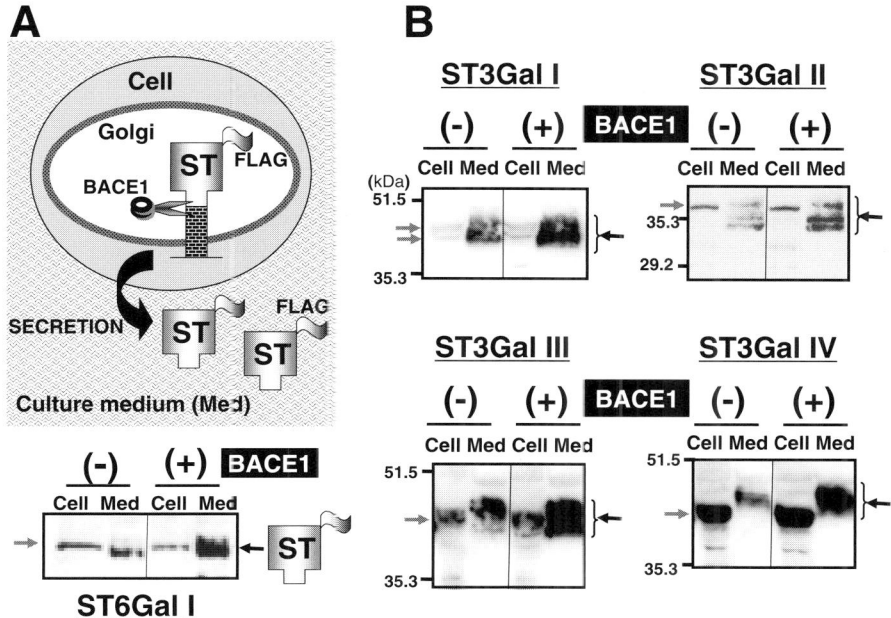

Fig. 1 A Cell-based assay for screening BACE1 substrates, e.g., FLAG-tagged ST6Gal I secretion is enhanced by BACE1 overexpression. B Each ST3Gal enzyme (ST3Gal I, II, III, or IV)-FLAG-pSVL was co-transfected into COS cells together with BACE1myc-pcDNA or an empty vector. The membrane-bound forms of ST-FLAG in the cell lysates (Cell) and soluble secreted forms of ST-FLAG in the media (Med) are indicated by *gray* and *black arrows*, respectively

Results

Secretions of ST3Gal family enzymes (ST3Gal I, II, III, and IV) were examined by cell-based assay (Fig. 1A) because some of these enzymes are found as soluble forms in extracellular fluids. As shown in Fig. 1b, overexpression of BACE1 significantly enhanced the secretions of all the ST3Gal enzymes, although the secretion enhancement was less than that of ST6Gal I. It was also noted that, without co-transfection of BACE1, soluble forms of all ST3Gal enzymes were detected, suggesting that the endogenous BACE1 is responsible for the secretion. Unlike the case of ST6Gal I, however, we did not detect direct cleavage of protein A-ST3Gal enzymes by purified BACE1-Fc in vitro (Kitazume et al. 2006). The result suggests that BACE1 does not cleave ST3Gal proteins directly but it indirectly affects ST3Gal secretion via another mechanism.

Comment

We speculate that BACE1 may activate a protease or proteases that are responsible for the cleavage and secretion of ST3Gal proteins. Alternatively, BACE1 could inactivate machinery for retention of ST3Gal proteins in the Golgi apparatus. The latter hypothesis appears to be supported by recent findings, in which a soluble form, but not membrane-bound form, of heparan sulfate 6-O-sulfotransferase is retained in the Golgi apparatus and the retention is abolished by overexpression of BACE1 in cultured cells (Nagai et al. 2007).

Acknowledgments This work was supported by NEDO of Japan. Kitazume was the recipient of grants from MEXT of Japan (Nos. 17046025 and 18570141). Hashimoto was the recipient of a grant from MHLW of Japan (No. 2006-Nanchi-Ippan-017).

References

Kitazume S, Tachida Y, Oka R, Shirotani K, Saido TC, Hashimoto Y (2001) Alzheimer's β-secretase, β-site amyliod precursor protein-cleaving enzyme, is responsible for cleavage secretion of a Golgi-resident sialyltransferase. Proc Natl Acad Sci USA 98:13554–13559

Kitazume S, Nakagawa K, Oka R, Tachida Y, Ogawa K, Luo Y, Citron M, Shitara H, Taya C, Yonekawa H, Pulson JC, Miyoshi E, Taniguchi N, Hashimoto Y (2005) In vivo cleavage of α2,6-sialyltransferase by Alzheimer's β-secretase. J Biol Chem 280:8589–8595

Kitazume S, Tachida Y, Oka R, Nakagawa K, Takashima S, Lee YC, Hashimoto Y (2006) Screening a series of sialyltransferases for possible BACE1 substrates Glycoconjugate J 23:437–441

Nagai N, Habuchi H, Kitazume S, Toyoda H, Hashimoto Y, Kimata K (2007) Regulation of heparan sulfate 6-*O*-sulfation by β-secretase activity. J Biol Chem M610691200; 282:14942–14951

Chondroitin Sulfate Proteoglycans: A Key Substance for Central Nervous System Injury Repair

Atsuhiko Oohira

Introduction

Research on proteoglycans originally advanced using connective tissues such as cartilage as a source. In the 1990s, it was found that multiple species of proteoglycans, both heparan sulfate proteoglycans (HSPGs) and chondroitin sulfate proteoglycans (CSPGs), existed in the central nervous system (CNS). Of these proteoglycans, some such as neurocan, phosphacan/RPTPβ, and neuroglycan C are predominantly expressed in the CNS. Since the CNS contains a very small amount of fibrous collagen, if any, proteoglycans are the major glycoconjugate in the microenvironment of neural cells.

Expression of an individual proteoglycan is speciotemporally regulated in the CNS, suggesting that proteoglycans play some roles in various cerebral areas and at various phases of cerebral development. In fact, evidence has been accumulated to show that neural proteoglycans are involved in a variety of cellular events including proliferation, adhesion, migration, differentiation, neurite elongation, and neuronal plasticity (Oohira 2007). In many cases, the glycosaminoglycan (GAG) moiety, rather than the core protein moiety, participates in the regulation of these developmental events of neural cells. This may be the reason why gene-targeting of a single proteoglycan core protein does not generally produce any marked CNS abnormalities: GAGs of other proteoglycan species compensate for the defective function of the knocked-out molecule.

When the CNS is injured, neuronal cells at the lesion site die and glial cells such as microglia, oligodendrocyte precursors, and astrocytes appear sequentially at and around the lesion site. Within a few days of lesion formation, astrocytes become the major cell type around the lesion site, and they actively synthesize and accumulate CSPGs in the extracellular matrix. The CSPG-rich glial tissue, the so-called glial scar, is considered to prevent the expansion of neuronal degeneration and aberrant innervation in the injured brain. However, it is also true that CSPGs are one of the major inhibitory molecules for axon regeneration in the CNS.

Principle of CNS Injury Repair

There are three phases in injury repair of the CNS (Fig. 1). In the first phase, neuronal cell death should be prevented. The concentration of glutamate, a typical excitatory neurotransmitter, transiently increases around the lesion site immediately following

---

Research Complex for the Medicine Frontiers, Aichi Medical University, Nagakute-cho, Aichi 480-1195, Japan
Phone: +81-52-264-4811, Fax: +81-561-61-1896
E-mail: atsu48@aichi-med-u.ac.jp

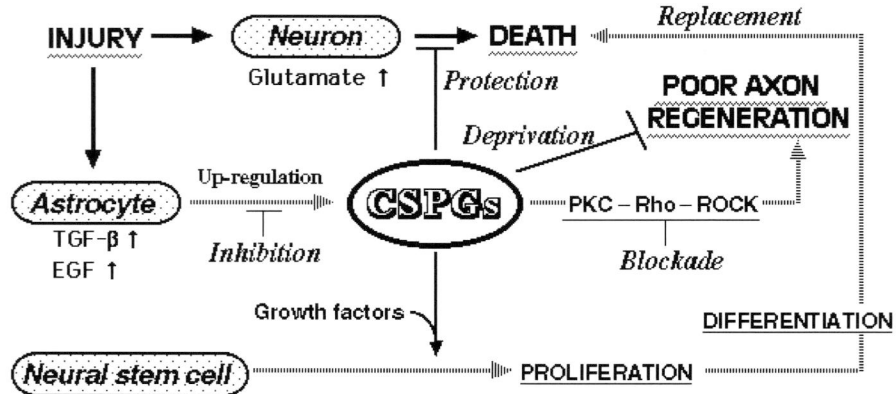

Fig. 1 CSPGs are a key substance for injury repair of the central nervous system. Some preparations of CS/CSPG protect neurons from glutamate-induced cell death, and promote proliferation of neural stem cells via interaction with heparin-binding growth factors. CSPGs up-regulated by astrocytes after injury prevent axonal regeneration through the Rho/ROCK signal transduction pathway. Deprivation of CSPGs from the lesion site and blockade of the signal pathway induce and promote axon regeneration

cerebral injury, and this leads to excitatory neuronal cell death. The excitatory death of neurons occurs in various neuronal diseases, as well as injury to the CNS, so that protection of neurons from glutamate-excitotoxicity is a common therapeutic step for neuronal degeneration.

In the second phase, degenerated neural cells should be replaced with functional cells by either cell transplantation or activation of endogenous neural stem cells. Bone marrow and cord blood would be practical sources of cells for transplantation because they contain a cell population that can differentiate into some neural cell types, including neurons.

In the third phase, a functional neural network should be reconstructed by either surviving cells or transplanted cells. It is well known that although neurons in the CNS have the potential to regenerate axons in themselves, axon regeneration is very poor and almost impossible in the CNS. This is attributable to various inhibitory molecules present in the microenvironment of the CNS, so that remodeling of the inhibitory microenvironment or deprivation of inhibitory molecules is necessary for induction and promotion of axon regeneration in the CNS.

Application of CSPG-Related Materials to CNS Injury Repair

Since CS is the major substance in the surroundings of neuronal cells, it is likely that a polysaccharide protects neurons from external harmful stimuli. Excitatory cell death can be induced in primary cultures of neurons by adding glutamate or glutamate-receptor agonists such as NMDA, and using this in vitro culture system, some preparations of CS and CSPG have been shown to support survival of neuronal cells. Recently, it was shown that polysialic acid also enhanced neuronal survival from glutamate-excitotoxicity, probably via inhibition of glutamate receptor activity. Negatively charged polysaccharides on cell surfaces may have a neuroprotective activity. Surprisingly and unexpectedly, CS-derived disaccharides produced by digestion with a CS lyase have been demonstrated to protect neurons in several in vivo rodent models of CNS neurodegeneration.

Neural stem cells that have self-renewal and multidifferentiation activities exist not only in the embryonic but also in adult brains, and are a promising material for cell supplementary therapy to treat neuronal degeneration in the CNS. Proliferation and differentiation of neural stem cells are regulated by both intrinsic factors such as transcription factors and extrinsic factors present in the cell environment. Some heparin-binding growth factors such as FGF-2, midkine, and pleiotrophin serve as extrinsic factors through interactions with HSPGs on the cell surface. It has been shown that CS with particular structural features also has a high affinity to these growth factors. Immunohistochemical studies revealed that several CSPGs such as neurocan, phosphacan/RPTPβ, neuroglycan C, and NG2 existed in the milieu of neural stem cells (Ida et al. 2006). CS preparations purified from developing rat brains actually promoted FGF-2-mediated proliferation of neural stem cells in culture, suggesting that interaction of CSPGs with some heparin-binding growth factors is implicated in the regulation of physiological events of neural stem cells.

Substrata containing CS at a low concentration promote neurite elongation, but those containing CS at a high concentration conversely inhibit neurite extension from cultured neurons (Oohira et al. 2000). Accordingly, efforts to reduce the concentration of CS around lesion sites in the CNS have been made to promote axon regeneration. Three trials outlined below have so far shown effective for regeneration of axons, even functional recovery, in experimental models of CNS injuries, mostly spinal cord injuries (Matsui and Oohira 2004): first, enzymatic removal of CS; second, inhibition of biosynthesis of CS/CSPGs; and third, blockade of the intracellular pathways of CS inhibitory signals (Fig. 1). Protease-free chondroitinase ABC supplied by Seikagaku Corporation, Tokyo, Japan, is often used to remove CS from glial scars. Decorin, a natural antagonist of TGF-β receptor, and a DNA enzyme designed to target a glycosyltransferase necessary for CS synthesis can be administered to lesion sites to reduce biosynthesis of CS/CSPGs. Since inhibitory signals for neuritogenesis are transduced intracellularly through the Rho/ROCK pathway, the inhibitors of this pathway such as Y-27632 are effective for promoting axon regeneration in the CNS.

Future Perspectives

Several lines of in vitro experiments have demonstrated that CSPGs have activities applicable to treatment of neurodegeneration. There are no practical treatments for neurodegeneration in lesions and diseases of the CNS at present, but some trials based on the outcomes of studies on neural CSPGs have proved effective prevention of neurodegeneration and promotion of axon regeneration in some experimental model animals. Combined treatment of neural stem cells with CSPG-related materials or CS lyases would synergistically intensify the therapeutic effects. Thus, CSPGs are really a key substance in not only development but also CNS injury repair.

References

Ida M, Shuo T, Hirano K, Tokita Y, Nakanishi K, Matsui F, Aono S, Fujita H, Fujiwara Y, Kaji T, Oohira A (2006) Identification and functions of chondroitin sulfate in the milieu of neural stem cells. J Biol Chem 281:5982–5991

Matsui F, Oohira A (2004) Proteoglycans and injury of the central nervous system. Congenit Anom (Kyoto) 44:181–188

Oohira A (2007) Multiple species and functions of proteoglycans in the central nervous system. In: Kamerling H (ed) Comprehensive glycoscience: from chemistry to systems biology, vol 4. pp 297–322, Elsevier, Oxford

Oohira A, Kushima Y, Tokita Y, Sugiura N, Sakurai K, Suzuki S, Kimata K (2000) Effects of lipid-derivatized glycosaminoglycans (GAGs): a novel prove for functional analyses of GAGs, on cell-to-substratum adhesion and neurite elongation in primary cultures of fetal rat hippocampal neurons. Arch Biochem Biophys 378:78–83

Section VIII
Quality Control of Proteins

A Cytoplasmic Peptide: *N*-Glycanase and ER-Associated Degradation

Tadashi Suzuki

Introduction

Peptide: *N*-glycanase (PNGase) hydrolyzes the amide bond between proximal GlcNAc and the linkage Asn residue of *N*-linked glycopeptides/glycoproteins, releasing free oligosaccharides bearing *N,N'*-diacetylchitobiose structure at their reducing termini (after spontaneous release of ammonia). This enzyme has been widely used as a tool reagent for analyzing the structure and functions of *N*-linked glycan chains on glycoproteins. The occurrence of cytoplasmic PNGase activity has been reported in a wide variety of eukaryotes (Suzuki et al. 2002 and references therein). The cytoplasmic PNGase was found to be quite distinct in terms of enzymatic properties from the "reagent" PNGases of plant and bacterial origin. A gene encoding the cytoplasmic enzyme, *PNG1*, was first identified in *Saccharomyces cerevisiae* (Suzuki et al. 2000). This gene has highly conserved its orthologues and can be found throughout eukaryotes (Suzuki et al. 2002).

Extensive studies during the last decade have clearly established that eukaryotic cells have a so-called ER-associated degradation (ERAD) system for elimination of newly synthesized malfolded/unassembled proteins. In this system, misfolded/unassembled (glyco)proteins are first "dislocated" or "retrotranslocated" from the ER into the cytosol. A number of ERAD substrates are known to be deglycosylated by the cytoplasmic PNGase during the degradation process (Tanabe et al. 2006 and references therein).

PNGase Assay: Principle

PNGase-catalyzed deglycosylation results not only in the removal of glycan(s) from the core protein/peptide but also in introducing negative charge(s) into the core protein/peptide by converting glycosylated Asn residue(s) into Asp residue(s). Therefore, to conclusively prove the PNGase activity, these two drastic changes in the substrates should be monitored (Table 1). For glycopeptide substrates, paper chromatograpghy is often used to detect the release of glycan from peptide, whereas paper electrophoresis is convenient to detect the introduction of negative charges into the core peptide (Suzuki 2005) (Fig. 1). The substrate glycopeptide can be radiolabeled at the N-terminal α-amino group by reductive methylation using [^{14}C]formaldehyde (Seko et al. 1991) to establish the very sensitive PNGase assay. The detailed method for the preparation of glycopeptide, as well as two analytical assays (paper chromatography and paper electrophoresis) has been described in detail (Suzuki 2005 and references therein).

Glycometabolome Laboratory, Systems Glycobiology Research Group, Frontier Research System, RIKEN Institute, 2-1 Hirosawa, Wako, Saitama 351-0198, Japan
Phone: +81-48-467-9626, Fax: +81-48-467-9628
E-mail: tsuzuki_gm@riken.jp

Table 1 Assay methods normally utilized for PNGase assay (Suzuki 2005 and references therein; Tanabe et al. 2006 and references therein)

	Removal of N-glycan	Introduction of negative charge(s) into core peptide/protein
Glycopeptide substrate	Paper chromatography	Paper electrophoresis
Glycoprotein substrate	SDS-polyacrylamide gel	Isoelectric focusing

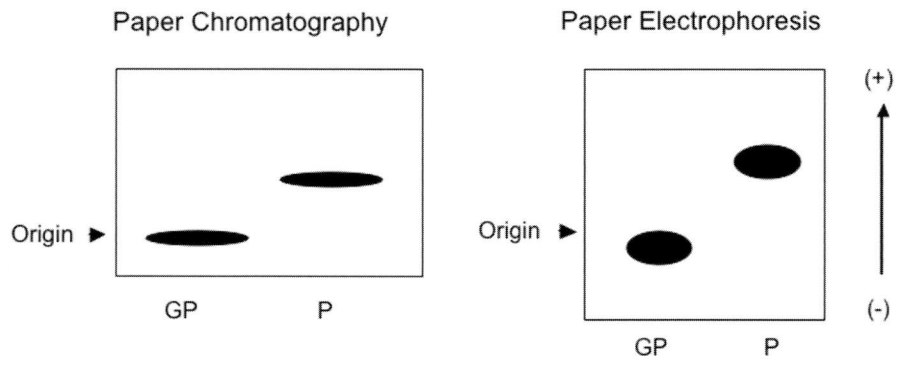

GP: glycopeptide
   ($[^{14}C]CH_3)_2$Leu-Asn(CHO)-Asp-Ser-Arg
P: deglycosylated peptide
   ($[^{14}C]CH_3)_2$Leu-Asn(CHO)-Asp-Ser-Arg

Fig. 1 Schematic representation of migration pattern of asialofetuin [^{14}C]glycopeptide I and its PNGase-deglycosylated product using two analytical methods (Suzuki 2005 and references therein)

The principle, i.e., two changes should be monitored to detect the PNGase activity, can also be applied to PNGase assay when glycoprotein substrates are used. This can be achieved by using SDS-PAGE for the detection of release of glycans from glycoproteins and isoelectric focusing (IEF) for the detection of charge changes (Tanabe et al. 2006).

For the detection of cytoplasmic PNGase activity using glycoprotein substrates, it has been shown that, although most of native glycoproteins are highly resistant to the cytoplasmic PNGase, extensively misfolded glycoprotein becomes susceptible to this enzyme. For in vitro deglycosylation analysis, S-alkylated bovine pancreatic ribonuclease B (RNase B; Sigma) turned out to be a good substrate (Suzuki 2005 and references therein).

Deglycosylation by the cytoplasmic PNGase seems a rate-limiting step in the ERAD reaction since the deglycosylated substrates can only be seen under conditions that the proteasome activity was compromised. Therefore, the inhibition of the proteasome activity is crucial for the detection of PNGase activity in vivo (Tanabe et al. 2006). In most cases SDS-PAGE was carried out for the detection of PNGase activity, but to distinguish the "deglycosylated form (Asp form)" from the "unglycosylated form (Asn form)", the confirmation of the negative charge introduction by isoelectric focusing is always desirable to demonstrate the PNGase activity. The detailed method for in vivo PNGase assay can be found elsewhere (Tanabe et al. 2006 and references therein). Theoretically, it is possible to assay the occurrence of deglycosylation reaction with Western blotting at steady state; however, pulse-chase experiments are always favored to confirm the deglycosylation reaction since it will allow one to detect the conversion of the glycosylated form to the deglycosylated form.

In Vivo PNGase Assay in *S. Cerevisiae* (Tanabe et al. 2006)

1. Prepare yeast cells harboring pRS315-PDG*Kar2-RTA*Δ*-(his)$_6$* plasmid.
2. Grow yeast cells in synthetic complete (SC) media lacking leucine.
3. Collect cells, prepare protein extract by a method of your choice (glass beads methods normally works fine).
4. Run SDS-PAGE gel; blot the gel to PDVF membrane for Western blotting.
5. Stain the gel with anti-(his)$_6$ antibody (Abcam, Cambridge, UK) or anti-RTA antibody (HyTest, Turku, Finland).

Comments

If cells have reduced/no PNGase activity, g1 form (glycosylated form) can be seen as a dominant form, whereas normally g0 form (unglycosylated/deglycosylated form) is seen as a major form. The difference is clearer for cells with stationary phase.

References

Seko A, Kitajima K, Inoue Y, Inoue S (1991) Peptide:*N*-glycosidase activity found in the early embryos of *Oryzias latipes* (Medaka fish): the first demonstration of the occurrence of peptide:*N*-glycosidase in animal cells and its implication for the presence of a de-*N*-glycosylation system in living organisms. J Biol Chem 266:22110–22114

Suzuki T (2005) A simple, sensitive in vitro assay for cytoplasmic deglycosylation by peptide:*N*-glycanase. Methods 35:360–365

Suzuki T, Park H, Hollingsworth NM, Sternglanz R, Lennarz WJ (2000) *PNG1*, a yeast gene encoding a highly conserved peptide:*N*-glycanase. J Cell Biol 149:1039–1052

Suzuki T, Park H, Lennarz WJ (2002) Cytoplasmic peptide:*N*-glycanase (PNGase) in eukaryotic cells: occurrence, primary structure, and potential functions. FASEB J 16:635–641

Tanabe K, Lennarz WJ, Suzuki T (2006) A cytoplasmic peptide:*N*-glycanase. Methods Enzymol 415:46–55

Ubiquitination of Glycoproteins in the Cytosol

Tadashi Tai

Introduction

The secretory proteins (membrane proteins, secretion proteins, and lysosomal enzymes) that are synthesized in the ribosome are translocated into the endoplasmic reticulum (ER). These nascent proteins are subjected to modification as the mature glycoproteins by the attachment of carbohydrates and move through the ER via the Golgi complex to their final destinations. The carbohydrate portions of glycoconjugates are implicated in many biological functions and are used as specific markers for cells. They are mainly expressed on the outside of cells. Carbohydrates, however, play an important role in the inside of cells (Helenius and Aebi 2001). N-linked glycoproteins are subjected to the quality control of glycoproteins through which aberrant proteins are distinguished from properly folded proteins and retained in the ER compartment. The quality control system includes the calnexin–calreticulin cycle, a unique chaperone system (Parodi 2000). Here, the author describes briefly the identification of N-glycan-binding proteins for E3 ubiquitin ligases and their perspectives in basic science as well as in clinical medicine.

N-Glycan-Binding Protein Assay

1. Prepare GlcNAc-terminated fetuin (GTF) by incubating asialofetuin (10 mg) with β-galactosidase (0.5 units) in 100 mM citrate/phosphate buffer (pH 6.5) at 37°C for 24 h.
2. Immobilize GTF to Affi-gel 15 (0.5 ml) by a manufacture procedure provided.
3. Homogenize 8-week-old ICR mouse brains in 10 volumes of TBS (20 mM Tris-HCl, pH 7.5, 150 mM NaCl) containing 5 mM $CaCl_2$ and protease inhibitor mixture.
4. Centrifuge the homogenate at 30,000 g for 30 min at 4°C.
5. Incubate the supernatant with GTF-immobilized beads for 18 h at 4°C.
6. Wash the beads with TBS containing 0.1% TritonX-100 (TBS-T).
7. Elute adsorbed proteins by 0.1 M chitobiose in TBS-T.
8. Separate the eluted proteins by SDS-polyacrylamide gel electrophoresis (SDS-PAGE), excise and digest with lysylendopeptidase, and determine their sequences by using a protein sequencer.

Comment

The eluted two proteins 45 and 23 kDa have been identified as Fbx2/NFB42 and Skp1, respectively, both of which are molecules associated with an ubiquitin ligase.

Seikagaku Corporation, Central Research Laboratories, 1253 Tateno 3-chome, Higashiyamato, Tokyo 207-0021, Japan
Phone: +81-42-563-5822, Fax: +81-42-563-5846
E-mail: tadashi.tai@seikagaku.co.jp

Principles

In an attempt to determine the biological roles of N-glycans in the central nervous system, mouse brain extracts were screened for proteins that bind to various sugar probes. We isolated an F-box protein, Fbx2 (NFB42/FBG1/Fbs1) that associated with the ubiquitin-proteasome system. We have found that (i) Fbx2 is expressed specifically on neuronal cells, (ii) Fbx2 binds to proteins attached to N-linked high-mannose oligosaccharide, (iii) pre-integrinβ1 is one of the Fbx2 target proteins, and (iv) the compound SCF^{Fbx2} may have a role in ER-associated degradation (ERAD) (Tai 2006). These findings clearly indicate that N-glycosylation acts as a targeting signal to eliminate intracellular glycoproteins in the cytosol.

It has recently been reported that Fbx2 belongs to a subfamily consisting of at least five homologous F-box proteins, Fbx6b (FBG2/Fbs2), Fbx30 (FBG3), FBG4, and FBG5 (Ilyn et al. 2002). We have determined that Fbx6b binds several N-linked glycoproteins, but other F-box proteins fail to bind any of the glycoproteins tested, and that Fbx2 is expressed only on mouse brain and testis, whereas Fbx6b is widely distributed in a variety of mouse tissues. Fbx6b protein exhibits 75% identity with Fbx2 (Tai 2006). These results suggest that there are a number of ubiquitin ligases that recognize N-glycans and that these ligases may have their specific functions in cells and tissues.

The ubiquitin-proteasome system is to degrade proteins by ATP-dependent manner, which differs from protease, a well-established protein degradation enzyme. Ubiquitin is a small protein containing 76-amino acids. Ubiquitin-mediated degradation of specific proteins has been found to contribute to the regulation of many cellular events, including maintenance of homeostasis such as stress responses and quality control of proteins and control of cell functions such as cell cycle, cell proliferation, and differentiation, apoptosis, signal transduction, control of transcription, and response of immune reaction (Hershko and Ciechanover 1998). Together with the ubiquitin-activating enzyme E1 and ubiquitin-conjugating enzyme E2, E3 ubiquitin ligases catalyze the ubiquitination of a variety of protein substrates for targeted-degradation via the 26S proteasome. E1 is a single molecule, whereas E2 and E3 are more than 10 and over 100 molecules, respectively. E3 ubiquitin ligases constitute a large protein family including SCF (Skp1, Cullin1, and F-box protein) (Fig. 1). F-box proteins are also diverse and pivotal in selecting target proteins for ubiquitination. Modification of target proteins is a prerequisite for the recognition by certain SCF-type E3 ubiquitin ligases and subsequent destruction. These include phosphorylation and proline hydroxylation.

Future Perspective

Abnormal protein aggregates have been detected in neuronal cells in common neurodegenerative disorders such as Alzheimer and Parkinson's diseases. Since most of these aggregates are poly-ubiquitinated, it is proposed that the failure of ERAD may be one of the major causes of the disorders. Several lines of evidence show that the ubiquitin ligase recognizing N-glycans, Fbx2, is expressed in neurons, especially in adult brain and may play a role for the quality control of glycoproteins. Thus, it is expected that further studies on Fbx2 and its homologues may develop and advance effective therapy for the degenerative disorders.

Unfolded or incompletely assembled proteins are degraded by the ERAD system. Some viruses also use the ERAD system for their protection from the host immune

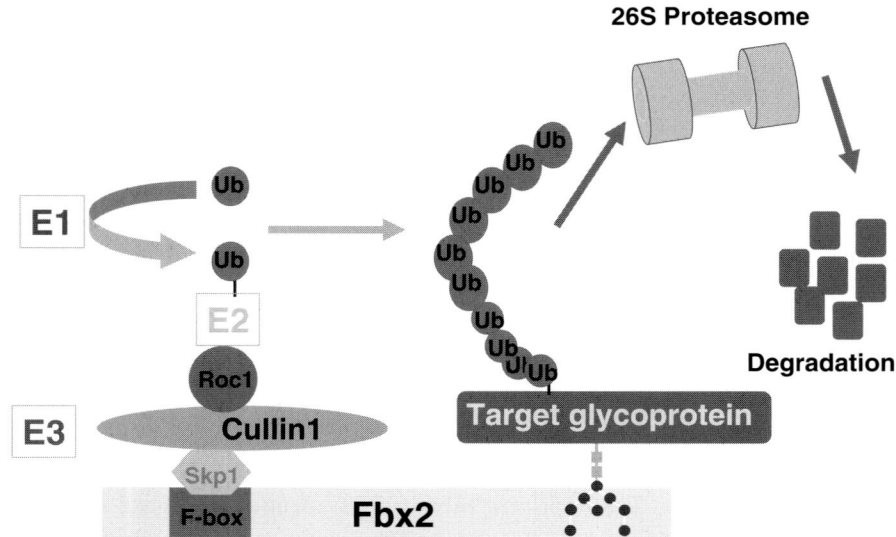

Fig. 1 A scheme of SCF complex-type ubiquitin ligase and the ubiquitin proteasome system. The ubiquitin system contains three enzymes, such as the activating enzyme, E1, the conjugating enzyme, E2, and the ligase, E3. The SCF-type E3 ubiquitin ligases consist of four molecules, such as Skp1, Cullin1, F-box protein, and Roc1. Fbx2, an F-box protein, recognizes N-linked high-mannose-type oligosaccharides. The compound catalyzes the ubiquitination of targeting protein substrate and polyubiquitinates a target protein that is degraded by the 26S proteasome

system. US1 and US2, both of which are the gene products of human cytomegalovirus, have functions of ERAD for heavy chain of MHC class 1. Similarly, mouse herpes virus, MK3, and human herpes virus, K3 and K5, induce the degradation of MHC class 1. Thus, the discovery of the ubiquitin ligases that recognize N-glycans is expected as a breakthrough for glycobiology and ubiquitin proteasome system as well as for the quality control of secretory proteins.

It has been shown that isolation and characterization of sugar-binding proteins are one of the most effective approaches for the analysis of biological functions in glycoconjugates. In fact, many sugar-binding proteins such as selectins and calnexins have been isolated and characterized in the past three decades. Further studies on sugar-binding proteins will be expected to reveal unknown and important functions of glycans in cells and tissues in the near future.

References

Helenius A, Aebi M (2001) Intracellular functions of N-linked glycans. Science 291:2364–2369
Hershko A, Ciechanover A (1998) The ubiquitin system. Annu Rev Biochem 67:425–479
Ilyn GP, Serandour AL, Pigeon C, Rialland M, Glaise D, Guguen-Guillouzo C (2002) A new subfamily of structurally related human F-box proteins. Gene 296:11–20
Parodi AJ (2000) Protein glycosylation and its role in protein folding. Annu Rev Biochem 69:69–93
Tai T (2006) Identification of N-glycan-binding proteins for E3 ubiquitin ligases. Methods Enzymol 415:20–30

Degradation of Misfolded Glycoproteins in the Endoplasmic Reticulum

Nobuko Hosokawa

Introduction

Secretory proteins and membrane proteins are synthesized in the endoplasmic reticulum (ER), and only correctly folded or properly assembled proteins are sorted to the early secretory pathway, whereas proteins that failed to obtain correct conformations are retained in the ER, followed by intracellular protein degradation named ER-associated degradation (ERAD). This cellular function is called ER quality control (ERQC) (Ellgaard et al. 1999; Trombetta and Parodi 2003) (Fig. 1). Most of the proteins synthesized in the ER are N-glycosylated, by the attachment of high mannose oligosaccharides ($Glc_3Man_9GlcNAc_2$) to the asparagine residues in the context of the consensus sequence of Asn-Xaa-Ser/Thr. The processing of the N-linked oligosaccharides plays an important role in the ERQC of glycoproteins. Removal and readdition of glucose at the terminus of the N-linked glycan contribute to the productive folding of glycoproteins in mammals, through the interaction with lectin chaperone calnexin and calreticulin. Trimming of the mannose residues, particularly from the middle branch of the N-linked glycan, is important for the entry into the ERAD pathway, but glycoproteins actually degraded seem to have $Man_{8-5}GlcNac_2$ glycans. We have cloned a mouse EDEM (ER-degradation enhancing α-mannosidase-like protein) (Hosokawa et al. 2001), and examined the function in glycoprotein ERAD.

Results

The mouse EDEM protein was cloned as a molecule whose mRNA expression was upregulated by ER stress caused by the accumulation of misfolded proteins in the ER (Hosokawa et al. 2001). This protein has a sequence similar to Class I α1,2-mannosidases (glycosylhydrolase family 47), but enzyme activity as a processing α-mannosidase was not detected in vitro. EDEM accelerates the ERAD of misfolded glycoproteins, but not that of non-glycosylated ERAD substrate. On the basis of these observations, EDEM is predicted to act as a lectin which discriminates misfolded glycoproteins destined for degradation, although its lectin activity is not experimentally proved. There is one homolog protein named as Htm1p/Mnl1p in yeast *Saccharomyces cerevisiae*, and recently, two mammalian EDEM homolog proteins are cloned and named as EDEM2 and EDEM3 (Fig. 2) (Hirao et al. 2006; Mast et al. 2005; Olivari et al. 2005). The amino acid identity among the glycosylhydrolase family 47 domain of EDEM1, 2, and 3 are approximately 45%, and it is about 33% between ER ManI and each of the three EDEM proteins. Both EDEM2 and EDEM3, similar to EDEM1, accelerate glycoprotein ERAD when

---

Department of Molecular and Cellular Biology, Institute for Frontier Medical Sciences, Kyoto University, 53 Kawahara-cho, Shogoin, Sakyo-ku, Kyoto 606-8397, Japan
Phone: +81-75-751-3849, Fax: +81-75-751-4646
E-mail: nobuko@frontier.kyoto-u.ac.jp

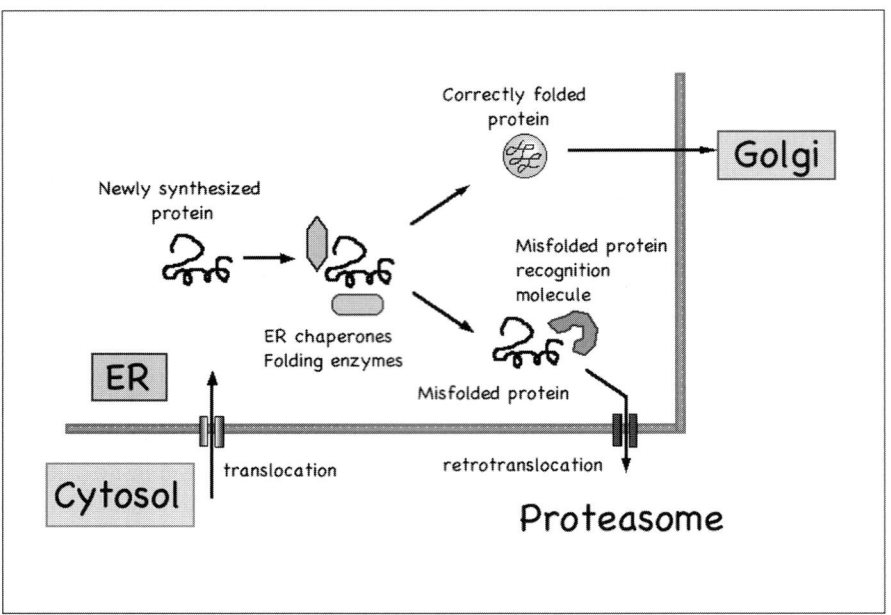

Fig. 1 Schema of ERQC and ERAD. Secretory and membrane proteins are synthesized in the ER, and by the assistance of folding enzymes and molecular chaperone proteins, nascent polypeptides acquire their mature conformation. Only correctly folded or properly assembled proteins are sorted out of the ER to the secretory pathway, whereas misfolded proteins are retained in the ER and subsequently degraded after retrotranslocation out of the ER, a mechanism known as ERAD

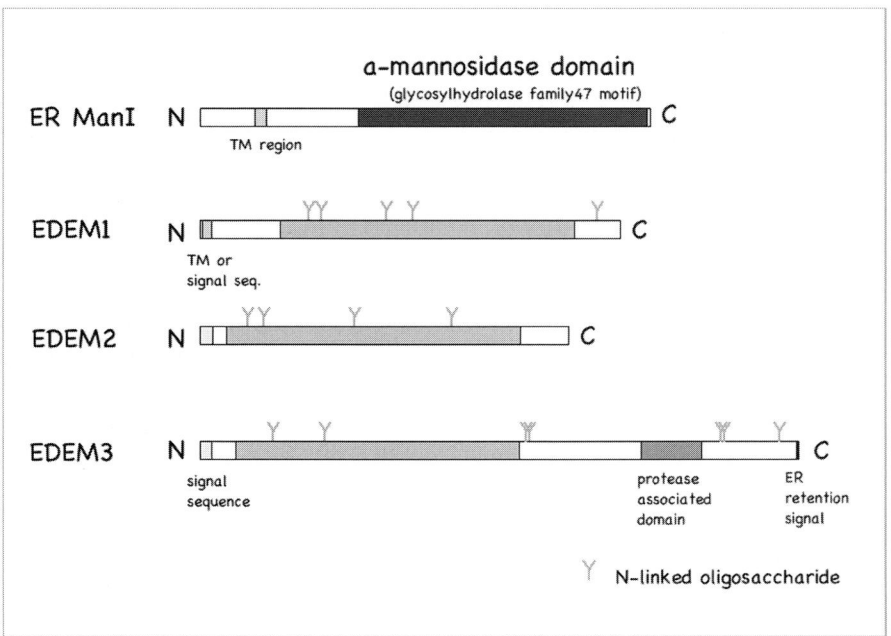

Fig. 2 Domain organization of ER ManI (ER α-mannosidase I) and EDEM proteins. Class I α1,2-mannosidases (glycosylhydrolase family 47) domains are *shaded*

transfected to mammalian cells. Functional analysis revealed that EDEM3 has a processing α-mannosidase activity in overexpressed cells, whereas EDEM2 does not. Furthermore, we have shown that the oligosaccharide structures on misfolded glycoproteins are different in cells overexpressing ER ManI, EDEM1, or EDEM3 (Hirao et al. 2006).

ER α-mannosidase I (ER ManI) is an enzyme which trims the mannose from the middle branch of the N-linked oligosaccharides, and this trimming acts as a timer for the misfolded glycoproteins sorted to the disposal pathway. Overexpression of ER ManI to cultured cells accelerates ERAD of glycoproteins by enhancing the mannose-trimming from the N-linked glycans. However, ER ManI does not discriminate the folding state of the substrates, i.e., ER ManI trims the mannose from correctly folded proteins as well as misfolded glycoproteins. Therefore, some molecules should discriminate the unfolded state of glycoproteins after ER ManI has trimmed the mannose. Recently, in yeast, Yos9p is reported to act as a lectin, discriminating the misfolded glycoproteins. The molecular mechanism how EDEM proteins act on the glycoprotein ERAD, as well as the detection and characterization of enzyme or lectin activity in vitro, is under investigation.

Pulse-Chase Analysis of ERAD Substrates

Transfection

1. Cells are plated on a dish approximately 16–24 h prior to transfection.
2. Plasmids encoding the ERAD substrates are transfected into cells using transfection reagent.
3. Approximately 24 to 48 h after transfection, cells are metabolically labeled.

Metabolic Labeling: [^{35}S]-Methionine/Cysteine

1. Cells are preincubated with medium lacking methionine/cystine supplemented with dialyzed fetal bovine serum for 20–30 min.
2. Pulse-label the cells with [^{35}S]-methionine/cysteine in fresh methionine/cystine-free medium.
3. Chase by replacing the medium to normal growth medium for the period desired.

Metabolic Labeling: [2-^{3}H]-Mannose

1. Cells are preincubated with medium containing 0.5–1 mM glucose for 0.5–1 h.
2. Pulse-label the cells with [2-^{3}H]-mannose in fresh medium containing 0.5–1 mM glucose.
3. Chase by replacing the medium to normal growth medium containing 4.5 g/l glucose (high glucose medium) supplemented with 5 mM mannose.

Immunoprecipitation

1. Cell lysates are mixed with specific antibodies and kept on ice for 1 h to overnight, depending on the antibodies.
2. Add Protein A- or Protein G-Sepharose beads and rotate at 4°C for 1–2 h.
3. Collect the beads by centrifugation, and wash with buffer containing high ionic strength.

4. Elute immunoprecipitates by boiling in Laemmli's buffer.
5. Electrophorese the immunoprecipitates on SDS-PAGE gel, and expose to phosphor image plate.

Comment

Following electrophoresis on SDS-PAGE gel, glycoproteins labeled with [2-^{3}H]-mannose can be blotted onto nylon membrane. The membrane is excised and N-glycans are released by EndoH digestion for further analysis of the N-linked oligosaccharide structures by HPLC.

References

Ellgaard L, Molinari M, Helenius A (1999) Setting the standards: quality control in the secretory pathway. Science 286:1882–1888

Hirao K, Natsuka Y, Tamura T, Wada I, Morito D, Natsuka S, Romero P, Sleno B, Tremblay LO, Herscovics A, Nagata K, Hosokawa N (2006) EDEM3, a soluble EDEM homolog, enhances glycoprotein endoplasmic reticulum-associated degradation and mannose trimming. J Biol Chem 281:9650–9658

Hosokawa N, Wada I, Hasegawa K, Yorihuzi T, Tremblay LO, Herscovics A, Nagata K (2001) A novel ER alpha-mannosidase-like protein accelerates ER-associated degradation. EMBO Rep 2:415–422

Mast SW, Diekman K, Karaveg K, Davis A, Sifers RN, Moremen KW (2005) Human EDEM2, a novel homolog of family 47 glycosidases, is involved in ER-associated degradation of glycoproteins. Glycobiology 15:421–436

Olivari S, Galli C, Alanen H, Ruddock L, Molinari M (2005) A novel stress-induced EDEM variant regulating endoplasmic reticulum-associated glycoprotein degradation. J Biol Chem 280:2424–2428

Trombetta ES, Parodi AJ (2003) Quality control and protein folding in the secretory pathway. Annu Rev Cell Dev Biol 19:649–676

Section IX
Golgi and Lysosomal Diseases

Retrograde Transport of Glycolipid-Bound Toxins

Ryo Misaki[1], Tomohiko Taguchi[2]

Introduction

Cholera toxin (CTX) produced by *Vibrio cholerae* is the virulence factor responsible for the massive secretory diarrhea. It belongs to the AB_5-subunit family of toxins. Five identical 11-kDa peptides (B-subunit) form a ring-like structure, through the hole of which A-subunit protrudes. The B-subunit is a lectin which binds to GM1 at the plasma membrane, whereas the A-subunit is an ADP-ribosyltransferase that can activate Gsα (heterotrimeric G protein). Other toxins, such as Shiga toxin (STX), bind other glycolipids, e.g., Gb3.

To exert toxic effect, CTX must cross lipid bilayer so that it can encounter Gsα in the cytosol. Remarkably, it is the endoplasmic reticulum (ER), not the plasma membrane, where the toxins penetrate into the cytosol.

Concept

It has recently been recognized that bacterial toxins exploit the retrograde membrane transport pathway equipped in the host cell. The first observation that the toxin moves retrogradely from the plasma membrane to the ER was made with STX (Sandvig et al. 1992). The drug brefeldin-A (BFA) was then found to inhibit the action of bacterial toxins. Although BFA is known to have many side effects, this result can be interpreted as the toxins traffic through the Golgi complex before reaching at the ER (Sandvig and van Deurs 2002). Recently, efforts have been made mainly on identifying the trafficking pathway from the plasma membrane to the Golgi complex. The group of Johannes found that STX passed through early endosomes before its arrival at the Golgi complex, but not via the pathway through late endosomes (Mallard et al. 1998).

The other advance has come from the recognition of other class of endosomes, "recycling endosomes", being the major sorting station for retro/anterograde trafficking (Ang et al. 2004). We have recently found that CTX/STX have to pass through recycling endosomes before their arrival at the Golgi complex. The overall retrograde traffic of toxins can then be outlined as "the plasma membrane, early endosomes, recycling endosomes, the Golgi complex, and the ER" (Fig. 1).

The next challenge would be summarized as follows. What can dictate toxins to go retrogradely to the ER? Do glycolipids play a role for this retrograde traffic all the way to the ER? What is the molecular mechanism underlying the sorting system operating

---

Department of Biochemistry, Osaka University School of Medicine, 2-2 Yamadaoka, Suita, Osaka 565-0871, Japan
Phone: +81-6-6879-3421, Fax: +81-6-6879-3429
E-mail: [1] misaki@biochem.med.osaka-u.ac.jp, [2] tom_taguchi@biochem.med.osaka-u.ac.jp

Fig. 1 The overall retrograde traffic of toxins. CTX moves retrogradely from the plasma membrane to the ER via endosomes (early endosomes and recycling endosomes) and the Golgi complex

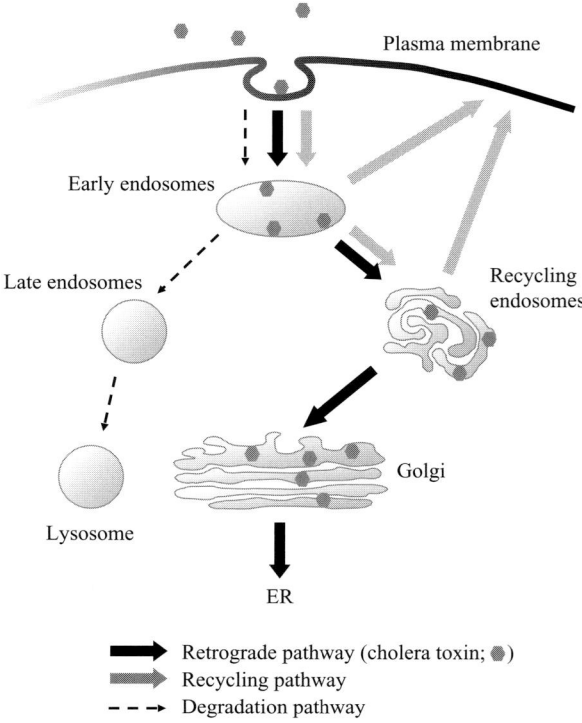

on each organelles (early endosomes, recycling endosomes)? How do glycolipids normally traffic without toxins?

Protocol for CTX Uptake Experiment

Cholera toxin B-subunit conjugated with Alexa fluorphore (CTxB-Alexa) is available from Invitrogen. COS-1 cells (green monkey kidney cells) are used in this protocol. All the experiments before cell fixation are performed at 37°C with 5% CO_2.

1. Culture cells on a cover slip or a bottom-glass dish suitable for subsequent fluorescent microscopy in Dulbecco's modified Eagle's medium (DMEM)/10% heat-inactivated fetal calf serum.
2. Add CTxB-Alexa (final concentration 1 μg/ml) to the medium and pulse for 5 min.
3. Chase for an appropriate time. CTxB can be usually found in recycling endosomes after 10–30 min and in the Golgi complex after 30—60-min chase.
4. Fix cells in 0.1 M sodium-phosphate buffer (pH 7.4) containing 4% paraformaldehyde. If necessary, indirect immunofluorescence protocol will be further applied so as to visualize organelle markers.

References

Ang AL, Taguchi T, Francis S, Folsch H, Murrells LJ, Pypaert M, Warren G, Mellman I (2004) Recycling endosomes can serve as intermediates during transport from the Golgi to the plasma membrane of MDCK cells. J Cell Biol 167:531–543

Mallard F, Antony C, Tenza D, Salamero J, Goud B, Johannes L (1998) Direct pathway from early/recycling endosomes to the Golgi apparatus revealed through the study of shiga toxin B-fragment transport. J Cell Biol 143:973–990

Sandvig K, van Deurs B (2002) Membrane traffic exploited by protein toxins. Annu Rev Cell Dev Biol 18:1–24

Sandvig K, Garred O, Prydz K, Kozlov JV, Hansen SH, van Deurs B (1992) Retrograde transport of endocytosed Shiga toxin to the endoplasmic reticulum. Nature 358:510–512

Degradation of Hyaluronan and Its Disorder

Masaki Yanagishita, Katarzyna Anna Podyma-Inoue

Introduction

Hyaluronan is distributed throughout extracellular matrices as an essential molecular component in animals (Hascall and Yanagishita 2007). It plays important biological roles in connective tissues, such as skin and articular cartilage, through its water-holding capacity and as a component of molecular scaffold. For example, hyaluronan is a major component of synovial fluid and is indispensable for the lubricating function, thus smooth mobility of the joint. It is also a major component of the vitreous fluid in the eye, maintaining necessary spherical shape of the eye. It has been demonstrated that the proper temporal and spacial production of hyaluronan is crucial during development for the proliferation, movement, and differentiation of cells and organogenesis. Hyaluronan is very actively metabolized in the body; of the total amount of approximately 15 g hyaluronan in a human with a body weight of 70 kg, about 5 g is degraded everyday, and the same amount replenished by biosynthesis (Stern 2004).

Concept

Hyaluronan is built by the repeating disaccharide structure $(4GlcA\beta 1\text{-}3GlcNAc\beta 1\text{-})n$. The number of repeats (n) is usually very large and varies widely, typically 10–40,000, making the size variation of hyaluronan between approximately 5,000 and 20,000,000 Da. Typical hyaluronan falls in the larger molecular size spectrum. The degradation of hyaluronan at a molecular level can be divided into two stages; first, by a group of enzymes called hyaluronidase (a family of endo-acetylhexosaminidases which are encoded by six genes in human, HAYL1, HYAL2, HYAL3, HYAL4, SPAM1, and PHYAL1, the last being a pseudogene) (Stern and Jedrzejas 2006). They cleave hyaluronan internally and generate fragments of varying sizes, theoretically down to as small as disaccharide size. The degradation occurs both extracellularly and intracellularly. An experimental protocol detecting hyaluronidase activity by zymography (Miura et al. 1995) is described in this chapter. Partial fragmentation of hyaluronan could also occur by non-enzymatic mechanisms involving free radicals. Fragments of hyaluronan are endocytosed through their binding to cell surface receptors, which include CD44 and Hyal-2 (one of the hyaluronidases, Fig. 1). Endocytosed hyaluronan fragments are eventually delivered to lysosomes and undergo the second-stage degradation by β-glucuronidase and β-acetylhexosaminidase (both are typical lysosomal exoglycosidases), finally releasing monosaccharides for reutilization. Detailed information on how hyaluronidases and exoglycosidases coordinate in hyaluronan degradation is not clear at this moment. Genetic defects of a

---

Biochemistry, Department of Hard Tissue Engineering, Graduate School, Tokyo Medical and Dental University, 1-5-45 Yushima, Bunkyo-ku, Tokyo 113-8549, Japan
Phone: +81-3-5803-5447, Fax: +81-3-5803-0187
E-mail: m.yanagishita.bch@tmd.ac.jp

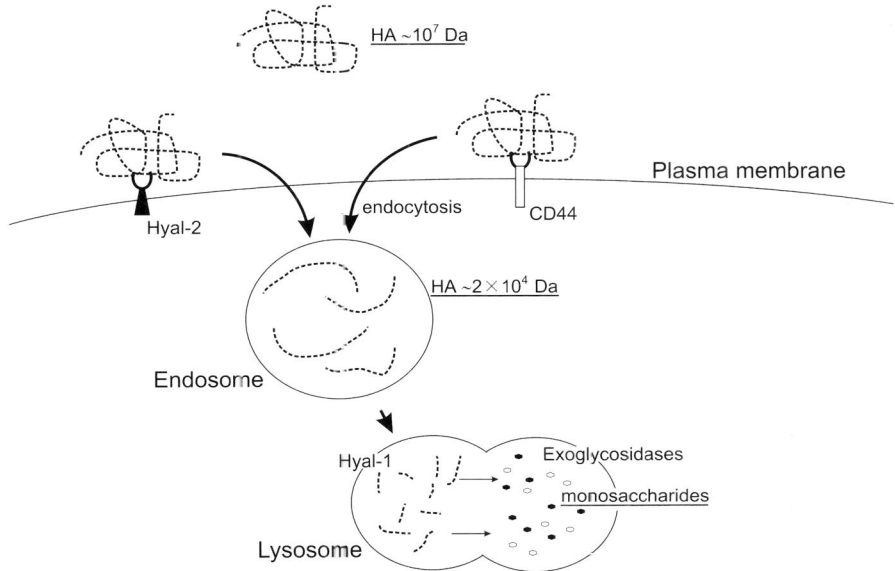

Fig. 1 Endocytosis and intracellular degradation of hyaluronan. Hyaluronan in the extracellular matrix (~10^7 Da) is endocytosed after binding to cell surface receptors such as CD44 and Hyal-2 (one of hyaluronidases). Details of endocytotic mechanisms are not known. However, Hyal-2 (a GPI-anchored protein) could use a mechanism involving caveola structure, which may be distinct from that used by internalization through CD44. Hyaluronan found in endosomes may have undergone fragmentation by actions of hyaluronidases. After being transported to lysosomes, hyaluronan is degraded by the concerted actions of Hyal-1 and exoglycosidases eventually, releasing component monosaccharides (i.e., glucuronic acid and N-acetylglucosamine)

hyaluronidase, β-glucuronidase or β-acetylhexosaminidase, leading to a classic lysosomal storage disease (Neufeld and Muenzer 2001) affecting hyaluronan degradation, have not been documented in human (Stern and Jedrzejas 2006); in the cases of hyaluronidase and β-acetylhexosaminidase, there are multiple enzymes with similar catalytic activity, and they could compensate the single enzyme defect. In the case of the genetic defect of β-glucuronidase, i.e., MPS VII, combined actions of hyaluronidases could, at least theoretically, degrade hyaluronan down to disaccharide size.

In contrast, enhanced degradation of hyaluronan is encountered in various clinical situations. Many of them are due to release of hyaluronidases from leukocytes in inflammatory processes. For example, degradation of hyaluronan in synovial fluid causes loss of lubricating function in arthritides, could lead to a severe damage to the cartilage tissue. For a therapeutic approach, intra-articular administration of a hyaluronan preparation could provide temporary relief from symptoms and some functional restoration, and has been used widely for the treatment of arthritides.

Recently, involvement of hyaluronan genes in various cancers, both as enhancers and suppressors, has been drawing attention. The elucidation of molecular mechanisms how they participate in the generation and progression of cancers will provide important tools for cancer prevention and treatment.

Protocol

Determination of hyaluronidase activity using hyaluronan-zymography:

1. Prepare the acrylamide gel ($8 \times 10 \times 0.1$ cm) of desired concentration impregnated with hyaluronan from rooster comb (final concentration 0.17 mg/ml).
2. Mix the samples with SDS-PAGE sample buffer containing no reducing reagent (20% glycerol, 63 mM Tris-HCl, pH 6.8, 2% SDS, 0.1% bromophenol blue) and apply directly on gel. Human serum (1 µl) can be used as a positive control for acidic hylauronidases.
3. Electrophorese at room temperature and at a constant current (20 mA/gel).
4. Wash the gel with 2.5% Triton X-100 for 1 h at room temperature.
5. Wash the gel with "incubation buffer" (e.g., 100 mM sodium formate, 30 mM NaCl, pH 4.0) for 10 min at room temperature.
6. Incubate the gel in "incubation buffer" for 16 h at 37°C.
7. Transfer the gel into a buffer (100 mM Tris-HCl, 20 mM NaCl, pH 8.0) containing 0.1 mg/ml pronase and incubate for 2 h at 37°C.
8. Fix the gel with 20% ethanol/10% acetic acid for 20 min at room temperature.
9. Stain the gel with 0.5% Alician Blue in 20% ethanol/10% acetic acid for 1 h at room temperature.
10. Destain the gel with 20% ethanol/10% acetic acid for 1 h at room temperature. Repeat this step once more.
11. Take a photograph and dry the gel.

Comment

The method allows the detection of hylauronidase activity in polyacrylamide gels after electrophoresis in native and SDS-containing buffers, providing information regarding the molecular weight of the enzyme, or optimum pH when buffers with varying pH are employed.

References

Hascall VC, Yanagishita M (eds) (2007) In: Hyaluronan today http://www.glycoforum.gr.jp/science/hyaluronan/hyaluronanE.html (Current in 2008)

Miura RO, Yamagata S, Miura Y, Harada T, Yamagata T (1995) Analysis of glycosaminoglycan-degrading enzymes by substrate gel electrophoresis (zymography). Anal Biochem 225:333–340

Neufeld EF, Muenzer J (2001) The mucopolysaccharidoses. In: Scriver CS et al (eds) The metabolic and molecular bases of inherited diseases, 8th edn. McGraw-Hill, New York, pp 3421–3452

Stern R (2004) Hyaluronan catabolism: a new metabolic pathway. Eur J Cell Biol 83:317–325

Stern R, Jedrzejas MJ (2006) Hyaluronidases: their genomics, structures, and mechanisms of action. Chem Rev 106:818–839

Recent Advances in Enzyme Replacement Therapy for Lysosomal Diseases

Kohji Itoh

Introduction

Lysosomal diseases are inherited metabolic disorders caused by gene defects of lysosomal enzymes and their cofactors, which are characterized by the accumulation of undegraded natural substrates, including glycoconjugates, in lysosomes. The patients develop quite heterogeneous and progressive manifestations, and most of them involve neurological symptoms. In recent years, enzyme replacement therapy (ERT) with recombinant lysosomal enzymes has been clinically available for several non-neurological diseases, such as Gaucher disease type 1, Fabry disease cardiac type, and Pompe disease, on the basis of the endocytotic mechanism through binding of the mannose-6-phosphate- and mannose residues-containing oligosaccharides attached to lysosomal enzymes to the cation-independent mannose-6-phosphate receptor (CI-M6PR) and mannose receptor (MR) as molecular targets in visceral organs of the patients as well as delivery to lysosomes. However, two major problems are present in the clinical application of the ERT for lysosomal diseases. One is the difficulty in producing a large amount of recombinant human enzymes inexpensively. The other is the inefficiency for neurological diseases because intravenously administered enzymes cannot be incorporated into the central nervous system (CNS) across the blood–brain barrier (BBB).

Research program "Development of recombinant lysosomal enzyme replacement therapy for brain diseases based on the specific functions of oligosaccharides" (2005–2009) in the JST CREST projects is proceeding to establish a massive production system of the human enzymes with human type-like oligosaccharides, which are expressed in the CHO cell line and the methylotrophic yeast mutants. Hyperfunctional recombinant enzymes are also designed in silico on the basis of amino acid substitutions to alter the subunit interaction as well as novel tag-fusion and conjugation techniques to deliver them across the BBB non-invasively into the CNS of the patients.

Concept

Tay-Sachs disease and Sandhoff disease are autosomal recessive GM2 gangliosidoses caused by the genetic defects of *HEXA* and *HEXB* encoding α- and β-subunits of the human β-hexosaminidase (Hex), respectively. These diseases are characterized by Hex isozyme deficiency and excessive accumulation of GM2 ganglioside (GM2) in the CNS,

---

Department of Medicinal Biotechnology, Graduate School of Pharmaceutical Sciences, The University of Tokushima, Sho-machi 1-78, Tokushima 770-8505, Japan;
CREST, JST, Chiyoda-ku, Tokyo 102-0075, Japan
Phone: +81-88-633-7290, Fax: +81-38-633-7290
E-mail: kitoh@ph.tokushima-u.ac.jp

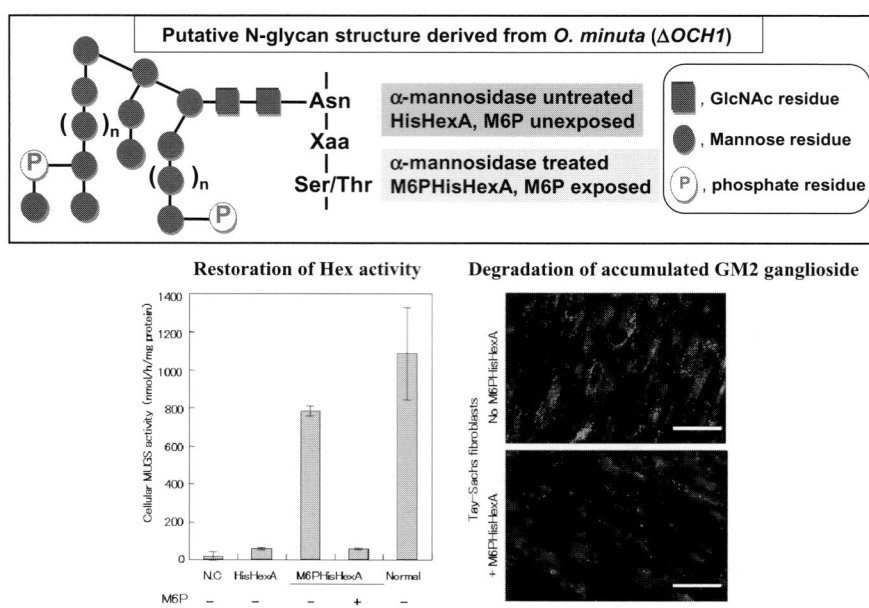

Scheme 1 Enzyme replacement effects of recombinant human HexA secreted from *O. minuta*

and the patients develop mainly neurological symptoms. Only the HexA isozyme (αβ heterodimer) can degrade GM2 in cooperation with GM2 activator protein.

To develop the ERT for these Hex deficiencies, the following approaches were examined (Sakuraba et al. 2006). (1) *HEXA* and *HEXB* were co-introduced into the CHO cells (Tsuji et al. 2005; Murata-Ohsawa et al. 2005; Itakura et al. 2006) and the methylotrophic yeast (*Ogataea minuta: Om*) mutant strain (Δ*OCH1*), and the stable transformants were established. The latter yeast strain produced and secreted a large amount of Hex isozymes (13 mg/l broth) with phosphomannosyl oligosaccharides (Scheme 1). The His-tagged HexA (HisHexA) was purified (1.9 mg), and then treated with α-mannosidase to expose the terminal mannose-6-phosphate residues in the oligosaccharides to obtain M6PHisHexA (0.83 mg). The purified M6PHisHexA was incorporated into the fibroblasts derived from patients with Tay-Sachs and Sandhoff diseases, and degraded GM2 accumulated in lysosomes (Scheme 1). A part of M6PHisHexA was delivered to the neonatal brains of Sandhoff disease model mice after intraperitoneal injection. (2) Introduction of *MNN4* gene encoding the positive regulator of α-2,6-mannosyltransferase into the *Om* strain stably expressing the human HexA was found to increase significantly the content of phosphomannosyl residues in the oligosaccharides and the therapeutic effects (Scheme 2). (3) Effects of fusion and conjugation of protein transduction domains (PTD) derived from Tat protein encoded by HIV gene and oligo-arginine sequences as well as brain-directed tag (peptide) sequences selected by phage display systems on the delivery across the BBB into the brain parenchyma are analyzed with Sandhoff disease model mice (Scheme 2).

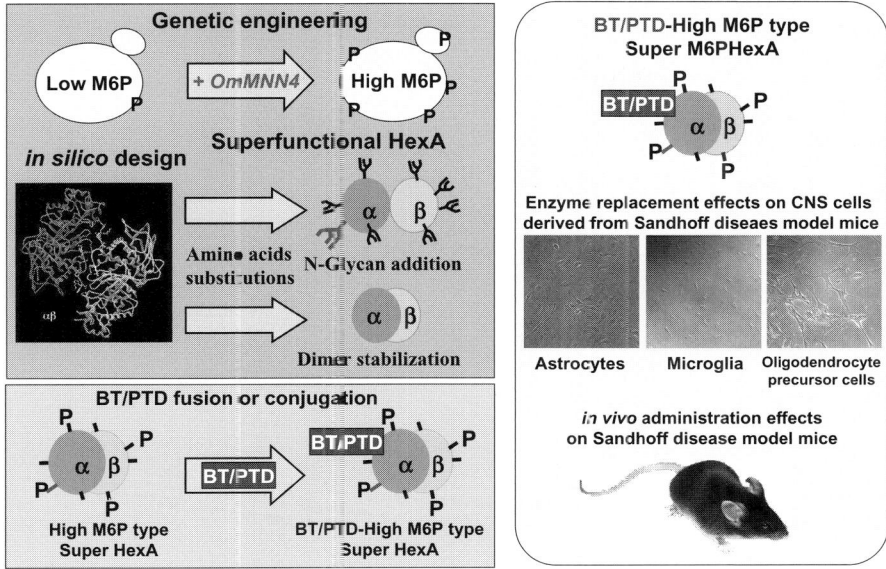

Scheme 2 Strategy for development of enzyme replacement therapy for Sandhoff disease

References

Itakura T, Kuroki A, Ishibashi Y, Tsuji D, Kawashita E, Higashine Y, Sakuraba H, Yamanaka S, Itoh K (2006) Inefficiency in GM2 ganglioside elimination by human lysosomal beta-hexosaminidase beta-subunit gene transfer to fibroblastic cell line derived from Sandhoff disease model mice. Biol Pharm Bull 29:1564–1569

Murata-Ohsawa M, Kotani M, Tajima Y, Tsuji D, Ishibashi Y, Kuroki A, Itoh K, Watabe, K, Sango K, Yamanaka S, Sakuraba H (2005) Establishment of immortalized Schwann cells from Sandhoff mice and corrective effect of recombinant human β-hexosaminidase A on the accumulated GM2 ganglioside. J Hum Genet 50:460–467

Sakuraba H, Sawada M, Matsuzawa F, Aikawa S, Chiba Y, Jigami Y, Itoh K (2006) Molecular pathologies and enzyme replacement therapies for lysosomal diseases. Curr Drug Targets Cent Nerv Syst Neurol Disord 5:401–413

Tsuji D, Kuroki A, Ishibashi Y, Itakura T, Itoh K (2005) Metabolic correction in microglia derived from Sandhoff disease model mice. J Neurochem 94:1631–1638

Section X
Infections

Developing an Assay System for the Hemagglutinin Mutations Responsible for the Binding of Highly Pathogenic Avian Influenza A Viruses to Human-Type Receptors

Yasuo Suzuki

Introduction

Influenza is one of the most widely distributed zoonotic infectious diseases in the world, and its pathogen, the influenza virus, is extremely mutable. Highly pathogenic H5N1 avian influenza viruses have become resident in poultry within Asia, Europe, Africa, and Middle East, and continue to pose a pandemic threat. The poultry infections in East and Central Asia and east Africa since 2004 have led to 380 confirmed cases of human H5N1 disease in 14 countries with 240 deaths until April 15, 2008 (WHO Web). However, the transmission of the virus from poultry to human remains inefficient and human-to-human transmission is so far, an infrequent event.

We found that the host cell membrane receptor for the influenza virus is based on the carbohydrate structures of sialyl lactosamine, sialic acid (Sia) α2-3 (6) galactose (Gal) β1-4 (3) N-acetylglucosamineβ1-. The hemagglutinin of avian and human influenza viruses distinguishes between the terminal Sia-Gal structure, (2-3, 2-6) of the sialyl lactosamine structure, and also the sialic acid molecular species such as N-acetylneuraminic acid (Neu5Ac) and N-glycolylneuraminic acid (Neu5Gc). In general, most avian influenza viruses bind preferentially to α2,3Sia (avian-type receptor), whereas human influenza viruses bind preferentially to α2,6Sia (human-type receptor). Switch from α2,3Sia to α2,6Sia receptor specificity is one of the most critical steps in the adaptation of avian viruses to human host, and a single amino acid substitution in the viral hemagglutinin (HA) changes the binding specificity to sialyl linkages (2-3) and (2-6), and resulted in the variation of host range, from birds to humans (Suzuki 2005, 2007).

Concept

The initial mutation of the highly pathogenic avian influenza viruses (H5N1) into the viruses which easily transmit among humans will primarily be an adaptation of receptor binding specificity of the H5N1 virus from α2,3Sia (bird type) to α2,6Sia (human type). A conversion from α2,3Sia to α2,6Sia recognition is believed to be one of the changes that must occur before avian influenza viruses can replicate efficiently in humans and acquire the potential to cause a pandemic. By identifying mutations in the receptor-binding hemagglutinin (HA) molecule that would enable avian H5N1 viruses to recognize human-type host cell receptors, it may be possible to predict (and thus to increase preparedness for) the emergence of pandemic viruses. Therefore, monitoring of the phenotypic mutation of virus receptor binding specificity which enable to ascertain the

Department of Biomedical Sciences, College of Life and Health Sciences, Chubu University, 1200 Matsumoto-cho, Kasugai, Aichi 487-8501, Japan

mutations of sialosugar chain receptor binding specificity of highly pathogenic avian viruses into the human type should be started globally. Recently, we developed several new assay systems for the receptor binding surveillance to detect the change of receptor binding specificity of the highly pathogenic avian influenza virus (H5N1) into human type. These methods are useful to detect the mutation of the host range of the highly pathogenic avian influenza virus (H5N1) into humans. Experimental conditions and several notable results have been published (Kobasa et al. 2004; Le et al. 2005; Yamada et al. 2006; Auewarakul et al. 2007).

References

Auewarakul P, Suptawiwat O, Kongchanagul A, Sangma C, Suzuki Y, Ungchusak K, Louisirirotchanakul S, Lerdsamran H, Pooruk P, Thitithanyanont A, Pittayawonganon C, Guo CT, Hiramatsu H, Jampangern W, Chunsutthiwat S, Puthavathana P (2007) An avian influenza H5N1 virus that binds to a human-type receptor. J Virol 81:9950–9955

Kobasa D, Takada A, Shinya K, Halfman P, Hatta M, Theriault S, Suzuki H, Nishimura H, Mitamura K, Sugaya N, Usui T, Murata T, Suzuki T, Suzuki Y, Feldman H, Kawaoka Y (2004) Enhanced pathogenicity of influenza A viruses possessing the haemagglutinin of the 1918 pandemic. Nature 431:703–707

Le MQ, Kiso M, Someya K, Sakai YT, Nguyen TH, Nguyen KHL, Pham ND, Ngyen HH, Yamada S, Muramoto Y, Horimoto T, Takada A, Goto H, Suzuki T, Suzuki Y, Kawaoka Y (2005) Isolation of drug-resistant H5N1 virus. Nature 437:1108

Suzuki Y (2005) Sialobiology of influenza: molecular mechanism of host range variation of influenza viruses. Biol Pharm Bull 28:399–408

Suzuki Y (2007) The highly pathogenic avian flu viruses and the molecular mechanism of the transmission of the viruses into humans. In: Kamerling H (ed) Comprehensive glycoscience from chemistry to systems biology, November issue. Elsevier Publishing book, Amsterdam pp 465–471

Yamada S, Suzuki Y, Suzuki T, Le MQ, Nidom CA, Sakai YT, Muramoto Y, Ito M, Kiso M, Horimoto T, Shinya K, Sawada T, Kiso M, Lin Y, Hay A, Haire LF, Stevens DJ, Russel RJ, Gambin SJ, Skehel JJ, Kawaoka Y (2006) Hemagglutinin nutations responsible for the binding of H5N1 influenza A viruses to human-type receptors. Nature 444:378–382

Helicobacter pylori Growth Assay for Glycans

Jun Nakayama

Introduction

Helicobacter pylori (*H. pylori*) is a Gram-negative microaerophilic bacterium isolated from the gastric mucosa of chronic gastritis patients (Warren and Marchall 1983). It is now well established that this microbe is associated with pathogenesis of gastric disorders such as chronic active gastritis, peptic ulcer, gastric cancer, and malignant lymphoma of the MALT type. Although *H. pylori* infects about half the world's population, only a fraction of infected patients develop severe gastric disease, suggesting the presence of a defense mechanism in the gastric mucosa itself. *H. pylori* colonizes surface mucous cells of the gastric mucosa and is rarely found in gland mucous cells such as mucous neck cells and pyloric gland cells located in deeper portions of the mucosa (Hidaka et al. 2001). *O*-glycans with terminal α1,4-linked GlcNAc residues [GlcNAcα1 $\rightarrow$ 4Galβ1 $\rightarrow$ 4GlcNAcβ1 $\rightarrow$ 6(GlcNAcα1 $\rightarrow$ 4Galβ1 $\rightarrow$ 3)GalNAcα $\rightarrow$ Ser/Thr] are unique to gland mucins secreted from gland mucous cells (Ishihara et al. 1996). Recently, we showed that α1,4-GlcNAc-capped *O*-glycans suppress *H. pylori* growth (Kawakubo et al. 2004). We further demonstrated that cholesterol-α-D-glucopyranoside (CGL), a major cell wall component of *Helicobacter* species, is important for *H. pylori* survival and that CGL biosynthesis was inhibited by exogenous α1,4-GlcNAc-capped *O*-glycans. More recently, we have successfully cloned cholesterol α-glucosyltransferase (CHLαGcT) from *H. pylori*, which is responsible for the CGL biosynthesis, and demonstrated that enzymatic activity of CHLαGcT is suppressed by a synthetic α1,4-GlcNAc-capped *O*-glycan (Lee et al. 2006). Here, I describe the procedure for a bacterial growth assay to test the effects of glycans on *H. pylori* growth.

H. Pylori Growth Assay Protocol

1. Streak a standard strain of *H. pylori* (ATCC43504; American Type Culture Collection, Manassas, VA, USA) from a frozen stock vial on a Trypticase Soy Agar II plate with 5% sheep blood (TSA II) (Nippon Becton Dickinson, Fukushima, Japan) and incubate at 35°C under 15% CO_2 with saturated humidity for 4–7 days.

2. Pick several colonies on the TSA II plate with a cotton-tipped swab and streak on another TSA II plate using the same culture conditions described for another 4 days.

3. Scrape all colonies appearing on the second plate using a cotton-tipped swab and suspend bacteria in 500 µl sterile saline. Then, inoculate bacteria in 50 ml of Brucella broth (Becton Dickinson Microbiology Systems, Sparks, MD, USA) supplemented

---

Department of Pathology, Shinshu University School of Medicine, Asahi 3-1-1, Matsumoto 390-8621, Japan
Phone: +81-263-37-3394, Fax: +81-263-37-2581
E-mail: jun@hsp.md.shinshu-u.ac.jp

with 10% horse serum at 35°C under 15% CO_2 for approximately 40 h with shaking (ca. 100 rpm) (pre-preculture).

4. Measure bacterial growth at OD 600 nm using a microplate spectrophotometer and adjust the concentration to 4×10^7 cells/ml with approximately 3 ml of Mueller–Hinton broth (Eiken Chemical, Tokyo, Japan) supplemented with 5.5% horse serum. Then, cultivate bacteria at 35°C under 15% CO_2 for 12–24 h with shaking (ca. 100 rpm) after motility of *H. pylori* is verified microscopically. (preculture).

5. Measure bacterial growth at OD 600 nm and adjust the concentration to 2×10^7 cells/ml with Mueller–Hinton broth supplemented with 5.5% horse serum. Then, apply 50 µl of the bacterial solution to individual wells of a 96-well plate after *H. pylori* motility is verified again under a microscope. In parallel, prepare glycans of interest dissolved in Mueller–Hinton broth supplemented with 5.5% horse serum and filter using a syringe filter. Then, add 50 µl of the glycan solution to bacterial solutions in the 96-well plate; the total volume of the mixed solution should be 100 µl.

6. Measure bacterial growth at OD 600 nm before culture and then cultivate at 35°C under 15% CO_2 and measure bacterial growth at OD 600 nm every day or every half-day for 4–4.5 days of culture.

Results

Since α1,4-GlcNAc-capped *O*-glycans are secreted from gland mucous cells of the gastric mucosa (Ishihara et al. 1996), we examined the effects of this unique glycan on *H. pylori* growth (Kawakubo et al. 2004). When *H. pylori* was cultured in the presence of α1,4-GlcNAc-capped *O*-glycans prepared from Lec2 cells cotransfected with plasmids encoding α1,4-*N*-acetylglucosaminyltransferase, core2 β1,6-*N*-acetylglucosaminyltransferase-I (C2GnT-I), and soluble CD43 (sCD43), *H. pylori* growth was suppressed in a dose-dependent manner (Fig. 1). However, such an inhibitory effect was not observed when a control *O*-glycan without terminal α1,4-linked GlcNAc secreted from Lec2 cells cotransfected with C2GnT-I and sCD43 cDNAs was used. Similar inhibitory effects were observed when pyloric gland cell-derived mucin having α1,4-GlcNAc-capped *O*-glycan isolated from the gastric mucosa or *p*-nitrophenyl αGlcNAc (GlcNAcα-PNP) were tested instead of recombinant α1,4-GlcNAc-capped *O*-glycan. These results indicate that α1,4-GlcNAc-capped *O*-glycans secreted from gastric gland mucous cells function as a natural antibiotic against *H. pylori*.

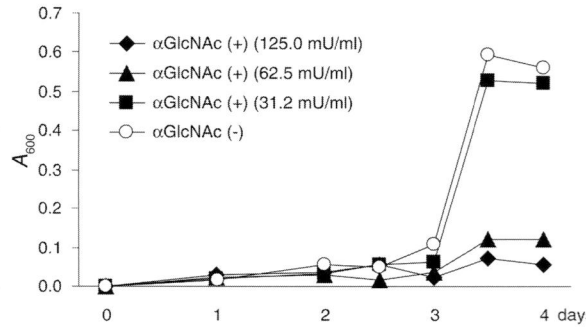

Fig. 1 Growth curves of *H. pylori* cultured with sCD43 with α1,4-GlcNAc-capped *O*-glycans [αGlcNAc (+)] or control sCD43 without α1,4-GlcNAc-capped *O*-glycans [αGlcNAc (−)]. One milliunit of αGlcNAc (+) corresponds to 1 µg of GlcNAcα-PNP (adapted from Kawakubo et al. 2004)

References

Hidaka E, Ota H, Hidaka H, Hayama M, Matsuzawa K, Akamatsu T, Nakayama J, Katsuyama T (2001) *Helicobacter pylori* and two ultrastructurally distinct layers of gastric mucous cell mucins in the surface mucous gel layer. Gut 49:474–480

Ishihara K, Kurihara M, Goso Y, Urata T, Ota H, Katsuyama T, Hotta K (1996) Peripheral α-linked *N*-acetylglucosamine on the carbohydrate moiety of mucin derived from mammalian gastric gland mucous cells: epitope recognized by a newly characterized monoclonal antibody. Biochem J 318:409–416

Kawakubo M, Ito Y, Okimura Y, Kobayashi M, Sakura K, Kasama S, Fukuda MN, Fukuda M, Katsuyama T, Nakayama J (2004) Natural antibiotic function of a human gastric mucin against *Helicobacter pylori* infection. Science 305:1003–1006

Lee H, Kobayashi M, Wang P, Nakayama J, Seeberger PH, Fukuda M (2006) Expression cloning of cholesterol α-glucosyltransferase, a unique enzyme that can be inhibited by natural antibiotic gastric mucin *O*-glycans, from *Helicobacter pylori*. Biochem Biophys Res Commun 349:1235–1241

Warren JR, Marchall BJ (1983) Unidentified curved bacilli on gastric epithelium in active chronic gastritis. Lancet 1:1273–1275

Binding Properties of *Clostridium botulinum* Type C Progenitor Toxin

Atsushi Nishikawa

Introduction

Clostridium botulinum has been classified into seven types, A to G, which are differentiated by the immunological specificity of the neurotoxin (NT) produced. NT is a high-potent inhibitor of the neurotransmitter release from the peripheral nerve terminus. Once localized in the cytoplasm of nerve cells, the NT functions by specifically cleaving one of the three different SNARE proteins essential for synaptic vesicle fusion (Lalli et al. 2003). The molecular masses of all types of these NTs are approximately 150 kDa. In the type C strain, two different sized progenitor toxins with molecular masses of 500 kDa (C16S) and 300 kDa (C12S) are produced. The C12S toxin consists of an NT and a non-toxic component having no hemagglutinin (HA) activity that is designated as non-toxic non-HA. The C16S toxin consists of C12S toxin and four kinds of different subcomponents (HAs) (Fig. 1). The previous data suggest that the HA subcomponents and non-toxic component assume to have the role of protecting the NT against acidity and proteases in the digestive tract. Furthermore, HAs seem to play an important role in transferring the toxin from digestive tract to circulating system. The HAs function as an adhesion in the attachment of 16S toxin to the microvilli of the intestine of a guinea pig (Fujinaga et al. 1997). Here, we introduce our assay methods of toxin binding ability to glycoprotein, and show the binding specificity of the C16S toxin to mucin.

The Binding Assay of C16S Toxin to Porcine Gastric Mucin

The binding ability of toxin to porcine gastric mucin (PGM) was measured as follows:
1. Two hundred microliters containing 2 ng of PGM was applied to PVDF membrane using a Bio-Dot SF microfiltration apparatus (Bio-Rad).
2. Following the membrane was blocked by 1% BSA, it was treated with 0.13 mg/ml of C16S toxin in PBST (PBS containing 0.1% of Tween-20) and various concentrations of sugars used as binding inhibitors of C16S toxin to PGM for 1 h.
3. The membrane was then washed by PBST, and it was incubated with anti-type C16S nontoxic components antibody for 1 h.
4. The C16S toxin binding was detected by an ECL Western blotting system, and visualized using a Lumino imaging analyzer LAS 3000 and Multi Gauge v2.1 software (Fuji Film).

Laboratory of Biochemistry, Department of Applied Biological Science, Tokyo University of Agriculture and Technology, 3-5-8 Saiwai-cho, Fuchu, Tokyo 183-8509, Japan
Phone: +81-42-367-5905, Fax: +81-42-367-5905
E-mail: nishikaw@cc.tuat.ac.jp

Fig. 1 Schematic supposed model of botulinum type C 16S toxin and 12S toxin. 12S toxin is a heterodimer of NT and non-toxic non-HA. 16S toxin seems to consist of a 12S toxin, six molecules of HA1, three of HA2, and three of HA3a and HA3b provided processing by protease after translation of HA3

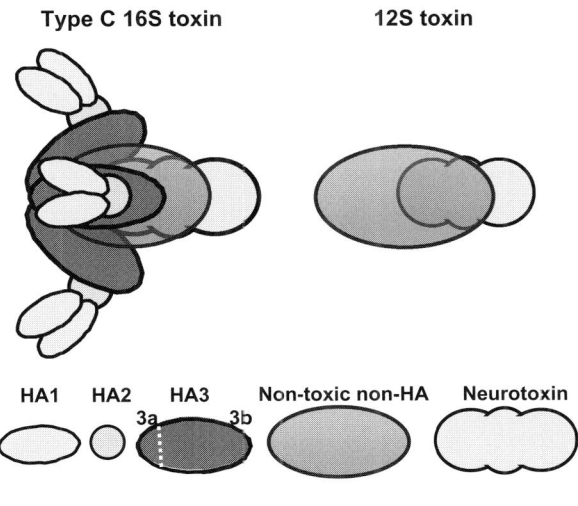

Fig. 2 Internalization of C16S toxin into HT-29 cells. **A** The human colon carcinoma cell line HT-29 cells were incubated with 10 nM each toxin at 37°C for 3 min and then washed with PBS. The cell-associated toxins were detected by immunofluorescence microscopy using rabbit anti-C7S antiserum and TRITC-labeled anti-rabbit IgG. **B** HT-29 cells were incubated with 40 nM C16S toxin at 37°C for 5 min, then washed, and incubated in toxin-free medium for 4 min. The cell-associated toxin and internalized toxin were analyzed by confocal laser scanning microscopy using rabbit anti-C non-toxic component anti serum and TRITC-labeled anti-rabbit IgG

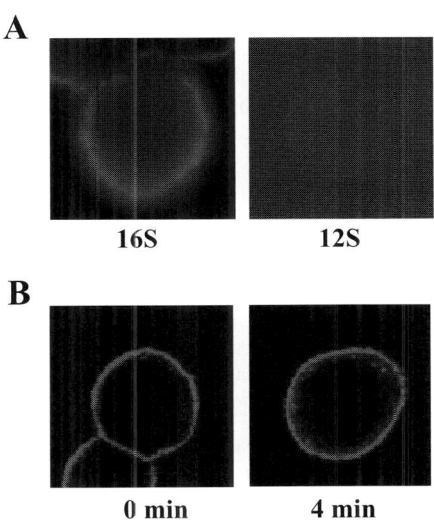

Results

As shown in Fig. 2, C16S toxin bound to the oligosaccharide of cell surface mucin and internalized into cells. Considering these phenomena and others, we proposed that the receptor on the HT-29 cell surface targeted by the C16S toxin is an O-linked sugar chain of mucin-like glycoproteins (Nishikawa 2004). Furthermore, we demonstrated that the C16S toxin is internalized into HT-29 cells with mucin-like glycoprotein via clathrin-mediated endocytosis (Uotsu et al. 2006).

The binding specificity of C16S toxin to porcine gastric mucin (PGM) was examined using some kinds of sugars as inhibitor (Fig. 3). Similar inhibition effects were observed with the addition of galactose and lactose, but glucose and N-acetylglucosamine did not prevent the toxin binding to PGM. On the other hand, sialic acid inhibited well the toxin

Fig. 3 Inhibitory effect of various sugars on the 16S toxin binding to PGM. The toxin binding rate to blotted PGM were plotted. Galactose (*open circle*), lactose (*square*), glucose (*closed circle*), and N-acetylglucosamine (*diamond*) used as binding inhibitors. Each point shows the mean ± SD of duplicate slots from three independent duplicate experiments

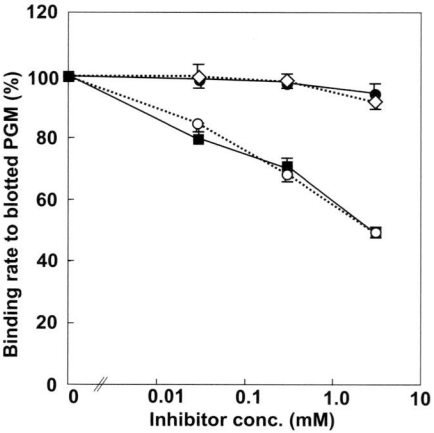

binding to bovine submaxillary mucin (data not shown) (Nakamura et al. 2007). From these results, it is suggested that the C16S toxin can recognize neutral and acidic sugars, which are reducing end sugar of oligosaccharide. C16S toxin composed of a lot of lectins, six molecules of HA1 and three of HA3, presumably that they are aligned together in the toxin binding in vivo, causing the broad admissible specificity for oligosaccharide structure being able to correspond to the various targets.

References

Fujinaga Y, Inoue K, Watanabe S, Yokota K, Hirai Y, Nagamachi E, Oguma K (1997) The haemagglutinin of *Clostridium botulinum* type C progenitor toxin plays an essential role in binding of toxin to the epithelial cells of guinea pig small intestine, leading to the efficient absorption of the toxin. Microbiology 143:3841–3847

Lalli G, Bohnert S, Deinhardt K, Verastegui C, Schiavo G (2003) The journey of tetanus and botulinum neurotoxins in neurons. Trends Microbiol 11:431–437

Nakamura T, Takada N, Tonozuka T, Sakano Y, Oguma K, Nishikawa A (2007) Binding properties of *Clostridium botulinum* type C progenitor toxin to mucins. Biochim Biophys Acta Gen Subj 1770:551–555

Nishikawa A, Uotsu N, Arimitsu H, Lee J-C, Miura Y, Fujinaga Y, Nakada H, Watanabe T, Ohyama T, Sakano Y, Oguma K (2004) The receptor and transporter for internalization of *Clostridium botulinum* type C progenitor toxin into HT-29 cells. Biochem Biophys Res Commun 319:327–333

Uotsu N, Nishikawa A, Watanabe T, Ohyama T, Tonozuka T, Sakano Y, Oguma K (2006) Cell internalization and traffic pathway of *Clostridium botulinum* type C neurotoxin in HT-29 cells. Biochim Biophys Acta Mol Cell Res 1763:120–128

Section XI
Cancer

Fucosylation and Cancer

Eiji Miyoshi

Introduction

Fucosylation is one of the most common modifications involving oligosaccharides on glycoproteins or glycolipids. Fucosylation comprises the attachment of a fucose residue to N-glycans, O-glycans, and glycolipids. O-Fucosylation, a special type of fucosylation, is very important for Notch signaling. The regulatory mechanisms for fucosylation are complicated. Many kinds of fucosyltransferases, the GDP-fucose synthesis pathway, and GDP-fucose transporter are involved in the regulation of fucosylation. Increased levels of fucosylation have been reported in a number of pathological conditions, including inflammation and cancer. Therefore, certain types of fucosylated glycoproteins such as AFP-L3 and several kinds of antibodies, which recognize fucosylated oligosaccharides such as sialyl Lewis a/x, have been used as tumor markers. Furthermore, fucosylation of glycoproteins regulates the biological functions of adhesion molecules and growth factor receptors. Changes in fucosylation could provide a novel strategy for cancer therapy. To determine levels of cellular fucosylation, lectin blot analyses using AAL (Aleuria aurantia lectin) and LCA (Lens culinaris agglutinin) are powerful tools.

AFP-L3, a Novel Tumor Marker for HCC

Fucose is one of the most interesting monosaccharides in sugar chains (Miyoshi et al. 1999). Hepatocellular carcinoma (HCC) is special because a high-risk group for HCC has developed. The incidence of HCC in chronic liver diseases such as hepatitis and liver cirrhosis is much higher (more than 1000-fold) than that in normal healthy controls. Chronic liver diseases can be easily diagnosed by blood tests and ultrasonography. It is very important to find an early tumor marker for HCC in patients with chronic liver diseases. One of the best-known tumor markers for HCC is alpha-fetoprotein (AFP), an oncofetal protein. Approximately 60% of HCC cases are seropositive for AFP, but a problem is that increased levels of AFP are also observed in certain cases of chronic liver diseases. AFP contains one N-linked sugar chain. Alpha1–6 fucosylation on the N-glycan on AFP can be detected by LCA (lens culinaris agglutinin) lectin affino-electrophoresis. Fucosylated AFP (AFP-L3) is a more specific tumor marker for HCC, which is negative for most benign liver diseases. Of all tumor markers, AFP-L3 is the most useful. In 2005, the Food and Drug Administration in the United States approved AFP-L3 as a tool for the detection of HCC.

Department of Molecular Biochemistry & Clinical Investigation, Osaka University Graduate School of Medicine, 1-7 Yamada-oka, Suita, Osaka 565-0871, Japan
Phone: +81-6-6879-2590, Fax: +81-6-6879-2590
E-mail: emiyoshi@sahs.med.osaka-u.ac.jp

Molecular Mechanism for Production of AFP-L3 in HCC

The purification and cDNA cloning of α1-6 fucosyltransferase (Fut8) from porcine brain and a human gastric cancer cell line were performed in 1996 (Uozumi et al. 1996). It was found that the expression of Fut8 in the liver was quite low and its expression was dramatically enhanced in hepatoma tissues, compared with the surrounding normal liver tissues in a rodent model. While the expression of Fut8 was low in normal human liver, similar to that in the rat liver, a high expression of Fut8 was observed in chronic liver diseases such as chronic hepatitis and liver cirrhosis. In contrast, GDP-fucose is a common donor substrate for fucosyltransferases. Levels of GDP-fucose in HCC tissues were significantly higher than that in the surrounding normal liver tissues. There are two synthetic pathways to produce GDP-fucose. One is a de novo pathway and the other is the salvage pathway. In the case of liver, the de novo pathway is dominant. The GDP-fucose synthase FX (human homologue of GDP-4-keto-6 deoxymannose-3) is a key enzyme in the de novo pathway, and expression of FX was significantly higher in HCC tissues than in surrounding liver tissues (Nada et al. 2003). Therefore, the molecular mechanisms underlying increases in fucosylation in HCC were found to be in part due to the enhancement of GDP-fucose synthesis pathway as well as the up-regulation of Fut8.

Physiological Function of Fut8 and Core-Fucose

Although Fut8 is not directly involved in the fucosylation of AFP in HCC, core-fucose, a product of Fut8, is very important in various biological phenomena. The overexpression of Fut8 in cancer cells changes their biological character in terms of metastasis, and the down-regulation of Fut8 suppresses the growth of cancer cells. These changes in cell phonotype are due to oligosaccharide modifications. Fut8 deficient mice died within 3 days after birth and the survivors showed severe growth retardation with emphysema-like lesions in their lungs. A deficiency of core-fucose on a TGFβ receptor deregulates its signal, leading to enhancement in the expression of several matrix metalloproteinases (MMPs). Some types of stress, such as smoking, induce the expression of several kinds of proteases. The signaling of TGFβ suppresses this induction of proteases. In the case of Fut8-deficient mice, this inhibitory signaling pathway is interrupted. The loss of core-fucose also influences the functions of many cell surface glycoproteins, including integrins and EGF-receptors.

Fucosylation Signal and Tumor Marker

Fucosylated haptoglobin was found to be a novel marker for pancreatic cancer (Okuyama et al. 2006). As preliminary experiments related to this finding, it was discovered that pancreatic cancer cells secrete many fucosylated proteins into a conditioned medium. Mass spectrometry analyses reveal that the fucosylation of haptoglobin in pancreatic cancer produces both α1-6 and α1-3/1-4 linkages. A question arises as to which organs produce fucosylated haptoglobin. Haptoglobin is normally produced in the liver, in which Fut8 is not expressed substantially. Haptoglobin is ectopically produced from infiltrated lymphocytes around tumors. To determine the origin of haptoglobin-producing organs, the site-directed identification of sugar chains might be useful. High amounts of fucosylated haptoglobin are present in the human bile. It had been found that in mice, glycoproteins in bile were highly fucosylated. It was then possible to hypothesize that fucosylation of hepatic glycoproteins could be a signal for secretion into a bile duct. To confirm this hypothesis, the oligosaccharide structures of human bile glycoproteins were analyzed in detail by 2D mapping of HPLC and mass-spectrometry. Furthermore, it was found that hepatic glycoproteins such as haptoglobin, $\alpha 1$ acid glycoproteins, and $\alpha 1$ anti-trypsin were not secreted into the bile in Fut8-deficient mice, suggesting that the fucosylation of oligosaccharides regulates the secretion of hepatic glycoproteins into bile ducts (Nakagawa et al. 2006). This hypothesis could be applied to AFP-L3 and HCC. Even if GDP-fucose levels are higher in HCC tissues as compared to their surrounding tissues, it is only two-fold higher. In contrast, the specificity of the AFP-L3 test for HCC diagnosis is approximately 95%. Another factor could also exist. If fucosylated AFP is secreted into bile in patients with chronic liver disease and this system is disrupted in HCC, the hypothesis proposed above would state the third mechanism underlying the production of fucosylated AFP in HCC.

Conclusion

Fucosylation is an interesting topic in cancer glycobiology and is also important in immunology. The regulatory mechanisms for the production of fucosylated proteins appear to be complicated. To determine the levels of cellular fucosylation, lectin blot analyses using AAL and LCA are powerful tools. The detailed procedure is described below.

Lectin Blot Analysis

1. Electrophorese 10–20 μg of cellular proteins or 0.5 μl of serum on 8%–12 % polyacrylamide gel.
2. Blot onto a nitrocellulose membrane or PVDF membrane.
3. Prehybridize with 3% bovine serum albumin.
4. Incubate with 1 μg/ml of biotinated AAL or AOL for 1 h at room temperature.
5. Wash with TBS (Tris-buffered saline) + 0.05% Tween-20 three times for 10 min each time.
6. Incubate with 1/2000 diluted horseradish peroxidase avidin D (usually prepared with ABC kit, VECTASTAIN).
7. Wash with TBS + 0.05% Tween-20 for 10 min each three times.
8. Develop with ECL kit.
9. Analyze results of lectin blot using CBB staining for best results.

References

Miyoshi E et al (1999) The α 1–6 fucosyltransferase gene and its biological significance. Biochem Biophys Acta 1473: 9–20
Nakagawa T et al (2006) Fucosylation of N-glycans regulates secretion of hepatic glycoproteins into bile ducts. J Biol Chem 281(40): 29797–806
Noda K et al (2003) Relationship between elevated FX expression and increased production of GDP-L-fucose, a common donor substrate for fucosylation in human hepatocellular carcinoma and hepatoma cell lines. Cancer Res 63(19): 6282–6289
Okuyama N et al (2006) Fucosylated heptoglobin is a novel marker for pancreatic cancer: A detailed analysis of the oligosaccharide structure and a possible mechanism for fucosylation. Int J Cancer 118 (11) 2803–2808
Uozumi N et al (1996) Purification and cDNA cloning of porcine brain GDP-L-Fuc: N-acetyl-β-D-glucosaminide $\alpha 1 \rightarrow 6$fucosyltransferase. J. Biol Chem 271(44): 27810–27817

Biological Significance of Mucins Produced by Epithelial Cancer Cells

Hiroshi Nakada

Background of Research

Since normal epithelial cells exhibit a clear polarity, synthesized mucins are transported to the apical cell surface and become secretory or membrane-bound glycoproteins. Mucins are major glycoprotein components of mucus, covering the luminal surfaces of the epithelial respiratory, gastrointestinal, and reproductive tracts. Upon malignant transformation, mucins are secreted into the tumor tissues and/or the bloodstream of cancer patients because of loss of the polarity of epithelial tissues. Qualitative changes of mucins are detected in O-glycans expressed on core proteins. Such aberrant O-glycans are generally called cancer-associated carbohydrate antigens. Many mAbs against cancer-associated carbohydrate antigens have been produced and used for the detection of tumor markers in the bloodstream and for the determination of epitopic structures. It has been reported that patients with a higher amount of mucins in their bloodstream have a lower 5-year survival rate. However, little is known regarding the biological significance of mucins. Since many lectins are found on immune cells, we predicted that mucins may interact with these lectins. Generally, a single carbohydrate chain binds to a lectin so weakly that the interaction may be biologically ineffective but carbohydrate chains clustered on the core protein like mucins could bind to multiple lectins, thus resulting in high-affinity binding. Mucins generally possess so many tandem repeats that if there is a single binding site in the tandem repeat unit, many lectins expressed on immune cells may be bridged. If a lectin plays a role in signal transduction, it is expected to mediate a strong signal in immune cells, maybe leading to an effect on immune function.

Current Studies and Perspectives

Activation of Monocytes/Macrophages Through Interaction of Mucins with Scavenger Receptors and Its Biological Significance (Inoue et al. 1999; Inaba et al. 2003; Sugihara et al. 2006; Yokoigawa et al. 2007)

In a study on the effect of mucins on immune cells, we found that mucins activate monocytes/macophages.

Department of Biotechnology, Faculty of Engineering, Kyoto Sangyo University, Kyoto, Japan

Purification of Mucins from Culture Medium or Xenograft of Carcinoma Cells

Xenograft
↓ homogenization in PBS, 0.02% NaN$_3$, 1 mM PMSF
↓ centrifugation
Supernatant   ⟨Cell culture medium⟩
↓   ↓
        ↓ gel filtration (Sepharose 6B)
        ↓ detection of mucins by dot blot analysis using mAb against carbohydrate antigens
      Mucin fraction
        ↓ precipitation with perchloric acid (0.1~0.5 M)
        ↓ centrifugation
      Supernatant
        ↓ neutralization with 1M Tris-HCl, pH 8.0
        ↓ dialysis
        ↓ reduction in 0.1 M Tris–HCl, pH 8.0, 2 mM PMSF,
                6 M guanidine-HCl, 0.1 M dithiothreitol, 2 h
        ↓ alkylation in 0.3 M iodoacetamide, 2 h in the dark.
        ↓ CsCl density gradient centrifugation
                4 M guanidine-HCl , 40 (w/w)% CsCl
                36,000 rpm, 72 h, 10°C (Hitachi, P40ST rotor)
   mucin

More than 20 kinds of mucin genes have been reported. The procedure as described above shows a general purification method of secretory mucins.

To detect the receptor involved, binding of ^{125}I-labeled mucins to human peripheral monocytes was examined in the presence of various substances. The binding activity was inhibited by mucins, fucoidan, and poly I but not by poly C or orosomucoid, suggesting that the scavenger receptor (SCR) is responsible for the binding. These results were confirmed using a stable SCR cDNA transfectant. It is well known that SCR can recognize a pattern of anionic charges on molecules such as acetyl LDL, poly I, and fucoidan (Kodama et al. 1990). Anionic charges due to sialic acid and sulfate on O-glycans of mucins may be recognized by the receptor.

It is well known that cyclooxygenase 2 (COX2) is induced in various types of cancers, and overproduced PGE$_2$ has various biological effects such as inhibition of apoptosis and immunosurveillance and promotion of tumor angiogenesis and invasion. Despite much evidence that COX2 overexpression is crucial for tumorigenesis and tumor growth, the mechanisms by which COX2 is originally induced in tumor tissues remain unresolved. We found that when human peripheral monocytes were cultured in the presence of mucins, the production of PGE$_2$ increased in a dose-dependent manner. The induction of COX2 mRNA and enzyme protein in the monocytes was confirmed, indicating that PGE$_2$ was overproduced through COX2 induction. Next, we examined the distribution of infiltrated macrophages, COX2 and mucin in human colon cancer tissues immunohistochemically. When mucins were present near the infiltrated macrophages, COX2 protein was detected in the macrophages. Thus, COX2 induction may take place in vivo through a similar mechanism.

Next, we performed similar experiments using mouse mammary adenocarcinoma cell lines and examined the effects of overproduced PGE_2 on tumor-bearing mice. Mouse mammary adenocarcinoma cell line, TA3-Ha and TA3-St may be a closely matched pair that could be exploited to compare tumor growth and the immunologic state in relation to mucins, because TA3-Ha cells produce a mucin named epiglycanin, whereas TA3-St cells, a subline, do not. The two sublines showed similar growth curves in vitro. In contrast, when injected subcutaneously into mice, the TA3-Ha tumors grew much faster than the TA3-St tumors did. When mouse macrophages were cultured in the presence of epiglycanin, COX2 was also induced, resulting in overproduction of PGE_2. Subcutaneous tumor tissues comprising TA3-Ha and TA3-St cells were examined immunochemically. Epiglycanin was detected in TA3-Ha tumor tissues but not in TA3-St tumor tissues. Similar numbers of macrophages had infiltrated into the two tumor tissues, and COX2 was detectable in TA3-Ha tumor tissues but only very slightly in TA3-St tumor tissues. Interestingly, COX2 was induced in most infiltrated macrophages in TA3-Ha tumor tissues. CD31-positive cells were also clearly observed in TA3-Ha tumor tissues but only slightly in TA3-St tumor tissues, indicating advanced angiogenesis in the TA3-Ha tumor tissues. Next, we compared the immunological states of TA3-Ha and TA3-St tumor-bearing mice. Splenic CD4 T cells prepared from tumor-bearing mice were treated with ionophore and breferdine A to pool cytokines intracellularly, and then the number of IL-4 or IFN-γ producing cells was determined. IFN-γ-producing T cells in TA3-Ha tumor-bearing mice were significantly reduced compared with control or TA3-St tumor-bearing mice, whereas IL-4 producing cells did not show any change irrespective of whether they were from control or tumor-bearing mice. These results are consistent with the report that the synthesis of Th1 cytokines is much more sensitive to inhibition by PGE_2 than that of Th2 cytokine production. Thus, this immune suppression and promotion of angiogenesis may together facilitate the progression of TA3-Ha xenografts. Taking these results and other reports together, we propose the following cascade in a tumor microenvironment (Fig. 1). Mucins are produced by cancer cells. Infiltrated macrophages are activated by the mucins through the SCR. PGE_2 secreted from the macrophages binds to the EP2 receptor present on cancer cells and/or other cells, resulting in further induction of COX2 and PGE_2 overproduction (Sonoshita et al. 2001). PGE_2 produced by various cells upregulates VEGF production and downregulates IL-12 production by macrophages in an autocrine manner, and IFN-γ production by Th1 cells in a paracrine manner. Thus, angiogenesis is promoted and tumor-rejection is prevented, leading to favorable conditions in epithelial cancer tissues for cancer cell growth.

Immune Suppression Through the Siglec Family

Siglecs (sialic acid binding Ig-like lectins) are characterized by an N-terminal V-set Ig domain that mediates sialic acid binding, followed by varying numbers of C2-set Ig domains. In human, there are 11 Siglecs, which are all expressed on immune cells except for Siglec4/MAG. Most Siglecs possess the immunoreceptor tyrosin-based inhibition motif (ITIM) in their cytoplasmic tails, which is considered to be involved in regulation of cellular activation within the immune system. However, signal transduction through Siglecs and its effect on immune cells have not been demonstrated, because endogenous ligands have not been fully characterized. Since Siglecs recognize sialic acid-containing glycoconjugates, it is expected that mucins produced by epithelial cancer cells can bind to Siglecs. Siglec 2 (CD22) is a B cell-specific glycoprotein that is associated with the

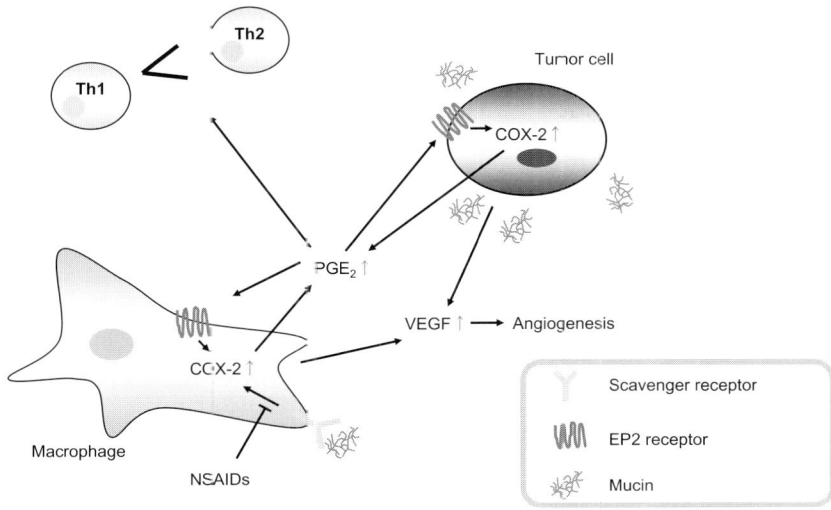

Fig. 1 Cascade originating from tumor-produced mucins in the tumor microenvironment

B cell receptor. Although the physiological significance of Siglec 2-mediated ligand binding remains unknown, many studies have suggested that Siglec 2 plays an important role in regulation of B cell function. In fact, B cells obtained from Siglec 2 gene knockout mice exhibit various biological changes.

First, we examined the binding of mucins to Siglec 2 expressed on B cells by means of FACS analyses and plate assaying. It was demonstrated that labeled MUC2 isolated from the culture medium of a human colorectal cell line, LS 180 cells, and bovine submaxillary mucin (BSM) binds to B cell lymphomas, Daudi cells, and stable Siglec 2 cDNA transfectants, and soluble recombinant Siglec 2 was also found to bind to these mucins in the plate assay. When the B cells were stimulated with the antigen in the presence of mucins, the phosphorylation of MAP kinase was reduced, indicating that mucins inhibit B cell signaling. The in vivo effects of mucins produced by epithelial tumor cells are now under investigation. This effect on B cells is not specific for these mucins, because Siglec 2 is known to recognize α2,6 linked sialic acid-containing carbohydrate chains including a sialyl Tn antigen, and sialyl Tn antigens are expressed in various types of human adenocarcinomas. Furthermore, since mucins have a variety of sialic acid-containing O-glycans including carbohydrate antigens, they may bind to many other Siglecs. Thus, mucins produced by epithelial cancer cells may downregulate immune functions in various immune cells.

Applications

We have demonstrated that overproduction of PGE_2 through SCR signal transduction caused immune regulation and angiogenesis, leading to favorable conditions for tumor growth and progression. As is obvious from the fact that COX2 inhibitors are potent drugs for decreasing the incidence and mortality of tumors, blocking of mucin-ligation to SCR may be applicable for the practical treatment of tumors. In fact, soluble recombinant SCR was effective for inhibiting the tumor growth in TA3-Ha bearing mice (Takagi et al., unpublished data). As for the function of the Siglec family, it is uncertain to what degree mucin-ligation affects the biological function of immune cells. It is attractive that this immune regulation system can be applied to modulate each immune cell, either elevating immune function in cancer patients or inhibiting it in patients with an autoimmune disease.

References

Inaba T, Sano H, Kawahito Y et al (2003) Induction of COX2 in monocyte/macrophages by mucins secreted from colon cancer cells. Proc Natl Acad Sci USA 100:2736–2741

Inoue M, Fujii H, Kaseyama H et al (1999) Stimulation of macrophages by mucins through a macrophage scavenger receptor. Biochem Biophys Res Commun 264:276–280

Kodama T, Treeman M, Rohrer L et al (1990) Type 1 macrophage scavenger receptor contains alpha helical and collagen-like coiled coils. Nature 343:531–535

Sonoshita M, Takaku K, Sasaki N et al (2001) Acceleration of intestinal polyposis through prostaglandin receptor EP2 in Apc (delta716) knockout mice. Nat Med 2:1048–1051

Sugihara I, Yoshida M, Shigenobu T et al (2006) Different progression of tumor xenografts between mucin-producing and mucin-non-producing mammary adenocarcinoma-bearing mice. Cancer Res 66:6175–6182

Yokoigawa N, Takeuchi N, Toda M et al (2007) Overproduction of PGE2 in peripheral blood monocytes of gastrointestinal cancer patients with mucins in their bloodstream. Cancer Lett 245:149–155

Regulation of Glycosyltransferases for O-Glycan Core Structure in Cancer and Their Relations to Metastasis

Toshie Iwai, Hisashi Narimatsu

Synthesis of many O-linked carbohydrate chains (O-glycan) is initiated by transferring GalNAc to the Ser/Thr residue in an amino acid sequence by polypeptide GalNAc transferase (pp-GalNAc-T) (Cheng et al. 2004). These O-glycans are called mucin-type carbohydrate chains, and most of them are found on glycoproteins called mucin. Mucin, which is the major constituent of mucus glycoproteins, not only protects the mucosal membrane from invasion of viruses and chemical substances but also is involved in various biological phenomena including cell adhesion, signal transduction, inflammation, and metastasis.

Structures of O-glycans are diverse and classified according to their root structure. Figure 1 shows structures of major O-glycans. As described above, synthesis of O-glycan is initiated by the addition of GalNAc to pp-GalNAc-T. The structure in which only GalNAc bounds to a Ser/Thr residue (Tn antigen) and the structure in which sialic acid bounds to Tn antigen (sTn antigen) with an α2-6 linkage have been known as the O-glycans which increase during the cancerous transformation. The core 1 structure and the core 3 structure are synthesized by the addition of Gal and GlcNAc, respectively, to a GalNAc residue with a β1-3 linkage. It is considered that the core 1 and core 3 structures are in competitive relationship to each other, because these structures are formed by the addition of Gal or GlcNAc to the same 3-position of a GalNAc residue. The core 2 and core 4 structures are synthesized by the addition of GlcNAc to the 6-position of the GalNAc residue at the root of the core 1 and core 3 structures, respectively. There have been various reports on synthetic enzymes of the core 1–4 structures including cloning of their genes. On the other hand, no study has been reported on the genes of synthetic enzymes of the core 5–8 structures, although their structures have been reported.

To date, an enzyme synthesizing the core 1 structure (C1Gal-T) and an enzyme synthesizing the core 3 structure (β3Gn-T6) have been reported (Ju et al. 2002, Iwai et al. 2002). Three enzymes have been known to synthesize the core 2 structure until now (C2GnT1, T2, and T3), and it has been reported that C2GnT2 also synthesizes the core 4 structure (Schwientek et al. 2000). Some studies reported that Cosmc acted as the chaperone of C1Gal-T, and coexistence of both proteins in the same cell is at least necessary to synthesize the core 1 structure (Ju and Cummings 2002, Kudo et al. 2002). C1Gal-T synthesizing the core 1 structure is expressed in almost all the tissues, but the core 3 synthase is expressed in the limited tissues including the epithelial cells of stomach, small intestine and large intestine. Detailed structural analysis has been performed on carbohydrate chains of O-glycans in the normal human large intestinal tissue,

---

Research Center for Medical Glycoscience, National Institute of Advanced Industrial Science and Technology (AIST), 1-1-1 Umezono, Tsukuba, Ibaraki 305-8568, Japan
E-mail: h.narimatsu@aist.go.jp

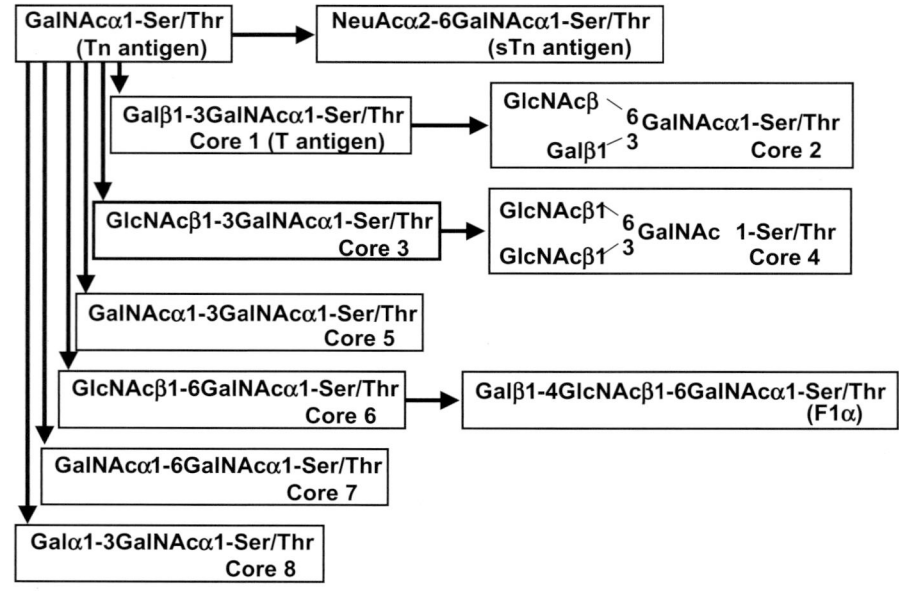

Fig. 1 Biosynthetic pathways for mucin type O-glycans

and it has been reported that most of the O-glycans have the core 3 structure at their root.

These core structures are involved in various phenomena in the body. During T-cell maturation in the thymus, ST antigen (NeuAcα2-3Galβ1-3GalNAc), the structure in which sialic acid is added to the core 1 structure, is synthesized after the temporal predominance of the core 1 structure. It has been reported that, as the sialylation required a sialyltransferase, ST3Gal I, the inactivation of the enzyme caused presentation of the naked core 1 structure on the surface of T cells, leading to the elimination of the T cells by apoptosis (Priatel et al. 2000). It has been also reported that lymphocyte rolling and recruitment of lymphocytes to inflammatory sites are reduced in C2GnT1-deficient mice, indicating profound involvement of the core 2 synthase in these phenomena (Ellies et al. 1998). It has been known that carbohydrate chains promoting metastasis, such as sialyl-Lewis A and sialyl-Lewis X, are synthesized at the end of core 2 structure, and core 2 synthase is involved in metastasis. Furthermore, the structure of O-glycan drastically changes during cancerous transformation. As described above, the core 3 structure is the major O-glycan in the normal human large intestine, but the core 3 structure drastically decreases and the core 1 structure increases instead during cancerous transformation. It is considered that, as the core 1 and core 3 structures exist in a competitive manner to each other in normal condition, the core 1 structure increases due to disappearance of Core 3 synthase during cancerous transformation. We prepared the antibody against Core 3 synthase (β3Gn-T6), and performed immunohistochemical staining of the stomach and large intestine with this antibody. The result demonstrated that β3Gn-T6 was expressed in the normal tissue, but completely disappeared in the cancerous tissue. Staining of the familial adenomatous polyposis tissue with this antigen demonstrated that β3Gn-T6 was expressed in the benign tumor tissue but completely disappeared in the malignant tissue. We also found that introduction of the β3Gn-T6 gene vector into the highly metastatic cell line dramatically lowered their migration and metastatic activities. In contrast to the expression pattern of β3Gn-T6, the cancer tissue was well stained with a peanut lectin which recognizes the core 1 structure, but the normal tissue was not stained with it (Iwai et al. 2005).

As described above, O-glycans are profoundly involved in important intercellular phenomena. Therefore, it is possible to develop various markers of diseases including cancer by using antibodies against carbohydrate chains and glycosyltransferases.

References

Cheng L, Tachibana K, Iwasaki H, Kameyama A, Zhang Y, Kubota T, Hiruma T, Tachibana K, Kudo T, Guo JM, Narimatsu H (2004) Characterization of a novel human UDP-GalNAc transferase, pp-GalNAc-T15. FEBS Lett 566(1–3):17–24

Ellies LG, Tsuboi S, Petryniak B, Lowe JB, Fukuda M, Marth JD (1998) Core 2 oligosaccharide biosynthesis distinguishes between selectin ligands essential for leukocyte homing and inflammation. Immunity 9(6):881–890

Iwai T, Inaba N, Naundorf A, Zhang Y, Gotoh M, Iwasaki H, Kudo T, Togayachi A, Ishizuka Y, Nakanishi H, Narimatsu H (2002) Molecular cloning and characterization of a novel UDP-GlcNAc:GalNAc-peptide β1,3-N-acetylglucosaminyltransferase (β 3Gn-T6), an enzyme synthesizing the core 3 structure of O-glycans. J Biol Chem 277(15):12802–12809

Iwai T, Kudo T, Kawamoto R, Kubota T, Togayachi A, Hiruma T, Okada T, Kawamoto T, Morozumi K, Narimatsu H (2005) Core 3 synthase is down-regulated in colon carcinoma and profoundly suppresses the matastatic potential of carcinoma cells. Proc Natl Acad Sci USA 102(12):4572–4577

Ju T, Cummings RD (2002) A unique molecular chaperone Cosmc required for activity of the mammalian core 1 β 3-galactosyltransferase. Proc Natl Acad Sci USA 99(26):16613–16618

Ju T, Brewer K, D'Souza A, Cummings RD, Canfield WM (2002) Cloning and expression of human core 1 β1,3-galactosyltransferase. J Biol Chem 277(1):178–186

Kudo T, Iwai T, Kubota T, Iwasaki H, Takayma Y, Hiruma T, Inaba N, Zhang Y, Gotoh M, Togayachi A, Narimatsu H (2002) Molecular cloning and characterization of a novel UDP-Gal:GalNAc(alpha) peptide β1,3-galactosyltransferase (C1Gal-T2), an enzyme synthesizing a core 1 structure of O-glycan. J Biol Chem 277(49):47724–47731

Priatel JJ, Chui D, Hiraoka N, Simmons CJ, Richardson KB, Page DM, Fukuda M, Varki NM, Marth JD (2000) The ST3Gal-I sialyltransferase controls CD8+ T lymphocyte homeostasis by modulating O-glycan biosynthesis. Immunity 12(3):273–283

Schwientek T, Yeh JC, Levery SB, Keck B, Merkx G, van Kessel AG, Fukuda M, Clausen H (2000) Control of O-glycan branch formation. Molecular cloning and characterization of a novel thymus-associated core 2 β1, 6-N-acetylglucosaminyltransferase. J Biol Chem 275(15):11106–11113

Roles of Carbohydrate-Mediated Cell Adhesion in Cancer Progression

Reiji Kannagi[1], Akira Seko[2]

Introduction

It has long been known that malignant transformation is associated with the appearance of abnormal carbohydrate determinants on the cell surface. The monoclonal antibody approach to cancer-associated antigens, which prevailed worldwide in the early 1980s, further facilitated research on carbohydrate determinants, because not a few monoclonal antibodies that had preferential reactivity to cancer cells turned out to recognize carbohydrate determinants. The serum diagnosis of tumors using cancer-associated carbohydrate determinants has been covered by the national medical insurance program in Japan over the past two decades, and is widely accepted by clinicians to be clinically beneficial (Kannagi et al. 2004).

The serum tumor diagnosis using these carbohydrate determinants is applied not only for initial diagnosis of cancers, but also for detection of recurrence following a surgical operation, and monitoring the therapeutic effect of radio- and chemotherapy. However, the pathophysiological significance of these determinants in cancer progression was not clear when they were first applied to serum diagnosis, nor was the underlying mechanism for their preferential appearance in malignant cells.

Principle

Involvement of Sialyl Lewis X/A Glycans in Hematogenous Metastasis of Cancer

It was not until the early 1990s that sialyl Lewis X (Phillips et al. 1990) and sialyl Lewis A (Takada et al. 1991) on the surface of cancer cells were found to serve as ligands for endothelial E-selectin. Sialyl Lewis A was shown to play a major role in the adhesion to the endothelium of cancer cells derived from the lower digestive organs, such as the colon and rectum, as well as from the pancreas and biliary tract. On the other hand, the sialyl Lewis X determinant was found to figure heavily in the adhesion of breast, ovarian, and pulmonary cancer cells. These cell-biological findings indicated that these carbohydrate determinants are not merely markers for cancers, but functional molecules involved in the pathophysiology of cancers. This immediately raised the possibility that the adhesion of cancer cells to the endometrium mediated by E-selectin and the specific carbohydrate ligands might be involved in the hematogenous metastasis of cancers (Fig. 1).

---

[1]Department of Molecular Pathology, Aichi Cancer Center, Aichi, Japan
Phone: +81-52-762-6111, Fax: +81-52-764-2973
E-mail: rkannagi@aichi-cc.jp

[2]Innovative Research Initiatives, Tokyo Institute of Technology, Tokyo, Japan
Phone: +81-45-921-4308, Fax: +81-45-921-4308
E-mail: kyamashi@bio.titech.ac.jp

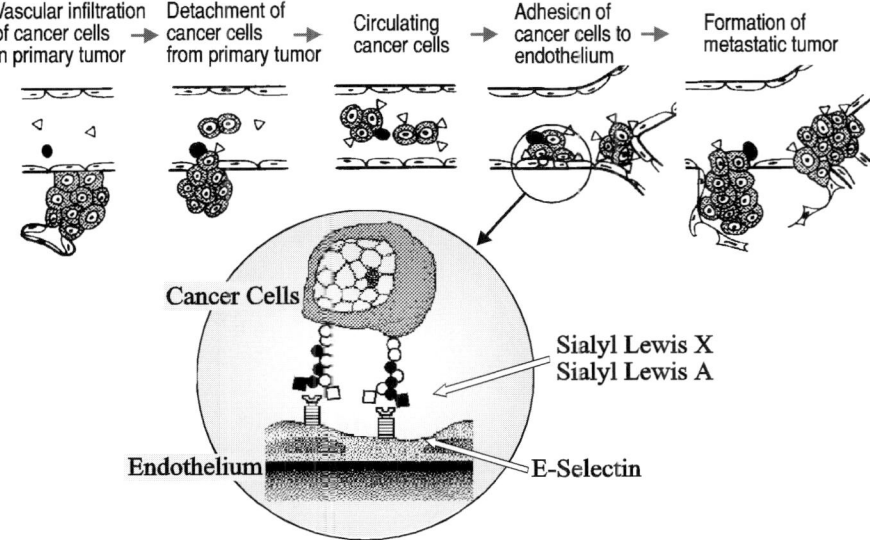

Fig. 1 Schematic representation of the complex, multistep process of hematogenous metastasis of cancer. The process starts with the intravasation of cancer cells into the bloodstream in the primary tumor lesion. Cancer cells then travel through the bloodstream, where they interact with various blood cells such as leukocytes and platelets, finally adhering to endothelial cells somewhere in the peripheral vessel walls. Cell adhesion mediated by the interaction of E-selectin on endothelial cells and sialyl Lewis X/A determinants on cancer cells is involved in this step

A lot of accumulated clinical statistics indicated a significantly higher risk of developing hematogenous metastasis for patients with cancer cells strongly expressing these carbohydrate ligands for selectins. Propitious results were reported for clinical trials that inhibited hematogenous metastasis by suppressing this cell adhesion system. The selectin-mediated cell adhesion is also known to facilitate tumor vascularization. This implies that inhibition of this cell adhesion system would result in suppression of the growth of primary tumors as well as in prevention of hematogenous metastasis.

Induction of Selectin Ligand Expression in Early-Stage Cancers

The expression of sialyl Lewis X and sialyl Lewis A is significantly enhanced in cancers as compared to nonmalignant epithelial cells. As for the mechanism of the preferential expression of these carbohydrate determinants in cancers, it has long been assumed that transcription of some genes for glycosyltransferases involved in their synthesis must be accelerated in cancers. However, the results obtained from studies in this area over the past decade indicated that this is not necessarily the case.

What has gradually become clear from the studies in the past decade is that the normal epithelial cells are, contrary to general expectations, equipped with sufficient amounts of enzymes and mRNAs required for the synthesis of sialyl Lewis X and sialyl Lewis A. The difference between normal epithelial cells and cancer cells is that only normal epithelial cells have additional enzymes to further modify and metabolize these determinants into carbohydrate determinants with more complex structures than sialyl Lewis X and sialyl Lewis A (Kannagi 2004; Kannagi et al. 2004) (Fig. 2).

Fig. 2 Induction of sialyl Lewis A or sialyl Lewis X expression in colon cancer as a result of *"incomplete synthesis."* *Ca* cancer cells, *N* nonmalignant epithelial cells. **a** Typical distribution pattern of sialyl Lewis X (*lower panel*) and sialyl 6-sulfo Lewis X (*upper panel*) determinants in colon cancer tissues, indicating that sialyl Lewis X determinant is preferentially expressed on cancer cells, while sialyl 6-sulfo Lewis X determinant is specifically localized in nonmalignant epithelial cells. Adopted from reference (Izawa et al. 2000). **b** Typical distribution pattern of sialyl Lewis A (*lower panel*) and disialyl Lewis A (*upper panel*) determinants in colon cancer tissues, indicating that the (mono-)sialyl Lewis A determinant is preferentially expressed on cancer cells, while the disialyl Lewis A determinant is specifically localized in nonmalignant epithelial cells. Adopted from reference (Miyazaki et al. 2004)

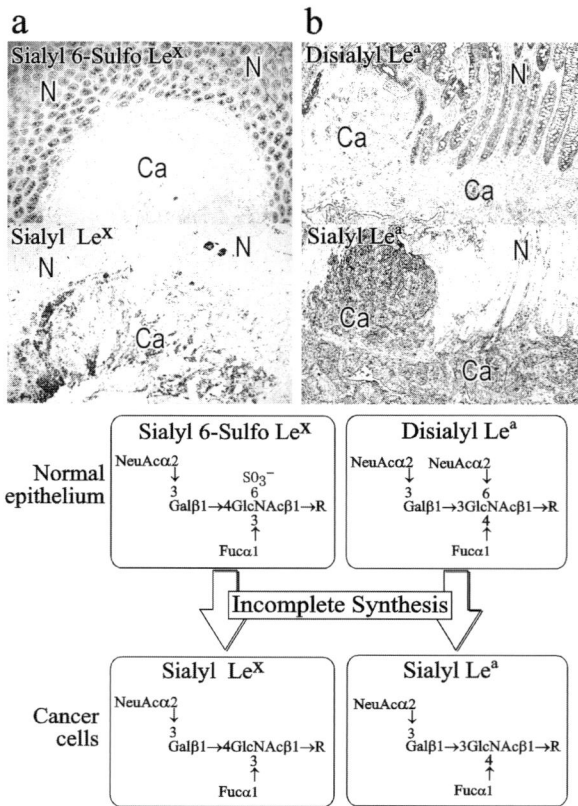

Disialyl Lewis A, a determinant having one more sialic acid residue than sialyl Lewis A, is a good example of carbohydrate determinants that are preferentially synthesized and expressed in normal epithelial cells (Miyazaki et al. 2004). In contrast, sialyl Lewis A that has a simpler structure accumulates in cancer cells, because the disturbed sialylation associated with malignant transformation hinders cancer cells from synthesizing the disialyl Lewis A determinant. Another example is sialyl 6-sulfo Lewis X, which has one sulfated residue attached to sialyl Lewis X. The sialyl 6-sulfo Lewis X determinant is preferentially expressed on normal epithelial cells (Izawa et al. 2000). Because of malfunction in sulfation accompanying malignant transformation, this determinant disappears in cancer cells, wherein the sialyl Lewis X determinant without sulfate residues takes its place. These examples fit well with the classical concept of *"incomplete synthesis,"* which was proposed in the early 1980s by Hakomori to characterize the cancer-associated alteration of cell-surface carbohydrates (Fig. 2).

Partial transcriptional impairment of glycogenes involved in the synthesis of complex carbohydrate determinants in normal epithelial cells thus leads to accumulation of cancer-associated carbohydrate determinants in malignant cells. An increasing line of evidence indicates that epigenetic silencing of glycogenes, which occurs at the early stage of carcinogenesis, is a predominant mechanism for this *incomplete synthesis*. The recent growth in our understanding of the relationship between chromatin organization and gene transcription has highlighted the importance of epigenetic mechanisms, such as DNA

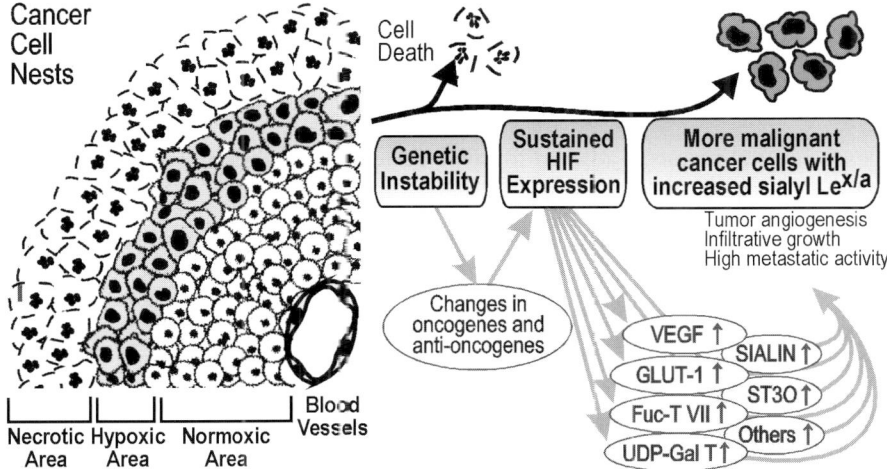

Fig. 3 Schematic illustration of hypoxia-induced cancer progression and sialyl Lewis X/A accumulation in locally advanced cancer nests. Expression of sialyl Lewis X/A is further enhanced during the course of hypoxia-induced cancer progression in locally advanced cancers through transcriptional induction of several glycogenes by a transcription factor HIF (hypoxia-inducible factor)

methylation and histone modification, in the initiation of human cancers. The increased sialyl Lewis X/A expression due to the *incomplete synthesis* mechanism is a link in the chains of these events. The inhibitors for DNA methylation or histone deacetylation, which have been developed to cope with cancer-associated epigenetic silencing, may well be effective also in correcting abnormal expression of sialyl Lewis X/A on cancer cells.

Further Acceleration of Sialyl Lewis X/A Expression in Advanced Stage Cancers

Malignant progression of cancers proceeds through accumulation of genetic abnormalities in locally advanced stage cancers, and cancer cell clones having more malignant characteristics are selected in accordance with the principle of survival of the fittest. Locally advanced tumor nests suffer from low oxygen supply, and the strongest natural selection occurs in terms of cell resistance to hypoxia. Cancer cell clones that become resistant to hypoxia by accumulating genetic changes preferentially survive and dominate the advanced cancer cell nests. Such cancer cells usually have a high and sustained expression of a transcription factor HIF-1 (hypoxia inducible factor-1).

Recently, we found that tumor hypoxia induces transcription of several glycogenes involved in sialyl Lewis X/A synthesis using a DNA array specially arranged for glycogenes by doctors, Kozutsumi and Suzuki at the Riken Institute (Koike et al. 2004). This transcriptional induction of glycogenes by hypoxia further accelerates expression of these glycans in advanced cancers (Fig. 3). This means that the more malignant cancer cells selected during cancer progression tend to express sialyl Lewis X/A determinants more strongly. These cells are at the highest risk of developing hematogenous metastasis in the terminal stages of cancers. This raises a possibility that therapeutic modalities targeted to HIF-1 may be useful in preventing hematogenous metastasis. Hypoxia induces drastic changes in cell surface glycoconjugates, some of which are useful for targeting therapy of hypoxia-resistant cancer cells (Yin et al. 2006).

Protocols

The experimental protocols adapted for hypoxia-induced adhesion of cancer cells are as follows (Koike et al. 2004).

1. Human colon cancer cell lines, SW480, C-1 and Colo201, were routinely maintained in Dulbecco's modified MEM (high glucose) supplemented with 10% FCS. For hypoxia experiments, the cells were cultured at 37°C with 5% CO_2/94% N_2/1% O_2 in a multigas incubator (Juji Field Inc., Tokyo, Japan) for 3–7 days in 24-well plates. Another plate for control was kept under normoxic conditions.

2. When a multigas incubator is not available, desferrioxamine (Sigma, St. Louis, MO, USA), a hypoxia-mimicking reagent, can be added to the culture medium at a final concentration of 100 μM (optimum concentration should be determined for each cell line).

3. BCECF-AM-labeled 300.19/E-selectin cells (1×10^6 cells/0.5 ml/well, murine B lymphoma 300.19 cells transfected with human E-selectin cDNA) were added to each well, and the 24-well plate was incubated on a rotating platform for incubation under shear (90 rpm with a rotary shaker R-20mini, Taitek Inc., Tokyo) for 20 min at room temperature.

4. After nonadherent cells were gently washed out three times with PBS, the adherent cells were lysed with 0.5% NP-40, and the attached cells were counted by measuring fluorescence intensity using an Arvo 1420 multilabel counter (Wallac, Gaithersburg, MD, USA).

Future Perspectives

There are many cancer-associated carbohydrate determinants other than sialyl Lewis A/X, awaiting future clinical application for diagnosis and therapy of cancers. As to O-glycans, the core-3 and core-4 structures are predominantly expressed in normal epithelial cells in the digestive organs, and the core-2 structure is carried in normal epithelial cells in other organs. Upon malignant transformation, determinants having simpler structures such as T-, Tn- and sialyl Tn-antigens appear in cancer cells. This line of studies was elaborated by Dr. Springer's group. This could also be regarded as an example of *incomplete synthesis* á la Hakomori, and therapeutic application of these determinants is now in progress in the European glycoconjugate consortium. With regard to *N*-glycans, it is well known that scientists in Japan have established the importance of increased GlcNAcβ1-6 branching and bisecting GlcNAc in CEA, hCG and γGTP. Regarding glycosaminoglycans, glypicans and syndecans are recently found to exhibit cancer-associated increase using DNA arrays.

A sulfotransferase, GlcNAc6ST-2, is known to be increased in colon and ovarian cancers (Figs. 4a, b). Its product is acknowledged to be preferentially expressed in colon cancers arising in the right hemicolon. Recently, Kanoh et al. (2006) found this enzyme to be a good marker for chemotherapy-resistant ovarian cancers. This enzyme is preferentially expressed in mucinous-, clear cell-, and papillary serous adenocarcinomas of the ovary (Fig. 4C shows an example of mucinous adenocarcinoma), which have a poor prognosis. In contrast, GlcNAc6ST-2 is much less frequent in solid serous adenocarcinoma

Fig. 4 An example of new tumor marker. Panels **a** and **b** show immunohistochemical staining of nonmalignant colonic epithelial cells (**a**) and colon cancers using a specific monoclonal antibody to the product of GlcNAc6ST-2. Panels C and D show staining of mucinous adenocarcinoma (**c**) and endometrioid adenocarcinoma (**d**) of the ovary using antibody directed to GlcNAc6ST-2 enzyme protein. (Adopted from Kanoh et al. 2006), *Bars* 50 μm

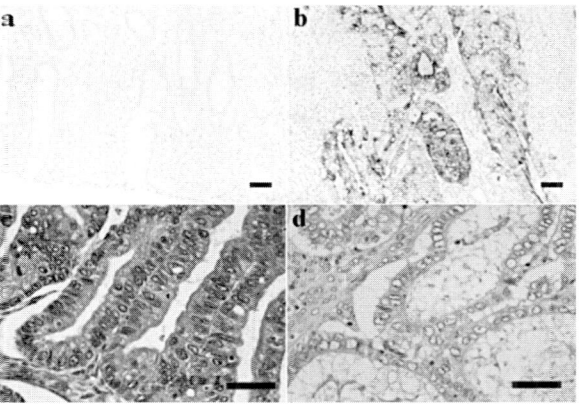

and endometrioid adenocarcinoma, which have a relatively good prognosis (Fig. 4D shows an example of endometrioid adenocarcinoma). Clinical application of this enzyme is now in progress by a scientist group at the Tokyo Institute of Technology.

References

Izawa M, Kumamoto K, Mitsuoka C, Kanamori A, Ohmori K, Ishida H, Nakamura S, Kurata-Miura K, Sasaki K, Nishi T, Kannagi R (2000) Expression of sialyl 6-sulfo Lewis x is inversely correlated with conventional sialyl Lewis x expression in human colorectal cancer. Cancer Res 60:1410–1416

Kannagi R (2004) Molecular mechanism for cancer-associated induction of sialyl Lewis X and sialyl Lewis A expression—the Warburg effect revisited. Glycoconj J 20:353–364

Kannagi R, Izawa M, Koike T, Miyazaki K, Kimura N (2004) Carbohydrate-mediated cell adhesion in cancer metastasis and angiogenesis. Cancer Sci 95:377–384

Kanoh A, Seko A, Ideo H, Yoshida M, Nomoto M, Yonezawa S, Sakamoto M, Kannagi R, Yamashita K (2006) Ectopic expression of *N*-acetylglucosamine 6-*O*-sulfotransferase 2 in chemotherapy-resistant ovarian adenocarcinomas. Glycoconj J 23:453–460

Koike T, Kimura N, Miyazaki K, Yabuta T, Kumamoto K, Takenoshita S, Chen J, Kobayashi M, Hosokawa M, Taniguchi A, Kojima T, Ishida N, Kawakita M, Yamamoto H, Takematsu H, Kozutsumi Y, Suzuki A, Kannagi R (2004) Hypoxia induces adhesion molecules on cancer cells—a missing link between Warburg effect and induction of selectin ligand carbohydrates. Proc Natl Acad Sci USA 101:8132–8137

Miyazaki K, Ohmori K, Izawa M, Koike T, Kumamoto K, Furukawa K, Ando T, Kiso M, Yamaji T, Hashimoto Y, Suzuki A, Yoshida A, Takeuchi M, Kannagi R (2004) Loss of disialyl Lewis[a], the ligand for lymphocyte inhibitory receptor Siglec-7, associated with increased sialyl Lewis[a] expression on human colon cancers. Cancer Res 64:4498–4505

Phillips ML, Nudelman E, Gaeta FCA, Perez M, Singhal AK, Hakomori S, Paulson JC (1990) ELAM-1 mediates cell adhesion by recognition of a carbohydrate ligand, sialyl-Le[x]. Science 250:1130–1132

Takada A, Ohmori K, Takahashi N, Tsuyuoka K, Yago K, Zenita K, Hasegawa A, Kannagi R (1991) Adhesion of human cancer cells to vascular endothelium mediated by a carbohydrate antigen, sialyl Lewis A. Biochem Biophys Res Commun 179:713–719

Yin J, Hashimoto A, Izawa M, Miyazaki K, Chen G-Y, Takematsu H, Kozutsumi Y, Suzuki A, Furuhata K, Cheng F-L, Lin C-H, Sato C, Kitajima K, Kannagi R (2006) Hypoxic culture induces expression of sialin, a sialic acid transporter, and cancer-associated gangliosides containing nonhuman sialic acid on human cancer cells. Cancer Res 66:2937–2945

Conditional Transgenic Mouse of Hyaluronan Synthase: A Potential Model of Advanced Breast Cancer

Naoki Itano

Introduction

Hyaluronan (HA) is a polysaccharide composed of repeating GlcNAcβ(1 $\rightarrow$ 4)-GlcAβ(1 $\rightarrow$ 3) disaccharide units that is synthesized by three members of the HA synthase, HAS1, HAS2, and HAS3. HA is a major constituent of the extracellular matrix (ECM), linking proteoglycans and other binding molecules into macromolecular aggregates. HA-rich ECM provides a favorable microenvironment for cell proliferation and migration by maintaining the turgidity and hydration of tissues, and also by activating intracellular signals through interaction with cell surface receptors. Recently, HA has attracted a great deal of attention because abnormal synthesis is often associated with cancer progression; increased synthesis of HA, and in many cases, upregulation of HAS, has been linked with malignant progression in certain types of human tumors, including breast cancer, where the level of HA is considered to be a reliable prognostic indicator. Additionally, ectopic expression of HAS genes and perturbation of endogenous HA function in several cancer cell lines have both suggested that accumulated HA stimulates the growth, survival, invasion, and metastasis of cancer cells. In the following study, a transgenic mouse model that allows overexpression of murine Has2 in the mammary glands under the control of Cre recombinase was generated to further investigate the roles of HA in carcinogenesis and cancer progression (Koyama et al. 2007).

Generation of Has2 Transgenic Animal and Backcross with Mammary Tumor Model

1. Murine Has2 cDNA is subcloned into multicloning sites of pCALNL5 expression vector (Kanegae et al. 1996) as depicted in Fig. 1.
2. The CALNL5-Has2 unit is excised from the vector by *Sal*I and *Sfi*I digestion and purified by agarose gel electrophoresis.
3. The purified CALNL5-Has2 fragment is microinjected into fertilized BALB/cCrSlc mouse eggs.
4. The generated Has2 conditional transgenic mice are backcrossed for nine generations to the mouse mammary tumor virus-Neu (MMTV-Neu) mammary tumor model (Charles River Laboratories International, Inc.Wilmington, MA, USA).

Department of Molecular Oncology, Division of Molecular and Cellular Biology, Institute on Aging and Adaptation, Shinshu University Graduate School of Medicine, Matsumoto, Nagano 390-8621, Japan
Phone: +81-263-37-2722, Fax: +81-263-37-2724
E-mail: itano@sch.md.shinshu-u.ac.jp

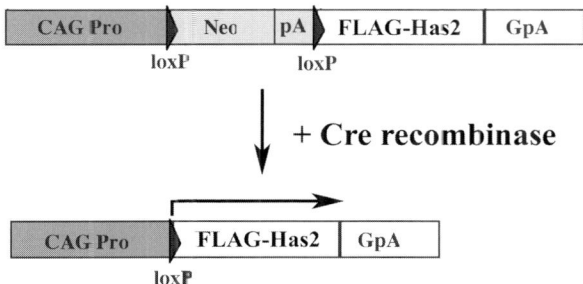

Fig. 1 Schematic illustrate of the transgenic construct. FLAG-tagged murine Has2 cDNA was positioned downstream of the transgene unit, which included a CAG promoter (Sakai and Miyazaki 1997) (CAG Pro), a *lox*P sequence, the *Neo*-resistance gene (Neo), the SV40 poly(A) signal (pA), and a second *lox*P sequence. Upon recognition of the *lox*P site, Cre recombinase deletes the Neo cassette along with one of the *lox*P sequences and then joins the CAG promoter and Has2 cDNA, leading to expression of Has2 mRNA

Table 1 Primer sequences of the oligonucleotides used for PCR

Primer	Sequence (5' → 3')
Has2 forward	5'-GACCTGGTGAGACAGAAGAGTCCC-3'
Has2 reverse	5'-TATATTAAAAGCCATCCAGTATCTCACG-3'
Cre forward	5'-GCGGTCTGGCAGTAAAAACTATC-3'
Cre reverse	5'-GTGAAACAGCATTGCTGTCACTT-3'
Neu forward	5'-GGAACCTTACTTCTGTGGTGTGAC-3'
Neu reverse	5'-TAGCAGACACTCTATGCCTGTGTG-3'

5. B6129-TgN (MMTV-cre)4Mam (MMTV-Cre) mice expressing Cre recombinase (Jackson Laboratories. Bar Harbor, ME, USA) are backcrossed for six generations to the MMTV-Neu mice
6. Has2:Neu bigenic mice bearing both *Has2* and *neu* transgenes are intercrossed to Cre: Neu bigenic mice bearing both *Cre* and *neu* transgenes.
7. The founder lineages with a different combination of three transgenes, MMTV-Neu (Neu), MMTV-Cre/MMTV-Neu (Cre:Neu), CAG-Neo-Has2/MMTV-Neu (Has2^{+Neo}), and CAG-Has2/MMTV-Cre/MMTV-Neu (Has2$^{\Delta Neo}$), are generated and genotyped by PCR analysis of genomic DNA.

Genotyping of Conditional Transgenic Mice

Potential founders were analyzed for the presence of the transgene by polymerase chain reaction (PCR) of mouse genomic DNA isolated from tail specimens.

1. Genomic DNA is extracted from the tail and analyzed by PCR using the oligonucleotide primers as described in Table 1.
2. The PCR condition for *Has2* and *neu* transgenes is as follows: 1 cycle at 94°C for 2 min; 35 cycles at 94°C for 45 s; 59°C for 1 min; 72°C for 1 min; 1 cycle at 72°C for 7 min.
3. The PCR condition for *Cre* transgene is as follows: 1 cycle at 94°C for 3 min; 35 cycles at 94°C for 30 s; 51°C for 1 min; 72°C for 1 min; 1 cycle at 72°C for 2 min.
4. PCR products are run on 2% agarose gels.

Results

We investigated the role of HA in carcinogenesis and cancer progression using the MMTV-Neu transgenic model of spontaneous breast cancer. Conditional transgenic (cTg) mice that express murine Has2 by Cre-mediated recombination were generated and crossed with MMTV-Neu mammary tumor models. In expressing Cre recombinase under the control of the MMTV promoter, the bigenic mice bearing *Has2* and *neu* transgenes exhibited elevated Has2 expression and HA production in the mammary tumors. The ratio of tumor-bearing mice to total mice was greater in the Has2-overexpression group, which was also associated with significantly faster tumor growth after appearance. Without the *neu* oncogene, however, Has2 cTg mice did not develop any visible tumors over a period of 1 year. Thus, overproduced HA is likely to synergistically enhance tumor development initiated by oncogenic alterations. Histologically, Has2 overexpressing tumors were poorly differentiated adenocarcinomas with numerous loosely cohesive tumor cells. As demonstrated by intense staining of type I collagen and fibronectin, prominent stromal reactions were induced within the tumors. Furthermore, microvessels frequently penetrated and accumulated into the stromal compartments.

Taken together, the transgenic mouse model may prove useful for the preclinical study of anti-cancer drugs targeting the host stromal reaction and tumor angiogenesis.

References

Kanegae Y, Takamori K, Sato Y, Lee G, Nakai M, Saito I (1996) Efficient gene activation system on mammalian cell chromosomes using recombinant adenovirus producing Cre recombinase. Gene 181:207–212

Koyama H, Hibi T, Isogai Z, Yoneda M, Fujimori M, AmanoJ, Kawakubo M, Kannagi R, Kimata K, Taniguchi S, Itano N (2007) Hyperproduction of hyaluronan in neu-induced mammary tumor accelerates angiogenesis through stromal cell recruitment: possible involvement of Versican/PG-M. Am J Pathol 170:1086–1099

Sakai K, Miyazaki J (1997) A transgenic mouse line that retains Cre recombinase activity in mature oocytes irrespective of the cre transgene transmission. Biochem Biophys Res Commun 237: 318–324

Section XII
Regeneration Medicine and Transplantation

Proteoglycans in Tissue Regeneration

Koji Kimata

Cells Used for the Regeneration

Three different types of cells are generally used: ES cells derived from the inner mass cells are known to have the ability to proliferate almost forever and differentiate into various cells with different functions (multi-differentiation activity). Stem cells which are thought to be precursor cells forming a tissue with specific function could be isolated from differentiating and differentiated tissues. Some of the differentiated cells such as epidermal cells could be cultured and proliferated with their specific differentiation functions.

Conditions for the Regeneration

Whatever cells are used, it is required for the regeneration to culture cells under the appropriate conditions. The cells need to adhere to proper extracellular substrate molecules. This requirement could correspond to microenvironments for the regeneration of cells and is considered the so-called "niche" which people recently use in those cases. The cells also need a suitable combination of cell growth factors with defined concentrations. It should be noted here that cell growth factors also need proper microenvironments for their activities to the cells. Therefore, we could say that cells having potentials of differentiation, microenvironments, and cell growth factors are three major requisites for cell regeneration (Fig. 1).

Participation of Sugar Chains in Regeneration

ES cells express on the cell surfaces embryoglycan having poly-N-acetyllactosamine which carries stage-specific embryonic antigen (SSEA)-1 antigen and has long been studied, TRA-1-60 which is a high molecular weight glycoprotein, and glycolipids of globo-series such as SSEA-3 and SSEA-4 antigens. Muramatsu (1988) greatly contributed to structural characterization of embryoglycan, and Hakomori (Eggens et al. 1989) also contributed in revealing its possible function that intermolecular interactions through the SSEA-1 antigen could affect adhesion among early embryonic cells. Those studies imply the first indication of the possibility that direct sugar–sugar interactions take part in cell adhesion. The above sugar antigens are now often used as markers for the selection or characterization of ES cells as well as some somatic stem cells. Further those may be essential for their long-lasting multiple-differentiating activities. ES cells also express the enzymes for the syntheses of heparan sulfate and chondroitin sulfate together with those of core proteins of some distinct proteoglycans (Kimata et al., unpublished observations) and so proteoglycans may also be involved in those activities. It has been

---

Aichi Medical University, Research Complex for the Medicine Frontiers, Aichi, Japan
Phone: +81-52-264-4811 ext. 1431, Fax: +81-561-61-1896

Fig. 1 Three essential factors for tissue regeneration

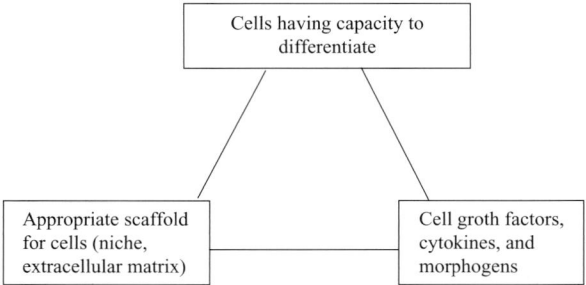

found recently that ES cells also express a series of molecules mediating BMP and Wnt signalings. Therefore, proteoglycans may be involved in those signalings. The reason will be stated later.

Application of Molecular Biology

Now that we knew genes for many molecules composed of microenvironments for regeneration as described above, we could take advantages of the information in order to have access to actual roles of microenvironments in regeneration. Gene knockout or mutation using homologous recombination techniques and gene silencing using RNAi or morpholine antisense RNA techniques are now available for this purpose. Hereinafter, we will see examples for such "loss-of-function" studies on proteoglycans later.

Cell Growth Factors and Heparan Sulfate Proteoglycans

Heparan sulfate proteoglycans (HSPG) are present ubiquitously on the cell surface and in the extracellular matrix. Syndecans and glypicans are two large families of HSPGs on the cell surfaces, and perlecan and agrin are major HSPGs in the extracellular matrix. Their HS chains interact with growth factors and morphogens such as FGFs, Wnts, BMPs, VEGFs, and Hh as well as receptors for those factors such as FGFR, and control activity, distribution, and stability of those factors through the interactions. The specificities of the interactions between HS and factors are at least in part due to the structure of HS chain characterized by the sulfation patterns and the isomer of hexuronic acid residues, GlcA or IdoA. The structures of HS chains are generated through the coordinate actions of HS modification enzymes, namely, N-deacetylase/N-sulfotransferases (NDSTs), C5-epimerase (HSepi), and 2-O-, 6-O-, and 3-O-sulfotransferases, following the biosynthesis of the HS backbone consisting of alternating glucuronic acid (GlcA) and N-acetylglucosamine (GlcNAc) residues. Recent loss-of-function studies on HS biosynthesis enzymes have demonstrated that HS has critical roles in the functions of cell growth factors necessary for developmental processes.

Wnt Signaling

In knockout mice of HS2ST, an enzyme catalyzing 2-O-sulfation of iduronic acid residues of heparan sulfate, the complete defect of kidney was detected and shown to be due to abnormal Wnt signal transduction causing failure of mesenchymal condensation surrounding ureteric buds and subsequently that of its branching (Bullock et al. 1998). The similar abnormality of kidney was also observed recently by the Swedish group in the heparan sulfate C5-epimerase knockout mice. These knockout mice can hardly synthesize heparan sulfate chains containing 2-O-sulfated IdoA (iduronic acid) residues, because HS2ST largely preferred IdoA residues to GlcA in HS as substrate. This may explain the reason for the above observation.

Mammals may have three genetically different HS6STs (−1, −2, and −3). Each shows distinct tissue distribution as well as different substrate specificity so as to yield HS chains having tissue-dependent sulfation patterns. Actually, KO mice of HS6ST-1 and HS6ST-2 showed largely different phenotypes in that the former mice are embryonic lethal with incomplete vasculogenesis in placenta. This abnormality may be caused by the down regulation of VEGF expression and fibronectin synthesis, in which abnormal Wnt2a signaling is involved.

FGF-Signaling

The local expressions and subsequent appropriate diffusions of FGF-10, FGF-8 and FGF-4 are essential for limb bud development and skeletal pattern formation in embryo. Actually FGF-10 knockout mice did not have any limbs. We have demonstrated that the siRNA inhibition of heparan sulfate 2-O-sulfotransferase caused structural defects in heparan sulfate, accompanied by the abnormal expression of those FGFs and the abnormal limb development (Kobayashi et al. 2007), which suggests that heparan sulfate proteolgycans are greatly involved in tissue morphogenesis.

Tissue Regeneration and Engineering, and Proteoglycans

Artificial Proteoglycans

We created new molecules by covalently conjugating glycosaminoglycans to phosphatidyl ethanolamine dipalmytoyl (PE) that could be considered as a substitute for the core proteins (Sugiura et al. 1993) (Fig. 2). These new molecules can easily bind not only to plastic surfaces of culture dishes and cell surfaces of living cells by their hydrophobic PE portions but also to a variety of glycosaminoglycan-binding molecules by the glycosaminoglycan portions. Thus, these glycosaminoglycan-PEs on plastic dish in culture mimic the proteoglycans in the extracellular matrix of tissues. Actually, glycosaminoglycan-PEs on the dish have the capacities to fix a variety of glycosaminoglycan-binding molecules to the surfaces of dishes. For example, heparin and/or heparan sulfate-PEs can fix FGFs as an active complex to the surfaces of plastic culture dishes.

Application of Artificial Proteoglycans to Tissue Engineering and the Regeneration

When we seeded liver parenchymal cells onto the chondroitin sulfate-PE-coated dish, the cells formed a cluster (we named these cell aggregates "spheroid") and kept their intrinsic differentiation function for a long time. Otherwise, cells adhered to the plastic surfaces firmly and lost such a function rapidly.

Akaike et al. in collaboration with Kobayashi's group synthesized β-galactose-conjugated stylene homopolymer (poly(*N*-*p*-vinylbenzyl-*O*-β-D-galactopyranosyl-D-gluconamide, PVLA) and used the culture substrate on plastic dishes for liver parenchymal

(A) 75PLC scaffold (from Honda et al. 2004)

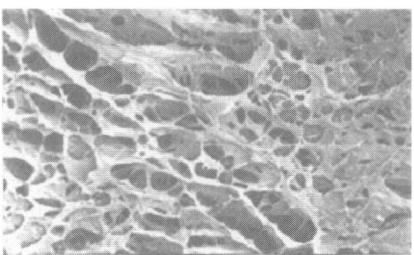

B) Artificial proteoglycans   (CS-PE) (from Sugiura et al. 1993)

Chondroitin 6-sulfate   **Phosphatidylethanolamine**

Fig. 2 Collagen-coated biodegradable polymer of L-lactide-ε-caprolactone, a new scaffold for tissue engineering for cartilage and bones (**A**), structure of CS-PE, an artificial proteoglycan (**B**)

cells. The idea was due to the observation that asialo-glycoprotein receptors on those cells specifically bind to β-galactose-residues. The cells on such a substrate were found to keep their differentiation function such as albumin synthetic activity (Cho et al. 1996). The results suggest the possible application of those substrate materials to the tissue engineering for artificial liver. The accumulation of such examples would lead to create new technology helpful for liver regeneration using stem cells.

"Niche" and Cell Growth Factors

Cell proliferation, cell motility, cell differentiation, cell morphology, and cell death are all needed for tissue regeneration, and those cell behaviors and properties are greatly dependent upon the substrates that the cells sit on and adhere to. A few decades earlier, in 1980, Yannas et al. developed artificial skin consisting of bilayers of a silicon sheet and a collagen-chondroitin sulfate C-sponge and used it for treatment of patients who got burned. The artificial skin brought about the skin reconstruction consisting of dermis and epidermis. Since then, many trials with some modification have been performed. For example, Kubo et al. reported that skin fibroblasts when grown on collagen/hyaluronan sponge gave a good cell sheet to treat skin ulcer because the cells releases some cell growth factors such as VEGF and FGF-2 which stimulate angiogenesis (Kubo and Kuroyanagi 2003).

Tissue engineering for cartilage and bones needs some idea to develop and/or maintain so-called a form. Collagen is usually degraded and metabolized rapidly, and so a form of the collagen sponge is fragile. We have developed a biodegradable hybrid polymer of L-lactide-ε-caprolactone in a 3:1 mixture, which is a plastic sponge as a new scaffold for such a purpose (Fig. 2). Actually, we have succeeded in making a cartilage tissue block by culturing rat costal chondrocytes in such a hybrid polymer block (Honda et al. 2004). In a series of experiments we are now trying to involve glycosaminoglycan-PE, artificial proteoglycans in such a biodegradable hybrid polymer, so that cell growth factors or morphogens are more accessible to their receptors and signaling cascades.

Conclusion

Tissue regeneration needs appropriate cell adhesion substrates and cell growth factors/morphogens as essential factors. Since we have succeeded in cloning genes for most of the enzymes involved in glycosaminoglycan synthesis, this would provide a profitable situation modifying glycosaminoglycan synthesis, so that these two essential factors for the regeneration could be modified by our hands.

References

Bullock SL, Fletcher JM, Beddington RS et al (1998) Renal agenesis in mice homozygous for a gene trap mutation in the gene encoding heparan sulfate 2-sulfotransferase. Genes Dev 12:1894–1906

Cho CS, Goto M, Kobayashi A et al (1996) Effect of ligand orientation on hepatocyte attachment onto the poly(N-p-vinylbenzyl-o-beta-D-galactopyranosyl-D-gluconamide) as a model ligand of asialoglycoprotein. J Biomater Sci Polym 7:1097–1104

Eggens I, Fenderson B, Toyokuni T et al (1989) Specific interaction between Lex and Lex determinants. A possible basis for cell recognition in preimplantation embryos and in embryonal carcinoma cells. J Biol Chem 264:9476–9484

Habuchi H, Nagai N, Sugaya N et al (2007) Mice deficient in heparan sulfate 6-O-sulfotransferase-1 exhibit defective heparan sulfate biosynthesis, abnormal placentation, and late embryonic lethality. J Biol Chem 282:15578–15588

Honda MJ, Yada T, Ueda M, Kimata K (2004) Cartilage formation by serial passaged cultured chondrocytes in a new scaffold: hybrid 75:25 poly(L-lactide-epsilon-caprolactone) sponge. J Oral Maxillofac Surg 62:1510–1516

Kobayashi T, Habuchi H, Tamura K, Ide H, Kimata K (2007) Essential role of heparan sulfate 2-O-sulfotransferase in chick limb bud patterning and development. J Biol Chem 282:19589–19597

Kubo K, Kuroyanagi Y (2003) Development of a cultured dermal substitute composed of a spongy matrix of hyaluronic acid and atelo-collagen combined with fibroblasts: fundamental evaluation. J Biomater Sci Polym Ed 14:625–641

Muramatsu T (1988) Developmentally regulated expression of cell surface carbohydrates during mouse embryogenesis. J Cell Biochem 36:1–14

Sugiura N, Sakurai K, Hori Y et al (1993) Preparation of lipid-derivatized glycosaminoglycans to probe a regulatory function of the carbohydrate moieties of proteoglycans in cell-matrix interaction. J Biol Chem 268:15779–15787

Stem Cell Glycobiology

Mitsuru Nakamura

Introduction

The field of "stem cell biology" is developing rapidly worldwide. However, research into "stem cell glycobiology" is yet to commence.

Meaning of Stem Cell Biology

Regenerative medicine was once thought to be a fantastic, futuristic notion. However, it actually started in 1957 with a clinical application of bone marrow allogenic transplantation: hematopoietic stem cell transplantation. Another form of tissue engineering-oriented regenerative medicine, intended for the loss of tissues that cannot be restored by cell transplantation therapy, originated from artificial skin therapy in the 1960s. Regenerative medicine involving both cell transplantation and tissue engineering is forecasted to become a worldwide market on the scale of tens of trillion yen.

Recently, research on the mechanism of stem cell self-renewal during normal development has advanced greatly. Signal transduction pathways in stem cell self-renewal have been revealed in hematopoietic and nerve tissues (Reya et al. 2001). The field of "stem cell biology" has emerged. When the signal pathways for self-renewal are controlled normally, tissues keep producing normal cells. In the hematopoietic system, 1×10^{16} blood cells are continuously produced from bone marrow-derived hematopoietic stem cells throughout a lifetime. While such systems are strictly controlled, the mechanisms are not fully understood. At present, one cannot control the proliferation and self-renewal of stem cells. Therefore, studies for ex vivo proliferation of tissue stem cells and ex vivo control of cell differentiation have started applying the outcomes of research on the mechanisms of stem cell self-renewal, growth, and differentiation.

How can glycobiology contribute to research on stem cell biology? First of all, sugar chains are possibly associated with stem cell self-renewal. Studying the mechanisms of glycobiology-associated stem cell self-renewal and control of differentiation may contribute to "stem cell biology." Second, it is necessary to search for sugar chain-associated markers of stem cells. A method of selecting stem cells prospectively has yet to be established. The most appropriate markers to discriminate stem cells from non-stem cells may be cell surface oligosaccharides. Possible sugar chain-related cell surface markers can be used for the enrichment of stem cells or exclusion of non-stem cells. Technologies for the purification and selection of stem cells will provide great practical convenience in the future. Even if "ex-vivo" stem cell proliferation becomes possible, it is impossible

---

Cell Regulation Analysis Team, Research Center for Medical Glycoscience, National Institute of Advanced Industrial Science and Technology, Function of Biomolecule, Medical Science for Control of Pathological Processes, Tsukuba University, Central-2, 1-1-1 Umezono, Tsukuba, Ibaraki 305-8568, Japan
Phone: +81-29-861-2745, Fax: +81-29-861-2744
E-mail: owl.nakamura@aist.go.jp

to differentiate all cells into the same lineage. There is no question that we need to establish selection technologies to separate desirable cells from nondesirable cells.

Somatic and Embryonic Stem Cells

There are two types of stem cells, distinguishable in terms of origin and differentiation potential: somatic or tissue stem cells and embryonic stem (ES) cells.

A. Somatic Stem Cells

Somatic stem cells studied extensively to date include hematopoietic, mesenchymal, and neural stem cells. However, the expression profiles of sugar chains in stem cells and the participation of these sugar chains in stem cell self-renewal and differentiation have not yet been fully elucidated. The reason for this is that each tissue contains very few somatic stem cells. A protocol for preparing side population cells from mouse bone marrow is presented in this section.

A rare example, in which sugar chains are associated with the mechanism of cell differentiation, is notch signaling during T cell development (Radtke et al. 2004). Another example is in vivo trafficking of hematopoietic stem cells via selectin-associated molecules (Simmons et al. 2001; Milinkovic et al. 2004; Xia et al. 2004). Even in the above two cases, however, the mechanisms are not fully elucidated. Some progress has been made in the study of neural stem cells. Galectin-1 was identified using mass-spectrometry as a soluble factor necessary for maintaining the self-renewal of neural stem cells (Sakaguchi et al. 2006). As for mesenchymal stem cells, the search for markers for selecting differentiated cells among human mesenchymal stem cells has started.

B. ES Cells

As there are ethical problems with the research using human ES cells, many scientists deal with murine ES cells (http://sweet.ccrc.uga.edu/glycomics/glycomics.php). However, there may be a limitation in using murine ES cells to find cell surface selection markers, as characteristic sugar chains are species-specific. Therefore, clarifying ethical issues is indispensable for the search of sugar chain-associated selection markers in human ES cells. Somatic stem cells are appropriate for finding sugar chain-associated selection markers. On the other hand, murine ES cells may be suitable for studying the mechanism of self-renewal and differentiation of stem cells. Typical examples of such research include Notch and EGF signaling.

Applications of Research Outcomes

Advances in regenerative medicine will be applicable to many diseases. At present, a number of diseases can only be treated with palliative therapy and many patients cannot expect a complete cure. Common examples are the so-called adult or geriatric diseases, lifestyle-related diseases, and diseases with wide-ranging loss of tissue or degenerated tissue. Such diseases will be treated using cell supplementation, stem cell transplantation, regeneration induced by possible medications, or reconstruction of the lost tissue or organ. Regenerative medicine covers a broad spectrum of diseases. Even at present, targets of hematopoietic stem cell transplantation therapy are thought to be not only hematologic diseases including leukemias and lymphomas but also autoimmune diseases and all solid malignancies in which chemotherapy is the first choice. Mesenchymal or other somatic stem cell transplantation therapies are targeted at myocardial infarctions, arterioscleroses, cerebrovascular disorders, diabetes, liver diseases, and intractable neuronal diseases including Parkinson's disease. Regenerative medicine may hold the key to the development of radical cures for many diseases. Similarly, advances in "stem cell glycobiology" will inevitably be applied to the treatment of intractable diseases. Stem cell research has become a worthy challenge.

Preparation of Side Population Cells from Mouse Bone Marrow

1. Take femora and tibiae from C57BL/6 mice and cut both ends. Insert a 25 G needle into the bones and squeeze out bone marrow cells with ice-cold PBS(-) into dishes.

2. After removing cellular clumps using a cell strainer (BD #352350), spin down the cells. Following hemolysis using lysis buffer for 8 min on ice, add cold PBS(-) gradually, and spin down the cells.

3. Resuspend the cells in cold PBS(-), count cell density, spin down, and again resuspend the cells in a warm staining medium to make a suspension of 10^6 cells/mL. Add Hoechst 33342 (5 μg/mL, final concentration) in the presence or absence of Verapamil (20 μg/mL, final concentration) to the suspension, and incubate the cells at 37°C for 90 min.

4. Stop the reaction by adding cold PBS(-) and spin down the cells.

5. Stain the cells with appropriate mAbs and 7AAD and analyze the expression of glyco-epitopes of interest using a flow cytometer.

References

Milinkovic M, Antin JH, Hergrueter CA et al (2004) CD44-hyaluronic acid interactions mediate shear-resistant binding of lymphocytes to dermal endothelium in acute cutaneous GVHD. Blood 103: 740–742

Radtke F, Wilson A, Mancini SJC et al (2004) Notch regulation of lymphocyte development and function. Nat Immunol 5:247–253

Reya T, Morrison SJ, Clarke MF et al (2001) Stem cells, cancer, and cancer stem cells. Nature 414:105–111

Sakaguchi M, Ishibashi S, Kuroiwa T et al (2006) A carbohydrate-binding protein, Galectin-1, promotes proliferation of adult neural stem cells. Proc Natl Acad Sci USA 2006 May 2, 103(18):7112–7117

Simmons PJ, Levesque J-P, Haylock DN (2001) Mucin-like molecules as modulators of the survival and proliferation of primitive hematopoietic cells. Ann NY Acad Sci 938:196–206

Xia L, McDaniel JM, Yago T et al (2004) Surface fucosylation of human cord blood cells augments binding to P-selectin and E-selectin and enhances engraftment in bone marrow. Blood 104: 3091–3096

Human Embryonic Stem (ES) Cells and Induced Pluripotent Stem (iPS) Cells Are Defined by Carbohydrate Markers

Akihiro Umezawa, Masashi Toyoda

Introduction

Embryonic stem (ES) cells and induced pluripotent stem (iPS) cells are sources for regenerative medicine and tissue replacement after injury or disease. The stem cells can be defined based on a distinctive set of cell surface carbohydrate markers. SSEAs are carbohydrate antigens associated with various core glycolipids and are routinely used for the characterization of human pluripotent cells. Human ES cells express SSEA-3 and SSEA-4 but not SSEA-1, while their differentiation is characterized by downregulation of SSEA-3 and SSEA-4 and an upregulation of SSEA-1 (Thomson et al. 1998). SSEA-3 (Galß-globoside) and SSEA-4 (sialyl-Galß-globoside) are epitopes localized at the cell surface that are associated with globoseries glycolipids (Kannagi et al. 1983). Undifferentiated, human ES cells also express the keratin sulphate-associated antigens TRA-1-60 and TRA-1-81 (Badcock et al. 1999). Interestingly, most of these markers are carbohydrates themselves or are closely related to the carbohydrates carried on glycosphingolipids and glycoproteins.

iPS cells are a type of pluripotent stem cell artificially derived from a non-pluripotent cell, typically an adult somatic cell, by inserting certain genes. In many respects, such as the expression of carbohydrate markers, teratoma formation, and differentiability, iPS cells are believed to be identical to natural pluripotent stem cells, such as ES cells. Human iPS cells express the markers specific to hESC, including SSEA-3, SSEA-4, TRA-1-60, TRA-1-81, and TRA-2-49/6E (Takahashi et al. 2007).

Procedure

Immunocytochemistry of SSEA-4 (Xu 2006)

Note: This protocol of immunocytochemistry is for cells grown in a 24-well tissue culture plate.

1. Wash the cells twice with 1 ml of PBS.
2. Fix the cells with 0.5 ml of 4% paraformaldehyde (prepared in PBS) for 20 min at room temperature.
3. Wash the cells twice with 1 ml of PBS for 5 min.
4. Block the cells with 0.5 ml of 0.1% BSA, 10% normal donkey serum in PBS at room temperature for 45 min.

Department of Reproductive Biology, National Institute for Child Health and Development, 2-10-1 Okura, Setagaya-ku, Tokyo, 157-8535, Japan
Phone: +81-3-5494-7047, Fax +81-3-5494-7048
E-mail: umezawa@nch.go.jp

Fig. 1 Detection of Tra-1-60 in embryonic stem cells

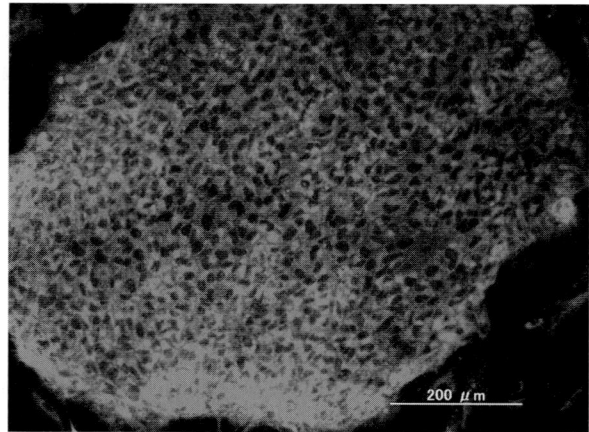

5. After blocking, incubate the cells with 300 µl/well of diluted primary antibody working solution overnight at 2°–8°C.
6. Wash the cells three times with 1 ml of PBS containing 1% BSA for 5 min.
7. Dilute the secondary antibody [e.g. Rhodamine Red-conjugated donkey anti-mouse IgG secondary antibody (Jackson Immunoresearch, Catalog # 715-295-150) for SSEA-4] according to the manufacturer's instructions in PBS containing 1% BSA. Incubate the cells with secondary antibody at 300 µl per well for 60 min at room temperature in the dark.
8. Wash the cells three times with 1 ml of PBS containing 1% BSA for 5 min.
9. Cover the cells with 1 ml of PBS and visualize with a fluorescence microscope.

List of Reagents

Antibodies against SSEA-1, SSEA-3, and SSEA-4 can be obtained from the Developmental Studies Hybridoma Bank (University of Iowa, Iowa City, IA) or from Chemicon (Temecula, CA), where antibodies against TRA-1-60 and TRA-1-81 are also available. For double staining, choose primary antibodies with different isotypes (e.g., IgG3-type SSEA-4 antibody together with IgM-type TRA-1-60 antibody).

References

Thomson JA, Itskovitz-Eldor J, Shapiro SS, Waknitz MA, Swiergiel JJ, Marshall VS, Jones JM (1998) Embryonic stem cell lines derived from human blastocysts. Science. 282(5391):1145–1147. Erratum in: Science, 282(5395):1827

Takahashi K, Tanabe K, Ohnuki M, Narita M, Ichisaka T, Tomoda K, Yamanaka S (2007) Induction of pluripotent stem cells from adult human fibroblasts by defined factors. Cell. 131(5):861–872

Draper JS, Pigott C, Thomson JA, Andrews PW (2002) Surface antigens of human embryonic stem cells: changes upon differentiation in culture. J Anat 200(3):249–258

Sato B, Katagiri YU, Miyado K, Akutsu H, Miyagawa Y, Horiuchi Y, Nakajima H, Okita H, Umezawa A, Hata J, Fujimoto J, Toshimori K, Kiyokawa N (2007) Preferential localization of SSEA-4 in interfaces between blastomeres of mouse preimplantaion embryos. Biochem Biophys Res Commun 364(4):838–343

Xu C (2006) Characterization and evaluation of human embryonic stem cells. Methods Enzymol 420:18–37

Glycoantigens in Xenotransplantation

Shuji Miyagawa

Introduction

The growing numerical gap between the number of patients and available human donor organs has led to a revival of interest in xenotransplantation. A number of trials are currently underway, especially pig islet transplantation, the aim of which is to develop transgenic and knockout pigs in which hyperacute rejection (HAR) and acute vascular rejection (AVR) can be overcome. These rejections are strongly related to species-specific glycoantigens.

The α-Gal epitope (Gal α1-3Galβ1-4 GlcNAc-R), which is biosynthesized by the action of α1,3 galactosyltransferase (α1,3GT), is closely associated with HAR in pig to human xenotransplantation. A variety of strategies have been pursued, in attempts to reduce or eliminate the α-Gal epitope from pig tissues. In addition to knocking out α1,3GT by nuclear transplantation techniques from the targeted somatic cells, other strategies, such as enzyme competition by α1,3GT with other glycosyltransferases and/or the control of sugar processing by glycosyltransferases, have provided some very interesting insights into the downregulation of such xenogenetic epitopes. Some of the strategies continue to pursue downregulating the non-α-Gal epitopes, include the Hanganutziu-Deicher (H-D) epitope, which contains N-glycolylneuraminic acid (NeuGc) and other currently unknown non-Gal epitopes, in transgenic animals after α1,3GT is eliminated.

In addition, the analysis of natural killer (NK) cell receptors, based on the importance of glycosyl epitopes in NK cell activity related to AVR is currently under study for downregulating the infectivity of porcine endogenous retrovirus (PERV) to human cells by modification of glycosyltransferases. The latter is also an important topic.

Concept

Substrate Competition in Core Glycosylation

In general, the regulation of sugar chain biosynthesis is under the control of glycosyltransferases, the substrate specificities of the enzymes and the localization of the enzymes in tissues and organelles. β-D-mannoside. β1,4-N-acetylglucosaminyltransferase III (GnT-III) catalyzes the branching of N-linked oligosaccharides, producing a bisecting N-acetylglucosamine (GlcNAc) residue. Once a bisecting GlcNAc residue is added to the core mannose by GnT-III, the action of other competitive enzymes such as α 3-D-mannoside β1,4-N-acetylglucosaminyltransferase IV (GnT-IV) and α-6-D-mannosideβ1,6-N-acetylglucosaminyltransferase V (GnT-V) is prevented from introducing any additional tri-antennary structures into the Golgi stack. As a result, it is likely that α-Gal

---

Division of Organ Transplantation, Department of Molecular Therapeutics, Osaka University Graduate School of Medicine, 2-2 Yamadaoka, Suita, Osaka 565-0871, Japan
Phone: +81-6-6879-3062, Fax: +81-6-6879-3069
E-mail: miyagawa@orgtrp.med.osaka-u.ac.jp

epitope levels are decreased because the bisecting GlcNAc suppresses further processing of complex-type sugar chains (Miyagawa et al. 2001).

Substrate Competition in Terminal Glycosylation

This strategy for downregulating the α-Gal epitope capitalizes on the enzymatic competition for terminal glycosylation between α1,3GT and other glycosyltransferases for the common acceptor substrate in the trans-Golgi network by masking the α-Gal epitope by fucosylation or sialylation, such as α1,2FT, α1,3 fucosyltransferase (α1,3FT), α2,3 sialyltransferase (α2,3ST), as well as α2,6ST.

Knock Out (KO)

Knocking out the pig α1,3GT gene not by homologous recombination in embryonic stem cells but by nuclear transplantation techniques from somatic cells, fetal fibroblasts, have been established at several institutes, include ours (Takahagi et al. 2005).

Regarding the xenoantigenicity to human serum remaining on cells from the α1,3GT-KO pig, no clear data have yet beenreported. In addition, other issues such as pig iGb3 synthase and the non-Gal epitope have become subjects of interest.

Glycosyltransferase Assays

The use of fluorescence-labeled oligosaccharides as acceptor substrates by a reversed-phase HPLC system equipped with a fluorescence detector.

Wash the cells twice with PBS and then centrifuge them at $1,500 \times g$ for 10 min.

Resuspend the pelleted cells in 100 ml of PBS and then lyse them by sonication.

In the activity assays for α2,6ST, α2,3ST, α1,2FT, and α1,3GT, use pyridylaminated lacto-N-neotetraose (LNnT-PA), a common acceptor substrate.

Incubate whole cell lysates at 37°C for 3 h in the reaction mixtures.

*Mixtures for α2,6ST, α2,3ST: 50 mM cacodylate buffer, 10 mM $MnCl_2$, 0.23% Triton X-100, 5 mM CMP-sialic acid as the donor substrate, and mM LNnT-PA as the acceptor, pH 6.8.

*Mixtures for α1,2FT: 13 mM potassium phosphate buffer, 0.1% Triton X-100, 13 mM phenyl-b-D-galactoside, 5 mM ATP, 1 mM GDP-fucose and 10 mM LNnT-PA, pH 6.1.

*Mixtures for α1,3GT; containing 10 mM 2-[4-(2-Hydroxyethyl)-1-piperazinyl] ethanesulfonic Acid (HEPES), pH 7.2, 20 mM UDP-Galactose, 10 mM $MnCl_2$, 33 mM NaCl, 3 mM KCl.

For the activity of GnT-III, react the cell lysates with 20 mM UDP-GlcNAc and 20 mM pyridylaminated agalacto biantennary oligosaccharide (GlcNAcβ1-2Manα1-6 (GlcNAcβ1-2Manα1-3) Manβ1-4GlcNAcβ1-4GlcNAc-PA) in the presence of 63 mM MES, 10 mM $MnCl_2$, 200 mM GlcNAc and 0.5% Triton X-100, pH 6.25.

Incubate the assay mixtures at 37°C for 3 h, and then terminate the reactions by heating in a boiling water bath for 5 min, followed by centrifugation of the samples at $15,000 \times g$ for 5 min.

Inject the resulting supernatants into a reversed phase HPLC equipped with a TSKgel column, ODS 80TM (4.6 250 mm^2).

Separate the product and the substrate isocratically with a 20 mM ammonium acetate buffer (pH 4.0) containing 0.01% n-butanol.

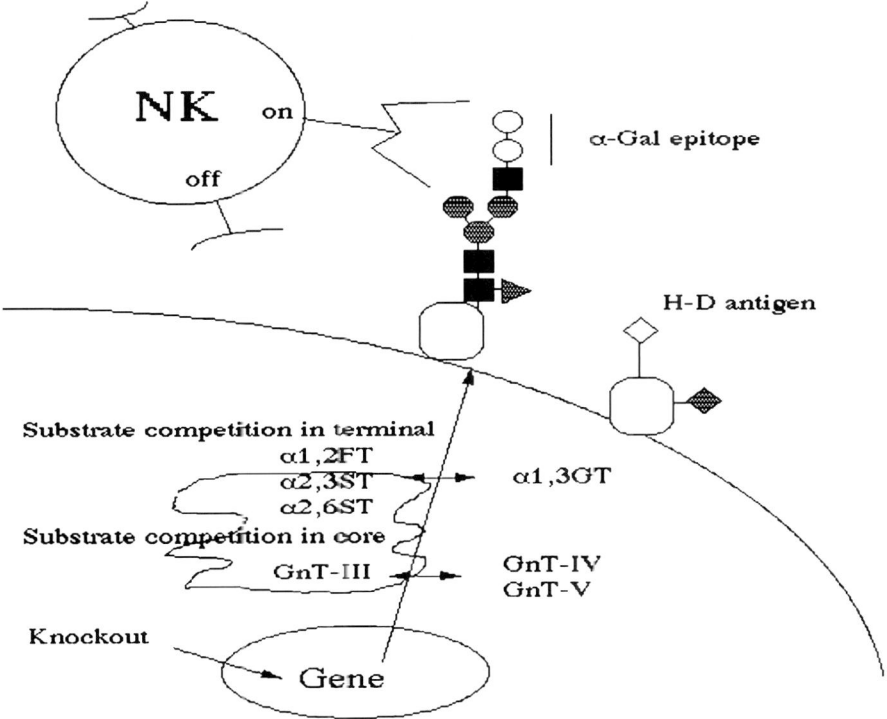

Fig. 1 Strategy for remodeling glycoantigens

Detect the fluorescence of the column elute with a fluorescence detector using excitation and emission wavelengths of 320 and 400 nm, respectively.

Express the specific activity of the enzyme as moles of product produced per hour of incubation per milligram of protein.

Determine protein concentrations using a BCA protein assay kit, using bovine serum albumin as a standard (Miyagawa et al. 1999, 2001).

Glycoantigens in Islets

The pig pancreas is considered to be the most suitable potential source of islets for xenotransplantation into patients with type I diabetes. The antigenicity of pig islets remains controversial. However, in our previous study, we clearly showed α-Gal expression in new bone pig islets but not in adult islets, via immunohistochemical analysis and the measurement of α1,3GT enzyme activity in the islets. We also clearly demonstrated the existence of the H-D antigen on both adult and new bone islets, which is in agreement with the conclusions reached in our previous PEC study (Omori et al. 2006).

Treatment with tunicamycin (TM), a specific inhibitor of the glycosylation of N-linked sugars, led to the drastic downregulation in human IgG and IgM deposition on pig islet cells, while the effect of PDMP (D-threo-1-phenyl-2-decan- oylamino-3-morpholino-1-propanol), a specific inhibitor of glycosylation of glycolipids, was much less. We also clearly demonstrated that nauraminidase treatment resulted in the clear downregulation of antigenicity without downregulation of the H-D antigen. These data indicate that pig sialic acids, in addition to not only the H-D antigen but also other sialic antigens, have antigenicity to human serum.

Enzyme Treatment

Tunicamycine: treat parental islets with 10 lg/ml tunicamycin, for 18 h in D-MEM containing 10% FBS.

PDMP: treat parental islets with 20 lMofthis reagent in D-MEM containing 10% FBS for 40 h, followed by FACS analysis.

Neuraminidase: treat parental islets with neuraminidase, in 0.2 m sodium acetate, pH 5.0, at 37°C for 3 h, and islets are then placed in phosphate-buffered saline again.

NK Cell Receptor

Increasing evidence has accumulated to suggest that NK cells play an important role in pig to human xenogenic cytotoxicity, whereas the molecular mechanism by which oligosaccharides are recognized by NK cells remains controversial. In our previous study, the remodeling of glycoantigen, especially the α-Gal epitope, by a glycosyltransferase, such as GnT- III, α2,3ST, and α1,2FT, affects the susceptibility of pig endothelial cells (PEC) to NK-mediated direct lysis in the absence of an antibody-mediated reaction (Miyagawa et al. 1999) (Fig. 1). Further studies to define the relation between the NK receptor and the xenogenic glycoantigen are needed.

PERV Infectivity and N-Linked Sugars

In pigs, at least several proviral copies of PERV-A, B, and C are present in the genome. In addition, recombination between PERV-A and PERV-C cannot be excluded. An analysis of O-glycans is less relevant to general retrovirus infectivity. On the other hand, most of the individual consensus N-linked glycosylation sites are indispensable for viral infectivity.

We addressed the effects of the remodeling of pig cell-surface glycoproteins, especially the high mannose type of N-glycan, on the susceptibility of PERV by the overexpression of α 1,2 mannosidase Ib (Man Ib), N-acetylglucosaminyltransferase I (GnT-I) and α-mannosidase II (Man II). The reduction in PERV infectivity by each enzyme was approximately 50%. It is possible that there is no specific limiting step in the N-linked sugar processing pathway and that each glycosyltransferase may function independently. Therefore, the individual over expression of each transferase is forced to the limit in suppressing PERV infectivity by reducing the high mannose type N-glycans on the PEC surface.

N-Linked Glyoprotein Processing Inhibitors

Two days before infection, add 10 µg/ml of Castanospermine (CS), 1 mM/ml of 1-Deoxynojirimycin (dNM), 1 mM/ml of 1-Deoxymannojirimycin (dMM), or 2 µg/ml of Swainsonine (SS) to the producer cell medium.

Continue the treatment for 18 h and then replace the medium with fresh D-MEM supplemented with 10% FBS.

Estimate the effect of N-glycosylation inhibition in a subsequent *Lac Z* assay (Hazama et al. 2003).

References

Hazama K, Miyagawa S, Miyazawa T, Yamada J, Tomonaga K, Ota M, Matsuda H, Shirakura R (2003) The significance of N-linked glycosylation in pig endogenous retrovirus (PERV) infectivity. Biochem Biophys Res Commun 310:327–333

Miyagawa S, Nakai R, Yamada M, Tanemura M, Ikeda Y, Taniguchi N, Shirakura R (1999) Regulation of natural killer cell-mediated swine endothelial cell lysis through genetic remodeling of a glycoantigen. J Biochem 126:1067–1073

Miyagawa S, Murakami H, Takahagi Y, Nakai R, Yamada M, Murase A, Koyota S, Koma M, Matsunami K, Fukuta D, Fujimura T, Shigehisa T, Okabe M, Nagashima H, Shirakura R, Taniguchi N (2001) Remodeling of the major pig xenoantigen by N-acetylglucosaminyltransferase III in transgenic pig. J Biol Chem 276:39310–39319

Omori T, Nishida T, Komoda, H, Fumimoto Y, Ito T, Sawa Y, Gao C, Nakatsu S, Shirakura R. Miyagawa S (2006) A study of the xenoantigenicity of neonatal porcine islet-like cell clusters (NPCC) and the efficiency of adenovirus-mediated DAF (CD55) expression. Xenotransplantation 11:237–246

Takahagi Y, Fujimura T, Miyagawa S, Nagashima H, Shigehisa T, Shirakura R, Murakami H (2005) Production of α 1,3-galactosyltransferase gene knockout pigs expressing both human decay-accelerating factor and N-acetylglucosaminyltransferase III. Mol Reprod Dev 71:331–338

Section XIII
Fertilization

Glycoconjugates in Spermatogenesis

Koichi Honke[1,2]

Introduction

Spermatogenesis occurs in the seminiferous tubules in the testis. Spermatogonia that are germline stem cells sit in a special place on the basement membrane of seminiferous tubules called niche. These spermatogonial stem cells differentiate into spermatocytes, ceasing proliferation. Subsequently, spermatocytes undergo meiosis and differentiate into haploid cells referred to as spermatids, migrating from the surrounding to the lumen of the seminiferous tubules on the stroma cells termed Sertoli cells with mutual interaction. After that, they mature into spermatozoa via morphogenesis and leave the seminiferous tubules for epididymis through the efferent ductules.

More than 90% of glycolipid in the testis consists of a unique glyceroglycolipid, seminolipid (Vos et al. 1994). Seminolipid is synthesized in spermatocytes and maintained in the subsequent germ cell stages (Ishizuka 1997). Its carbohydrate moiety is the same as that of sulfatide and is biosynthesized via sequential reactions catalyzed by common enzymes: ceramide galactosyltransferase (CGT, EC 2.4.1.45) and cerebroside sulfotransferase (CST, EC 2.8.2.11). Gangliosides are differentially distributed in mouse testis, namely "b" series gangliosides, such as GD1b and GT1b, are mainly expressed in germ cells, whereas "a" series gangliosides, such as GM1 and GD1a, are located in Leydig cells and Sertoli cells, respectively (Takamiya et al. 1998).

Results

In CGT-null mice, neither galactosylalkylacylglycerol (GalEAG) nor seminolipid is synthesized, and spermatogenesis is arrested prior to the meiosis (Fujimoto et al. 2000). On the other hand, primary spermatocytes seem to be normal in CST-null mice, but spermatogenesis is blocked at the metaphase of the first meiosis (Honke et al. 2002). The arrested stage in the germ cell differentiation of CST-null mice appears to be somewhat later than that in CGT-null mice, suggesting that both GalEAG and seminolipid are successively involved in the genetic program of spermatogenesis in the same order as their biosynthesis. Since these glycolipids are expressed on the cell surface of primary spermatocytes from the end of the leptotene stage or the zygotene stage and later, the interaction between Sertoli cells and spermatocytes, which is known to be important for their differentiation, may be disrupted in these knockout mice. To address the issue of which side of cell function, germ cells or Sertoli cells, is deteriorated in these mutant mice, it was examined whether spermatogenesis is restored or not after testis germ-cell trans-

---

[1] Department of Biochemistry, Kochi University Medical School, Kohasu, Oko-cho, Nankoku, Kochi 783-8505, Japan
Phone: +81-88-880-2313, Fax: +81-88-880-2314
E-mail: khonke@kochi-u.ac.jp

[2] CREST, Japan Science and Technology Agency, Japan

Fig 1 Glycans essential for spermatogenesis. Spermatogenesis is a precisely regulated process in which the germ cells closely interact with Sertoli cells. Seminolipid and a GlcNAc-terminated and triantennary and fucosylated N-glycan structure that are expressed on the germ cells play a key role in this germ cell–Sertoli cell interaction. In addition, complex gangliosides are involved in the transport of testosterone into the seminiferous tubules, which is required for differentiation of Sertoli cells

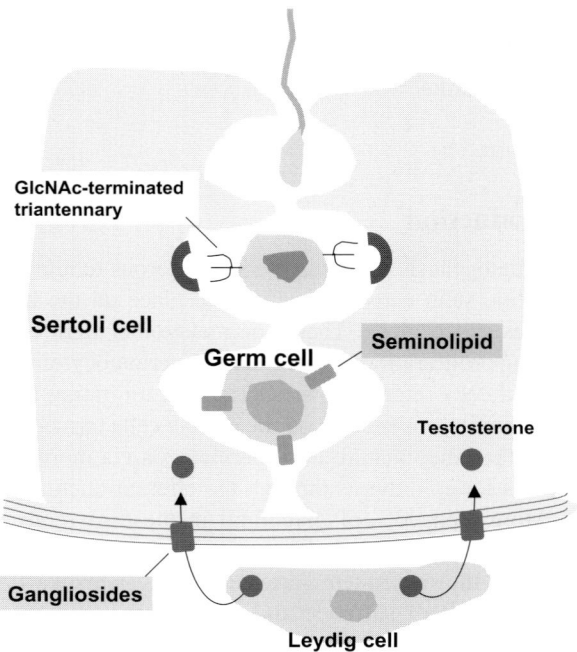

plantation, in which wild-type spermatogonial stem cells were injected into the seminiferous tubules of CST-null mice (Zhang et al. 2005). As a result, spermatogenesis was restored, proving that seminolipid is essential for germ cell function in spermatogenesis (Fig. 1). This functional result is consistent with the expression distribution of seminolipid.

Ganglioside GM2/GD2 synthase-knockout mice lack all complex gangliosides, and the males are sterile due to an arrest of spermatogenesis (Takamiya et al. 1998). The serum testosterone level is markedly low in the mutant mice, although the testosterone production in the Leydig cells is preserved. Testosterone is accumulated in Leydig cells, suggesting a defect in the testosterone transport from Leydig cells into serum and into Sertoli cells (Fig. 1). In fact, transport of intratesticularly injected testosterone is suppressed in GM2 synthase-null males.

The α-mannosidase IIx (MX) was identified as the gene product of α-mannosidase II (MII)-related gene. MX apparently plays subsidiary role for MII in many cell types, as N-glycan patterns of MX-null mouse tissues are not altered significantly. However, MX-null male mice are infertile due to a failure of spermatogenesis (Akama et al. 2002). The MX-null mice revealed that a specific N-glycan structure that is N-acetylglucosamine-terminated and has a fucosylated triantennary structure plays an important role in the adhesion between germ cells and Sertoli cells (Fig. 1).

References

Akama TO, Nakagawa H, Sugihara K, Narisawa S, Ohyama C, Nishimura S, O'Brien DA, Moremen KW, Millan JL, Fukuda MN (2002) Germ cell survival through carbohydrate-mediated interaction with Sertoli cells. Science 295:124–127

Fujimoto H, Tadano-Aritomi K, Tokumasu A, Ito K, Hikita T, Suzuki K, Ishizuka I (2000) Requirement of seminolipid in spermatogenesis revealed by UDP-galactose: ceramide galactosyltransferase-deficient mice. J Biol Chem 275:22623–22626

Honke K, Hirahara Y, Dupree J, Suzuki K, Popko B, Fukushima K, Fukushima J, Nagasawa T, Yoshida N, Wada Y, Taniguchi N (2002) Paranodal junction formation and spermatogenesis require sulfoglycolipids. Proc Natl Acad Sci USA 99:4227–4232

Ishizuka I (1997) Chemistry and functional distribution of sulfoglycolipids. Prog Lipid Res 36:245–319

Takamiya K, Yamamoto A, Furukawa K, Zhao J, Fukumoto S, Yamashiro S, Okada M, Haraguchi M, Shin M, Kishikawa M, Shiku H, Aizawa S, Furukawa K (1998) Complex gangliosides are essential in spermatogenesis of mice: possible roles in the transport of testosterone. Proc Natl Acad Sci USA 95:12147–12152

Vos JP, Lopes-Cardozo M, Gadella BM (1994) Metabolic and functional aspects of sulfogalactolipids. Biochim Biophys Acta 1211:125–149 (review)

Zhang Y, Hayashi Y, Cheng X, Watanabe T, Wang X, Taniguchi N, Honke K (2005) Testis-specific sulfoglycolipid, seminolipid is essential for germ cell function in spermatogenesis. Glycobiology 15:649–654

Recognition Mechanism of Egg and Sperm Based on Sugar Chains

Midori Matsumoto

Introduction

Recognition between sperms and eggs is required for species maintenance in organisms. Eggs are covered in an egg extracellular matrix, which contains considerable amounts of glycoprotein. This matrix usually possesses a site that prevents the recognition of sperms, which protect the egg against an attack from foreign bodies such as bacteria. The sperm is induced by a component of the egg extracellular matrix and is attached to the latter. The sperm increases the $[pH]_i$ and $[Ca^{2+}]_i$ on receiving a signal from the egg extracellular matrix and evokes exocytosis of the acrosomal vesicle at the tip of the sperm head (acrosome reaction). It then passes through the egg extracellular matrix and the vitelline membrane, and arrives at the egg cell membrane. Only the new sperm cell membrane, which is detached due to the exocytosis, can fuse with the egg cell membrane. A sugar chain is known to play an important role in the acrosome reaction and in the binding between the egg and the sperm.

List

Mammals

Among vertebrates, the mouse is the main animal used for studying fertilization mechanisms. The mammalian egg extracellular matrix is known as the zona pellucida (ZP). The mouse ZP (mZP) is composed of three major glycoproteins: mZP1, mZP2, and mZP3. The mZP1 helps in organizing the mZP2 and mZP3 molecules, and the mZP3 is responsible for primary sperm–egg binding and is also the inducer of the acrosome reaction. The mZP2 binds to the sperm after the occurrence of acrosome reaction. O-linked glycans present in mZP3 are important in sperm receptor activity; however, N-linked glycans do not affect the activity. The structures of the major mZP3 O-linked glycans are now known. These glycans are core type 2 sequences terminated in sialic acid, lacNAc (Galβ1-4GlcNAc), lacdiNAc (Gal-NAcβ1-4GlcNAc), Galα1-3Gal, and NeuAcα2-3[GalNAcβ1-4]Galβ1-4 (Dell et al. 2003). Core type 1 O-glycans are also present, of which some are terminated with sialic acid, β-linked N-acetylhexosamine, or NeuAcα2-3[GalNAcβ1-4]Galβ1-4. Furthermore, this study demonstrates that mouse eggs expressing human ZP3 bind to mouse sperm but not to human sperm, indicating that human ZP3 (huZP3) requires the same O-linked glycans as that of native mZP3. Recently, it was demonstrated that mouse eggs lacking core-1-derived O-glycans are

---

Department of Biosciences and Informatics, Faculty of Science and Technology, Keio University, 3-14-1 Hiyoshi, Kohoku-ku, Yokohama 223-8522, Japan
Phone: +81-45-566-1774, Fax: +81-45-566-1448
E-mail: mmatsumo@bio.keio.ac.jp

efficiently fertilized and bind to sperm well. Additional removal of complex and hybrid N-glycans in double mutant T-synF/FMgat1F/F:ZP3Cre females also resulted in fertilized eggs (Williams et al. 2007). Thus, neither Gal nor GlcNAc on N- or O-glycans of the mouse ZP function as essential sperm receptors. Further, since the precise glycan structures involved in such interactions remain unknown, the study of the recognition mechanism of egg and sperm in mouse remains unknown.

Echinoderm—Sea Urchin

The ligand from the egg jelly binds to the sperm and induces the acrosome reaction in the eggs of the echinoderm sea urchin. The sulfates, the polysaccharides of the egg jelly come into contact with the sperm membrane; these polysaccharides have been characterized, and they induce the acrosome reaction in a species-specific manner. In *Strongylocentrotus purpuratus*, the polysaccharide is a fucose sulfate polymer (FSP). The species specificity of the interaction is determined by the glycoside linkage of the polymer and the pattern of sulfation of the sugar residues (Viela-Silva et al. 2002). *S. franciscanus* and *S. purpuratus* belong to the same genus. They differ in the position of the sulfate residue linked to their polysaccharides. In the sea urchin *S. franciscanus*, sulfation of the polysaccharide occurs at the 2-position of fucose, whereas it occurs at the 4-position in *S. purpuratus*. Each sperm recognizes this difference and induces a species-specific acrosome reaction. A sea urchin egg jelly receptor (suREJ1) to which FSP binds on the sperm membrane is present, and an anti-suREJ1 antibody can induce the acrosome reaction. The N-terminal end of SuREJ1 contains 2 carbohydrate recognition domains (CRDs), a superfamily of immunoglobulin-type C domains. It is assumed that CRD can contribute to the binding between suREJ1 and FSP, and it is known to be a domain that plays an important role in cell recognition, natural immunity, and cell signaling. It is analyzed that the CRD of suREJ1-3 has been analyzed in six species of sea urchins in terms of evolution, and it was detected that CRD has a high evolution speed. It was also demonstrated that the gene was controlled by positive selection (Darwinian selection; Mah et al. 2005).

Echinoderm—Sea Star

The egg jelly of starfish contained three molecules that are involved in the acrosome reaction. In the case of *Asterias amurensis*, the Co-ARIS of steroidal saponin and peptide asterosap of 34 amino acid residues assists in the reactions of the acrosome reaction-inducing substance (ARIS), which is a large and highly sulfated glycoprotein conjugate. The activity of the sugar chain is maintained by pronase digestion but is lost on hyper-iodine acid oxidation. Ten repeats of the 5 sugar chains -4) β-Xyl$p$(1-3)-α-Gal$p$ (1-3)-α -Fuc$p$(4-SO$_3^-$)(1-3)-α -Fuc$p$(4-SO$_3^-$)(1-4)-α-Fuc$p$ (1-form the smallest activity unit. *A. amurensis* ARIS induces the acrosome reaction at the same level (intra-species level) for the sperm of a starfish belonging to Asteriidae (same genus as *A. amurensis*) but not for the sperm of a starfish belonging to a different genus (Nakachi et al. 2006). A similar result was obtained with the egg jelly from four species of starfish belonging to Asteriidae, as well as with the ARIS of *A. amurensis*. The egg jelly from Asteriidae contains the sugar repeat, including the same sulfate as that present in ARIS; however, this repeat is absent in the egg jelly of other starfish. It is believed that the sperm recognizes the structure of the sugar chain of a specific ARIS at the genus level.

From these data, we can examine whether the sugar chain and sulfate residues of the sugar repeats contribute to species differentiation by reproductive isolation.

Ascidian

In the case of the egg and sperm binding, sugar chain-modifying enzymes, which usually act as lectins, bind the sperm to the egg. The binding with the fucose ends of the vitelline membrane and the fucosidase of the sperm surface is essential for fertilization in the protochordates *Ciona intestinalis* and *Halocynthia roretzi*. Further, the binding of the GlcNAc end of the vitelline membrane and the hexosaminidase of the sperm surface is necessary for fertilization in *Phallusia mammillata*.

Comments

Several million species inhabit the earth; therefore, their gametes should be able to recognize another. The sugar chain is probably the best tool for recognition between gametes in organisms.

A Cleavage of C–C Bonds in Sugar Chain by Per-Iodidate Oxidation

The size distribution of radiolabeled HA synthesized in the cell-free reaction is analyzed by agarose gel electrophoresis.

1. A dried sample of sugar chain (3 mg) was dissolved in 1 ml of 0.1 M ammonium acetate buffer containing 9.0 M urea at pH 4.0. $NaIO_4$ (0.5 ml of a 1% solution).
2. Isopropyl alcohol (6 µl) was added to the sample solution and stirred in the dark at room temperature for 2 h.
3. The reaction mixture was stored in the dark for 7 days at 4°C.
4. The reaction was quenched by the addition of ethylene glycol, and the mixture was dialyzed against water.
5. The nondialyzable fraction was freeze-dried.
6. The product was reduced with $NaBH_4$ (1 ml of a 1% solution) for 16 h at room temperature.
7. The product was dialyzed against water and then hydrolyzed with 0.1 N trifluoroacetic acid (0.5 ml) at 100°C for 30 min.
8. After evaporation to dryness, the mixture was chromatographed on a column of Bio-Gel P-2 using 0.1 M NaCl as the eluent.
9. The tri-, di-, and monosaccharide fractions obtained were desalted by passing through Dowex 50W.
10. The samples were hydrolyzed with 2 N trifluoroacetic acid at 120°C for 1 h and then analyzed for their sugar compositions using a Dionex HPAEC-PAD.

References

Dell A, Chalabi S, Easton RL, Haslam SM, Sutton-Smith M, Patankar MS, Lattanzio F, Panico M, Morris HR, Clark GF (2003) Murine and human zona pellucida 3 derived from mouse eggs express identical O-glycans. Proc Natl Acad Sci 100:15631–15636

Mah SA, Swanson WJ, Vacquier VD (2005) Positive selection in the carbohydrate recognition domains of sea urchin sperm receptor for egg jelly (suREJ) proteins. Mol Biol Evol 22:533–541

Nakachi M, Moriyama H, Hoshi M, Matsumoto M (2006) Acrosome reaction is subfamily specific in sea star fertilization. Dev Biol 298:597–604

Viela-Silva AC, Castro MO, Valente AF, Biermann CH, Mourao PA (2002) Sulfated fucans from the egg jellies of the closely related sea urchins *Strongylocentrotus droebachiensis* and *Strongylocentrotus pallidus* Ensure species-specific fertilization. J Biol Chem 277:379–387

Williams SA, Xia L, Cummings RD, McEver RP, Stanley P (2007) Fertilization in mouse does not require terminal galactose or N-acetylglucosamine on the zona pellucia glycans. J Cell Sci 120: 1341–1349

Section XIV
Development, Differentiation, and Morphogenesis

Functional Analysis of Sugar Chains Using a Genome-Wide RNAi System in *Drosophila*

Shoko Nishihara

Introduction

Drosophila melanogaster was first used for genetic studies by Thomas Hunt Morgan in 1908. *Drosophila* has since become one of the most important model organisms for developmental biology studies in addition to those of classical and molecular genetics. Recently, effective analysis of glycan functions in vivo has been achieved using *Drosophila*. Flies with mutations of some glycosyltransferases display distinctive phenotypes, and these mutant flies have been invaluable in clarifying the role of glycosylation in Wg/Wnt, Hh, FGF, and Notch signaling in vivo. These data emphasize the crucial importance of protein glycosylation for many biological processes and indicate that glycobiology will provide a new focus of investigation for many aspects of Biology and Medicine.

Glycosylation is one of the most important post-translational modifications along with phosphorylation. Consequently, elucidation of the overall function of glycans at the cellular or whole body level is a vital issue for investigation following completion of many genome projects. The glycosylation of proteins and lipids is performed in the Golgi apparatus by glycosyltransferases, which transfer sugars from nucleotide sugars to acceptor substrates. Glycosyltransferases are responsible for synthesizing the huge diversity of complex oligosaccharides. Consequently, to achieve an overview of glycan functions in the biological processes within the whole organism, we developed a genetic approach in *Drosophila* that uses a heritable and inducible RNA interference (RNAi) system for glycosyltransferases (Nishihara et al. 2004).

Principle

Drosophila *Glycosyltransferases*

More than 160 human glycan-related transferases (GRTs), including glycosyltransferases and sulfotransferases, have been cloned and their activities identified. Using these human GRT sequences, we performed a BLAST search of the *Drosophila* genome database and identified 91 putative GRTs, including glycosyltransferases and sulfotransferases. Of these, the activities of 38 glycosyltransferases and 3 sulfotransferases had already been identified.

Laboratory of Cell Biology, Department of Bioinformatics, Faculty of Engineering, Soka University, 1-236 Tangi-cho, Hachioji, Tokyo 192-8577, Japan
Phone: +81-426-91-8140, Fax: +81-426-91-8140
E-mail: shoko@scc1.t.soka.ac.jp

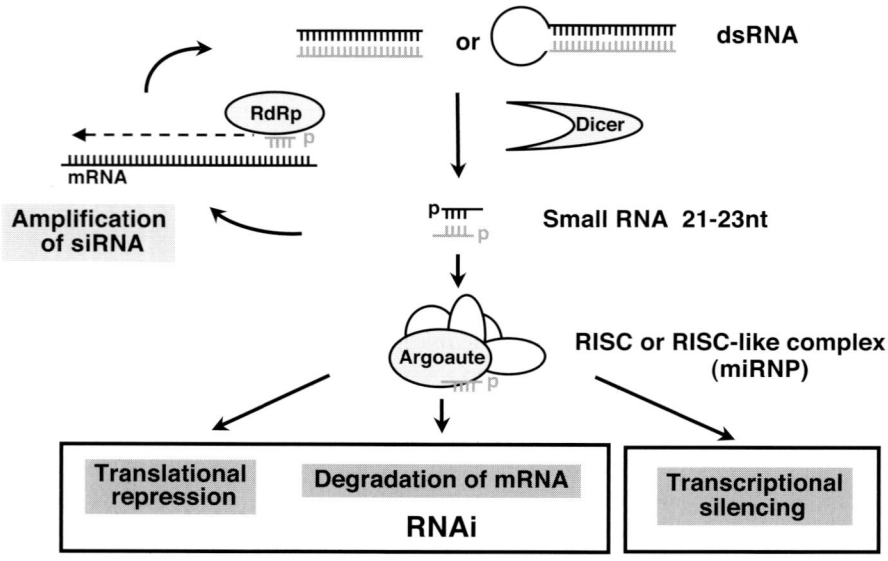

Fig. 1 Mechanism of RNA interference (*RNAi*). Long double-stranded RNA (*dsRNA*) is cleaved by Dicer into 21–23 nt small RNA duplexes, termed small interfering RNAs (*siRNAs*). These small RNA duplexes are incorporated into either a multimeric protein complex called an RNA-induced silencing complex (*RISC*), or an RISC-like complex. These complexes induce three different types of RNA silencing: translational repression, degradation of mRNA, and transcriptional silencing. Degradation of mRNA is generally termed RNAi. During RNAi, the RISC recognizes and specifically cleaves a target RNA complementary to the guide strand of the siRNA. MicroRNAs are incorporated into an RISC-like micro-ribonucleoprotein particle, miRNP, and induce translational repression. RNAi and translational repressions are types of post-transcriptional gene silencing. RISC can also induce transcriptional gene silencing

RNA Interference (RNAi)

The RNAi is an evolutionarily conserved phenomenon in which double-stranded RNA (dsRNA) induces gene silencing (Fig. 1). In the initiation step, a long dsRNA is processed by the RNase III enzyme Dicer into 21–23 nt small RNA duplexes containing 2-nt 3′ overhangs (small interfering RNAs, siRNAs). These siRNAs are incorporated into a multimeric protein complex termed the RNA-induced silencing complex (RISC). The RISC can recognize and specifically cleave a target RNA complementary to the guide strand of the siRNA. The last step is so rigorous that a single base mismatch between the target RNA and the guide strand of the siRNA is enough to stop destruction of the target RNA. This "degradation of mRNA" is termed RNAi, one type of post-transcriptional gene silencing (PTGS). In many model organisms such as *Drosophila*, long dsRNAs efficiently induce gene-specific silencing, whereas only siRNAs can efficiently suppress the expression of the corresponding gene in mammalian cells. *Caenorhabditis elegans* and plants have a pathway termed "amplification of siRNAs" in which an RNA-dependent RNA polymerase (RdRp) reduces the specificity of RNAi. This pathway is absent in *Drosophila* and humans.

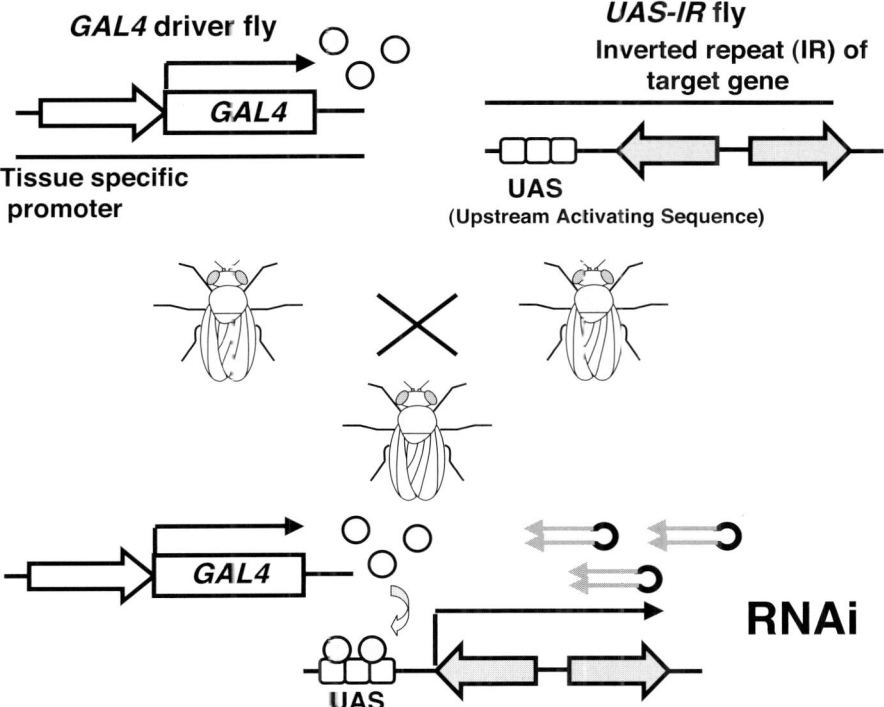

Fig. 2 Schema of the heritable and inducible RNAi system. Two transgenic fly stocks, *GAL4* driver and *UAS-IR*, are used in this system. The *GAL4* driver fly has a transgene containing the yeast transcriptional factor *GAL4*, the expression of which is under the control of a tissue-specific promoter. The *UAS-IR* fly has a transgene containing an inverted repeat (*IR*) of the target gene ligated to the UAS, a target of GAL4. In the F_1 generation of these flies, the dsRNA of the target gene is expressed in a tissue-specific manner and induces gene silencing

Drosophila Heritable and Inducible RNA Interference (RNAi) System

Genome-wide analysis of glycan functions, i.e., of glycosyltransferases, requires a convenient reverse genetics approach. Recently, two novel mutagenesis methods have been developed: gene targeting by homologous recombination and gene silencing by RNAi. The former approach suffers from the fact that it requires a considerable time to carry out targeted gene replacement by homologous recombination. Consequently, this approach is not suitable for identifying the functions of large numbers of genes. In contrast, RNAi is a powerful reverse genetics tool for studying many gene functions in model organisms including *Drosophila*. The large dsRNAs efficiently induce gene-specific silencing without interferon responses. We therefore established an inducible RNAi knockdown system in *Drosophila* for the functional analysis of glycans using the GAL4-upstream activating sequence (UAS) system (Takemae et al. 2003; Nishihara et al. 2004).

A schema of this approach is shown in Fig. 2 (Takemae et al. 2003; Nishihara et al. 2004). Two transgenic fly lines, *GAL4* driver and *UAS-inverted repeat* (*UAS-IR*), are used. The *GAL4* driver fly has a transgene containing the yeast transcription factor *GAL4*, the expression of which is under the control of a tissue-specific promoter. Currently,

4,300 *GAL4* driver fly lines are available from *Drosophila* stock centers. The *UAS-IR* fly has a transgene containing an IR of the target gene ligated to the UAS, a target of GAL4. Approximately 500 bps of target gene sequences are selected by dsCeck (http://dscheck.rnai.jp/), highly sensitive off-target search software for dsRNA-mediated RNAi, to avoid the off-target effects. In the F_1 generation of these flies, the dsRNA of the target gene is expressed under the control of a tissue-specific promoter to induce gene silencing. The strength of the RNAi effect depends on the temperature at which the F_1 progeny is raised, because binding of GAL4 to UAS is temperature dependent. Stronger RNAi effects are expected in F_1 progeny raised at 28°C compared to 25°C (Ichimiya et al. 2004). Using this system, we can induce tissue-specific gene silencing and analyze gene function, even for lethal null mutations. The *Drosophila* heritable and inducible RNAi system provides an excellent resource for analysis of glycan functions.

Procedure

RNAi Flies

RNAi fly lines are obtained as follows:

1. Select about 500 bp of the target gene sequence using dsCheck (http://dscheck.rnai.jp/); this highly sensitive search software is used to avoid off-target effects using dsRNA-mediated RNAi.

2. Amplify the target gene sequence by PCR, using a cDNA library derived from *Drosophila melanogaster* or an expression sequence tag clone. The following PCR primer set should be used: 5′ primer, AAGGCCTACATGGCCGGACCG *plus* 21 bp of a target gene sequence that includes an SfiI site and a CpoI site; 3′ primer, AATCTAGAGGTACC *plus* 21 bp of a target gene sequence that includes a KpnI site and an XbaI site.

3. The PCR fragment obtained is then inserted as an inverted repeat (IR) sequence into the pUAST-R57 vector (accession number: AB233207) using the above enzyme sites. pUAST-R57 is the modified pUAST vector for cloning IRs of the target gene sequence. A fragment comprising exons 5 to 7 of the fly Ret oncogene is placed between the insertion sites of the IRs. For details of the cloning strategies, refer to the web site http://www.shigen.nig.ac.jp/fly/nigfly/about/aboutRnai.jsp.

4. Introduce the constructed vector into *Drosophila* embryos of the w^{1118} mutant stock that was used to establish the *UAS-IR* transgenic fly lines. For details of the mating scheme, refer to the web site http://www.shigen.nig.ac.jp/fly/nigfly/about/aboutRnai.jsp.

5. Mate flies from each *UAS-IR* fly line with flies from the appropriate *GAL4* driver fly lines. Raise the F_1 progeny at 25°C or 28°C and assess their phenotypes.

Application

We have applied the *Drosophila* heritable and inducible RNAi system to three kinds of glycosyltransferases: proteoglycan β1,4galactosyltransferase I (Takemae et al. 2003), protein *O*-mannosyltransferase (POMT) 1 and 2 (Ichimiya et al. 2004). The experiments clearly demonstrated that RNAi specifically disrupts expression of target genes in this *Drosophila* RNAi system, and that this disruption produces flies with abnormal phenotypes. In addition, we also applied this system to two kinds of PAPS transporters, *sll*

(Kamiyama et al. 2003) and *PAPST2* (Goda et al. 2006). It is possible to identify potential interactions between two molecules without recourse to biochemical analysis: genetic interaction between two molecules indicates that these molecules may be involved in the same signaling pathway in vivo. Genetic interactions between *POMT 1* and *2*, and between the newly identified *PAPST2* and *heparan sulfate 6-O-sulfotransferase* and *heparan sulfate 3-O-sulfotransferase B* indicate that these gene products cooperate with each other to synthesize the products, *O*-mannosyl glycans (Ichimiya et al. 2004) and heparan sulfate proteoglycans (Goda et al. 2006), respectively. For more detailed description of the data, refer to references (Kamiyama et al. 2003; Takemae et al. 2003; Ichimiya et al. 2004; Goda et al. 2006).

References

Goda E, Kamiyama S, Uno T, Yoshida H, Ueyama M, Kinoshita-Toyoda A, Toyoda H, Ueda R, Nishihara S (2006) Identification and characterization of a novel *Drosophila* 3'-phosphoadenosine 5'-phosphosulfate transporter. J Biol Chem 281:28508–28517

Ichimiya T, Manya H, Ohmae Y, Yoshida H, Takahashi K, Ueda R, Endo T, Nishihara S (2004) The twisted abdomen phenotype of *Drosophila* POMT1 and POMT2 mutants coincides with their heterophilic protein *O*-mannosyltransferase activity. J Biol Chem 279:42638–42647

Kamiyama S, Suda T, Ueda R, Suzuki M, Okubo R, Kikuchi N, Chiba Y, Goto S, Toyoda H, Saigo K, Watanabe M, Narimatsu H, Jigami Y, Nishihara S (2003) Molecular cloning and identification of 3'-phosphoadenosine 5'-phosphosulfate transporter. J Biol Chem 278:25958–25963

Nishihara S, Ueda R, Goto S, Toyoda H, Ishida H, Nakamura M (2004) Approach for functional analysis of glycan using RNA interference. Glycoconj J 21:63–68

Takemae H, Ueda R, Okubo R, Nakato H, Izumi S, Saigo K, Nishihara S (2003) Proteoglycan UDP-galactose:beta-xylose β1,4-galactosyltransferase I is essential for viability in *Drosophila melanogaster*. J Biol Chem 278:15571–15578

Functional Glycomics at the Level of Single Cells: Studying Roles of Sugars in Cell Division, Differentiation, and Morphogenesis with 4D Microscopy

Souhei Mizuguchi, Katsufumi Dejima, Kazuko H. Nomura, Daisuke Murata, Kazuya Nomura

Introduction

The nematode *Caenorhabditis elegans* is the first multicellular organism whose complete cell lineage, cellular anatomy, and neural wiring diagram of its nervous system are determined. Because the complete DNA sequence of its genome is also available, it is easy to isolate deletion mutants from TMP (trimethylpsoralen)/UV mutagenized library of animals by PCR screening with carefully designed primer sets (cf. http://shigen.lab.nig.ac.jp/c.elegans/ChangeLocale.do?lang=en&url=home). Mutant strains can be stored frozen indefinitely, and this makes this model organism the only multicellular organism compatible with high throughput biologic study. In addition, genes defined by sequence similarity to any known genes can be easily identified and mutated, and the phenotypes of mutant animals can be studied at the level of single cells as the body of the nematode is transparent.

We are making use of advantages of this excellent model organism to determine functions of carbohydrates and aiming to reveal unknown roles of sugar-related molecules in fundamental biologic processes. Many glycosyltransferases and sugar modifying molecules are common to human and the worm. With the aid of bioinformatics, we pinpointed 145 different worm glycogenes orthologous to human glycogenes including various glycosyltransferases and proteoglycan-related genes. To understand roles of carbohydrates in development and morphogenesis, ablation of gene functions by RNAi or deletion mutagenesis is the method of choice. RNAi or deletion mutagenesis of 44% of the 145 glycogenes resulted in abnormal phenotypes. Abnormality due to ablation of gene functions can be often spotted by four-dimensional (4D) microscopy which can capture and visualize information throughout the volume of a living three-dimensional (3D) worm as it changes over time (four-dimensional imaging) (Thomas and White 2000). Here, we would like to describe essential knowledge to start 4D microscopy to monitor development of worm embryos and to detect abnormalities in the gene knocked-out/down worms in a laboratory. For other protocols for *C. elegans* research, readers should consult an excellent online protocol resource WormBook (http://www.wormbook.org). Also indispensable is the database WormBase (http://www.wormbase.org/) and you can also use it as an excellent portal site for the world-wide nematode databases and resources.

Department of Biological Sciences, Kyushu University Graduate School, Hakozaki 6-1-1, Fukuoka 812-8581, Japan
Phone: +81-92-642-4613, Fax: +81-92-642-4613
E-mail: knomuscb@mbox.nc.kyushu-u.ac.jp

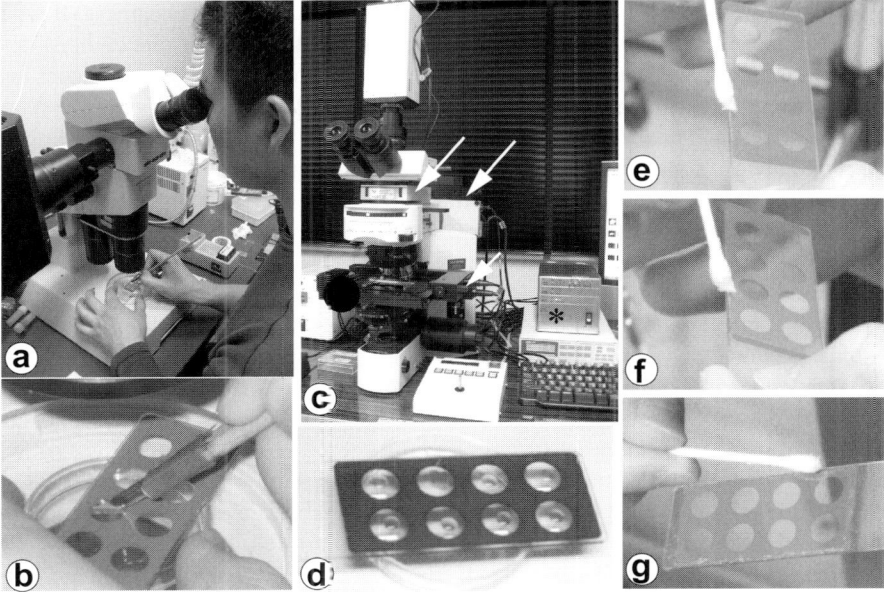

Fig. 1 Sample preparation for 4D (four-dimensional) microscopy (**a, b**) Dissection of gravid worms to obtain fertilized eggs on an eight-well printed slide glass. **c** Our 4D microscope system. Excitation/emission filter wheel (shutter is in excitation filter wheel, *white arrows*), shutter (not shown) and z-axis motor are controlled by MAC5000 controller system (*asterisk*). The XY stage (*short white arrow*) can be moved manually by the joystick or automatically via the joystick controller box (Nihon Molecular Device Corp) under the MAC5000 while 4D imaging. **d** In each well, there are 1–50 eggs. Adjust the liquid volume of the droplet to about 2.6 μl with M9 before a coverslip is placed. (**e–g**) Sealing the coverslip with white petrolatum. This is an important step for successful 4D imaging

Procedure

Software and Hardware for 4D Microscopy

We use MetaMorph Software (Molecular Devices Corp., USA) for controlling fluorescence/DIC (Nomarski differential interference contrast) erecting microscope (Fig. 1c) to capture and analyze 4D images (for a free software on 4D microscope control, see http://www.loci.wisc.edu/4d/ or Thomas and White 2000). During 4D image acquisition, images are transferred to an internal hard disk drive (HDD) and stored for further analysis. Since transfer of image data to DVD or CD takes longer time, their use is not recommended for real time 4D image acquisition and storage. Our computer for 4D microscopy is a Dell Precision 360 computer (Pentium 4 CPU 2.6 GHz) with 2 GB memory and 600 GB HDD. Any computer with GHz Pentium III or 4 (or higher CPU), and with GB RAM and 100 GB or larger HDD is suitable for 4D analysis.

Digital cooled CCD camera used is PCO SensiCam QE (PCO. imaging, Germany). Any equivalents such as CoolSNAP HQ (Roper, USA) or ORCA AG (Hamamatsu Photonics, Japan) are appropriate for 4D imaging. Usually, it is useless to record areas without embryos, and in our sytem, 600 × 500 pixels (i.e., ~100 × 80 μm) can cover a whole embryo. Thus, the field of view is usually restricted to this value, which is about one-fifth of the whole pixels (1376 × 1040 pixels) of the cooled CCD camera. If we use 600 × 500 pixel recording area, and if we record 15 focal planes for one embryo, about 8.5 MB hard disk space is needed. Thus, if we record 10 embryos (15 focal planes each) at 10 min interval for 10 h, we need $8.5 \times 10 \times 60 = 5,100$ MB disk space for the recording. Hard disk space needed for 4D recording is strictly proportional to recording area (pixels), number of focal planes along z-axis, number of image capture (recording intervals and time). If we use all the pixels of the camera (1376 × 1040) in the above experiment, we need about 25 GB hard disk space. We use several cassette type hard disk drives connected to the computer for 4D imaging.

Our microscope is an Olympus BX51 fluorescence/DIC microscope equipped with MAC 5000 controller system (Ludl Electronic Products Ltd., USA) which controls a z-axis motor, two filter wheels with shutters, and an automatic XY stage. The speed of filter wheel change should be about 100 ms, and switching of shutters should be completed within 5 ms. Minimizing optical damages and heat damages from the light source (UV and halogen lamp) is the most critical point for successful 4D imaging. To avoid killing embryos by these damages, you should optimize duration of illumination as well as shutter/filter change speed for your own 4D system. The sample stage we use is either BioPrecision XY stage (Ludl, USA) which is controlled by MAC5000 or MD-X60Y30100S-META which is controlled by its own controller (Nihon Molecular Devices Corp, Japan) (Fig. 1c). For observing early embryonic cells, we usually use 40× dry objective (N.A. 0.7–0.85). We also use 63× (dry) or 100× (oil immersion) objectives for higher magnification. Oil immersion objectives can be used, but during optical sectioning of the sample along the z-axis, the coverslip often slightly moves and embryos sometimes drift off. If embryos are placed on agar pad (2% agar pad on a slide glass) (Hamahashi et al. 2005), this movement seldom occurs due to the firm attachment of embryos to agar surface. We do not like distortion of eggs adhered to the agar pad, and usually use an eight-well printed slide glass instead (see below).

For 4D microscopy, recently introduced rotating disc confocal system is also very useful. Our system can be upgradable with a Nipkow dual rotating-disk confocal scanner (CSU10; Yokogawa Electric, Tokyo, Japan) for 4D confocal imaging.

Mounting Embryos for 4D Microscopy

We do not recommend hypochlorite treatment of adult worms to obtain eggs for 4D microscopy. The brood size of an adult worm raised from an egg obtained by the treatment is often reduced, and it is possible that various stress responses induced by this treatment may make the interpretation of gene disruption results difficult and confusing. Instead, we prefer manual dissection of uterus of gravid hermaphrodite to obtain intact eggs as described below.

1. Fill the two wells of an eight-well printed microscope slide glass (Matsunami Glass, Japan or Erie Scientific, USA) with 6 and 4 µl of M9, respectively.

(M9: mix KH_2PO_4 3 g, Na_2HPO_4 6 g, NaCl 5 g in water and mess up to 1,000 ml, then autoclave it and add 1 ml of sterile 1 M $MgSO_4$)

2. Pick one or a few gravid hermaphrodites and put them into the well with 6 µl M9.

3. Stir the well with a worm picker and wash the worm to remove *Escherichia coli*. Pick the worms and put them into the well with 4 µl M9.

4. Under dissecting microscope, cut the uterus of the worm with sterile surgical blade (carbon steel No. 15, Feather Safety Razor Co., Japan). At least 10–20 eggs can be obtained from one hermaphrodite (Fig. 1a, 1b).

5. Pipette 2 μl M9 to the empty wells of the slide glass or of a new eight-well printed slide glass. First, suck a few microliter of M9 into a 10 μl micropipette tip (Sorenson or Gilson, USA) and exhale the content into the well. Then, using this tip, move eggs to the new wells. (If you worry about damages caused by disrupted adult worms, you may move eggs to another new well.) Adjust the volume of M9 in a well to 2.6 μl to avoid spilling of the liquid when covered with a coverslip (24 mm × 50 mm., Matsunami Glass, Japan). (Fig. 1d)

Steps 1 to 5 should be completed quickly lest the water droplet in the well quickly evaporates (e.g., the second well from the top left in Fig. 1d).

6. Put the cover slip carefully and confirm the presence of eggs in each well. Usually, more than 10 eggs were put into each well, and in each experiment selected 5–10 eggs were observed with 4D microscopy.

7. Seal the cover slip and the slide with white petrolatum by using a cotton swab (Fig. 1e–g). This is the critical step for the experiment. Be careful not to touch spilled white petrolatum with objectives. If the sealing is successful, you may observe swimming wild type worms after 1 week! If the sealing is not perfect, water evaporates and embryos move with the evaporating water and often disappear from the field of view. The resulted high salt concentration kills the worm during 4D examination. Sealing with nail-polish is not suitable for 4D imaging because it is toxic.

Results

Among the 145 glycogenes tested, knocking out of genes related to proteoglycan synthesis showed the most spectacular phenotypes. In *C. elegans*, chondroitin is the major glycosaminoglycan, and nonsulfated chondroitin consists of a major part of its glycosaminoglycans. There is only one chondroitin synthesizing enzyme *sqv-5* (*C. elegans* chondroitin synthase: *ChSy*) in the nematode. Heparan sulfate is the minor component of glycosaminoglycan chains, and *rib-1* and *rib-2* are the genes responsible for heparan sulfate synthesis in the nematode. Four-dimensional microscopy of RNAi-treated worms revealed that chondroitin is involved in embryonic cytokinesis and chromosome partition (Fig. 2; Mizuguchi et al. 2003). The method also revealed that enzymes involved in heparan sulfate proteoglycan synthesis or sulfation are indispensable in embryos after

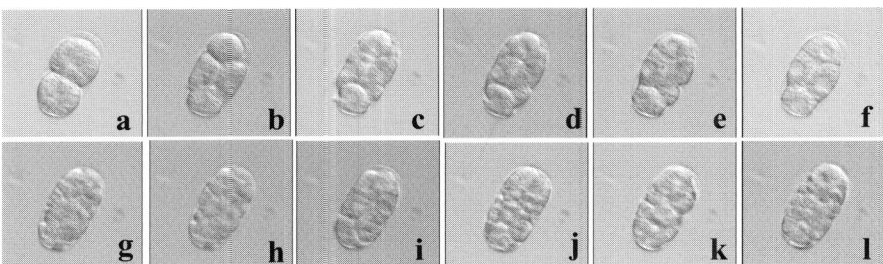

Fig. 2 Four-dimensional microscopy reveals abnormal cell division in chondroitin synthase (*sqv-5*) knocked down embryos (RNAi). Apparent cell number increases and decreases during time development (a-l). Modified from Mizuguchi et al. (2003)

gastrulation and not in early embryonic cell division (Kitagawa et al. 2007; Dejima et al. 2006).

References

Dejima K, Seko A, Yamashita K, Gengyo-Ando K, Mitani S, Izumikawa T, Kitagawa H, Sugahara K, Mizuguchi S, Nomura K (2006) Essential roles of 3'-phosphoadenosine 5'-phoshosulfate synthase in embryonic and larval development of the nematode *Caenorhabditis elegans* J Biol Chem 281:11431–11440

Hamahashi S, Onami S, Kitano H (2005) Detection of nuclei in 4D Nomarski DIC microscope images of early *Caenorhabditis elegans* embryos using local image entropy and object tracking. BMC Bioinformatics 6:125

Kitagawa H, Izumikawa T, Mizuguchi S, Dejima K, Nomura KH, Egusa N, Taniguchi F, Tamura J, Gengyo-Ando K, Mitani S, Nomura K, Sugahara K (2007) Expression of *rib-1*, a *Caenorhabditis elegans* homolog of the human tumor suppressor EXT Genes, is indispensable for heparan sulfate synthesis and embryonic morphogenesis. J Biol Chem 282:8533–8544

Mizuguchi S, Uyama T, Kitagawa H, Nomura KH, Dejima K, Gengyo-Ando K, Mitani S, Sugahara K, Nomura K (2003) Chondroitin proteoglycans are involved in cell division of *Caenorhabditis elegans*. Nature 423:443–448

Thomas CF, White JG (2000) Acquisition, display, and analysis of digital three-dimensional time-lapse (four-dimensional) data sets using free software applications. Methods Mol Biol 135:263–276

Studies on Functions of Notch O-Fucosylation in *Drosophila*

Tomonori Ayukawa, Kenji Matsuno

Introduction

Notch signaling is an evolutionarily conserved mechanism that regulates a broad spectrum of cell-specification events through local cell–cell communication. *Notch* encodes a single-pass transmembrane receptor protein with 36 epidermal growth factor-like (EGF) repeats. The EGF-like repeats of the Notch extracellular domain, which contain a consensus sequence, are modified by the O-linked tetrasaccharide Sia-α2,3-Gal-β1,4-GlcNAc-β1,3-Fuc (Fig. 1A). A GDP-fucose protein O-fucosyltransferase1 catalyzes this O-linked fucosylation (Wang et al. 2001). In *Drosophila*, this enzyme is encoded by *O-fut1*.

Protein fucosylation requires GDP-fucose as a donor of fucose. Two pathways for the synthesis of GDP-fucose, which are the salvage and de novo pathways, have been identified in mammals (Fig. 1H). However, the genes encoding the essential enzymes for the salvage pathway are not found in the *Drosophila* genome, suggesting that GDP-fucose is supplied only from the de novo pathway in this organism. Therefore, if we disrupted the de novo pathway of GDP-fucose synthesis in *Drosophila*, we could study the consequences of GDP-fucose starvation, which results in lacking of fucose glycan synthesis.

The GDP-fucose transporter is a nucleotide-sugar transporter, classified as belonging to solute carrier family 35 (SLC35). The transporter is predicted to span the Golgi membrane 10 times and couples the import of GDP-fucose into the Golgi lumen with the export of GMP into the cytoplasm. In the Golgi GDP-fucose is used by specific fucosyltransferases to add fucose to a variety of glycoproteins and glycolipids. Since fucosyltransferases utilize GDP-fucose, the fucose donor which is synthesized in the cytosol, the uptake of GDP-fucose into the Golgi is thought to be a critical step for the fucosylation event (Fig. 1H). Recently, the gene whose defect is responsible for Congenital Disorders of Glycosylation (CDG) IIc was cloned by the complementation of cells derived from CDG IIc patients and was found to encode a GDP-fucose transporter. CDG IIc, also termed LADII, is a rare recessive syndrome characterized by growth and mental retardation and severe immunodeficiency with marked neutrophilia.

Results

A *Drosophila* O-FucT-1 homolog, encoded by *O-fut1*, catalyzes the O-fucosylation of Notch (Fig. 1A). We isolated an *O-fut1* mutant (Sasamura et al. 2003). Genetic analysis

---

Department of Biological Science and Technology, Tokyo University of Science, 2641 Yamazaki, Noda, Chiba 278-8510, Japan
Phone: +81-4-7122-9714, Fax: +81-4-7122-1499
E-mail: matsuno@rs.noda.tus.ac.jp

of this *O-fut1* mutant revealed that *O-fut1* is generally required for Notch signaling (Sasamura et al. 2003). For example, embryos lacking *O-fut1* functions showed hyperplasia of neurons, which is also observed in Notch mutant embryos (Fig. 1B–D). We also found that the functions of *O-fut1* are highly specific to Notch signaling (Sasamura et al. 2003). Our biochemical analysis revealed that O-fut1 is required for the binding between Notch and its ligands (Sasamura et al. 2003). It was recently demonstrated that O-fut1 acts as a chaperon for Notch in the endoplasmic reticulum and is required for Notch to exit the endoplasmic reticulum (Okajima et al. 2005). We reported that O-fut1 has additional functions in the endocytic transportation of Notch. O-fut1 was indispensable for the constitutive transportation of Notch from the plasmamembrane to the early endosome, which we showed was independent of O-fut1's O-fucosyltransferase activity (Sasamura et al. 2007). We also found that O-fut1 promoted the turnover of Notch, which consequently downregulated Notch signaling (Sasamura et al. 2007). We proposed that O-fut1 is the first example, except for ligands, of a molecule that is required extracellularly for receptor transportation by endocytosis.

We studied developmental defects in mutant of *GDP-mannose 4,6 dehydratase* (*Gmd*) and *GDP-4-keto-6-deoxy-D-mannose 3,5-epimerase/4-reductase* (*Gmer*), which encode enzymes essential for de novo synthesis of GDP-fucose. The GDP-D-mannose 4,6-dehydratase (Gmd) gene encodes the first enzyme of the de novo GDP-fucose synthesis pathway. We generated two null-mutant alleles of *Gmd* (Sasamura et al. 2007). The concentration of GDP-fucose in lysates prepared from the third-instar larvae of *Gmd* mutants was determined biochemically (Sasamura et al. 2007). The GDP-fucose level in the Gmd^{H78} and Gmd^{H43} heterozygotes were approximately half that of wild-type larvae, and it was not detectable in the Gmd^{H78} and Gmd^{H43} homozygotes (Sasamura et al. 2007). In addition, staining with Aleuria Aurantia lectin (AAL), which recognizes α-1,3- and β-1,6-linked fucose, was reduced in *Gmd* mutant compared with that of wild type (Fig. 1E and F). We found that Notch signaling was disrupted in *Gmd* mutant (Sasamura et al. 2007). The defects of Notch signaling in the wing discs of *Gmd* mutant were fully rescued by the expression of *Gmd*, respectively (Sasamura et al. 2007).

Recently, the gene responsible for CDG IIc was found to encode a guanosine diphosphate (GDP)-fucose transporter (Fig. 1H). We investigated the possible cause of the developmental defects in CDG IIc patients using a *Drosophila* model. Biochemically, we demonstrated that a *Drosophila* homolog of the GDP-fucose transporter, the Golgi GDP-fucose transporter (*Gfr*), specifically transports GDP-fucose in vitro (Ishikawa et al. 2005). To understand the function of the *Gfr* gene, we generated null mutants of *Gfr* in *Drosophila*. Our phenotype analyses revealed that Notch signaling was deficient in these *Gfr* mutants (Ishikawa et al. 2005). These phenotypes were rescued by the human GDP-fucose transporter transgene (Ishikawa et al. 2005). GDP-fucose is known to be essential for the fucosylation of *N*-linked glycans and for *O*-fucosylation, and Notch is modified by both of fucosylation. Our results suggest that Gfr is involved in the fucosylation of *N*-linked glycans on Notch and its *O*-fucosylation, as well as those of bulk proteins (Ishikawa et al. 2005). In *Gfr* mutant embryos, AAL staining was reduced (Fig. 1G). However, despite the essential role of Notch *O*-fucosylation during development, the *Gfr* homozygote was viable (Sasamura et al. 2003; Ishikawa et al. 2005; Fig. 1). Thus, our results also indicate that the *Drosophila* genome encodes at least another GDP-fucose transporter that is involved in the *O*-fucosylation of Notch. Finally, we found that mammalian *Gfr* is required for Notch signaling in mammalian cultured cells (Ishikawa et al. 2005). Therefore, our results implicate reduced Notch signaling in the pathology of CDG IIc.

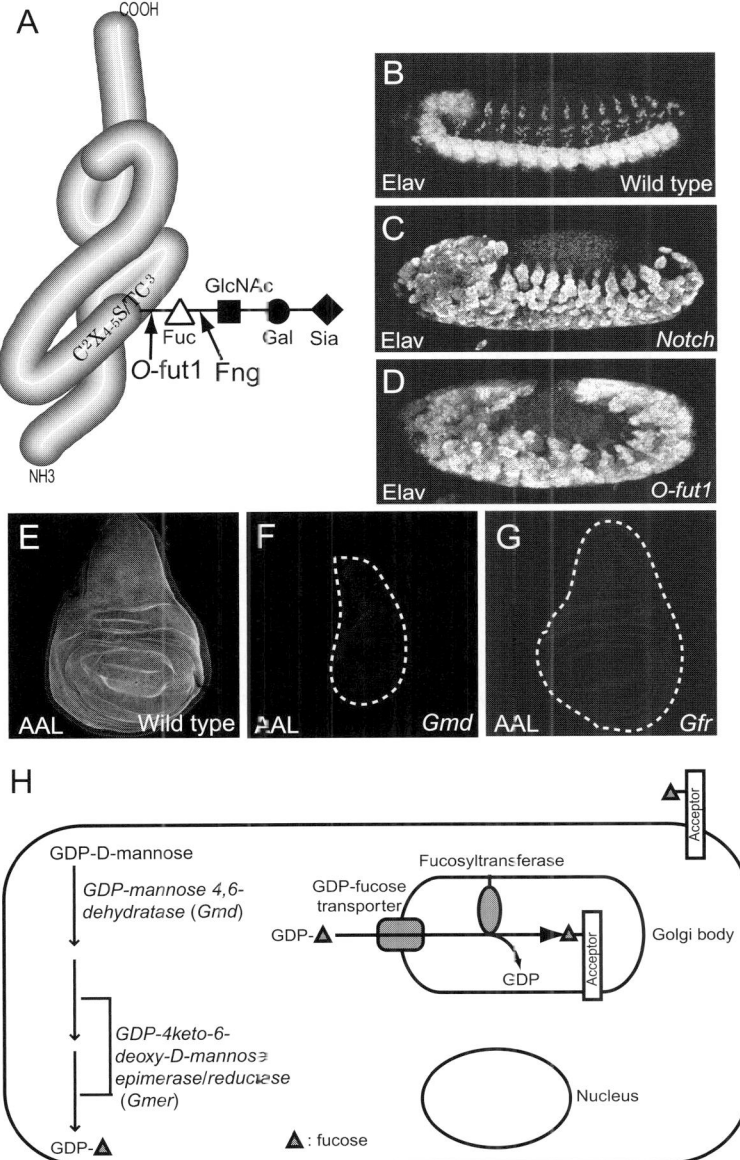

Fig. 1 Roles and mechanisms of Notch O-fucosylation in *Drosophila*. (**A**) O-fucosylation of Notch EGF repeats is essential for Notch signaling. *Left* An O-fucose glycan is attached to the EGF repeats of Notch. The modifications of Fucose and GlcNAc are catalyzed by O-fut1 and Fringe. (**B–D**) Drosophila embryos of wild type (**B**), *Notch* homozygote (**C**), and *O-fut1* maternal mutant (**D**): The embryos are stained with an antibody recognizing neuronal cells. Note that *Notch* and *O-fut1* mutant embryos show excess neurons, indicating that Notch signaling is defective. (**E–G**) AAL staining of wild type (**E**) and *Gmd* (**F**) and *Gfr* mutant wing discs. (**H**) GDP-fucose transporter(s) is required for protein O-fucosylation. DGP-fucose is synthesized in cytoplasm and transported into the lumen of Golgi by GDP-fucose transporter(s). In the lumen of Golgi, protein fucosylations catalyzed by Fucosyltransferase occur

Aleuria Aurantia Lectin (AAL) Staining of Drosophila Tissue

To detect fucose glycans in *Drosophila* tissues, fixed tissues are stained with ALL using the following method.

1. Dissect out the imaginal discs from the third instar larvae and fix them in PLP [2% paraformaidehyde, 0.01 M $NaIO_4$, 0.075 M lysine, 0.037 M $NaPO_4$ (pH 7.2)] for 40 min on ice.
2. Wash the sample three times with PBS-DT [0.3% (w/v) deoxycholic acid sodium salt monohydrate, 0.3%TritonX (v/v)] for 5 min at room temperature.
3. Block the sample with 1% bovine serum albumin (in PBS) for 1 h at room temperature.
4. Incubate the sample with 1 μg/ml biotin-conjugated AAL (in 1% BSA) for 2 h at room temperature.
5. Wash the sample three times with PBS-DT for 5 min at room temperature.
6. Incubate the sample with streptavidin-conjugated Alexa 555 (1 : 500 in 1% BSA) for 1 h at room temperature.
7. Wash the sample four times with PBS-DT for 5 min at room temperature.

Mount samples with 90% glycerol and obtain the image with a confocal microscope.

References

Ishikawa HO, Higashi S, Ayukawa T, Sasamura T, Aoki K, Ishida N, Sanai Y and Matsuno K (2005) A *Drosophila* model of congenital disorder of glycosylation IIc implicates the deficiency of Notch signaling in its pathogenesis. Proc Natl Acad Sci USA 102:18532–18537

Okajima T, Xu A, Lei L, Irvine K (2005) Chaperone activity of protein *O*-Fucosyltransferase 1 promotes Notch receptor folding. Science 307:1599–1603

Sasamura T, Sasaki N, Miyashita F, Nakao S, Ishikawa HO, Ito M, Kitagawa M, Harigaya K, Spana E, Bilder D, Perrimon N, Matsuno K (2003) *neurotic*, a novel maternal neurogenic gene, encodes an *O*-fucosyltransferase that is essential for Notch–Delta interactions. Development 130:4785–4795

Sasamura T, Ishikawa HO, Sasaki N, Higashi S, Kanai M, Nakao S, Ayukawa T, Aigaki T, Noda K, Miyoshi E, Taniguchi N, Matsuno K (2007) The *O*-fucosyltransferase *O*-fut1 is an extracellular component that is essential for the constitutive endocytic trafficking of Notch in *Drosophila*. Development 134:1347–1356

Wang Y, Shao L, Shi S, Harris RJ, Spellman MW, Stanley P, Haltiwanger RS (2001) Modification of epidermal growth factor-like repeats with *O*-fucose. Molecular cloning and expression of a novel GDP-fucose protein *O*-fucosyltransferase. J Biol Chem 276:40338–40345

Membrane Microdomain as a Platform of Carbohydrate-Mediated Interactions During Early Development of Medaka Fish

Tomoko Adachi, Chihiro Sato, Ken Kitajima

Introduction

Membrane microdomains, which are often called lipid rafts, are a platform of signal transduction in various biological processes like recognition and cell growth, because they contain not only receptors for ligands and growth factors, but also downstream signal transducer molecules on the same microdomains. Membrane microdomains are rich in cholesterol and glycosphingolipids and concentrate on particular proteins and carbohydrates in the restricted membrane area. Our group has recently demonstrated that membrane microdomains isolated from sperm function as a platform not only for signal transduction in sperm activation (Ohta et al. 1999, 2000), but also for sperm–egg binding at fertilization of sea urchin (Maehashi et al. 2003). In fertilization, sperm binding to the vitelline coat is the first step of the sperm–egg binding and at least two kinds of interactions occur. One is a protein–protein interaction between bindin and the sperm binding protein (SBP) localized in vitelline coat. The other is a carbohydrate–protein interaction between the glycolipids of sperm microdomain and the SBP (Maehashi et al. 2003). The most striking feature is that these two interactions concomitantly occur on the membrane microdomains. Furthermore, the sperm microdomains also contain transducer proteins like Src-family kinase and adenylate cyclase (Ohta et al. 2000). Therefore, co-localization of those proteins and carbohydrates involved in cell adhesion with those transducer proteins involved in signal transduction in the microdomains suggests the important roles of the membrane microdomain in carbohydrate- and protein-mediated interactions in the sperm–egg binding.

To demonstrate the importance of membrane microdomains as a platform for carbohydrate-mediated interactions in other cell–cell interactions than the sea urchin sperm–egg binding, we focus on epiboly, in which blastomeres actively migrate from the animal to vegetal poles of embryos at blastula and gastrula stages in medaka, *Oryzias latipes*.

Formation of Membrane Microdomains is Crucial for Cell Migration in Gastrula Embryos

In early development, the cell surface carbohydrates spatiotemporally change to possibly mediate cell–cell interactions. For example, in mouse morula stage embryos, the Le^x-glycan (Galβ1–4(Fucα1–3)GlcNAcβ1-) is transiently expressed on the surface of blastomeres and involved in compaction, at which blastomeres tightly contact with each

---

Bioscience and Biotechnology Center, and Graduate School of Bioagricultural Sciences, Nagoya University, Chikusa, Nagoya 464-8601, Japan
Phone: +81-52-789-4297, Fax: +81-52-789-5228
E-mail: kitajima@agr.nagoya-u.ac.jp

other at least through its carbohydrate–carbohydrate interaction (Fenderson et al. 1990). In medaka, Le^x-glycans are also expressed on the surface of early embryos, including gastrula stage embryos (Sasado et al. 1999). We thus hypothesize that the Le^x-glycan-mediated interactions occur through membrane microdomains of the blastomeres. We first isolated and characterized membrane microdomains from medaka embryos during epiboly. The isolated microdomains are characterized by enrichment of cholesterol and sphingomyelin as well as by the exclusive occurrence of Le^x-containing glycolipids. In addition, the high-molecular weight (>350,000) glycoproteins containing Le^x-glycans are also enriched in the microdomain (Adachi et al. 2007). These glycoproteins appear to contain huge glycan chains corresponding to embryoglycans in mammalian embryos. Furthermore, cadherin and transducer proteins such as cSrc kinase are also enriched in the microdomain (in a preliminary report in Adachi et al. 2002). These results suggest that the membrane microdomain may function as a platform of carbohydrate- and protein-mediated cell–cell interactions as well as of signal transductions in medaka embryos.

We assessed the importance of the membrane microdomain formation in the progress of epiboly using methyl β-cyclodextrin (MBCD) and C2-ceramide that disrupt microdomains through different mechanisms (Adachi et al. 2007). Both reagents efficiently disrupt the microdomain structure and concomitantly impaired epiboly. Notably, when embryos pretreated with MBCD (a cholesterol-binding molecule) are treated with exogenously added cholesterol, the embryos reconstitute the microdomain and completely restore a normal progress of epiboly. Thus, normal or impaired development is reversibly controlled by the cholesterol-dependent formation or disruption of microdomains (Fig. 1). In most cases, microdomain-disrupted embryos show detachment of cells from the blastoderm. These results suggest that the microdomains in epiboly mediate cell adhesion and movements between blastodermal cells.

Membrane Microdomain as a Platform for the Le^x-Mediated Interaction in Gastrula Stage Embryos

To elucidate an underlying mechanism for membrane microdomains-mediated cell–cell interactions, we further examined if isolated microdomains have the ability to bind with each other (preliminary results shown in Adachi et al. 2002). Most interestingly, binding and inhibition experiments demonstrate that the isolated microdomains bind with each other in an Le^x-glycan-dependent manner. Furthermore, the microdomain–microdomain binding also depends on cadherin, as revealed by the inhibition experiment using an inhibitory peptide of homophilic binding of cadherin. Thus, both Le^x-glycan- and cadherin-mediated interactions are involved in the microdomain-mediated cell–cell interactions (Fig. 2).

Procedures

1. Collect embryos (1 g) from medaka fish, *Oryzias latipes*, which were bred in aquariums under a 14-h light/10-h dark cycle at 27°C.
2. Homogenize the embryos in 5 ml of PBS containing 5 mM EDTA and protease inhibitors (4 milliunits leupeptin and 12.6 milliunits aprotinin) at 0°C with a mortar, and remove the fertilization envelope.
3. Transfer the homogenate into a centrifuge tube.
4. Centrifuge at 200,000 × *g* at 4°C for 30 min and discard the supernatant.

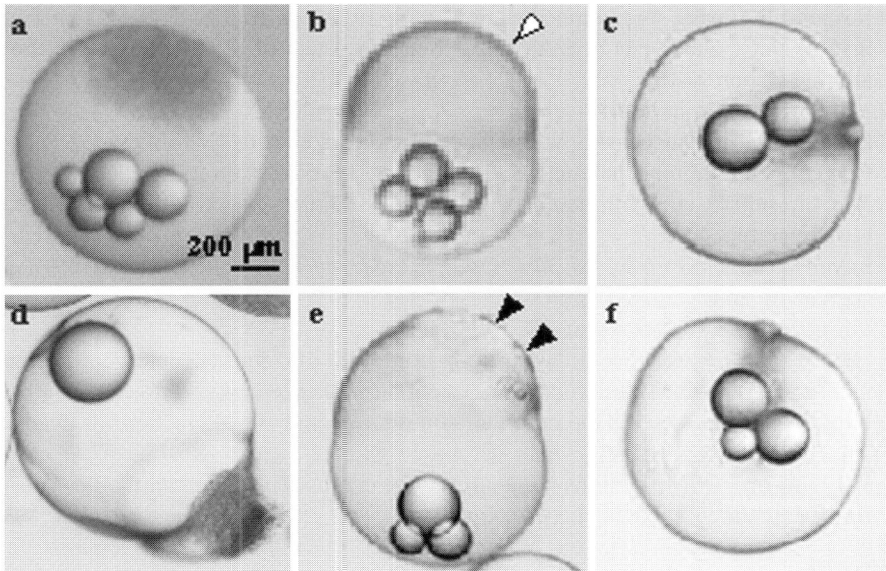

Fig. 1 Formation of membrane microdomains is crucial for cell adhesion and migration in epiboly of medaka gastrula embryos. Dechorionated embryos were untreated (**a, b, c**) or treated with MBCD at 3 mM (**d**) and 2 mM (**e**) at 27°C for 15 min. **a** Stage 13, **b** stage 15, **c** stage 19. After the 15-min-pulse-treatment with 3 mM MBCD, the blastodermal cells were peeled off from the animal pole (**d**). In some cases, embryos appeared to be normal immediately after the 15-min-pulse-treatment, but most cells were detached from the blastoderm 1 h after the pulse-treatment (**e**). The *filled* and *open arrowheads* indicate abnormal and the corresponding normal sites, respectively. **f** After the 15-min-pulse-treatment, the embryos were further incubated with cholesterol-MBCD complex (1 mM cholesterol) for 45 min at stage 19. The scale bar, 200 μm

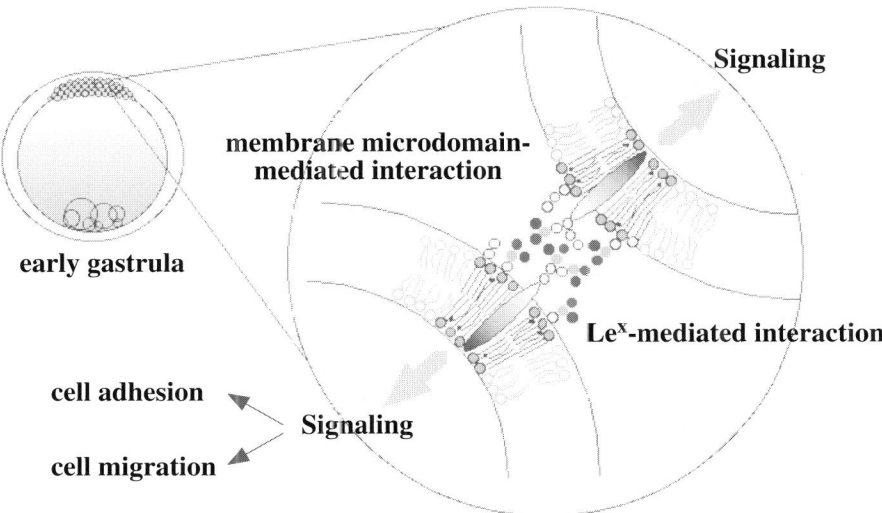

Fig. 2 Membrane microdomains as a platform for the Le^x-mediated interaction in gastrula stage embryos

5. Suspend the pellet (membrane fraction) with 1.0 ml of 10 mM Tris–HCl (pH 7.5), 0.15 M NaCl, and 5 mM EDTA (TNE) containing the protease inhibitors (see above) and 1% Triton X-100.
6. Stand on ice for 20 min.
7. Homogenize the suspension with Dunce homogenizer with 10 strokes.
8. Centrifuge the homogenate at $1,300 \times g$ at 4°C for 5 min to remove debris
9. Mix the supernatant with an equal volume of 85% (w/v) sucrose in TNE in a new centrifuge tube.
10. Layer successively 6 ml of 30% (w/v) sucrose in TNE and 3.5 ml of 5% (w/v) sucrose in TNE.
11. Centrifuge at $200,000 \times g$ at 4°C for 18 h
12. Collect 1 ml fractions from the top to bottom. The low density, detergent-insoluble membrane (LD-DIM), microdomain, fraction can be seen as a light-scattering band. Pool the 3–5 fractions as the microdomain fraction and 10–11 fractions as detergent soluble membrane (DSM).
13. LD-DIM and DSM fractions are analyzed for components using antibodies specific for certain carbohydrates and proteins.

Concept

On the basis of experiments on at least two different biological processes, the sea urchin sperm–egg binding and the epiboly of medaka gastrula embryos, our concept that membrane microdomains are a platform for carbohydrate-mediated interactions is demonstrated. Notably, the carbohydrate-mediated interactions are accompanied by cell adhesion protein-mediated interactions in the membrane microdomain-mediated cell–cell bindings. Our concept predicts that we could find various unknown carbohydrate-mediated interactions if we focused on isolated membrane microdomains in various cell adhesion events. An important future subject is, however, to elucidate how the protein- and carbohydrate-mediated interactions co-operate and link to the following signal transduction at the membrane microdomains.

References

Adachi T, Sato C, Matsuda T, Totani K, Murata T, Usui T, Kitajima K (2002) Isolation and roles in cell–cell interaction of the membrane microdomains (lipid rafts) of developing embryos of medaka. Zool Sci 19:1444

Adachi T, Sato C, Kitajima K (2007) Membrane microdomain formation is crucial in epiboly during gastrulation of medaka. Biochem Biophys Res Commun 358:848–853

Fenderson BA, Eddy EM, Hakomori S (1990) Glycoconjugate expression during embryogenesis and its biological significance. Bioessays 12:173–179

Maehashi E, Sato C, Ohta K, Harada Y, Matsuda T, Hirohashi N, Lennarz WJ, Kitajima K (2003) Identification of the sea urchin 350-kDa sperm-binding protein as a new sialic acid-binding lectin that belongs to the heat shock protein 110 family: implication of its binding to gangliosides in sperm lipid rafts in fertilization. J Biol Chem 278:42050–42057

Ohta K, Sato C, Matsuda T, Toriyama M, Lennarz WJ, Kitajima K (1999) Isolation and characterization of low density detergent-insoluble membrane (LD-DIM) fraction from sea urchin sperm. Biochem Biophys Res Commun 258:616–623

Ohta K, Sato C, Matsuda T, Toriyama M, Vacquier VD, Lennarz WJ, Kitajima K (2000) Co-localization of receptor and transducer proteins in the glycosphingolipid-enriched, low density, detergent-insoluble membrane fraction of sea urchin sperm. Glycoconj J 17:205–214

Sasado T, Kani S, Washimi K, Ozato K, Wakamatsu Y (1999) Expression of murine early embryonic antigens, SSEA-1 and antigenic determinant of EMA-1, in embryos and ovarian follicles of a teleost medaka (Oryzias latipes). Dev Growth Differ 41:293–302

Functions of Heparan Sulfate Proteoglycans in Morphogenesis

Hiroko Habuchi, Koji Kimata

Introduction

Heparan sulfate proteoglycans (HSPG) are present ubiquitously on the cell surface and in the extracellular matrix including the basement membranes. There are two families of HSPGs in the cell surface, syndecans and glypicans. HSPGs in the extracellular matrix are mainly perlecan and agrin. These HSPGs bear HS chains through serine residues in the core proteins. Their HS chains interact with divergent bioactive ligands such as growth factors and morphogens (FGFs, Wnts, BMPs, VEGFs, Hh, etc) and their receptors (FGFR, etc), extracellular matrix molecules (collagen, fibronectin, laminin, etc), proteases, etc. Such interactions modulate the activity, distribution, concentration, and stability of ligands. Thereby, HS chain plays important roles in a variety of developmental, morphogenetic and pathogenic processes. The specificities of the interactions between HS and ligands are thought to be due, at least in part, to the structure of HS characterized by the sulfation patterns and the distributions of isomers of hexuronic acid residues, GlcA or IdoA. The fine and divergent structures of HS are generated in the Golgi apparatus through the coordinate actions of HS modification enzymes, namely, N-deacetylase/N-sulfotransferases (NDSTs), C5-epimerase (Hsepi) and 2-O-, 6-O-, and 3-O-sulfotransferases, following the biosynthesis of the HS backbone structure comprising alternating glucuronic acid (GlcA) and N-acetylglucosamine (GlcNAc) residues (Fig. 1) (Habuchi et al. 2004). Further modification of HS occurs at the cell surface by the action of 6-O-endosulfatase. These modification enzymes, with the exception of Hsepi and HS2ST, exist as multiple isoforms that differ in substrate specificity and expression patterns. Recent loss-of-function studies on HS biosynthesis enzymes have demonstrated that HS has critical roles in developmental processes. Studies on mice deficient in the HS chains have demonstrated that HS is essential for gastrulation in early embryonic development. In addition, the inactivation of the genes involved in HS modification reactions leads to distinct phenotypes, suggesting that the specific structure of HS regulates the activity of growth factors and morphogens distinctly in vivo.

In Table 1, the aberrant phenotype and abnormal signaling of HS-modification enzyme genes-targeted animals are summarized (Bulow and Hobert 2006; Kimata et al. 2007). Among them, here we focus below on the NDST-deficeint mice and HS6ST-deficient mice.

NDST Knockout Mouse

NDST is a bifunctional enzyme composed of a single polypeptide that catalyzes the two reactions, N-deacetylation and N-sulfation of GlcNAc residue in heparin and heparan

---

Institute for Molecular Science of Medicine, Aichi Medical University, Nagakute, Aichi 480-1195, Japan
Phone: +81-52-264-4811. Fax: +81-561-633-532
E-mail: kimata@amugw-aichi-med-u.ac.jp

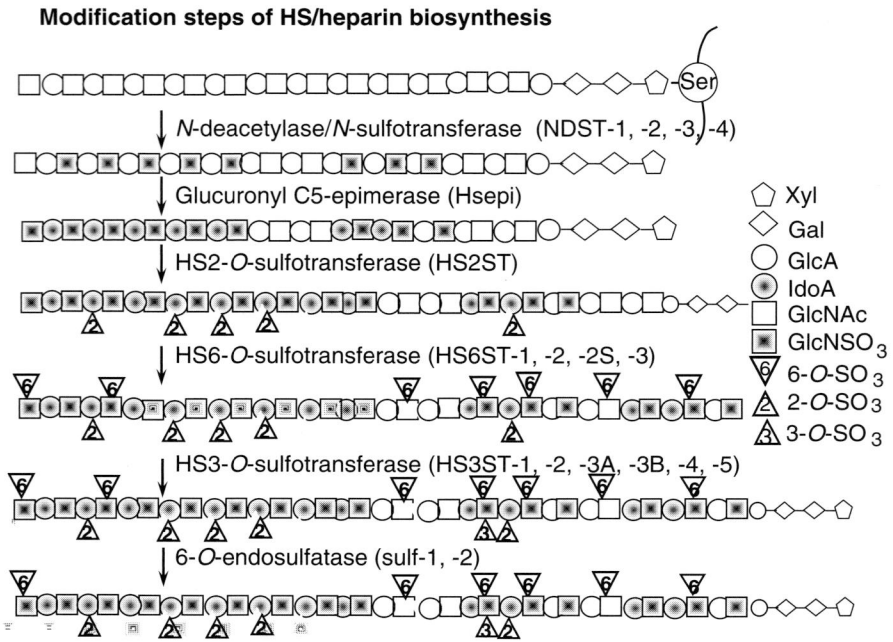

Fig. 1 Modification steps of HS/heparin biosynthesis

sulfate. There are four NDST isozymes in mammals. Most NDST1-deficient embryos survive until or just before birth. However, they turn out to be cyanotic and die neonatally in a condition resembling respiratory distress syndrome. Lungs from these newborn NDST1 null mice are severely atelectatic. This appears to be caused by reduced secretion of Surfactant protein A and B. In addition, some of NDST1-deficient embryos arrest their development in utero between E10 and E18.5 and exhibit severe developmental defects of the forebrain and forebrain-derived structures, including cerebral hypoplasis, lack of olfactory bulbs, eye defects, axon guidance errors and neural crest-derived facial structures. Fine structures of HS prepared from NDST1 null embryo are altered; the relative amount of N-sulfated disaccharides and 2-O-sulfated disaccharides markedly decrease in the mutant relative to the wild type. Therefore, heparan sulfate properly synthesized is required for the normal development of the brain and face. The phenotype of severely affected NDST1$^{-/-}$ embryos strongly resembles mouse embryos carrying neural crest specific deletions of sonic hedgehog (Shh) and Fgf8. Indeed, the genetic interaction between Ndst1 and Shh was demonstrated. All these results suggest that impaired signaling via Shh contributes to the observed phenotypes in NDST1$^{-/-}$ embryos (Grobe et al. 2002).

Heparan sulfate is a potential ligand for P- and L-selectin, two key molecules involved in the adhesion of platelets and leukocytes to inflamed endothelium. NDST1gene was inactivated in lung microvascular endothelial cells and lymphocytes using cre-loxP system regulated under the promoter of *Tek*, which encodes Tie2. Mutant mice developed normally but showed impaired neutrophil infiltration. Bone marrow transplantation experiments to generate chimeric mice demonstrated that impaired infiltration is observed only in endothelial NDST1-deficient mice, but not in leukocyte. Thus, endothelial cell expression of heparan sulfate is essential at least for three stages of the inflammatory

Table 1 Mutants and their phenotypes of heparan sulfate (HS)/heparin biosynthesis enzymes genes

Enzyme and function	Phenotype	Abnormal signaling molecule
Human (Homo sapience)		
EXT-1 and EXT-2, HS copolymerase	Hereditary multiple exostoses (HME), an autosomal dominant mode of inheritance	
Mouse (Mus musculus)		
EXT-1, HS copolymerase	Systemic null mice: Embryonic death by E8.5.	Ihh
	Mice deficient in the nervous system: Neonatal death. Malformation in the caudal midbrain-cerebral region, an abnormally small cerebral cortex, and the absence of the olfactory bulbs.	Fgf, Wnt, Slit
EXT-2, HS copolymerase	Null mice: Growth arrest after E6.0. Heterozygous mice: Ectopic bone growths (exostoses). Multiple abnormalities in cartilage differenciation.	Ihh
NDST-1, HS N-deacetylase/ N-sulfotransferase	Neonatal death. Pulmonary atelectasis. Severe developmental defects of the forebrain and forebrain-derived structures.	FGF, Shh
	Pericyte detachment and delayed migration	PDGF-BB
NDST-2, HS N-deacetylase/ N-sulfotransferase	Defects in connective tissue-type mast cells	
Hsepi, glucuronyl-C5 epimerase	Neonatal death. Renal agenesis, lung defects, and skeletal malformation.	
HS2ST, HS 2-O-sulfotransferase	Neonatal death. Bilateral renal agenesis. Defects of the eye and skeleton	
HS6ST-1, HS 6-O-sulfotransferase	Most null mice: Embryonic death during E15.5 and the perinatal stage. Abnormal angiogenesis in labyrinthine zone	VEGF
	Axon guidance error at the optic chiasm	Slit
HS3ST-1, HS 3-O-sulfotransferase	Normal hemostasis. Genetic background-specific lethality and intrauterine growth retardation	
Drosophila melanogaster		
Sulfateless, (sfl), N-deacetylase/ N-sulfotransferase	Abnormal segment polarity	FGF
Tout velu (ttv), heparan sulate co-polymerase (EXT-1)	Abnormal segment polarity	Hh, Wg
Sister of tout velu (sotv), heparan sulfate co-polymerase (EXT-2)	Abnormal segment polarity	Hh, Wg
Brother of tout velu (botv), heparan sulfate GlcNAc transferase-1 (EXTL3)	Abnormal segment polarity	Hh, Wg
Hs2st, heparan sulfate 2-O-sulfotransferase		
Hs6st, heparan sulfate 6-O-sulfotransferase	Abnormal formation of the trachea	FGF
Hs3st-B, heparan sulfate 3-O-sulfotransferase	Neurogenic phenotypes that are characteristic of notch signaling mutants	Notch

Table 1 (continued)

Enzyme and function	Phenotype	Abnormal signaling molecule
Caenorhabditis. elegans		
Rib-2, heparan sulfate GlcNAc transferase-I and –II (EXTL-3)	Early developmental defects and stops the development during the gastrulation stage	
Hsepi, glucuronyl-C5 epimerase	axonal and cellular guidance defects in specific neuron classes	Slit, Kal-1
Hs2st, heparan sulfate 2-O-sulfotransferase	axonal and cellular guidance defects in specific neuron classes	Slit
Hs6st, heparan sulfate 6-O-sulfotransferase	axonal and cellular guidance defects in specific neuron classes	Slit, Kal-1
Zebrafish		
Dackel, HS co-polymerase (Ext-2)	Dorsal RGC axons missort in the optic tract	Slit, FGF
Boxer, HS co-polymerase (Extl-3)	Dorsal RGC axons missort in the optic tract	Slit, FGF
Hsepi A, B, glucuronyl-C5 epimerase	Abnormal dorsal-ventral axis formation	
Hs6st-2, heparan sulfate 6-O-sulfotransferase	Destruction of muscle development and abnormal vascular development	

response: by acting as a ligand for L-selectin during neutrophil rolling; in chemokine transcytosis; and by binding and presenting chemokines at the luminal surface of the endothelium.

The conditional NDST1 knockout mice in liver showed reduced uptake of triglyceride-rich lipoproteins indicating that hepatic heparan sulfate play a crutial role in the clearance of hepatic lipoprotein particles.

NDST2-deficient mice have no obvious developmental abnormality. However, these transgenic mice cannot produce heparin, and posses abnormal mast cells in their skin and peritoneal cavities due to an inability to store certain neutral protease (mMCP-4, -5, -6, and mMC-CPA)/proteoglycan complexes in their secretory granules.

HS6ST Knockout Mice

Heparan sulfate 6-O-sulfotransferases exist in three isoforms (HS6ST-1, -2 and -3) and one alternatively spliced form (HS6ST-2S) in mammals, and the individual isoforms exhibit a characteristic preference in their substrate specificities for the uronic acid residue neighboring the N-sulfoglucosamine. HS6ST-1 predominantly sulfates the IdoA-GlcASO$_3$ unit. On the other hand, HS6ST-2 transfers sulfate preferentially to the GlcA-GlcNSO$_3$ unit and the IdoA(2SO$_4$)-GlcNSO$_3$ unit to generate trisulfated disaccharide units in HS. HS6ST-3 has intermediate properties between HS6ST-1 and HS6ST-2. During organogenesis, *HS6ST-1* is predominantly transcribed in epithelial and neural-derived tissues, whereas *HS6ST-2* is more mesenchymal. *HS6ST-3* appears at later stages and in a more restricted manner.

Most of HS6ST-1-null mice died between E15.5 and the perinatal stage, and those mice that survived were considerably smaller than their wild-type littermates (Habuchi et al. 2007). Some of these HS6ST-1-null mice exhibited development abnormalities, and histochemical and molecular analyses of these mice revealed an ~50% reduction in the number of fetal microvessels in the labyrinthine zone of the placenta relative to that in the wild-type mice. VEGF-A mRNA and protein were reduced modestly in the placental labyrinthine zone of HS6ST-1-null mice; HS-dependent defect in cytokine signaling probably contributes to increased embryonic lethality and decreased growth. Biochemical studies of the HS chains isolated from various organs of our HS6ST-1-null mice revealed a marked reduction of GlcNAc($6SO_4$) and Hex-GlcNSO$_3$($6SO_4$) levels and a reduced ability to bind Wnt2. Pratt et al. reported the generation of HS6ST-1$^{-/-}$ mice using a gene-trap approach. The homozygous mice exhibit navigation errors of retinal ganglion cell (RGC) axons from each eye. It appears to be due to the abnormal response of the navigating growth cone to Slit protein, because HS6ST-1$^{-/-}$ RGCs in vitro are less sensitive to repulsion by Slit2-expressing cells. Therefore, 6-O-sulfated HSPG on the growth cone surface regulates the response of the growth cone to Slit2. Thus, despite the presence of three closely related 6-O-sulfotransferase genes in the mouse's genome, HS6ST-1 is the primary one used in HS biosynthesis in most tissues.

Preparation and Analysis of Disaccharide Compositions of HS

1. Various tissues are defatted by homogenizing in acetone and then dried.
2. Each tissue is suspended in 10 volumes of 0.2 M NaOH per wet tissue weight, incubated for 16 h at room temperature, neutralized with 4 M acetic acid, and then incubated at 37°C for 2 h with DNase I, RNase A in 50 mM Tris–HCl (pH8.0) containing 10 mM MgCl$_2$. Proteinase (1/200 of tissue wet weight, Actinase E; Kaken-Seiyaku Corp.) is added and incubation is continued at 37°C for 16 h.
3. The reaction is stopped by heating to 100°C for 5 min and the samples are centrifuged at 13,000 rpm for 10 min to remove any insoluble materials.
4. The supernatants are diluted with an equal volume of 20 mM Tris–HCl buffer (pH 7.2) and loaded on a DEAE-Sephacel column equilibrated with the same buffer (usually 0.5 ml DEAE-Sephacel for tissue of less than 200 mg).
5. The columns are washed with 10 column volumes of buffer containing 0.2 M NaCl and then eluted with 4 column volumes of 2 M NaCl in Tris–HCl buffer.
6. The eluates are mixed with 2.5 volumes of cold 95% (v/v) ethanol containing 1.3% (w/v) potassium acetate and 1 mM EDTA, and then the GAGs are recovered by centrifugation.
7. An aliquot of the GAG is digested at 37°C for 2 h with a mixture of 0.2 mU of heparitinase I, 0.1 mU of heparitinase II and 0.2 mU of heparinase in 50 µl of 50 mM Tris–HCl buffer (pH7.2), 1 mM CaCl$_2$ and 5 µg bovine serum albumin.
8. After filtration of the digests with Ultrafree-MC (5 kDa molecular weight cutoff filter: Millipore Corp.), the unsaturated disaccharide products in the filtrates are analyzed by reverse-phase ion-pair chromatography using Senshu Pak column Docosil with a fluorescence detector according to Toyoda's method (Toyoda et al. 2000) with slightly modified elution conditions.

References

Bulow HE, Hobert O (2006) The molecular diversity of glycosaminoglycans shapes animal development. Annu Rev Cell Dev Biol 22:375–407

Grobe K, Ledin J, Ringvall M, Holmborn K, Forsberg E, Esko JD, Kjellen L (2002) Heparan sulfate and development: differential roles of the *N*-acetylglucosamine *N*-deacetylase/*N*-sulfotransferase isozymes. Biochim Biophys Acta 1573:209–215

Habuchi H, Habuchi O, Kimata K (2004) Sulfation pattern in glycosaminoglycan: does it have a code? Glycoconj J 21:47–52

Habuchi H, Nagai N, Sugaya N, Atsumi F, Stevens RL, Kimata K (2007) Mice deficient in heparan sulfate 6-*O*-sulfotransferase-1 exhibit defective heparan sulfate biosynthesis, abnormal placentation and late embryonic lethality. J Biol Chem 282:15578–15588

Kimata K, Habuchi O, Habuchi H, Watanabe H (2008) Knockout mice and proteoglycans. In: Kamerling JP et al (eds) Comprehensive glycoscience 3:159–191

Toyoda H, Kinoshita-Toyada A, Selleck SB (2000) Structural analysis of glycosaminoglycans in *Drosophila* and *Caenorhabditis elegans* and demonstration that tout-velu, a Drosophila gene related to EXT tumor suppressors, affects heparan sulfate in vitro. J Biol Chem 275:2269–2275

Fukutin and Fukuyama Congenital Muscular Dystrophy

Motoi Kanagawa[1], Tatsushi Toda[2]

Introduction

Recent genetic and biochemical studies have revealed that mutations in (putative) glycosyltransferases and subsequent abnormal glycosylation of dystroglycan are associated with several forms of congenital muscular dystrophies (Kanagawa and Toda 2006). Fukuyama congenital muscular dystrophy (FCMD), the second most common childhood muscular dystrophy in Japan, is one of the congenital muscular dystrophies displaying glycosylation defects of α-dystroglycan. The gene responsible for this disease is *fukutin*, and its protein product is assumed to participate in cellular glycosylation events. FCMD is characterized by severe congenital muscular dystrophy, abnormal neuronal migration associated with mental retardation and epilepsy, and frequent eye abnormalities. Therefore, fukutin-dependent glycosylation of α-dystroglycan plays crucial roles in structural/functional maintenance of skeletal muscle, central/peripheral nervous system, and eye. α-Dystroglycan, a highly glycosylated protein, forms a protein complex with β-dystroglycan, and the dystroglycan complex links laminin in the extracellular matrix to the cellular actin cytoskeleton. Abnormal glycosylation of α-dystroglycan results in a severe reduction in laminin-binding activity, and thus disruption of the interaction between dystroglycan and laminin caused by *fukutin* mutations is believed to be the major cause of FCMD.

Fukutin

The gene responsible for FCMD was identified by positional cloning and its protein product was named as fukutin (Kobayashi et al. 1998). The major mutation in FCMD is a 3-kb retrotransposon insertion in the 3' noncoding region of the *fukutin* gene. Some FCMD patients carrying point-mutation(s) in the *fukutin* gene have also been reported. The function of fukutin is still unknown; however, computer analysis has predicted that fukutin has homology to enzymes that modify glycolipids and glycoproteins. Fukutin, a 461-amino-acid protein with a predicted molecular weight of 53.7 kDa, is a type-II membrane protein and possesses a DXD motif. Fukutin has been shown to localize to the Golgi apparatus. In FCMD muscle, α-dystroglycan is hypoglycosylated and its ligand-binding activities are dramatically decreased. These findings indicate that fukutin is involved in the glycosylation pathway of α-dystroglycan. A recent study has demonstrated the possibility that fukutin interacts with and regulates POMGnT1, which synthesizes *O*-mannosylglycan on α-dystroglycan (for details on POMGnT1 see Chapter by

---

[1,2] Division of Clinical Genetics, Department of Medical Genetics, Osaka University Graduate School of Medicine, 2-2-E9 Yamadaoka, Suita, Osaka 565-0871, Japan
Phone: +81-6-6879-3381, Fax: +81-6-6879-3389
E-mail: [1] kanagawa@clgene.med.osaka-u.ac.jp, [2] toda@clgene.med.osaka-u.ac.jp

T. Endo and H. Manya and Chapter by T. Endo, in this volume). Generation of fukutin chimeric mice confirmed the essential role of fukutin-dependent glycosylation of dystroglycan in the maintenance of muscle integrity, cortical histiogenesis, and ocular development. Targeted homozygous mutation of the *fukutin* gene in mice leads to embryonic lethality prior to the development of skeletal muscle and mature neurons. Detailed analyses of *fukutin*-null embryos suggested that fukutin is necessary for the maintenance of basement membrane function during early embryonic development.

Abnormal Glycosylation of Dystroglycan in FCMD

α/β-Dystroglycan is encoded by a single mRNA and is post-translationally cleaved into two subunits. α-Dystroglycan is a receptor for laminin, a major component of the extracellular matrix, and O-glycosylation of α-dystroglycan is essential for its laminin-binding activity. α-Dystroglycan is anchored on the plasma membrane through the non-covalent interaction with β-dystroglycan. β-Dystroglycan is a transmembrane protein and interacts intracellularly with dystrophin. Therefore, the dystroglycan complex acts as a linkage between the extracellular matrix and the dystrophin-actin cytoskeleton across the muscle plasma membrane (Kanagawa and Toda 2006). This interaction is believed to provide mechanical integrity to the muscle cell membrane. Since fukutin defects result in aberrant glycosylation of dystroglycan with reduced ligand-binding activity, the dystroglycan-mediated linkage between the extracellular matrix and the cytoskeleton is likely to be disrupted in FCMD, which renders the muscle cell membrane to be fragile and more susceptible to mechanical stress by muscle contraction, eventually leading to muscle cell wasting.

Abnormal glycosylation of α-dystroglycan can be tested by immunofluorescence or Western blotting analysis using monoclonal antibodies that recognize the functionally glycosylated form of α-dystroglycan (IIH6 and VIA4–1) and polyclonal antibodies that recognize the α-dystroglycan core peptide (GT20ADG). Abnormally glycosylated α-dystroglycan loses reactivity against IIH6 and VIA4–1, and shows a reduced molecular weight that can be detected by Western blotting with the α-dystroglycan core antibodies (Michele et al. 2002). A monoclonal antibody to β-dystroglycan (8D5) can be used to confirm the amount of dystroglycan protein expressed.

Glycotherapy for FCMD

Since defects in laminin-binding glycans on α-dystroglycan are a cause of FCMD, therefore complementing glycans, having affinity to laminin, on α-dystroglycan could be a therapeutic strategy of FCMD. LARGE, a putative glycosyltransferase involved in the functional glycosylation of dystroglycan, is a potential target for glycotherapy. Overexpression of LARGE produces hyperglycosylated α-dystroglycan with increased ligand-binding activity in cells from FCMD patients (Barresi et al. 2004). Although the detailed mechanism of LARGE-dependent hyperglycosylation of α-dystroglycan is still unclear, this finding may lead to a novel strategy to treat FCMD and related congenital muscular dystrophies by bypassing the alternative dystroglycan-laminin linkage. Thus, glycotherapies and treatments aimed at modulating the expression or the activity of LARGE may be a future therapeutic option for glycosyltransferase-deficient muscular dystrophies.

Fig. 1 Laminin overlay assay and Western blotting analysis of dystroglycan. WGA-purified preparations from mouse skeletal muscle extracts were analyzed by laminin overlay assay (LM OL) and Western blotting. IIH6 and 8D5 are monoclonal antibodies to α-dystroglycan and β-dystroglycan, respectively

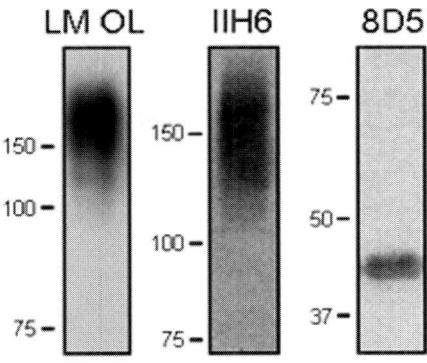

Laminin Overlay Assay

Laminin-binding activity of α-dystroglycan can be determined by the laminin overlay assay (Fig. 1) and the solid-phase laminin-binding assay (Michele et al. 2002). In this chapter, the protocol of the laminin overlay assay is described (for the solid-phase laminin-binding assay, see Michele et al. 2002).

Laminin-binding buffer (LBB): 10 mM triethanolamine-HCl (pH 7.4), 140 mM NaCl, 1 mM $CaCl_2$, 1 mM $MgCl_2$

1. Subject tissue/cell extracts or dystroglycan preparations to SDS-PAGE (dystroglycan can be enriched by WGA- or Con A-affinity chromatography).
2. Transfer separated proteins to a PVDF membrane and block the blot with 5% skim milk/LBB at room temperature.
3. Briefly wash the blot with 3% BSA/LBB.
4. Incubate the blot with 1–10 nM laminin in 3% BSA/LBB at 4°C for >12 h.
5. After washing with 5% skim milk/LBB, incubate the blot with an anti-laminin antibody at room temperature for several hours.
6. After washing with 5% skim milk/LBB, incubate the blot with the appropriate HRP-conjugated secondary antibody at room temperature for 1 h.
7. After washing with LBB, develop the blot by enhanced chemiluminescence.

Comment

Laminin-binding activity of dystroglycan varies in different tissues and cell types. Skeletal muscle dystroglycan can be used as a positive control (~150 kDa).

Acknowledgments The work from our laboratory was supported by the 21st Century COE program (Integrated functional analyses of disease-associated sugar chains and proteins) from the Ministry of Education, Culture, Sports, Science, and Technology of Japan.

References

Barresi R, Michele DE, Kanagawa M, Harper HA, Dovico SA, Satz JS, Moore SA, Zhang W, Schachter H, Dumanski JP, Cohn RD, Nishino I, Campbell KP (2004) LARGE can functionally bypass alpha-dystroglycan glycosylation defects in distinct congenital muscular dystrophies. Nat Med 10:696–703

Kanagawa M, Toda T (2006) The genetic and molecular basis of muscular dystrophy: roles of cell-matrix linkage in the pathogenesis. J Hum Genet 51:915–926

Kobayashi K, Nakahori Y, Miyake M, Matsumura K, Kondo-Iida E, Nomura Y, Segawa M, Yoshioka M, Saito K, Osawa M, Hamano K, Sakakihara Y, Nonaka I, Nakagome Y, Kanazawa I, Nakamura Y, Tokunaga K, Toda T (1998) An ancient retrotransposal insertion causes Fukuyama-type congenital muscular dystrophy. Nature 394:388–392

Michele DE, Barresi R, Kanagawa M, Saito F, Cohn RD, Satz JS, Dollar J, Nishino I, Kelley RI, Somer H, Straub V, Mathews KD, Moore SA, Campbell KP (2002) Post-translational disruption of dystroglycan-ligand interactions in congenital muscular dystrophies. Nature 418:417–422

Section XV
Muscle Dystrophy and Carbohydrate Disorders of Glycosylation

Congenital Muscular Dystrophies Due to the Glycosylation Defect

Tamao Endo

Introduction

Muscular dystrophies are genetic diseases that cause progressive muscle weakness and wasting. The causative genes of several muscular dystrophies have been identified in the past 15 years. The best known is the one described by Duchenne that results from mutations in the gene encoding a protein called dystrophin. Another subclass is congenital muscular dystrophies, where muscle weakness is apparent at birth or shortly afterward. The dystrophin–glycoprotein complex (DGC), a multisubunit complex comprising peripheral and integral membrane glycoproteins, links the cytoskeleton to the extracellular matrix. Dystroglycan is a central component of the DGC, a protein assembly that plays a critical role in a variety of muscular dystrophies. Recent data suggest that the aberrant protein glycosylation of a specific glycoprotein, α-dystroglycan, is the primary cause of some forms of congenital muscular dystrophy. Here, I describe the glycobiology of muscular dystrophy.

Incomplete Glycosylation-Induced Dystroglycanopathies

Dystroglycan is encoded by a single gene and is cleaved into two proteins, α-dystroglycan and β-dystroglycan, by posttranslational processing. α- and β-Dystroglycan constitute a membrane-spanning complex that connects the extracellular matrix to the cytoskeleton. α-Dystroglycan is heavily glycosylated and its sugars have a role in binding to laminin, neurexin and agrin. We previously demonstrated that the main glycan of α-dystroglycan was O-mannosylglycan which is a ligand of laminin. We identified and characterized two glycosyltransferases (POMGnT1 and POMT1/2) involved in the biosynthesis of O-mannosylglycans (see Chapter by T. Endo and H. Manya, this volume). Then we found that these two enzymes are responsible for congenital muscular dystrophies, muscle–eye–brain disease (MEB: OMIM 253280) and Walker–Warburg syndrome (WWS: OMIM 236670). Recent data suggest that aberrant glycosylation of α-dystroglycan is the cause of other four muscular dystrophies, e.g., Fukuyama-type congenital muscular dystrophy (FCMD: OMIM 253800), congenital muscular dystrophy type 1C (MDC1C: OMIM 606612), limb-girdle muscular dystrophy 2I (LGMD2I: OMIM 607155), and congenital muscular dystrophy type 1D (MDC1D: OMIM 608840) (Table 1). These are named as dystroglycanopathies because these are caused by incomplete-glycosylation of dystroglycan. The details of FCMD are described in Chapter by M. Kanagawa and T. Toda, this volume.

Glycobiology Research Group, Tokyo Metropolitan Institute of Gerontology, Foundation for Research on Aging and Promotion of Human Welfare, 35–2 Sakae-cho, Itabashi-ku, Tokyo 173-0015, Japan
Phone: +81-3-3964-3241, Fax: +81-3-3579-4776
E-mail: endo@tmig.or.jp

Table 1 Possible muscular dystrophies caused by defective glycosylation of α-dystroglycan

Condition	Gene	Protein function	Gene locus
Muscle–eye–brain disease (MEB)	POMGnT1	O-mannosyl glycan GlcNAc transferase	1p33
Walker-Warburg syndrome (WWS)	POMT1	Protein	9q34.1
	POMT2	O-mannosyl transferase	14q24.3
Fukuyama congenital muscular dystrophy (FCMD)	Fukutin	Putative glycosyl-transferase	9q31
MDCIC and limb–girdle muscular dystrophy 2I(LGMD2I)	Fukutin-related protein (FKRP)	Putative glycosyl-transferase	19q13.3
MDCID	Large	Putative glycosyl-transferase	22q12.3-13.1

The human *POMGnT1* gene is located at 1p33 and is responsible for MEB. MEB is an autosomal recessive disorder characterized by congenital muscular dystrophy, ocular abnormalities, and brain malformation (type II lissencephaly). Patients with MEB show severe cerebral and ocular anomalies, but some patients reach adulthood. MEB has been observed mainly in Finland. We screened the *POMGnT1* gene for mutations in patients with MEB and identified 13 mutations in these patients. To confirm that the observed mutations are responsible for the defects in the synthesis of O-mannosyl glycan, we expressed all of the mutant proteins and found that none of them had enzymatic activity (Yoshida et al. 2001; Manya et al. 2003). These findings indicate that MEB is inherited as a loss of function of the *POMGnT1* gene. Additionally, we found a selective deficiency of glycosylated α-dystroglycan in MEB patients. This finding suggests that α-dystroglycan is a potential target of POMGnT1 and that hypoglycosylation of α-dystroglycan may be a pathomechanism of MEB. MEB muscle and brain phenotypes can be explained by a loss of function of α-dystroglycan due to abnormal O-mannosylation.

WWS is another form of congenital muscular dystrophy that is characterized by severe brain malformation and eye anomalies. Patients with WWS are severely affected from birth and usually die within their first year. Recently, 20% of WWS patients were found to have mutations in *POMT1*. POMT1 is highly expressed in fetal brain, testis, and skeletal muscle, which are the affected tissues in WWS. Mutations have also been found in the *POMT2* gene in this syndrome. Together, the known defects of these two genes can account for approximately one third of WWS cases. This suggests that other as yet unidentified genes are responsible for WWS. To confirm that the observed *POMT1* mutations are responsible for the defects in the synthesis of O-mannosyl glycan, we expressed the mutant POMT1 proteins and found that none of them had enzymatic activity (Akasaka et al. 2004, 2006). These findings indicate that WWS is inherited as a defect of O-mannosylation. In WWS patients, as in MEB patients, the glycosylated α-dystroglycan was selectively deficient in skeletal muscle. WWS and MEB are clinically similar, but WWS is more severe than MEB. The difference of severity between the two diseases may be explained as follows. If POMGnT1, which is responsible for the formation of the GlcNAcβ1–2Man linkage of O-mannosyl glycans is nonfunctional, then only O-mannose residues may be present on α-dystroglycan in MEB. On the other hand, *POMT1* or *POMT2* mutations cause complete loss of O-mannosyl glycans in WWS because POMT1 and 2 are responsible for the formation of the O-mannosyl linkage. It is possible that attachment of the single mannose residue on α-dystroglycan is responsible for the difference in clinical severity of WWS and MEB.

Defective glycosylation of α-dystroglycan has also been implicated in congenital muscular dystrophy type 1C (MDC1C), which is caused by the defect of fukutin-related protein (*FKRP*), a homolog of *fukutin*. MDC1C is characterized by severe muscle weakness and degeneration, and cardiomyopathy. Mental retardation and cerebellar cysts have been observed in some cases. Allelic mutations in the *FKRP* gene also cause a milder and more common form of muscular dystrophy called LGMD2I, which is frequently associated with cardiomyopathy and shows variable onsets ranging from adolescence to adulthood. Patients with mutations in the *FKRP* gene invariably exhibit a reduced expression of glycosylated α-dystroglycan, which is strongly correlated with disease severity. Although the function of FKRP is unknown, FKRP has been suggested to be a Golgi-resident protein and to be involved in the glycosylation of α-dystroglycan as a glycosyltransferase or a kind of modulator.

The gene *large*, which is mutated in the myodystrophy (Large^{myd}) mouse, encodes a putative glycosyltransferase. However, its biochemical activity has not been confirmed. The Large^{myd} mouse shows a progressive muscular dystrophy, ocular defects, and a central nervous system phenotype characterized by abnormal neuronal migration and disruption of the basal lamina. The Large^{myd} mouse shows hypoglycosylation of α-dystroglycan in muscle and brain. A recent study described a patient with congenital muscular dystrophy, profound mental retardation, white matter changes, and subtle structural abnormalities in the brain and a reduction of immunologically detectable α-dystroglycan. The patient was found to have mutations in the *LARGE* gene. This type of muscular dystrophy was named MDC1D.

A recent study revealed that LARGE can functionally compensate α-dystroglycan glycosylation defects in many congenital muscular dystrophies, including FCMD, MEB, and WWS. LARGE may also affect an alternative glycosylation pathway of α-dystroglycan instead of the O-mannosylation pathway. A homolog of *LARGE* (*LARGE 2* or *GYLTL1B*) has been shown to hyperglycosylate α-dystroglycan and cause an increase in laminin binding.

In summary, hypoglycosylated α-dystroglycan in the muscle membrane has greatly reduced affinities for laminin, neurexin and agrin. These findings suggest that defective glycosylation of α-dystroglycan due to the primary genetic defects of glycosyltransferases may be the common denominator causing muscle cell degeneration and abnormal brain structure in the diseases listed in Table 1. However, the function of these proteins, with the exception of POMGnT1, and POMT1/2, is largely unknown. Elucidation of the intrinsic characteristics of these gene products will improve our understanding of the pathomechanisms of these complicated diseases. Since hypoglycosylation of α-dystroglycan is a common feature in these diseases, α-dystroglycan may be a potential target of new glycotherapeutics strategy for muscular dystrophy in the future.

Acknowledgments The work from our laboratory summarized here was supported by Research Grants for Nervous and Mental Disorders (17A-10) from the Ministry of Health, Labour and Welfare of Japan, and for Scientific Research on Priority Area (14082209) from the Ministry of Education, Culture, Sports, Science and Technology of Japan.

References

Akasaka-Manya K, Manya H, Endo T (2004) Mutations of the *POMT1* gene found in patients with Walker–Warburg syndrome lead to a defect of protein *O*-mannosylation. Biochem Biophys Res Commun 325:75–79

Akasaka-Manya K, Manya H, Nakajima A, Kawakita M, Endo T (2006) Physical and functional association of human protein *O*-mannosyltransferases 1 and 2. J Biol Chem 281:19339–19345

Manya H, Sakai K, Kobayashi K, Taniguchi K, Kawakita M, Toda T, Endo T (2003) Loss-of-function of an *N*-acetylglucosaminyltransferase, POMGnT1, in muscle–eye–brain disease. Biochem Biophys Res Commun 306:93–97

Yoshida A, Kobayashi K, Manya H, Taniguchi K, Kano H, Mizuno M, Inazu T, Mitsuhashi H, Takahashi S, Takeuchi M, Hermann R, Straub V, Talim B, Voit T, Topaloglu H, Toda T, Endo T (2001) Muscular dystrophy and neuronal migration disorder caused mutations in a glycosyltransferase, POMGnT1. Dev Cell 1:717–724

Molecular Diagnosis of Congenital Disorders of Glycosylation

Yoshinao Wada

Introduction

Congenital disorders of glycosylation (CDG) represent a group of diseases affecting N-linked glycosylation pathways and require structural analysis of glycoproteins or their glycans for decisive diagnosis. CDG are classified into two groups: the classical type of CDG (CDG-I) results from deficiencies in the early glycosylation pathway for biosynthesis of lipid-linked oligosaccharide and its transfer to proteins in endoplasmic reticulum, while the CDG-II diseases are caused by defects in the subsequent processing steps. Different types of disorders in each class are indicated by a small letter code (a, b, c, etc.) consistent with the chronological order in which the defective gene was identified, and 18 disorders have been documented to date (Freeze and Aebi 2005). Isoelectric focusing (IEF) of serum transferrin (Tf) has been used for laboratory screening of CDG. Serum Tf has two disialo-biantennary N-glycans as the major oligosaccharide at Asn-432 and Asn-630, and a defect in the glycan moiety accompanies sialic acid deficiency causing a shift of the protein pI values. Recently, mass spectrometry (MS) has been utilized to detect a change in the molecular mass by 2,200 Da per glycan set in CDG-I or a smaller change due to the glycan processing defects in CDG-II. A program for molecular diagnosis has started in Japan according to the protocols in Scheme 1 (Wada 2006).

Isoelectric Focusing

1. Mix 5 µL serum (or plasma) with 35 µL of 6 mM ferrous ammonium sulfate.
2. Incubate at room temperature for 1 h to load ions completely to Tf molecules.
3. Apply 10 µL sample solution to IEF, pH 4.0–6.5, precast gel (Ampholine PAGplate, GE Healthcare) for 4,000 V h.
4. Fix in 15% (w/v) trichloroacetic acid.
5. Wash the gel in ethanol/acetic acid/H$_2$O (25:8:68, v/v/v).
6. Stain with 0.1% (w/v) Coomassie Brilliant Blue R-250 dissolved in ethanol/acetic acid/H$_2$O (9:2:9, v/v/v).
7. Destain in ethanol/acetic acid/H$_2$O (25:8:68, v/v/v) (Fig. 1).

Mass Spectrometric Analysis of Transferrin

1. Prepare anti-Tf immunoaffinity gel by coupling rabbit anti-human transferrin IgG (Dako) with NHS-activated Sepharose (GE Healthcare).

Osaka Medical Center and Research Institute for Maternal and Child Health,
840 Murodo-cho, Izumi, Osaka 594-1101, Japan
Phone: +81-725-57-4105, Fax: +81-725-57-3021
E-mail: waday@mch.pref.osaka.jp

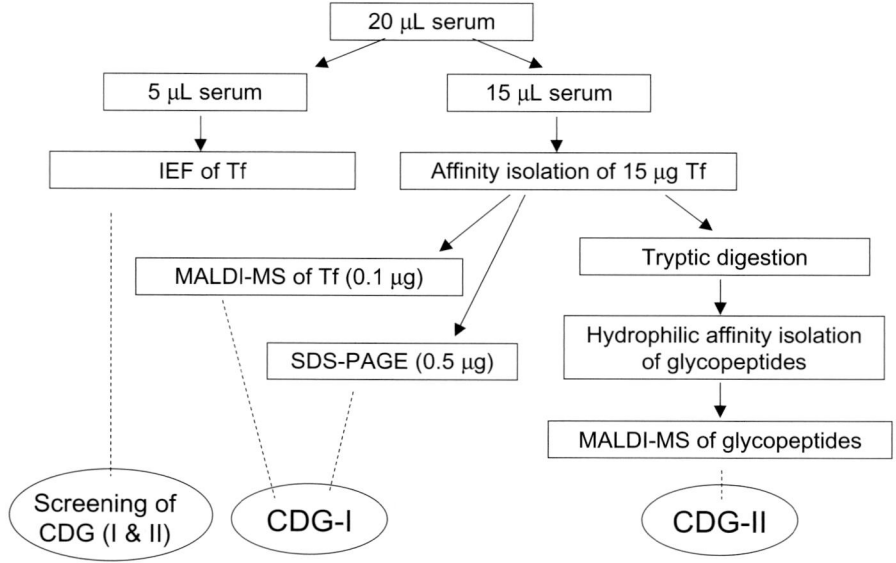

Scheme 1 Flow chart of molecular diagnosis of CDG

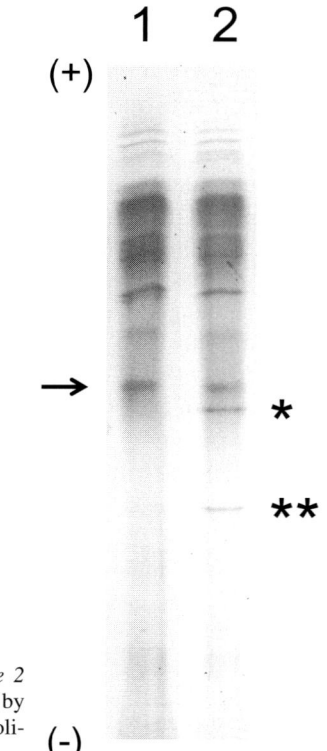

Fig. 1 Isoelectric focusing of serum. *Lane 1* healthy control, *lane 2* CDG-I patient. Normal and aberrant Tf molecules are indicated by *arrow* and *asterisks*, respectively. One (*asterisk*) or two (*asterisks*) oligosaccharide chains are missing in the patient Tf

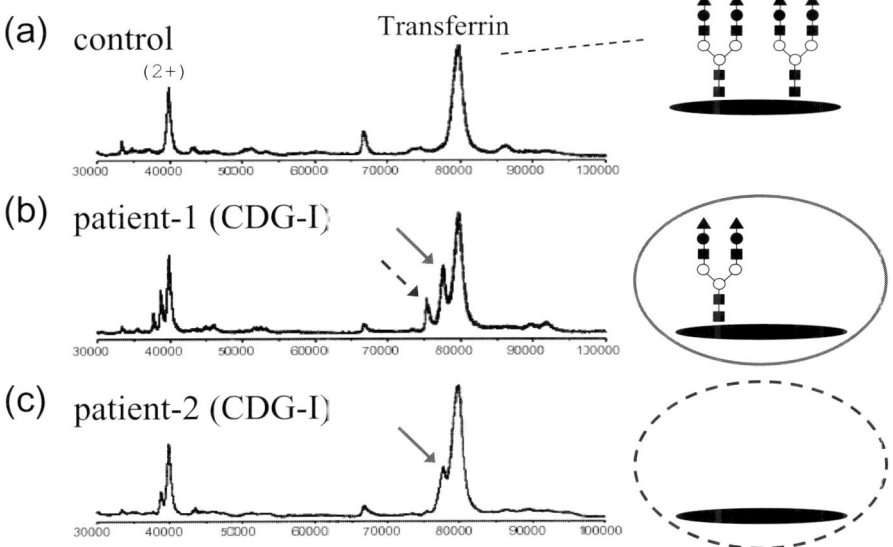

Fig. 2 MALDI TOF mass spectra of Tf. a Healthy control. b, c) CDG-I patients

2. Incubate 15 μL serum (or plasma) and 5 μL anti-Tf gel in 0.5 mL phosphate-buffered saline (PBS) at room temperature for 3 h or at 4°C overnight.
3. Wash the gel with PBS, three times.
4. Elute Tf in 100 μL of 0.2 M glycine HCl, pH 2.7.
5. Apply one-tenth part of the eluent to a reversed phase device such as Ziptip (Millipore) for desalting, according to manufacturer's instruction (another part is reserved for glycopeptide analysis).
6. Elute Tf from Ziptip in acetonitrile/H_2O/trifluoroacetic acid (70:30:0.1, v/v/v).
7. Mix 1 μL eluent with 1 μL saturated sinapinic acid in acetonitrile/H_2O/trifluoroacetic acid (30:70:0.1, v/v/v) on the sample target of MALDI mass spectrometer (Fig. 2).

Mass Spectrometric Analysis of Glycopeptides

1. Add 0.4 mL of 6 M guanidium HCl, 0.25 M Tris–HCl, pH 8.6 to the eluent (90 μL) from immunoaffinity gel.
2. Add 5 mg of dithiothreitol and incubate at 50°C for 3 h for reduction.
3. Add 10 mg of iodoacetamide and incubate in the dark at room temperature for 30 min for S-alkylation (S-carbamidomethylation).
4. Exchange the solvent by gel filtration with NAP-5 (GE Healthcare) equilibrated with 0.05 N HCl.
5. Add one-tenth volume of 1.5 M Tris–HCl, pH 11, to raise the solution pH to 8.2.
6. Digest with trypsin.
7. Follow protocol in Chapter by Y. Wada, in *Experimental Glycoscience: Glycochemistry*, for isolation of glycopeptides and mass spectrometry (Fig. 3).

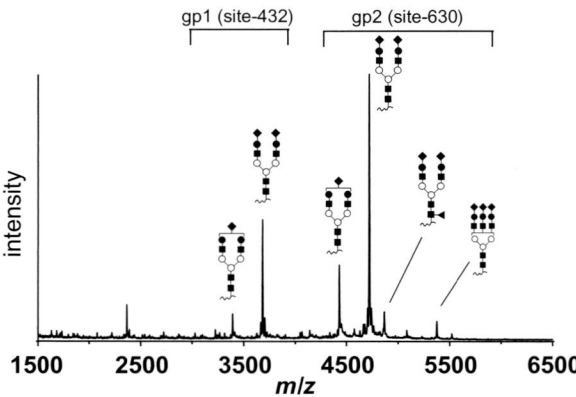

Fig. 3 MALDI TOF mass spectrum of Tf glycopeptides. Site-specific glycoforms at Asn-432 (*gp1*) and Asn-630 (*gp2*) are indicated above individual signals

References

Freeze HH, Aebi M (2005) Altered glycan structures: the molecular basis of congenital disorders of glycosylation. Curr Opin Struct Biol 15:490–498

Wada Y (2006) Mass spectrometry for congenital disorders of glycosylation, CDG. J Chromatogr B Anal Technol Biomed Life Sci 838:3–8

Research in Japan Has Contributed to the Understanding of GPI Anchor Deficiency

Yoshiko Murakami, Taroh Kinoshita

Introduction

Glycosylphosphatidylinositol (GPI) anchors more than 200 surface proteins to the plasma membrane. Therefore, GPI deficiency causes severe effects. Twenty-six genes involved in GPI biosynthesis have been cloned and among them *PIGA* carries the first step. The finding, that a *Piga* knockout mouse is lethal, indicates that GPI is essential for ontogenesis. Even the tissue-specific knockout of the *Piga* gene causes severe defects, e.g., a keratinocyte-specific knockout causes death soon after birth and an oocyte-specific knockout causes infertility. These lines of evidence suggest that only partial GPI deficiency causes disease. Among 26 GPI biosynthesis genes, 4 genes, *DPM1*, *DPM2*, *DPM3* and *SL15* (*MPDU1*), are also involved in biosynthesis of *N*-glycan. A defect in these genes causes congenital deficiencies of glycosylation (CDG). Deficiencies of *DPM1* and *SL15* (*MPDU1*) have been reported and called CDG-Ie and CDG-If, respectively. Both of these deficiencies are partial and the symptoms of these patients seem to be mainly caused by defective N-glycosylation.

Discovery of Inherited GPI Deficiency

Recently, our lab and colleagues in England identified a novel disease, caused by GPI deficiency, inherited in an autosomal recessive manner and characterized by portal vein thrombosis and seizures (Almeida et al. 2006). In two unrelated families, we found an identical point mutation (C > G) at position 270 in the promoter region of the *PIGM* gene, which encodes a mannosyltransferase essential for the addition of the first mannose to a GPI intermediate. This point mutation disrupts binding of the transcription factor Sp1 to its promoter motif. As a result, transcription of *PIGM* is severely reduced in patients leading to partial GPI deficiency. Expression levels of *PIGM* differ among cell lineages, indicating tissue-specific regulation of *PIGM* expression. For example, expression of GPI-anchored proteins on red blood cells is only slightly decreased, which is in contrast to a severe defect on granulocytes and fibroblasts. To our surprise, these patients have no congenital, morphological, and developmental abnormalities. Various stimulations during embryogenesis may have compensated the decreased basal promoter activity caused by the disruption of Sp1 binding. Analysis of the *PIGM* promoter region of the affected cells showed that this point mutation results in reduced histone acetylation. Treatment with Na butyrate, a histone deacetylase inhibitor, restores histone acetylation, *PIGM* expression and the surface expression of GPI-anchored proteins in vitro as well

---

Department of Immunoregulation, Research Institute for Microbial Diseases, Osaka University, 3-1 Yamada-oka, Suita Osaka 565-0871, Japan
Phone: +81-6-6879-8328, Fax: +81-6-6875-5233
E-mail: tkinoshi@biken.osaka-u.ac.jp

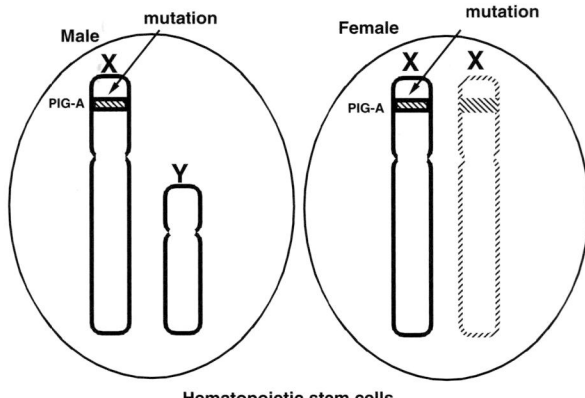

Fig. 1 One somatic mutation in *PIGA* gene generates a GPI-deficient hematopoietic stem cell in both males and females

as in vivo. Indeed, butyrate treatment is very effective to relieve the affected child from intractable seizures.

This inherited disease is completely different from paroxysmal nocturnal hemoglobinuria (PNH), in which hemolysis is a major symptom. Hemolysis is caused by complement owing to the defective expression of GPI-anchored complement regulatory proteins, which protect red blood cells. In inherited GPI deficiency, as mentioned above, expression of GPI-anchored proteins on erythrocytes is mostly normal, and thus, there are no episodes of hemolytic attack. Hence, this new disease may be passed over without being checked for the expression of GPI-anchored proteins.

An Acquired GPI Deficiency—PNH

PNH is a well-known acquired GPI deficiency, which affects only hematopoietic cells. Intravascular hemolysis, bone marrow failure and venous thrombosis are triad of symptoms of PNH. Since complement regulatory factors such as CD59 and decay accelerating factors (DAF) are GPI-anchored proteins, these factors are defective on the affected red blood cells from PNH patients, which leads to destruction of red blood cells upon activation of complement. Nakakuma, Luzzatto and our laboratory have shown that the affected cells from patients with PNH are defective in the first step of GPI anchor biosynthesis. In 1993, we cloned the gene involved in the first step of GPI biosynthesis and named it *PIGA* (Miyata et al. 1993). We then showed that *PIGA* is responsible for GPI deficiency in PNH (Takeda et al. 1993). *PIGA*, which is localized on Xp22.1, encodes the catalytic subunit of the *N*-acetylglucosamine transferase that mediates the first step. We demonstrated that a somatic mutation in *PIGA* occurs in one or a few hematopoietic stem cells.

Since *PIGA* is X-linked, only one hit of somatic mutation to the male X-chromosome or to active X-chromosomes in female cells is sufficient to cause GPI deficiency (Fig. 1). More than 200 cases of PNH patients have been analyzed and almost all were shown to be caused by *PIGA* mutation. The reason for this uniformity is that all other genes involved in GPI biosynthesis are autosomal and hence require two inactivating mutations to cause GPI deficiency, but such frequency is extremely low. This is quite different from CDG, for which many different genes involved in the biosynthesis of *N*-glycan are responsible. The basis of this difference is that PNH is caused by a somatic mutation whereas CDG is caused by congenital abnormalities. Indeed, inherited GPI deficiency is caused by mutation in the autosomal gene, *PIGM*.

Unique Pathogenesis of PNH

The pathogenesis of PNH is still not well understood. *PIGA* somatic mutation occurs in only one or a few hematopoietic cells and mutant cells are clonally expanded. Several lines of evidence indicate that *PIGA* mutation alone does not cause clonal expansion. It is well known that PNH is often associated with aplastic anemia, an autoimmune disease against hematopoietic stem cells It has been proposed that the *PIGA* mutant clone is positively selected by autoimmune attack to hematopoietic stem cells. We have shown that GPI-negative cells are less susceptible than GPI-positive cells to the attack of cytotoxic T cells, and that the percentage of GPI-negative cells increases under immunological selection in a mouse model (Murakami et al. 2002). Current studies are focused on the identification of autoantigens and the basis of relative resistance of *PIGA* mutant cells to autoimmunity. A further problem in PNH pathogenesis is that PNH cells in patients with aplastic anemia usually show limited expansion, suggesting that other events are required for full expansion of mutant clones. We recently reported two PNH patients who have a chromosomal 12 abnormality only in the PNH clone. In these PNH cells, 3′UTR of the *HMGA2* gene was disrupted (Inoue et al. 2006). *HMGA2* encodes an architectural transcriptional factor and is mainly expressed during the embryonic stage. Ectopic expression of the *HMGA2* transcript with 3′ truncation owing to a loss of 3′UTR is reported to cause several benign tumors, such as uterine myoma and lipoma. In two patients, 3′UTR disruption was associated with ectopic expression of HMGA2 in bone marrow cells, strongly suggesting that the PNH cells-acquired characteristics of the benign tumor. We intend to analyze other PNH patients to determine whether we can generalize this idea.

Thus, PNH shows a very unique pathogenesis in which red blood cells derived from fully expanded PNH clones undergo hemolysis upon infection owing to their own dysregulation of the complement system.

References

Almeida AM, Murakami Y, Layton DM, Hillmen P, Sellick GS, Maeda Y, Richards S, Patterson S, Kotsianidis I, Mollica L et al (2006) Hypomorphic promoter mutation in *PIGM* causes inherited glycosylphosphatidylinositol deficiency. Nat Med 12:846–851

Inoue N, Izui-Sarumaru T, Murakami Y, Endo Y, Nishimura J, Kurokawa K, Kuwayama M, Shime H, Machii T, Kanakura Y et al (2006) Molecular basis of clonal expansion of hematopoiesis in 2 patients with paroxysmal nocturnal hemoglobinuria (PNH). Blood 108:4232–4236

Miyata T, Takeda J, Iida Y, Yamada N, Inoue N, Takahashi M, Maeda K, Kitani T, Kinoshita T (1993) Cloning of PIG-A, a component in the early step of GPI-anchor biosynthesis. Science 259:1318–1320

Murakami Y, Kosaka H, Maeda Y, Nishimura J, Inoue N, Ohishi K, Okabe M, Takeda J, Kinoshita T (2002) Inefficient response of T lymphocytes to GPI-anchor-negative cells: implications for paroxysmal nocturnal hemoglobinuria. Blood 100:4116–4122

Takeda J, Miyata T, Kawagoe K, Iida Y, Endo Y, Fujita T, Takahashi M, Kitani T, Kinoshita T (1993) Deficiency of the GPI anchor caused by a somatic mutation of the *PIGA* gene in paroxysmal nocturnal hemoglobinuria. Cell 73:703–711

Section XVI
Lifestyle-Related Diseases

Aberrant Expression of Sialidase in Cancer and Diabetes

Taeko Miyagi

Introduction

Altered sialylation is associated with malignant properties such as invasiveness and metastasis. To understand the causes and the consequences of aberrant forms, our studies have focused on sialidases responsible for the removal of sialic acids from glycoproteins and glycolipids. Mammalian sialidases have been classified into four types (designated as Neu1, Neu2, Neu3, and Neu4) differing in subcellular localization and enzymatic properties. We found that the four types behave in different manners during carcinogenesis (Miyagi et al. 2004). Among the sialidases, Neu3 is a key enzyme for ganglioside degradation and is unique in being localized mainly in the plasma membrane and in hydrolyzing specifically gangliosides, thereby probably participating in cell surface events including neuronal differentiation and transmembrane signaling. The human Neu3 ortholog, NEU3, is indeed upregulated in various cancers where it suppresses apoptosis. Its overexpression in mice results in a diabetic phenotype. Herein molecular mechanisms and significance of aberrant expression of NEU3 are briefly introduced.

Principles

When human sialidase (NEU3) expression was measured in colon cancers, mRNA levels were found to be increased by 3- to 100-fold as compared to adjacent non-tumor mucosa, this being associated with significant elevation of sialidase activity (Kakugawa et al. 2002). In situ hybridization showed high sialidase expression in epithelial elements of adenocarcinomas. To examine the significance of increased expression, cultured human colon cancer cells were treated with sodium butyrate, and changes during differentiation and apoptosis were observed. The NEU3 level was downregulated by the treatment while NEU1 was upregulated. Transfection of the *NEU3* gene into cancer cells inhibited apoptosis accompanied by increased Bcl-2 and decreased caspase expression, while knock down with a short interfering RNA resulted in acceleration of apoptosis in cancer, but not in normal cells. NEU3 was also found to contribute the expression of malignant properties, including cell invasion and motility, by affecting integrin-mediated signaling associated with the extracellular matrix and by activating IL-6-mediated signaling via the PI3K/Akt cascade in renal cancers (Ueno et al. 2006). Interestingly, NEU3 overexpression enhanced Ras activation with consequent stimulation of ERK and Akt but no change in JNK and p38. On the other hand, NEU3 siRNA inhibited Ras activity with down-regulation of ERK and Akt (Wada et al. 2007). These results indicate that NEU3 plays an important role in cancer-cell survival, increased expression promoting

---

Division of Biochemistry, Miyagi Cancer Center Research Institute, Natori, Miyagi 981-1293, Japan
E-mail: miyagi-ta173@pref.miyagi.jp

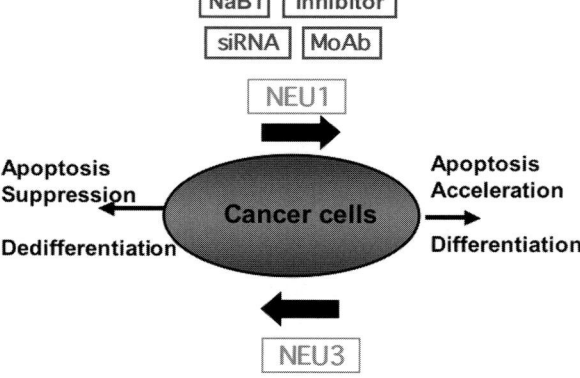

Fig. 1 Up-regulation of NEU3 in cancer cells and the schematic representation of possible targeted therapy for cancer

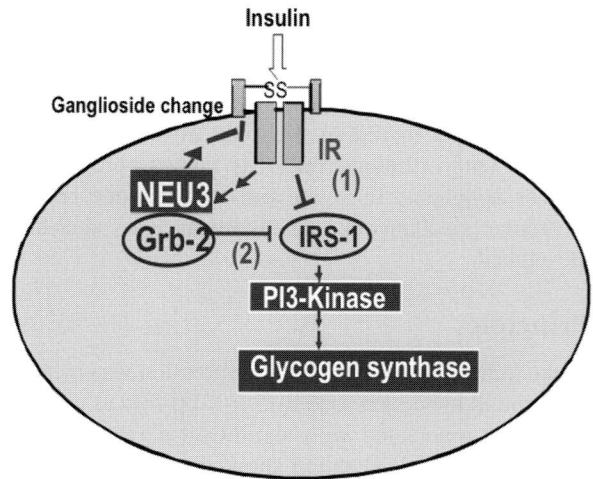

Fig. 2 Schematic model of the hypothetical role of NEU3 in insulin signaling. In the upstream pathway: *(1)* NEU3 down-regulates IR phosphorylation through modulation of gangliosides, and in the downstream pathway: *(2)* NEU3 influences signaling negatively through interaction with Grb2

malignancy by modulating Ras-mediated signaling. NEU3 may thus have utility for gene-based therapy of various human cancers, as proposed in Fig. 1.

To investigate the effects of NEU3 overexpression in vivo, transgenic mice were generated. These mice developed diabetic phenotype by 18–22 weeks of age, associated with hyperinsulinemia, islet hyperplasia and an increased β-cell mass (Sasaki et al. 2003). Compared to the wild type, insulin-stimulated phosphorylation of the insulin receptor and insulin receptor substrate I was significantly reduced, leading to retardation in post receptor insulin signaling (Fig. 2). In response to insulin, NEU3 was found to undergo tyrosine-phosphorylation and associate with Grb2 protein, as assessed by immunoprecipitation with anti-NEU3 monoclonal antiboby, thus being activated and negatively regulate insulin signaling. In fact, accumulation of GM1 and GM2, possible sialidase products in transgenic tissues, has been found to cause inhibition of IR phosphorylation in vitro, and blocking of the association with Grb-2 resulted in reversion of impaired insulin signaling. The data indicate that NEU3 is a novel molecule involved in insulin signaling, and the mice can serve as a valuable model for human diabetes mellitus. In addition, genetic polymorphisms of NEU3 are associated with the pathogenesis of type 2 diabetes mellitus in Japanese suggesting that SNPs might contribute to insulin resistance.

Procedures

In Situ Hybridization

1. Generate sense and antisense probes using DIG-labeled UTP (Roche Molecular Biochemicals, Mannheim) and a Bluescript plasmid containing human NEU3 cDNA fragment (nucleotides 1–928).
   (a) Linearize the plasmids by cutting with *Xho*I and *Xba*I, and transcribe by T3 and T7 RNA polymerases, respectively, in the presence of digoxygenin-labeled UTP (Roche Molecular Biochemicals, Mannheim).
   (b) Cut the probes into approximately 300-bp fragments by alkaline treatment before use.
2. Slice the tissues on a Micron cryostat at 20°C, and thaw-mount onto silane-coated slides.
3. Fix the sections in a mixture of 4% paraformaldehyde and 0.5% glutaraldehyde/PBS, and incubate with 10 µg/ml proteinase K for 30 min at 37°C.
4. Treat the sections with 0.25% acetic anhydride for 10 min at room temperature, and dehydrate and prehybridize them for 1 h at 48°C in a hybridization buffer consisting of 50% deionized formamide, 4 × SSC, 20% dextran sulphate, 1 × Denhardt's solution, and 0.25 mg/ml yeast tRNA.
5. Incubate them with an antisense riboprobe (or a sense probe as a negative control) at 48°C overnight in a humidified box.
6. Incubate them successively in 2 × SSC at 48°C for 15 min twice, in 2 × SSC with 100 µg/ml RNaseA at 37°C for 30 min, and in 0.1 × SSC at room temperature for 10 min.
7. Detect the positive signals by immunoassay with anti-digoxygenin–alkaline phosphatase conjugate and the color substrates, NBT/X-phosphate.

Preparation of Monoclonal Antibody to Human Sialidase NEU3

1. Prepare the particulate fractions of NEU3-transfected COS cells by centrifugation of the crude homogenates at 100.000 × g for 1 h at 4°C.
2. Emulsify 2 mg protein of the above particulate fractions with incomplete adjuvant, and inject every week into the foot pads of 6-week-old female BALB/c mice for 5 weeks followed by a booster injection.
3. Harvest the spleens and bilateral inguinal and parietal abdominal lymphnodes of immunized mice.
4. Fuse B cells with the mouse myeloma X 63 Ag8.653 (The Cancer Cell Repository, Tohoku University) in 50% polyethylene glycol 1000 (Wako) and culture for a week.
5. Screen hybridomas by examining the specific immunoreactivity of the supernatant of their culture medium in ELISA system (ZYMED).
   (a) Precoat the polystylene microtiter plates with the particulate fractions of the NEU3 transfectants or those of the vehicle-transfected cells.
   (b) Incubate with the supernatants of hybridomas at 37°C for 1–2 h, then with alkaline phosphatase-conjugated goat anti-IgG.

(c) Detect positive clones by color development, and then remove the clones, double-positive with NEU3 and the vehicle-transfected cells to exclude the false positives.
6. Screen hybridomas by FACS analysis.
    (a) Incubate the hybridomas with the NEU3-transfectants or vehicle-transfected cells in the presence of 0.5% saponine.
    (b) Incubate the hybridomas with anti-mouse F(ab')$_2$ fragments of IgG conjugated with fluorescein.
7. Screen the positive cells by dilution to obtain single specific clones.
8. Determine the subclass of monoclonal antibody obtained above with a Mouse isotyping kit.

Future Perspectives

From the observations described above, increased expression of NEU3 may be essential for the survival of various cancer cells. This sialidase, therefore, could be a useful target for cancer diagnosis and therapy. As shown in the schema in Fig. 1, discovery of specific inhibitors for NEU3, or use of small interfering RNAs or specific antibodies, opens up potential applications in cures for cancer. Furthermore, transgenic mice may serve as a valuable model for human diabetes mellitus. It is of obvious interest that NEU3 overexpression causes disturbance in both insulin and apoptosis signaling, given the significant correlation between certain types of cancer development and diabetes demonstrated by epidemiological studies.

References

Kakugawa Y, Wada T, Yamaguchi K, Yamanami H, Ouchi K, Sato I, Miyagi T (2002) Up-regulation of plasma membrane-associated ganglioside sialidase (Neu 3) in human colon cancer and its involvement in apoptosis suppression. Proc Natl Acad Sci 99:10718–10723

Miyagi T, Wada T, Yamaguchi K, Hata K (2004) Sialidase and malignancy: a minireview. Glycoconjugate J 20:189–198

Sasaki A, Hata H, Suzuki S, Sawada M, Wada T, Yamaguchi K,.Obinata M, Tateno H, Suzuki Y, Miyagi T (2003) Overexpression of plasma membrane-associated sialidase attenuates insulin signaling in transgenic mice. J Biol Chem 278:27896–27902

Ueno S, Saito S, Wada T, Yamaguchi K, Satoh M, Arai Y, Miyagi T (2006) Plasma membrane-associated sialidase is up-regulated in renal cell carcinoma and promotes the interleukin-6-induced apoptosis suppression and cell motility. J Biol Chem 281:7756–7764

Wada T, Hata K, Yamaguchi K, Shiozaki K, Koseki K, Moriya S, Miyagi T (2007) A crucial role of plasma membrane-associated sialidase (NEU3) in the survival of human cancer cells. Oncogene 26:2483–2490

Insulin Resistance and Type 2 Diabetes as Microdomain Disease: Implication of Ganglioside GM3

Jin-ichi Inokuchi, Kazuya Kabayama

Introduction

Caveolae are a subset of membrane microdomains (lipid raft) particularly abundant in adipocytes. Critical dependence of the insulin metabolic signal transduction on caveolae/microdomains in adipocytes has been demonstrated. These microdomains can be biochemically isolated with their detergent insolubility and were designated as detergent resistant microdomains (DRM). Gangliosides are known as structurally and functionally important components in microdomains. We demonstrated that increased GM3 expression was accompanied in the state of insulin resistance in mouse 3T3-L1 adipocytes induced by TNFα and in the adipose tissues of obese/diabetic rodent models such as Zucker *fa/fa* rats and *ob/ob* mice (Tagami et al. 2002). We examined the effect of TNFα on the composition and function of DRM in adipocytes and demonstrated that increased GM3 levels result in the elimination of insulin receptor (IR) from the DRM while caveolin and flotillin remain in the DRM, leading to the inhibition of insulin's metabolic signaling (Kabayama et al. 2005). These findings are further supported by the report that mice lacking GM3 synthase exhibit enhanced insulin signaling (Yamashita et al. 2003). Thus, we present a new pathological feature of insulin resistance in adipocytes induced by TNFα.

Lists

1. Summary on the evidence implicating caveolae microdomains is critical for proper compartmentalization of IR in caveolae to execute the successful insulin metabolic signaling in adipocytes (Table 1) (Inokuchi and Kabayama 2007).

2. Reduction of GM3 level ameliorates insulin resistance induced by TNFα (Tagami et al. 2002).

Fully differentiated 3T3-L1 adipocytes were treated with 0.1 nM TNFα for 96 h as described previously (Guo and Donner 1996). TNFα induced a moderate decrease of insulin-stimulated phosphorylation of the IR and a more pronounced inhibition of insulin-promoted phosphorylation of IRS-1 without affecting expression of either IR or IRS-1 (Fig. 1a). Marked accumulation of GM3 occurred with TNFα treatment (Fig. 1b). D-PDMP, a well-known inhibitor of glucosylceramide synthase lowered GM3 content in 3T3-L1 adipocytes with a concomitant increase of tyrosine phosphorylation of IRS-1 in

---

Division of Glycopathology and CREST, Japan Science and Technology Agency, Institute of Molecular Biomembranes and Glycobiology, Tohoku Pharmaceutical University, 4-4-1 Komatsushima, Aoba-ku, Sendai, 981-8558, Japan
E-mail: jin@tohoku-pharm.ac.jp

Table 1 Localization of insulin receptor in caveolae microdomains is essential for insulin's metabolic signaling

1. Direct binding of IR and caveolin-1
IR has caveolin binding domain
Coimmunoprecipitation of IR and caveolin
2. Colocalization of IR and caveolin-1
IR and caveolin in light-density fractions by sucrose density floatation assay
Fluorescence microscope
Electron microscope
3. Insulin signaling via caveolae
Stimulation of caveolin-1 tyrosine phosphorylation by insulin
Caveolin-deficient mice show insulin resistance due to accelerated degradation of IR in adipose tissue
Cholesterol depletion disrupts caveolae and metabolic signaling of insulin
Increased GM3 eliminates IR from DRMx and inhibits IR–IRS-1 signaling

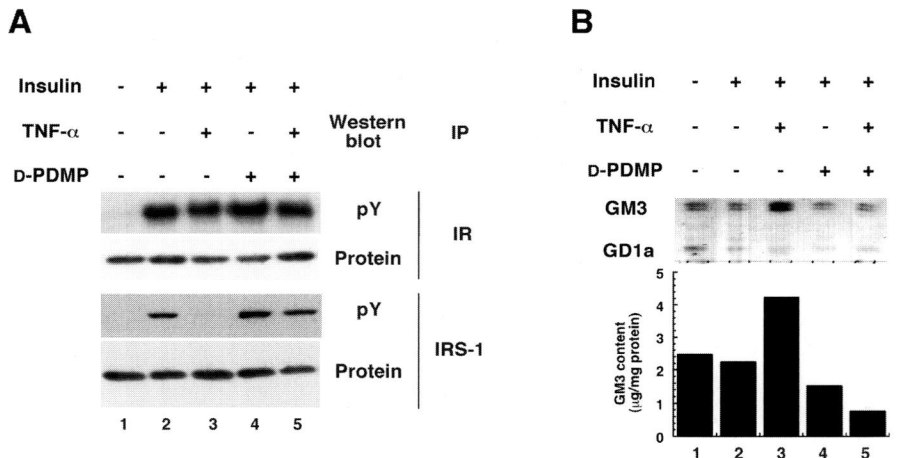

Fig. 1 TNFα increases the expression of GM3 and prevention of GM3 synthesis reverses TNFα induced suppression of insulin signaling in adipocytes (Tagami et al. 2002)

response to insulin stimulation (Fig. 1A), suggesting that D-PDMP and its analogs may open a new therapeutic strategy for insulin resistance and type 2 diabetes.

3. Dissociation of IR from caveolae microdomains in a state of insulin resistance.

Adipocytes, untreated or treated with 0.1 nM TNFα for 96 h in the presence or absence of 20 µM D-PDMP, were lysed with buffer containing 0.08% Triton X-100 and subjected to a sucrose density gradient floatation assay. GM3 was preferentially localized at the DRM in both normal and TNFα-treated 3T3-L1 adipocytes. Remarkably, the accumulation of GM3 observed in the DRM was twofold higher in the TNFα-treated cells (Fig. 2a). Next, each fraction was subjected to SDS-PAGE and immunoblotted with antibodies against IRβ, caveolin-1 and flotillin. The DRMs were able to hold the IR, although in the TNFα-treated cells the IR tended to shift to higher density fractions without affecting the localization of caveolin and flotillin in DRM (Fig. 2B). After GM3 depletion by D-PDMP, the dissociation of IR from the DRM was effectively blocked (Fig. 2C). These results demonstrate the selective elimination of IR from the DRM owing

Insulin Resistance and Type 2 Diabetes as Microdomain Disease: Implication of Ganglioside

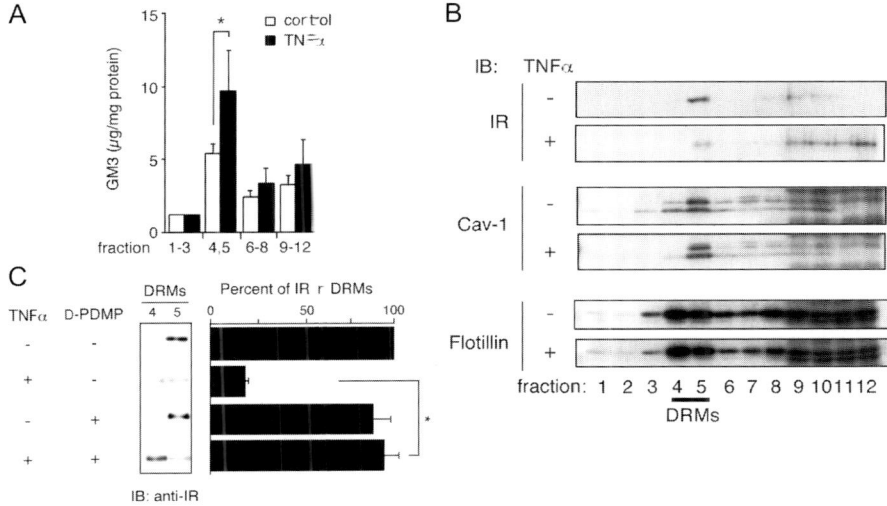

Fig. 2 Elimination of IR from the DRM owing to over accumulation of GM3 (Kabayama et al. 2005)

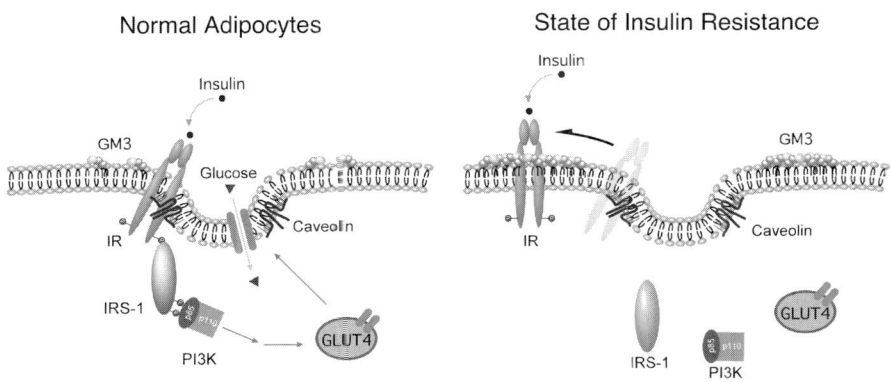

Fig. 3 Proposed model of caveolae microdomains in the state of insulin resistance in adipocytes (Kabayama et al. 2007)

to over accumulation of GM3 in adipocytes under a chronic state of TNFα-induced insulin resistance.

Comments

The current view of microdomains in the state of insulin resistance is depicted in Fig. 3. We propose a working hypothesis, "Life style-related diseases such as type 2 diabetes are membrane microdomain disorders caused by aberrant expression of glycosphingolipids". Further work is in progress to elucidate the mechanisms for the interactions of the ganglioside GM3, IR and caveolin-1 in the microdomains (Kabayama et al. 2007).

Protocols

Elimination of Insulin Receptor from DRMs in the State of Insulin Resistance

All steps were carried out at 4°C.

1. Murine 3T3-L1 preadipocytes were differentiated, maintained and then incubated with or without 0.1 nM TNFα for 96 h.
2. Cells were then washed with PBS, added 2 ml of TNE buffer (10 mM Tris–HCl, pH 7.5, 150 mM NaCl, 5 mM ethylenediamine tetra acetic acid) containing protease inhibitors and 0.08% Triton X-100, and allow it to stand for 30 min to lyse.
3. Lysates were centrifuged for 5 min at 1,300×g to remove large cellular debris, and the 2 ml of supernatants were transferred to ultracentrifuge tubes, and diluted with 2 ml of 85% (w/v) sucrose in TNE buffer.
4. In the ultracentrifuge tubes, the diluted supernatants were overlaid with 4 ml 30% sucrose (w/v) in TNE buffer, then with 4 ml 5% sucrose (w/v) in TNE buffer. The samples were centrifuged at 198,000×g for 18 h in a swinging bucket rotor.
5. One-ml of fractions were collected from the top for immunoblotting by anti-IRβ antibody or antibodies for raft (microdomain) marker proteins (Kabayama et al. 2005).

Comment

In the presence of TNFα, IR was selectively eliminated from the DRMs and no distinct differences in the expression levels of cholesterol and no distribution of caveolin, flotillin, or fyn were observed (Kabayama et al. 2005).

References

Guo D, Donner DB (1996) Tumor necrosis factor promotes phosphorylation and binding of insulin receptor substrate 1 to phosphatidylinositol 3-kinase in 3T3-L1 adipocytes. J Biol Chem 271:615–618

Inokuchi J, Kabayama K (2007) Receptor modifications in glycobiology. In: Kamerling JP (ed) Comprehensive glycoscience, 3:733–744, Elsevier, Amsterdam

Kabayama K, Sato T, Kitamura F, Uemura S, Kang BW, Igarashi Y, Inokuchi J (2005) TNFalpha-induced insulin resistance in adipocytes as a membrane microdomain disorder: involvement of ganglioside GM3. Glycobiology 15:21–29

Kabayama K, Sato T, Saito K, Loberto N, Prinetti A, Sonnino S, Kinjo M, Igarashi Y, Inokuchi J (2007) Dissociation of the insulin receptor and caveolin-1 complex by ganglioside GM3 in the state of insulin resistance. Proc Natl Acad Sci USA 104:13678–13683

Tagami S, Inokuchi J, Kabayama K, Yoshimura H, Kitamura F, Uemura S, Ogawa C, Ishii A, Saito M, Ohtsuka Y, Sakaue S, Igarashi Y (2002) Ganglioside GM3 participates in the pathological conditions of insulin resistance. J Biol Chem 277:3085–3092

Yamashita T, Hashiramoto A, Haluzik M, Mizukami H, Beck S, Norton A, Kono M, Tsuji S, Daniotti JL, Werth N, Sandhoff R, Sandhoff K, Proia RL (2003) Enhanced insulin sensitivity in mice lacking ganglioside GM3. Proc Natl Acad Sci USA 100:3445–3449

Section XVII
IgA Nephropathy

Incompleteness of O-glycans at the Hinge Region of IgA1 and IgA Nephropathy

Hitoo Iwase[1], Yoshiyuki Hiki[2]

Introduction

IgA nephropathy (IgAN) was first reported as a disease by Berger. The definite diagnosis of IgAN can be made by detection of deposited IgA in the glomerular mesangial region by renal biopsy. IgAN was well known as one of the most common primary glomerulonephritis forms in the world. The frequency of IgAN in renal diseases is considered to be 20–30% in Asia and Europe. Among IgAN patients, about 40% of the patients in Japan develop renal failure within 20 years. The patient is often left without treatment because the symptoms were usually slight (protein urea and episodes of hematuria). Glomerular deposited IgA was dominantly the IgA1 subclass. The disease was thought to occur due to the deposition of an immune complex. Participation of many antigens, such as a food antigen, virus and bacteria, were reported as candidates. In addition, the implication of an abnormality in the molecular structure, especially a hinge mucin-type sugar chain, of IgA1 was reported mainly based on our research results in this article (Fig. 1).

Mucin-Like Structure of Hinge Portion of IgA1

Human IgA is composed of IgA1 and IgA2 subclasses. About 90% of serum IgA is the IgA1 subclass. Among them, IgA1 has five mucin-type sugar chains on the hinge region of the heavy chain, but this chain was absent in IgA2. The sugar chain is composed of GalNAc (Tn antigen), Galβ1,3GalNAc (TF antigen), sialyl Tn and sialyl TF antigen (Mattu et al. 1998).

Presence of Incompletely Glycosylated IgA1 and Anti-Hinge IgG Antibody Against Its Naked Hinge Portion in Sera

An antibody raised against the synthetic hinge peptide (Anti-sHP) was specifically reacted with artificially deglycosylated IgA1. The content of the aberrant IgA1 detected by the anti-sHP antibody was significantly higher in the sera of an IgAN patient than in a healthy control and other renal disease patients (Kokubo et al. 1999). The activity of anti-sHP IgG antibody was also detected in patient sera. The antibody activity was significantly higher in IgAN than in other renal diseases. A gender difference in anti-sHP IgG activity was also detected in IgAN patients. This was significantly higher in female patients than that in male patients. Because the frequency of the incidence of IgAN is

---

[1] Department of Biochemistry, Kitasato University, 1-15-1 Kitasato, Sagamihara, Kanagawa 228-8555, Japan
Phone: +81-42-778-9267, Fax: +81-42-778-8441
E-mail: iwaseh@med.kitasato-u.ac.jp

[2] Division of Nephrology, Fujita Health University, Aichi, Japan

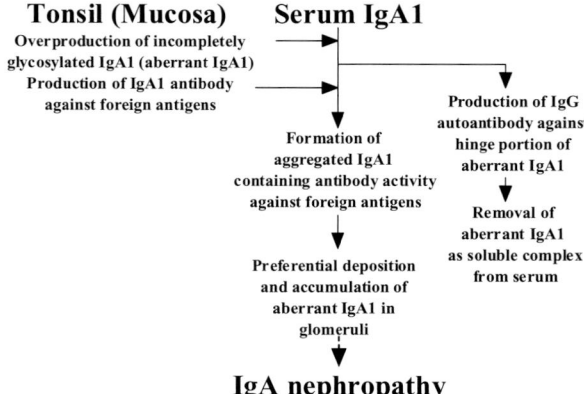

Fig. 1 Proposed mechanism for glomerular deposition of incompletely glycosylated IgA1

known to be slightly higher in males, this gender difference might indicate a protective meaning to remove incompletely glycosylated IgA1 from the patient's serum (Nakamura et al. 2004).

Self-Aggregation of Incompletely Glycosylated IgA1

From the analyses of heat-labile IgA1 and heat-stable IgA1, it was found that the heat-stable IgA1 was rich in sialylated mucin-type sugar chains. An aggregation phenomenon by the treatment of IgA1 with neuraminidase was also shown. Thus, artificial deglycosylation of IgA1 could induce its self-aggregation. Although there is some aggregated IgA even in control serum, almost all of the aggregated IgA was composed of the IgA1 subclass. Aberrant IgA1 could be separated from the serum by asialo-, agalacto-IgA1-Sepharose affinity chromatography using its self-aggregation characteristics. Concentration of the aberrant IgA1 was ~15 µg/ml serum, and its content was significantly higher in IgAN patients. The compositional similarity was reported between immunoglobulins binding to asialo-, agalacto-IgA1-Sepharose and those deposited in glomeruli of IgAN patients (Takatani et al. 2004). It was also found that the artificially deglycosylated IgA1 preferentially bound to the matrix proteins, such as fibronectin, laminin and type IV collagen.

Preferential Glomerular Deposition of Deglycosylated IgA1 in Animal Experiments

IgA1 from an IgAN patient was subfractionated by ion-exchange chromatography and jacalin affinity chromatography. Each subfraction was injected into the aorta of a rat kidney. Among the subfractions, a relatively neutral and jacalin-high-affinity fraction was preferentially deposited into the rat glomeruli. Analysis of the IgA1 subfraction indicated incompletely glycosylated IgA1. Meanwhile, artificially deglycosylated IgA1 was also injected into the rat kidney as above. Preferential accumulation of asialo-, agalacto-IgA1 in rat glomeruli became obvious, and simultaneously, infiltration of polymorphonuclear cells into the glomeruli was also observed (Sano et al. 2002). Although these are animal experiments, incompleteness of the sugar chain of IgA1 seemed to accelerate its accumulation in the glomeruli. Recently, the importance of polymeric IgA (secretory IgA) in sera was reported for the pathogenesis of IgAN.

Analyses of Hinge Glycopeptide Derived from IgA1

Hinge glycopeptide (33mer) was prepared from IgA1 by its S-pyridylethylation, and subsequent tryptic digestion. It could be analyzed by a mass spectrometric technique, such as matrix-assisted laser-desorption ionization time-of-flight mass spectrometry (MALDI-TOFMS), electro-spray-ionization mass spectrometry (ESI-MS) and surface-enhanced laser desorption-ionization (SELDI)-TOFMS (Odani et al. 2000). It was also advantageous to use capillary electrophoresis for analysis of the microheterogeneity of the IgA1 hinge glycopeptide having multiple O-linked oligosaccharides. MALDI-TOFMS analysis of the hinge glycopeptide derived from deposited IgA1 indicated that the incompletely glycosylated IgA1 would play a crucial role in its glomerular deposition (Hiki et al. 2001). In SELDI-TOFMS analysis, the jacalin-, PNA-, and VVL-immobilized ProteinChips could provide 13, 4, and 2 peaks, respectively, from the hinge glycopeptide mixture. Thus, SELDI-TOFMS analysis indicated the possibility that the glycopeptide having an IgAN specific glycoform could be selected by the appropriate ligand-coupled proteinChip array.

Method of Preparation of Hinge Glycopeptide from Human Serum IgA1

1. Serum (10 ml) was applied to jacalin-agarose column (10 ml, Vector Laboratories), and the column was washed with five times of column volume of J-Buffer (0.175 M Tris–HCl, pH 7.5 containing 0.02% sodium azide).
2. IgA1 was eluted from the column with the J-Buffer containing 0.8 M of galactose (50 ml), and the IgA1 fraction was dialyzed against distilled water and then lyophilized.
3. About 1 mg of IgA1 was desalted on a PD-10 column, and the desalted sample was dissolved in 0.5 ml of 0.4 M Tris–HCl buffer, pH 8.6, containing 6 M guanidine–HCl and 0.2 M EDTA.
4. To dissociate the disulfide bond, 5.4 µl of DTT solution (200 mg/ml) was added with stirring.
5. After heating at 50°C for 4 h, 1.6 ml of 4-vinyl pyridine was added and the reaction mixture was allowed to stand for 90 min at room temperature.
6. The reaction was terminated by the addition of 50 µl of 2.0 M formic acid, and the S-pyridylethylated α1 chain was separated by HPLC on a Cosmosil 5C4-300 column (Nacalai Tesque; 4.6 × 150 mm) by a linear gradient of 0–90% acetonitrile in 0.1% TFA.
7. About 160 µl of 50 mM Tris–HCl, pH 8.0, containing 2.0 M urea, 20 µl of trypsin (10 µg/20 ml) and 20 µl of 0.1 M $CaCl_2$ were added to about 0.5 mg of heavy chain and the mixture was incubated overnight at 37°C.
8. The trypsin digest was made up to 1 ml by adding 0.8 ml of J-Buffer, and the hinge glycopeptide was separated using jacalin-agarose column (2 ml) as above by eluting with 0.1 M melibiose.
9. The hinge glycopeptide (33mer) was further purified by HPLC method as mentioned above.

Comment

To obtain shorter hinge glycopeptide (25mer), purified 33mer glycopeptide was further treated in thermolysin solution (0.1 mg/ml of 0.1 M ammonium bicarbonate).

Tonsillar (or Mucosal) Tissue as a Possible Origin of Incompletely Glycosylated IgA1

The production mechanism of incompletely glycosylated IgA1 is still unclear. There are some reports on the relationship between tonsillectomy and IgA nephropathy (Xie et al. 2003). They reported that a tonsillectomy was effective in the long-term renal survival of an IgAN patient. Biochemical analysis of IgA1 derived from tonsil tissue and tonsillar lymphocytes was carried out. Overproduction of IgA1 bearing an asialo-TF antigen in tonsillar tissue from an IgAN patient and enrichment of asialo-, agalacto-IgA1 in the tonsillar lymphocytes from an IgAN patient were reported (Itoh et al. 2003; Horie et al. 2003). The distribution ratio of IgA1/IgA2 in the tonsillar extracts was 61/39, differing from that of serum IgA1, and the overproduction of IgA1 owing to its disordered balance in the tonsil tissue of some IgAN patients was also observed.

Participation of Glycosyltransferases in IgAN

The participation of glycosyltransferases in IgAN was summarized in Chapter by M. Asano and Chapter by Y. Narimatsu and H. Narimatsu, this volume.

References

Hiki Y, Odani H, Takahashi M, Yasuda Y, Nishimoto A, Iwase H, Shinzato T, Kobayashi Y, Maeda K (2001) Mass spectrometry proves under O-glycosylation of glomerular IgA1 in IgA nephropathy. Kidney Int 59:1077–1085

Horie A, Hiki Y, Odani H, Yasuda Y, Takahashi M, Kato M, Iwase H, Kobayashi Y, Nakashima I, Maeda K (2003) IgA1 molecules produced by tonsillar lymphocytes are under-O-glycosylated in IgA nephropathy. Am J Kidney Dis 42:486–498

Itoh A, Iwase H, Takatani T, Nakamura I, Hayashi M, Oba K, Hiki Y, Kobayashi Y, Okamoto M (2003) Tonsillar IgA1 as a possible source of hypoglycosylated IgA in the serum of IgA nephropathy patients. Nephrol Dial Transplant 18:1108–1114

Kokubo Y, Hiki Y, Iwase H, Tanaka A, Nishikido J, Hotta K, Kobayashi Y (1999) Exposed peptide core of IgA1 hinge region in IgA nephropathy. Nephrol Dial Transplant 14:81–85

Mattu TS, Pleass RJ, Willis AC, Kilian M, Wormald MR, Lellouch AC, Rudd PN, Woof JM, Dwek RA (1998) The glycosylation and structure of human serum IgA1, Fab, and Fc regions and the role of N-glycosylation on Fc alpha receptor interactions. J Biol Chem 273:2260–2272

Nakamura I, Iwase H, Arai K, Nagai Y, Toma K, Katsunata T, Hiki Y, Kokubo T, Sano T, Kobayashi Y (2004) Detection of gender difference and epitope specificity of IgG antibody activity against IgA1 hinge portion in IgA nephropathy patients by using synthetic hinge peptide and glycopeptides probes. Nephrology 9:26–30

Odani H, Hiki Y, Takahahsi M, Nishimoto A, Yasuda Y, Iwase H, Shinzato T, Maeda K (2000) Direct evidence for decreased sialylation and galactosylation of human serum IgA1 Fc O-glycosylated hinge peptides in IgA nephropathy by mass spectrometry. Biochem Biophys Res Commun 271:268–274

Sano T, Hiki Y, Kokubo T, Iwase H, Shigematsu H, Kobayashi Y (2002) Enzymatically deglycosylated human IgA1 molecules accumulate and induce inflammatory cell reaction in rat glomeruli. Nephrol Dial Transplant 17:50–56

Takatani T, Iwase H, Itoh A, Nakamura I, Hayashi M, Sakamoto H, Kamata K, Kobayashi Y, Hiki Y, Okamoto M, Higashihara M (2004) Compositional similarity between immunoglobulins binding to asialo-, agalacto-IgA1-Sepharose and those deposited in glomeruli in IgA nephropathy. J Nephrol 17:679–686

β4-Galactosyltransferase Deficiency and IgA Nephropathy

Masahide Asano

Introduction

Immunoglobulin A nephropathy (IgAN) is the most common glomerulonephritis worldwide, and about 40% of the patients progress to renal failure 20 years after the onset of the disease in Japan. IgAN is characterized by glomerular mesangial expansion and IgA deposition. However, the pathological molecular mechanisms remain to be fully elucidated. Pathological roles for the abnormal serum IgA, especially abnormal galactosylation and sialylation of O-glycans of IgA1, have been proposed. We have reported that β1,4-galactosyltransferase-I-deficient (β4GalT-I$^{-/-}$) mice spontaneously developed human IgAN-like disease with IgA deposition and expanded mesangial matrix (Nishie et al. 2007). This is the first report to demonstrate that genetic remodeling of protein glycosylation causes IgAN.

Results

β4GalT-I$^{-/-}$ mice showed severe albuminuria and some of them also showed hematuria. Histological examination of the kidney revealed that β4GalT-I$^{-/-}$ mice developed IgAN-like disease consistent with the pathological diagnosis of human IgAN, including IgA deposition, expanded mesangial matrix and electron-dense deposits in the paramesangial regions. IgA was predominately deposited in the mesangial area of the glomeruli with IgG, IgM and C3. About 75% of glomeruli in β4GalT-I$^{-/-}$ mice were already affected at younger age (8–13 weeks) and the ratio of glomeruli with global lesion increased with age. Glomerular sclerosis was also observed at 8 weeks of age and progressed with age. Furthermore, the β4GalT-I$^{-/-}$ mice showed high serum IgA levels with increased polymeric forms as observed in IgAN patients. However, hematuria was observed in only some of the β4GalT-I$^{-/-}$ mice, which is not consistent with human IgAN. The difference might be explained by reduced inflammatory responses attributable to the impaired biosynthesis of selectin ligands in the β4GalT-I$^{-/-}$ mice (see Chapter by M. Asano, this volume), because hematuria is induced by inflammation.

As expected, MALDI-TOF MS analysis revealed that β4-galactosylation on the N-glycans of the serum IgA of the β4GalT-I$^{-/-}$ mice was completely absent. Aberrant O-glycosylation in the hinge region of human IgA1 has been suggested to be involved in the pathogenesis of human IgAN. Although N-glycans, but not O-glycans, are attached to mouse IgA (Fig. 1), pathological roles of aberrant galactosylation on IgA in the development of IgAN in mice are suggested. Thus, we suppose that carbohydrates of IgA

Advanced Science Research Center, Kanazawa University, 13-1 Takara-machi, Kanazawa 920-8640, Japan
Phone: +81-76-265-2460, Fax: +81-76-234-4240
E-mail: asano@kiea.m.kanazawa-u.ac.jp

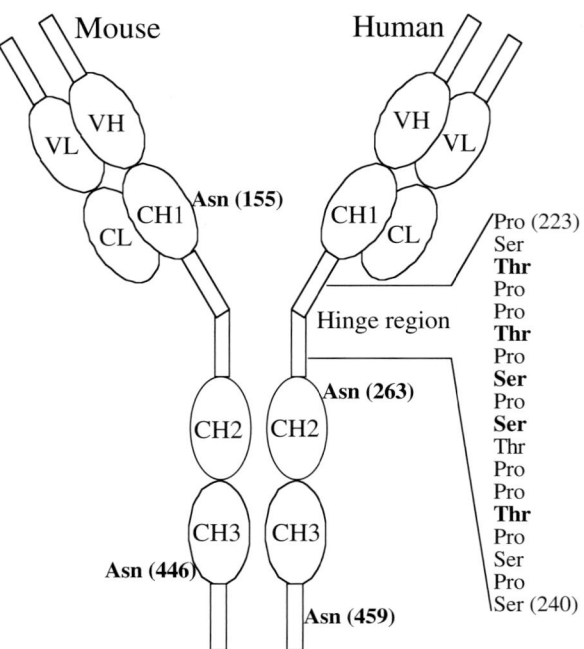

Fig. 1 Mouse (*left*) and human (*right*) IgA molecules with *O*- and *N*-glycosylation sites. Both mouse and human IgA have two *N*-glycosylation sites and human IgA1 alone has five *O*-glycosylation sites on the hinge region

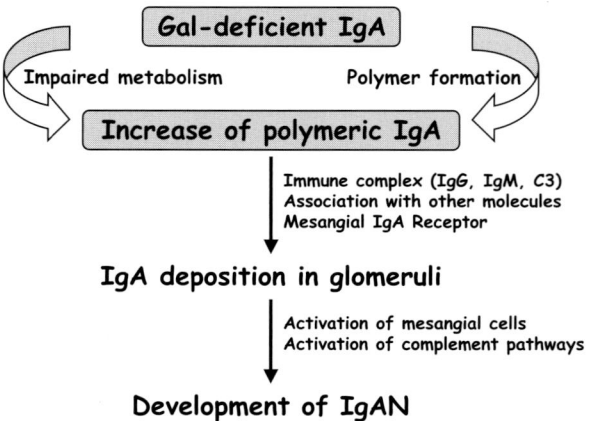

Fig. 2 Proposed molecular mechanisms on the development of IgAN caused by Gal-deficient IgA

might be involved in the development of IgAN both in humans and mice independent of whether they are *O*-glycans or *N*-glycans (Fig. 2).

Histological Analysis

1. Fix kidneys in 10% neutral buffered formalin, dehydrate, and embed them in paraffin according to standard procedures.
2. Deparaffinize paraffin sections (4–6-μm thick), rehydrate, and subject them to H&E, PAS, and Masson-trichrome staining using standard methods.

Immunofluorescence Microscopy

1. Prepare frozen kidney sections (5-μm thick) using cryostat.
2. Wash the slides in phosphate-buffered solution (PBS) and treat them with SuperBlock blocking buffer (Pierce Biotechnology, Rockford, IL, USA).
3. Incubate the slides at 4°C overnight with goat anti-mouse-IgA, -IgG, -IgM, or -C3 Fab fragment.
4. Rinse the slides with PBS/0.1% Tween-20.
5. Incubate the slides at room temperature for 40 min with fluorescein isothiocyanate (FITC)-conjugated rabbit anti-goat IgG antibody.
6. Wash the slides with PBS/0.1% Tween-20.
7. Observe immunofluorescence signals under immunofluorescence microscopy.

Electron Microscopy

1. Fix kidneys with glutaraldehyde and osmium tetroxide.
2. Embed the kidneys in Epon 812 (Oken Shoji, Tokyo, Japan).
3. Slice them into 0.1-μm sections.
4. Double-stain the sections with uranyl acetate and lead citrate.
5. Examine the sections under the electron microscope (Hitachi H-600, Hitachi, Tokyo, Japan).

Comments

β4GalT-I$^{-/-}$ mice developed an IgAN-like disease consistent with the pathological diagnosis of human IgAN, including mesangial IgA deposition, mesangial matrix expansion, and electron dense deposits in the paramesagial regions. Protocols described here are to examine histopathological diagnosis of IgAN.

Reference

Nishie T, Miyaishi O, Azuma H, Kameyama A, Naruse C, Hashimoto N, Yokoyama H, Narimatsu H, Wada T, Asano M (2007) Development of immunoglobulin A nephropathy-like disease in β-1,4-galactosyltransferase-I deficient mice. Am J Pathol 170:447–459

Glycosyltransferases Involved in the Biosynthesis of IgA O-Glycans

Yoshiki Narimatsu, Hisashi Narimatsu

Background and Significance of the Study

IgA nephropathy (IgAN) is a chronic inflammatory renal disease caused by the deposition of IgA1 subclass in the glomerular mesangium area, and one-third patients with IgAN could progress to end-stage renal failure. It is proposed that IgA molecules from patients with IgAN have incomplete carbohydrate chains in the O-glycan structure at the hinge region of IgA, though there is no clear evidence for this proposal. The sialyl-T (ST) antigen is a O-glycan having the following simple structure: SAα2,3Galβ1,3GalNAcα-Ser/Thr. IgA1 subclass bears at least five ST antigen in the hinge regions. It has been reported that the IgA1 molecule from the patients lacks a galactose or sialic acid residue at the end of the ST structure. The incomplete synthesis of O-glycans may induce physical change in the IgA molecule, which in turn makes IgA easy to deposit in the renal glomeruli. However, we found that only about 1% of IgA obtained from the patients contained incomplete carbohydrate chains in its molecule. Definitive diagnosis of IgAN is now being done by a renal biopsy, which imposes burden on the patients. Thus, it is required to develop novel methods for the diagnosis. Identification of glycosyltransferases involved in the biosynthesis of IgA O-glycan may lead us to elucidate cause of the disease and development of the diagnosis and therapeutic method.

Present Status and Future Direction

It has been elucidated that pp-GalNAc-T2 is the enzyme which first transfers GalNAc to the Ser/Thr residue in the hinge region of IgA molecules (Iwasaki et al. 2003). Then, the enzyme which transfers Gal to form Galβ1,3GalNAcα-Ser/Thr (core 1 structure) has been found and named C1Gal-T1 (Ju et al. 2002). C1Gal-T1 requires a specific chaperone, Cosmc, for its folding (Ju and Cummings 2002). The enzyme which transfers sialic acid to the core 1 structure is ST3Gal I (see Chapter by S. Tsuji, this volume).

References

Iwasaki H, Zhang Y, Tachibana K, Gotoh M, Kikuchi N, Kwon YD, Togayachi A, Kudo T, Kubota T, Narimatsu H (2003) Intiation of O-glycan synthesis in IgA1 hinge region is determined by a single enzyme, UDP-N-acetyl-α-D-galactosamine:polypeptide N-acetylgalactosaminyltransferase 2. J Biol Chem 278(8):5613–5621

Ju T, Cummings RD (2002) A unique molecular chaperone Cosmc required for activity of the mammalian core 1 β3-galactosyltransferase. Proc Natl Acad Sci USA 99(26):16613–16618

Ju T, Brewer K, D'Souza A, Cummings RD, Canfield WM (2002) Cloning and expression of human core 1 β1,3-galactosyltransferase. J Biol Chem 277(1):178–186

Research Center for Medical Glycosciences, National Institute of Advanced Industrial Science and Technology (AIST), 1-1-1 Umezono, Tsukuba, Ibaraki 305-8568, Japan
E-mail: h.narimatsu@aist.go.jp

Section XVIII
Growth Factor Receptors

Importance of Sugar Chains in the Function of Growth Factor Receptors

Yoshitaka Ikeda

Introduction

Growth of normal cells in our body is strictly controlled on demand, and proceeds, e.g., under conditions of wound healing and recovery from inflammation of tissues. Usually, normal cells do not freely grow without appropriate external signals. Such signals are often mediated by proteins known as growth factors. Although these proteins cannot translocate across the plasma membrane, integral membrane receptor proteins receive the extracellular signals through binding growth factors and transmit the signal of growth into the intracellular space. Many receptor proteins are glycosylated, and the addition of sugar chains appears to confer expression of functional receptors. If the receptors do not correctly trigger transmembrane signaling, the proteins may instead activate the intracellular signaling pathways, in spite of the absence of an extracellular signal, causing abnormal cell growth. Thus, loss of normal receptor function could lead to a malignant alteration of cells.

Epidermal growth factor (EGF) stimulates cell growth by binding to EGF receptor localized on the cell surface. The EGF receptor is a single-transmembrane spanning protein, and is composed of extracellular-, transmembrane- and intracellular-domains. EGF, as a ligand for the receptor, first binds to the extracellular domain, and then induces the conformational change of the receptor protein, which leads to dimerization of the receptor. The resulting close proximity of the two intracellular domains permits mutual phosphorylation of tyrosine residues in the domains by intrinsic kinase activity. This process is known as autophosphorylation of the receptor and is the first step in intracellular growth signaling.

It has been found that N-glycosylation plays an essential role in the function of the EGF receptor (Tsuda et al. 2000; Wang et al. 2001). The loss of the specific sugar chain in the receptor impairs the EGF-directed dimerization of the receptor. As reported, even in the absence of EGF, the mutant receptor, which lacks only a single N-linked sugar chain, spontaneously clusters to form a homo-oligomer, which eventually causes receptor autophosphorylation (Tsuda et al. 2000). These findings suggest that alteration of sugar chains in the growth factor receptor can potentially disrupt regulation of cell growth, although it may or may not transform the cells into malignant cells.

Division of Molecular Cell Biology, Department of Biomolecular Sciences, Saga University, Faculty of Medicine, 5-1-1 Nabeshima, Saga 849-8501, Japan
Phone: +81-952-34-2190, Fax: +81-952-34-2189
E-mail: yikeda@med.saga-u.ac.jp

Concept

It is well known that genetic alterations of genes encoding growth factor receptors yield mutant receptors that activate intracellular signal transduction pathways for cell growth without appropriate extracellular stimuli. Such mutated genes are often referred as oncogenes and have been extensively investigated. As described above, however, normal functions of the receptors are conferred not only by polypeptides but also by added sugar chains. Thus, sugar chains in the glycoproteins are as important as the polypeptide portion, in terms of expression of functional receptors. Further study on the involvement of sugar chains in the receptor function will enhance their importance.

References

Tsuda T, Ikeda Y, Taniguchi N (2000) The Asn-420-linked sugar chain in human epidermal growth factor receptor suppresses ligand-independent spontaneous oligomerization. Possible role of a specific sugar chain in controllable receptor activation. J Biol Chem 275:21988–21994

Wang XQ, Sun P, O'Gorman M et al (2001) Epidermal growth factor receptor glycosylation is required for ganglioside GM3 binding and GM3-mediated suppression [correction of suppression] of activation. Glycobiology 11:515–522

Analysis of *N*-glycan of Growth Factor Receptors

Motoko Takahashi

Introduction

Changes in oligosaccharide structure are associated with many physiological and pathological events. A number of membrane bound receptors for growth factors and cytokines are glycosylated, and oligosaccharide moieties are crucial for the regulation of some of those receptors. Strategies for examining the function of oligosaccharides in glycoproteins in culture cells are somewhat limited. One involves eliminating the oligosaccharide moiety from the target proteins by introducing a mutation in the oligosaccharide binding sites of the protein. The other is, in order to determine the role of each structure, manipulation of the oligosaccharides via the introduction of glycosyltransferases or by using RNAi. The drawback of the latter strategy is that it is difficult to identify the key target protein as the change in glycosyltransferase levels can alter the oligosaccharides in a variety of proteins.

The procedures below are utilized for analysis of growth factor receptors such as epidermal growth factor receptor (EGFR) as well as other membrane proteins such as Fas, LDL receptor-related protein-1 or E-cadherin.

Determination of Cell-Surface Expression Level of the Target Protein

One of the most important function of glycosylation is regulation of the sorting of glycoproteins. It is possible that localization of the target protein is altered as changes in glycosylation occur. Cell-surface expression levels of target protein can be determined by a combination of cell-surface biotinylation and immunoprecipitation.

1. The cells are washed twice with ice-cold PBS supplemented with 0.1 mM $CaCl_2$ and 1 mM $MgCl_2$ [PBS (+)], and reacted with 0.2 mg/ml of sulfo-NHS-biotin (Pierce) diluted in the same solution for 30 min on ice.

2. The reaction is terminated by washing the cells in Tris–saline once and PBS (+) twice.

3. The cells are solubilized with lysis buffer [20 mM Tris–HCl, pH 7.4, 150 mM NaCl, 5 mM EDTA, 1% (w/v) Nonidet P-40, 10% (w/v) glycerol, 5 mM sodium pyrophosphate, 10 mM NaF, 1 mM sodium orthovanadate, 10 mM β-glycerophosphate, 1 mM phenylmethylsulfonyl fluoride, 2 μg/ml aprotinin, 5 μg/ml leupeptin, and 1 mM dithiothreitol], incubated for 30 min at 4°C, and centrifuged at 15,000 rpm for 10 min at 4°C to obtain the supernatant.

Department of Biochemistry, Sapporo Medical University School of Medicine, South-1 West-17, Chuo-ku, Sapporo 060-8556, Japan
Phone: +81-11-611-2111, Fax: +81-11-611-8556
E-mail: takam@sapmed.ac.jp

4. To immunoprecipitate the target protein, 15 µl of Protein G Sepharose (GE Healthcare Biosciences) and 4 µg of an antibody are added to 1 ml of the cell lysate and incubated with agitation for 2 h~ overnight at 4°C.

5. The samples are centrifuged at 10,000 rpm for 30 s at 4°C and the supernatant is discarded.

6. The gel is washed with 1 ml of lysis buffer four times by repeating the centrifugation step as above.

7. The sample buffer of SDS-PAGE is added to the sample, boiled for 5 min and centrifuged to obtain the supernatant.

8. The samples are subjected to SDS-PAGE and transferred to PVDF membrane.

9. The membrane is blocked with 3% BSA (w/v) and the biotinylated target protein is visualized by ABC kit (Vectastain) and ECL kit (GE healthcare Biosciences).

Examination of N-glycan Modification of the Target Protein by Glycopeptidase F Digestion

The issue of whether the protein is modified by *N*-glycan can be determined by glycopeptidase F digestion and by examining the changes in molecular weight. Glycopeptidase F can cleave all types of *N*-glycans at GlcNAc-Asn bonds in glycopeptides or glycoproteins, unless the Asn is located at N-termini or C-termini, or unless the *N*-glycan contains α1,3-linked core fucose.

1. The cells are solubilized with lysis buffer [20 mM Tris–HCl, pH 7.4, 150 mM NaCl, 5 mM EDTA, 1% (w/v) Nonidet P-40, 10% (w/v) glycerol, 5 mM sodium pyrophosphate, 10 mM NaF, 1 mM sodium orthovanadate, 10 mM β-glycerophosphate, 1 mM phenylmethylsulfonyl fluoride, 2 µg/ml aprotinin, 5 µg/ml leupeptin, and 1 mM dithiothreitol], incubated for 30 min at 4°C and centrifuged at 15,000 rpm for 10 min at 4°C to obtain the supernatant.

2. To immunoprecipitate the target protein, 15 µl of Protein G Sepharose (GE Healthcare Biosciences) and 4 µg of an antibody are added to 1 ml of the cell lysate and incubated with agitation for 2 h~ overnight at 4°C.

3. The samples are centrifuged at 10,000 rpm for 30 s at 4°C and the supernatant is discarded.

4. The gel is washed with lysis buffer four times by repeating the centrifugation step as above.

5. The immunoprecipitated samples are boiled in 0.1 M 2-mercaptoethanol and 0.1% SDS for 10 min and centrifuged to obtain the supernatant.

6. The samples are incubated with glycopeptidase F (final 40 mU/ml) in the buffer containing 60 mM Tris–HCl (pH 8.6) and 0.8% NP-40 for 16 h at 37°C.

7. The samples are subjected to SDS-PAGE and Western blot using an antibody for the target protein. The status of *N*-glycan modification is estimated from the changes in molecular weight.

The immunoprecipitation step (Steps 2–4) can be omitted. When using glycopeptidase F from Takara Bio Inc., the buffers are supplemented and the manufacturer's instruction can be followed for the digestion step.

Lectin Blot Analysis of the Target Protein

The precise analysis of N-glycan is done by NMR or MS/MS, but other procedures are also available for simple analysis. Lectin blot analysis is one of the choices. The biotinylated lectins such as E_4-PHA, L_4-PHA, ConA, LCA and AAL are available from Seikagaku Corp. and the information on the specificity of each lectin can be obtained from the catalog or textbooks.

1. The cells are solubilized with lysis buffer [20 mM Tris–HCl, pH 7.4, 150 mM NaCl, 5 mM EDTA, 1% (w/v) Nonidet P-40, 10% (w/v) glycerol, 5 mM sodium pyrophosphate, 10 mM NaF, 1 mM sodium orthovanadate, 10 mM β-glycerophosphate, 1 mM phenylmethylsulfonyl fluoride, 2 μg/ml aprotinin, 5 μg/ml leupeptin, and 1 mM dithiothreitol], incubated for 30 min at 4°C, and centrifuged at 15,000 rpm for 10 min at 4°C to obtain the supernatant.
2. The samples are subjected to SDS-PAGE and transferred to PVDF membrane.
3. The membrane is blocked with 3% BSA (w/v) and then incubated with 1 μg/ml of biotinylated lectin for 1 h at room temperature.
4. The lectin reactive proteins are visualized by ABC kit (Vectastain) and ECL kit (GE healthcare Biosciences).

Applications

It has been reported that initial N-glycosylation is required for the proper sorting of EGFR to the membrane as well as for ligand binding. By introducing mutations in the putative oligosaccharide binding sites, we have found that N-glycans in domain III of ErbB receptors play important roles in dimerization (Tsuda et al. 2000; Takahashi et al. 2004; Yokoe et al. 2007). We also introduced GnT-III into the cells and examined the effect on EGF signaling. Cell-surface expression levels of EGFR were not changed, and the oligosaccharides of EGFR were successfully modified (Figs. 1, 2). The rate of EGF-induced EGFR endocytosis was revealed to be increased in the GnT-III transfectants, leading to the up-regulation of downstream ERK phosphorylation (Sato et al. 2001). It could be assumed that the enhancement of internalization of EGFR is caused by the reduction of interaction of N-glycan of EGFR with lectin-like molecules on the cell surface. A recent study showed that ganglioside GM3 inhibits EGFR tyrosine kinase through binding

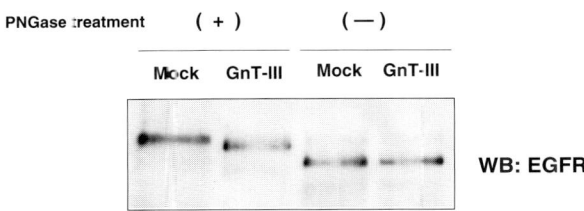

Fig. 1 Glycopeptidase F treatment of EGFR. Whole cell lysates from mock and GnT-III transfectants were treated with glycopeptidase F, and subjected to Western blot using an anti-EGFR antibody. The difference in molecular weight between mock and GnT-III transfectants without glycopeptidase F treatment seems to attribute to the difference in molecular weight of oligosaccharides, since the difference was diminished when the samples were treated with glycopeptidase F

Fig. 2 Lectin blot analysis of EGFR. EGFR was immunoprecipitated from 500 μg of whole cell lysate of mock and GnT-III transfectants and subjected to 8% SDS-PAGE and transferred to nitrocellulose membranes, which were probed with E_4-PHA

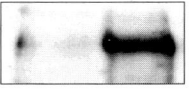

IP: α EGFR
Lectin: E_4-PHA

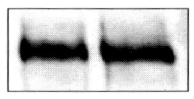

IP: α EGFR
WB: α EGFR

Mock GnT-III

to N-glycan with multivalent GlcNAc termini, also indicating the importance of the interaction of N-glycan of receptors and cell-surface molecules for signal transduction (Yoon et al. 2006).

References

Sato Y, Takahashi M, Shibukawa Y, Jain SK, Hamaoka R, Miyagawa J, Yaginuma Y, Honke K, Ishikawa M, Taniguchi N (2001) Overexpression of N-acetylglucosaminyltransferase III enhances the epidermal growth factor-induced phosphorylation of ERK in HeLaS3 cells by up-regulation of the internalization rate of the receptors. J Biol Chem 276:11956–11962

Takahashi M, Tsuda T, Ikeda Y, Honke K, Taniguchi N (2004) Role of N-glycans in growth factor signaling. Glycoconj J 20:207–212

Tsuda T, Ikeda Y, Taniguchi N (2000) The Asn-420-linked sugar chain in human epidermal growth factor receptor suppresses ligand-independent spontaneous oligomerization. Possible role of a specific sugar chain in controlable receptor activation. J Biol Chem 275:21988–21994

Yokoe S, Takahashi M, Asahi M, Lee SH, Li W, Osumi D, Miyoshi E, Taniguchi N (2007) The Asn418-linked N-glycan of ErbB3 plays a crucial role in preventing spontaneous heterodimerization and tumor promotion. Cancer Res 67:1935–1942

Yoon SJ, Nakayama K, Hikita T, Handa K, Hakomori SI (2006) Epidermal growth factor receptor tyrosine kinase is modulated by GM3 interaction with N-linked GlcNAc termini of the receptor. Proc Natl Acad Sci USA 103:18987–18991

The Midkine Family and Its Receptors

Kenji Kadomatsu

Midkine (MK) and pleiotrophin (PTN) comprise a family, the MK family, which is distinct from other growth factor families. The MK family is evolutionarily conserved, and it is studied in a variety of animals. The C-terminal domain of MK contains cluster I, a cluster of basic amino acids. Cluster I is important for binding to heparin and chondroitin sulfate.

The MK family members strongly bind to Syndecan (a heparan sulfate proteoglycan) and PTPζ (a chondroitin sulfate proteoglycan and a protein tyrosine phosphatase). Chondroitinase treatment of PTPζ, a mutation in cluster I of MK, or chondroitin sulfate E significantly suppresses cell migration mediated via MK and PTPζ.

It is also known that the MK family members bind to anaplastic lymphoma kinase (ALK), LDL receptor-related protein (LRP), integrin $\alpha_4\beta$ and integrin $\alpha_6\beta_1$. Thus, it is thought that a receptor complex involving integrin, LRP and PTPζ mediates the signals of the MK family.

The MK family plays important roles in (1) the nervous system (injury, etc.), (2) cancer, and inflammation (nephritis, vascular diseases, rheumatoid arthritis, etc.). Therefore, studies on the MK family will contribute to understanding of the pathogenesis of and to development of therapeutic strategies for these diseases. As mentioned above, E-type chondroitin sulfate suppresses MK-mediated cell migration, suggesting that sugars are candidate therapeutics if MK is a molecular target for the therapy for a disease. In addition, a soluble fragment of a receptor could specifically block the functions of the MK family.

Studies on the MK family may also contribute to elucidation of the action mechanisms of functional sugar chains. For example, cluster I of MK is localized at a particular site, i.e., not exposed to the opposite site. Therefore, it is interesting what the function of the opposite site is. In addition, the MK family members bind to two glycosaminoglycans, i.e., heparan sulfate and chondroitin sulfate. It remains unknown whether this differential binding has different biological significance. Studies on the MK family may shed light on the biological roles of glycosaminoglycans.

Department of Biochemistry, Nagoya University Graduate School of Medicine, Nagoya, Japan

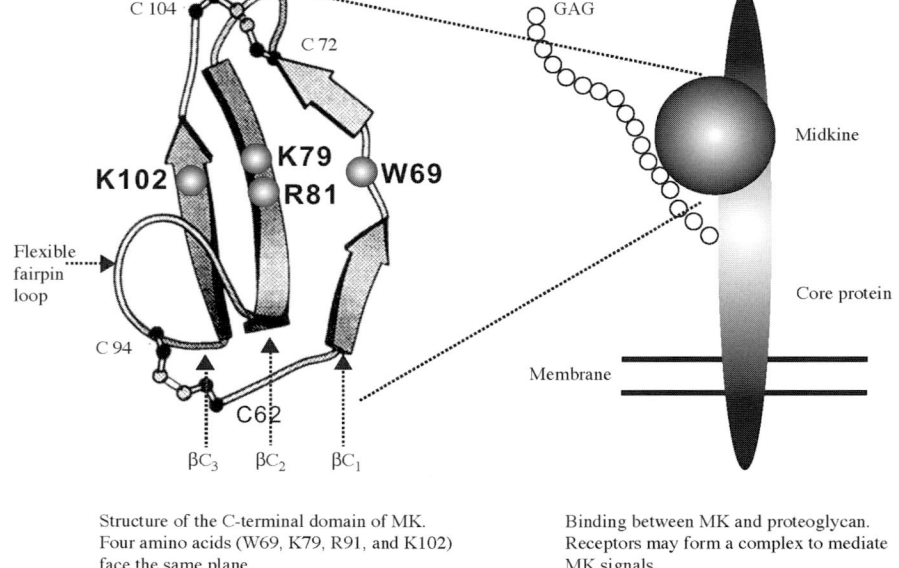

Structure of the C-terminal domain of MK. Four amino acids (W69, K79, R91, and K102) face the same plane.

Binding between MK and proteoglycan. Receptors may form a complex to mediate MK signals.

Fig. 1

Midkine Detection

Extraction of Midkine from Tissues and Detection by Western Blotting

1. Homogenize a tissue sample (0.1 g) with 1 ml of lysis buffer (1% Triton X100, 10 mM Tris–HCl pH 7.4, 150 mM sodium chloride with protease inhibitors) on ice.
2. Centrifuge the lysate at 10,000×g for 30 min at 4°C.
3. Add 40 µl of 50% slurry heparin Sepharose to the supernatant and rotate the mixture for 1 h at 4°C. [Before adding heparin Sepharose, take an aliquot (e.g., 100 µl) for an internal control, such as β actin expression].
4. Wash the heparin Sepharose three times with the lysis buffer by centrifugationg at 1,000 rpm for 10 seconds.
5. Further, wash the heparin Sepharose beads three times with the lysis buffer containing 500 mM sodium chloride.
6. Add 100 µl of sample buffer for SDS PAGE and apply 20 µl to SDS PAGE.
7. Perform Western blotting as usual.

Comment

The molecular weight of midkine is 13 kDa, but its band on Western blot appears around 17 kDa because midkine is rich in basic amino acids. Since non-specific bands often appear around 17 kDa, it is strongly recommended that heparin Sepharose extraction should be done. The affinity between midkine and heparin is very strong; 500 mM sodium chloride cannot release the binding between the two molecules but can wash out other molecules bound to heparin with low affinities.

Reference

Kadomatsu K, Muramatsu T (2004) Midkine and pleiotrophin in neural development and cancer. Cancer Lett 204:127–143

N-glycans Regulate Integrin α5β1 Functions

Jianguo Gu, Yuya Sato, Tomoya Isaji

Introduction

Integrin, which is formed by α and β subunits, is cell surface transmembrane glycoprotein that functions as an adhesion receptor. Integrin engagement during cell adhesion leads to intracellular phosphorylation, such as the phosphorylation of the focal adhesion kinase, thereby regulating gene expression, cell growth, differentiation and survival. Integrin is a major carrier of *N*-glycans. A growing body of evidence indicates that the *N*-glycosylation of integrin α5β1 plays crucial roles in hetero-dimer formation and in its biological functions (Gu and Taniguchi 2004). In fact, α5β1 integrin contains 26 potential *N*-linked glycosylation sites, 14 in the α5 subunit and 12 in the β1 subunit. To determine which of the *N*-glycosylation sites on the α5 subunit are essential for these functions, we sequentially mutated one or combined asparagines (*N*) residues in the putative *N*-glycosylation sites to glutamine (Q) residues as shown in Fig. 1, then transfected these mutant genes into α5-deficient Chinese hamster ovary (CHO) cells (CHO-B2) for cell spreading assay (Isaji et al. 2006).

Cell Spreading Assay

1. Coat 24-wells cell culture plate with 10 µg/ml human serum fibronectin in PBS for overnight at 4°C.
2. Block the plate with 1% bovine serum albumin (BSA) in PBS for 1 h at 37°C.
3. Detach the cells with trypsin containing 1 mM EDTA, stop trypsin digestion with serum containing DMEM and resuspend the cells in 600 µl serum-free DMEM with 0.1% BSA at 4×10^4 cells/ml.
4. Incubate the cell on fibronectin-coated cell culture plate for 20 min at 37°C
5. Remove nonadherent cells by washing with PBS and fix the cells with 3.7% paraformaldehyde in PBS.
6. Observe adhered cells by phase contrast microscopy.

Results

The activities for cell spreading and migration for the α5 subunit carrying only three potential *N*-glycosylation sites (3–5 sites) on the β-propeller were comparable to those of the wild type. In contrast, mutation of these three sites resulted in a significant decrease in cell spreading as well as in functional expression, although the total expression level of 3–5 deleted mutants on the cell surface was comparable to that of the wild type. The

---

Division of Regulatory Glycobiology, Institute of Molecular Biomembrane and Glycobiology, Tohoku Pharmaceutical University, 4-4-1 Komatsusima, Aoba-ku, Sendai, Miyagi 981-8558, Japan
Phone: +81-22-727-0216, Fax: +81-22-727-0078
E-mail: jgu@tohoku-pharm.ac.jp

Fig. 1 Schematic illustration of potential N-glycosylation sites on the integrin α5 subunit

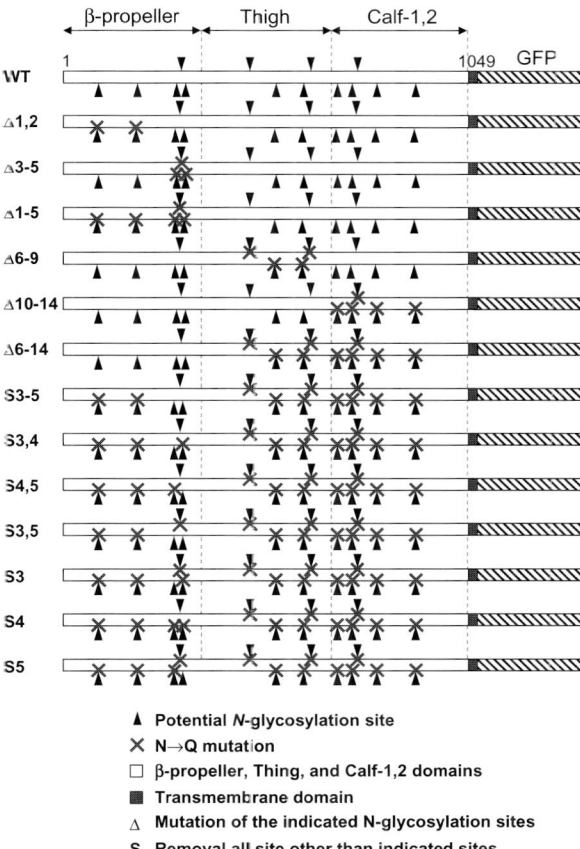

site5 is the most important site for its expression on the cell surface, but the S5 mutant did not show any biological functions. Taken together, these data reveal that the N-glycosylation on the β-propeller domain of the α5 subunit is essential for heterodimerization and biological functions of α5β1 integrin, and might also be useful for studies of the molecular structure (Isaji et al. 2006). Such regulation of integrin α5 subunit by N-glycans may also occur in other α subunits such as α3. Since both α3 and α5 subunits could be modified by glycosyltransferases, GnT-III, GnT-V or Fut8, to regulate their biological functions in a similar manner (Zhao et al. 2006a, b).

References

Gu J, Taniguchi N (2004) Regulation of integrin functions by N-glycans. Glycoconj J 21:9–15

Isaji T, Sato Y, Zhao Y, Miyoshi E, Wada Y, Taniguchi N, Gu J (2006) N-glycosylation of the beta-propeller domain of the integrin alpha5 subunit is essential for alpha5beta1 heterodimerization, expression on the cell surface, and its biological function. J Biol Chem 281:33258–33267

Zhao Y, Itoh S, Wang X, Isaji T, Miyoshi E, Kariya Y, Miyazaki K, Kawasaki N, Taniguchi N, Gu J (2006a) Deletion of core fucosylation on alpha3beta1 integrin down-regulates its functions. J Biol Chem 281:38343–38350

Zhao Y, Nakagawa T, Itoh S, Inamori K, Isaji T, Kariya Y, Kondo A, Miyoshi E, Miyazaki K, Kawasaki N et al (2006b) N-acetylglucosaminyltransferase III antagonizes the effect of N-acetylglucosaminyltransferase V on alpha3beta1 integrin-mediated cell migration. J Biol Chem 281:32122–32130

Section XIX
Blood Groups and Blood Cells

Blood Group Antigens: Blood Group Carbohydrate Antigens

Koichi Furukawa, Koichi Iwamura, Keiko Furukawa

Introduction

The most popular blood group antigen system, ABO antigens had been known to consist of alpha1,2-fucosylated galactose substituted with alpha1,3-*N*-acetylgalactosamine (A) or galactose (B). These biochemical results were clearly confirmed and further investigated by the molecular cloning of blood group A synthase cDNA (Yamamoto et al. 1990a, 1990b). It was demonstrated that blood group A synthase gene is located at the human 9th chromosome ABO locus with blood group B synthase as an allelic gene having 7 nucleotides mutation in the coding region (Fig. 1). Blood group O gene is also present at the same locus with no enzymatic activity due to one nucleotide deletion. As for P1/P/pk/p blood group system, P/pk/p blood group system has been clarified to be based on globo-series glycosphingolipids, i.e., P is globoside (Gb4, globotetraosylceramide) (Okajima et al. 2000), pk is Gb3 (globotriaosylceramide) that is accumulated in the deficiency of Gb4 synthase, and p means individuals lacking the activity of Gb3 synthesis (Furukawa et al. 2000; Table 1). P1 is, in turn, alpha4Gal- structure substituted on a neolacto-series core structure. There has been a long-time controversy on the identity between P1 synthase and Gb3 synthase (Kojima et al. 2000; Table 2). Iwamura et al. (2003) demonstrated P1 is synthesized by Gb3 synthase, suggesting that there should be differences in the transcription efficiency of Gb3 synthase gene between P1 and P2 individuals.

Results

Investigation of genetic mechanisms for the ABO blood group system revealed that phenotypic differences in the major blood group are regulated by subtle mutations at the ABO gene locus. Differences between A and B individuals are only 4 amino acids in the coding region (1,062 bp) of the blood group A (and B) synthases, and only two of them are essential for the donor sugar specificity. Blood group O individuals have only one nucleotide deletion leading to inactivation of the enzyme. On the basis of this fundamental information, a number of additional mutations in A synthase gene have been reported. As for P1/P/pk/p blood group system, mechanisms for regulating blood group P/pk/p have been clearly explained. Mutations in Gb3 synthase gene coding region in Swedish and Japanese p individuals were reported, showing heterogeneous point mutations. These results indicated that point mutations in Gb3 synthase occurred much later than those in the ABO synthase gene. Mutations in Gb4 synthase gene in pk individuals were also reported. For the synthesis of P1 blood group antigen, it was clearly demonstrated that

---

Department of Biochemistry II, Nagoya University Graduate School of Medicine, 65 Tsurumai, Showa-ku, Nagoya 466-0065, Japan
Phone: +81-52-744-2070, Fax: +81-52-744-2069
E-mail: koichi@med.nagoya-u.ac.jp

ABO Blood group antigens

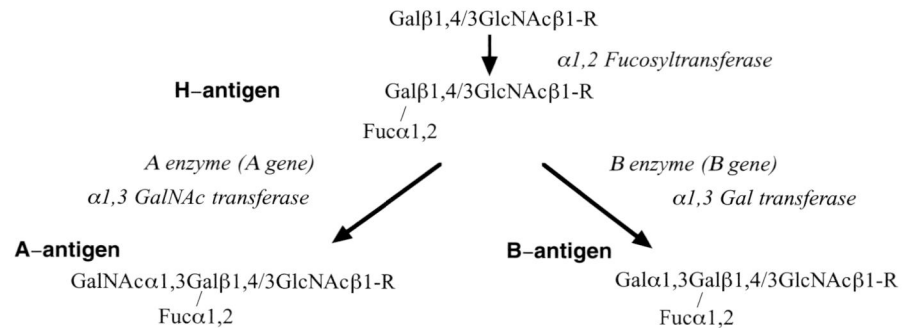

Fig. 1 ABO blood group antigen: phenotypes and genotypes
Alpha1,2fucosylated galactose is a common precursor named H antigen. Blood group A gene codes alpha1,3-N-acetylgalactosaminyltransferase and B gene codes alpha1,3-galactosyltransferase are allelic genes to each other

Table 1 Structures of P1/P/pk/p antigens

Antigens	Structures
P antigen (Gb4)	GalNAcβ1,3Galα1,4Galβ1,4Glc-Cer
Pk antigen (Gb3)	Galα1,4Galβ1,4Glc-Cer
p individuals (no Gb3)	Galβ1,4Glc-Cer
P1 antigens	Galα1,4Galβ1,4GlcNAcβ1,3Galβ1,4Glc-Cer
Paragloboside	Galβ1,4GlcNAcβ1,3Galβ1,4Glc-Cer

Table 2 P1/P2 system and P and pk system

P1/P2	P/pk
Population	
P1(P1+): 80%	P and pk: almost
P2 (P1−): 20%	p < 1/100,000
Chromosome assignment	
P1 synthase: 22q12-ter	Gb3 synthase: 22q13.2
Correlation	
p individuals: all P2	

Gb3 synthase was able to induce P1 antigen expression. More over, differences between P1 and P2 were due to the different degree of expression levels of P1 antigen or cripticity of the antigen. Eventually, majority of P2 individuals could synthesize the Gb3 antigen with lower efficiency than P1 individuals. Real P2 completely lacking P1 were of p type, corresponding to the story that P1 antigen is synthesized by Gb3 synthase. Mechanisms for the differences in the efficiency of Gb3 synthase (P1 antigen synthesis) between P1 and P2 are not clear at this moment, although Iwamura et al. (2003) suggested that single nucleotide polymorphism (SNP) in the regulatory region of the Gb3 synthase gene could be the causal factor.

Protocols

1. Gb3 synthase enzyme assay was as described by Kojima et al.
2. The enzyme activity of α1,4Gal-T to generate P1 structure was measured using membrane fractions prepared by nitrogen capitation apparatus.
3. The reaction mixture for the assay contained the following in a volume of 50 µl: 0.2 mM UDP-Gal (Sigma), UDP-[^{14}C]Gal(2.5×10^5 dpm) (PerkinElmer Life Science), 2.5 µg of nLc4, 20 µM CDP-choline, and 100 µg of phosphatidylglycerol (Sigma), and after evaporation, 20 mM galactonolactone (Sigma), 0.3% Triton X-100, 250 µg of α-lactalbumin, and membrane fraction containing 100 µg of protein.

The products were isolated by a C18 SepPak cartridge (Waters) and analyzed by TLC and autofluorography.

4. Reaction products were detected by TLC-immunostaining using anti-P1 monoclonal antibody.

Comments

On the polymorphism in the 5'-upstream region of Gb3 synthase, there are, at least, two papers from foreign countries following our report.

References

Furukawa K, Iwamura K, Uchikawa M, Sojka BN, Wiels J, Okajima T, Urano T, Furukawa K (2000) Molecular basis for the p phenotype: identification of distinct and multiple mutations in the a1,4-galactosyltransferase gene in Swedish and Japanese individuals. J Biol Chem 275:37752–37756

Iwamura K, Furukawa K, Uchikawa M, Birgitta NS, Kojima Y, Wiels J, Urano T, Furukawa K (2003) The blood group P1 synthase gene is identical to the Gb3/CD77 synthase gene: a clue to the solution of the P1/P2/p puzzle. J Biol Chem 278:44429–44438

Kojima Y, Fukumoto S, Furukawa K, Tetsuya O, Wiels J, Yokoyama K, Suzuki Y, Ohta M, Furukawa K (2000) Molecular cloning of Gb3/CD77 synthase, a glycosyltransferase that initiate the synthesis of globo-series glycosphingolipids. J Biol Chem 275:15152–15156

Okajima T, Nakamura Y, Uchikawa M, Haslam DB, Numata S, Furukawa K, Urano T, Furukawa K (2000) Expression cloning of human globoside synthase cDNAs: identification of β3 Gal-T3 as UDP-N-acetylgalactosamine: globotriaosylceramide β1,3-N-acetylgalactosaminyl-transferase. J Biol Chem 275:40498–40503

Yamamoto F, Marken J, Tsuji T, White T, Clausen H, Hakomori S (1990a) Cloning and characterization of DNA complementary to human UDP-GalNAc: Fuc alpha 1-2Gal alpha 1-3GalNAc transferase (histo-blood group A transferase) mRNA. J Biol Chem 265:1146–1151

Yamamoto F, Clausen H, White T, Marken J, Hakomori S (1990b) Molecular genetic basis of the histo-blood group ABO system. Nature 345:229–233

Bombay, Lewis and Ii Blood Group Antigens on Human Erythrocytes

Hisashi Narimatsu

Background and Significance of the Study

Relationship Between Lewis and ABO Blood Group Systems

Representatives of blood groups determined by carbohydrate antigens are Lewis, ABO, P, and Ii blood group systems. The synthetic pathway of carbohydrate antigens of ABO and Lewis blood groups is shown in Fig. 1. The type of a carbohydrate blood group antigen is determined by the genetic difference of the carbohydrate synthesized by glycosyltransferases in individuals, and the difference occurs by loss of activity or change in the substrate specificity of glycosyltransferases, mainly due to the point mutation in their gene.

Blood group antigens occur not only on erythrocytes but also in various tissues in the body. Lewis blood group antigens are highly expressed in almost all of epithelial tissues in the digestive organs. As shown in Fig. 1, there are two types of α1,2-fucosyltransferase, which synthesize the H antigen (O-type antigen) in human; these are FUT1 (H) and FUT2 (Se: secretor enzyme). FUT1, but not FUT2, is expressed in erythropoietic cells including erythroblasts. On the other hand, FUT2 is strongly expressed, whereas FUT1 is slightly expressed in the digestive organs, i.e., colorectal and stomach tissues. FUT1 is the enzyme synthesizing the H antigen on erythrocytes, and FUT2 is the major enzyme, which synthesizes it in the digestive organs. *FUT1* and *FUT2* genes exist in the chromosome 19q13.3. As *FUT1*, *FUT2*, and *Sec1* genes lie very proximal to each other, their haplotypes are almost always inherited in a linking manner. *Sec1* gene has already become an inactive pseudogene in human.

Mutant alleles of *FUT1* gene have been rarely and sporadically found. The mutant alleles named *h1*, *h2*, *h3*, *h4*, and *h5* were found in Japanese (Kaneko et al. 1997). Amount of the H antigen on erythrocytes is determined corresponding to the activity of these mutant enzymes in individuals. Mutant alleles of *FUT2* gene have been considerably frequently found (Narimatsu et al. 1998). The major inactive alleles found in Japanese are named *sej* and *se5*. The enzyme encoded by *sej* retained about 3% of activity compared to wild one, but that encoded by *se5* completely lost the activity (Kudo et al. 1996). Mutant alleles of *FUT2(Se)* gene are common in Asian, but the different mutant alleles are spreading in Caucasian, indicating that the inactivation of *FUT2(Se)* gene occurred after the divergence of both ethnic groups. The Bombay blood type which was first found in the individuals in Bombay, India who did not have ABO system antigens not only on erythrocytes but also in other tissues. As both *FUT1(H)* gene and *FUT2(Se)* gene are completely inactivated in these individuals, the H antigen is not expressed in any tissues in the body (Kaneko et al. 1997).

Research Center for Medical Glycosciences, National Institute of Advanced Industrial Science and Technology (AIST), 1-1-1 Umezono, Tsukuba, Ibaraki 305-8568, Japan
E-mail: h.narimatsu@aist.go.jp

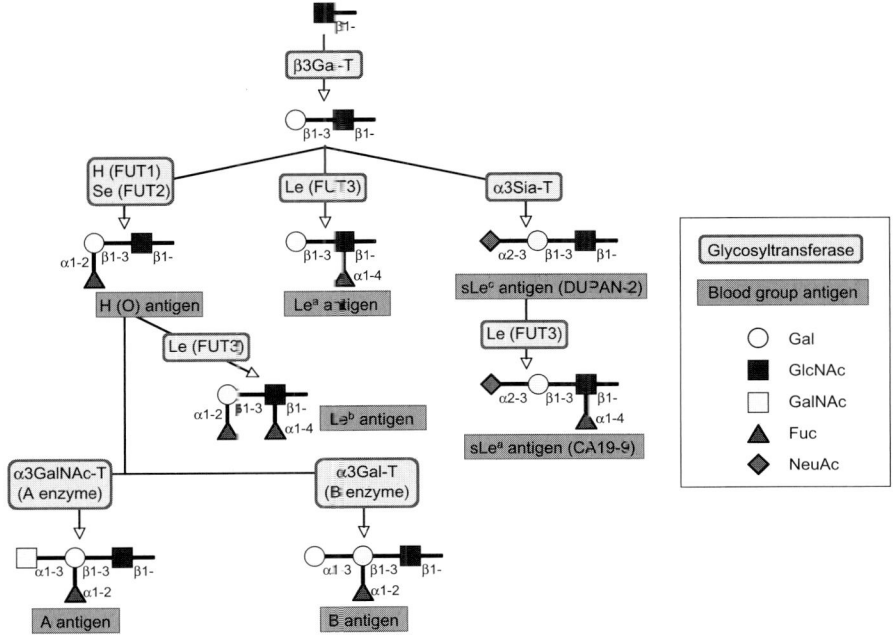

Fig. 1

The Lewis b antigen is synthesized by the action of FUT3(Le), α1,3/4-fucosyltransferase, on the H antigen. *FUT2* and *FUT3* genes are expressed in the epithelium of the digestive tract but are not expressed on erythrocytes (Nishihara et al. 1999). Lewis glycolipid antigens are first synthesized in the digestive epithelium, secreted into the blood, absorbed on the surface of erythrocytes, and then function as Lewis blood group antigens. The phenotype of individuals with both active FUT2 and FUT3 is Le (a–b+), that of individuals with active FUT3 and defective FUT2 is Le (a+b–), and that of individuals with defective FUT3, irrespective of activity of FUT2, is Le (a–b–) (Nishihara et al. 1994). The Le (a+b+) individuals possess the active FUT3 and the weakened activity of FUT2 by point mutations, i.e., *sej* allele. We previously reported mutant genes of FUT2 and FUT3 in Japanese (Narimatsu et al. 1998). We also found that measured values of an important tumor marker, CA19-9, are greatly influenced by the genotype of FUT2 and FUT3. Thus, care should be taken for clinical diagnosis (Narimatsu et al. 1998).

Ii System of Blood Group

Ii system blood group antigens are distributed in almost all the organs in the body. As demonstrated in our previous studies, three types of Ii blood group genes (IGnT glycosyltransferase genes) exist on the chromosome 6p24. These genes are named IGnT1, IGnT2, and IGnT3, respectively (Inaba et al. 2003). As the first exon of each gene is different, tissue distribution of gene expression differs each other. However, the second and third exons are common to the three genes. The three enzymes exhibit almost the same activity to synthesize I antigen. IGnT1 is expressed in the neuronal tissues, IGnT2 is mainly expressed in leukocytes, the digestive tract and the mammary gland, and IGnT3 is expressed on erythroblasts. It has been demonstrated that the enzyme expressed in each tissue synthesize I antigen, respectively. We found a Japanese pedigree in which the three enzymes lack their activity due to the mutation of the second or third exon which are shared by three enzymes. It has been reported that I antigen disappeared in the whole body of these individuals, and they developed congenital cataract. Cataract may be induced by the defect of IGnT2. On the other hand, Yu LC et al. reported the pedigree in which the activity of IGnT3 was lost due to the mutation of the first exon of IGnT3 (Yu et al. 2003). In these individuals, I antigen on erythrocytes disappeared, but that is expressed in other organs. These findings have resolved the mystery on the Ii blood group antigens.

References

Inaba N, Hiruma T et al (2003) A novel I-branching β-1,6-N-acetylglucosaminyl transferase involved in human blood group I antigen expression. Blood 101:2870–2876

Kaneko M, Nishihara S et al (1997) Wide variety of point mutations in the H gene of Bombay and para-Bombay individuals that inactivate H enzyme. Blood 90:839–849

Kudo T, Iwasaki H et al (1996) Molecular genetic analysis of the human Lewis histo-blood group system. II. Secretor gene inactivation by a novel single missense mutation A385T in Japanese nonsecretor individuals. J Biol Chem 271:9830–9837

Narimatsu H, Iwasaki H et al (1998) Lewis and secretor gene dosages affect CA19-9 and DU-PAN-2 serum levels in normal individuals and colorectal cancer patients. Cancer Res 58:512–518

Nishihara S, Narimatsu H et al (1999) Molecular behavior of mutant Lewis enzymes in vivo. Glycobiology 9:373–382

Nishihara S, Narimatsu H et al (1994) Molecular genetic analysis of the human Lewis histo-blood group system. J Biol Chem 269:29271–29278

Yu LC, Twu YC et al (2003) The molecular genetics of the human I locus and molecular background explain the partial association of the adult i phenotype with congenital cataracts. Blood 101:2081–8

Blood Cells in Glycobiology

Mitsuru Nakamura

Introduction

In this section, blood cells are discussed from the viewpoint of so-called white blood cell differentiation antigens and glycobiology. Blood types, hemolytic diseases, hematopoietic stem cells and so on are summarized elsewhere.

Meaning of Blood Cell Glycobiology

The pathophysiological value of blood is very high. It is relatively easy to obtain blood samples and to prepare them for experiments. Therefore, in medical research, blood is best used. Cluster designation (CD) plays an important role in the classification of leukocyte cell surface antigens and their antibodies. CD numbers are assigned in international workshops (HLDA: human leukocyte differentiation antigens) for human leukocyte surface antigens. The workshops have been held at intervals of 3–4 years since 1982 (Mason et al. 2001; Mason et al. 2002; http://www.mh-hannover.de/aktuelles/projekte/hlda7/hldabase/select.htm; http://www.bc-cytometry.com/Data/cytoinfo/CD%20Chart%20kaisetsu.htm). In the eighth workshop (HLDA 8 December 2004, Adelaide), 99 antigens were added and CD antigens were expanded to CD1a–CD339 (Zola et al. 2005; http://www.hlda8.org/HLDAOverview.htm; http://www.bc-cytometry.com/Data/cytoinfo/CD%20Chart%20HLDA8.htm). The identification of cell lineages of leukocytes, their differentiation stages, their functional subpopulations and new surface antigens has been achieved using CD-related monoclonal antibodies. The systematic classification of leukocyte surface differentiation antigens and their antibodies has been attempted on the basis of a standardized understanding of HLDA. At the beginning, the classification was limited to the surface antigens of leukocytes and their progenitors. Now, it has expanded also to the surface differentiation antigens of platelets, erythrocytes and endothelial cells. Furthermore, it is applied also to stromal cells or intracellular antigens. Therefore, a change of name from HLDA to HCDM (human cell differentiation molecules) is currently being considered. Molecules that have high variability and polymorphisms are excluded from the CD classification; T cell receptors, immunoglobulins, MHC complexes, etc are excluded.

Among CD antigens, there are several sugar chain antigens and cell adhesion molecules that designate sugar chains as ligands. Most cell surface antigens are glycoproteins. The role of oligosaccharide antigens as cell differentiation markers needs to be revealed. In addition, functional elucidation of the cell-to-cell interaction between oligosaccharide antigens and related molecules is important. Glycobiology is indispensable in the research

---

Cell Regulation Analysis Team, Research Center for Medical Glycoscience, National Institute of Advanced Industrial Science and Technology, Function of Biomolecule, Medical Science for Control of Pathological Processes, Tsukuba University, Central-2, 1-1-1 Umezono, Tsukuba, Ibaraki 305-8568, Japan
Phone: +81-29-861-2745, Fax: +81-29-861-2744
E-mail: owl.nakamura@aist.go.jp

of differentiation antigens. A protocol for an in vivo cell migration assay of human leukemia cells is presented in this section.

CD Antigens in Blood Cell Glycobiology

CD antigens related to sugar chains and glycoproteins are summarized in Table 1. There are five main CD antigen families with which functions are examined as cell adhesion

Table 1 CD antigens (sugar chain-related molecules)

CD15	LeX		
CD15s	Sialyl-LeX		
CD15u	Sulfated-LeX		
CDw17	LacCer		
CD22	Siglec-2		
CD23	C-type lectin	FcεRII	Receptor for Ig-Fcε
CD33	Siglec-3		
CD34	Sialomucin	CD62L-ligand	Hematopoietic stem cell marker
CD43	Sialomucin	Leukosialin	
CD44/CD44S	Hyaladherin		
CD44R	Hyaladherin		
CD56	N-CAM		
CD57	HNK1		
CD60a	GD3		
CD60b	9-O-Ac-GD3		
CD60c	7-O-Ac-GD3		
CD62E	E-Selectin		
CD62L	L-Selectin		
CD62P	P-Selectin		
CD65	Poly-N-acetyllactosamine		
CD65s	sialylated poly-N-acetyllactosamine		
CD68	Sialomucin	Macrosialin	
CD69	C-type lectin		
CD72	C-type lectin		
CD75	**Lactosamine**		
CD75s	Sialylated lactosamine		
CD77	Gb3		
CD94	C-type lectin		
CD138	Syndecan-1		
CD141	C-type lectin	Thrombomudulin	
CD147	Basigin		
CD159a	C-type lectin		
CD161	C-type lectin		
CD162	Sialomucin	P-selectin glycoprotein ligand-1	CD62P-ligand
CD162R	Sialomucin	CD162-KS-GalNAc epitope	KS = keratan sulfate
CD164	Sialomucin		
CD168	RHAMM (receptor for hyaluronan mediated motility)	HA-binding protein	HA = hyaluronic acid

Table 1 (continued)

CD169	Siglec-1		
CD170	Siglec-5		
CD173	ABH type-2		
CD174	LeY		
CD175	Tn		
CD175s	Sialyl-Tn		
CD176	T	Thomsen-Friedenreich	
CD205	C-type lectin		
CD206	C-type lectin		
CD207	C-type lectin		
CD209	C-type lectin		
CD222	M6P-R	Mannose-6-phosphate receptor	
CD227	MUC1		
CD235a	Glycophorin A	MN Ag	
CD235b	Glycophorin B	Ss Ag	
CD235ab	Glycophorin A/B crossing epitope		
CD236	Glycophorin C/D crossing epitope		
CD236R	Glycophorin C	Dh Ag	Ge Ag
CD248	Endosialin	TEM1 (tumor endothelial marker 1)	
CD299	L-SIGN (liver-specific ICAM3-grabbing nonintegrin)	DC-SIGN-related	
CD301	MGL (macrophage galactose-type C-type lectin)		
CD302	Type-I transmembrane C-type lectin receptor DCL-1		
CDw327	Siglec-6		
CDw328	Siglec-7		
CDw329	Siglec-9		
(CD330)	Siglec-10 (reserved)		

Glycoproteins except for glycophorin and GPI anchor proteins are omitted

molecules: (1) a selectin family (CD62E, CD62L, CD62P), and selectin ligands (CD15s, CD15u, CD65s), (2) sialomucin family (CD34, CD43, CD68, CD162, CD162R, CD164), (3) immunoglobulin superfamily (IgSF), (4) integrin family, and (5) CD44 family. There are many reports on these molecules in the field of immunology, hematology, and cancer cell biology. Glycosyltransferases hold key roles in biosynthesis of oligosaccharide CD antigens and of oligosaccharide portions of CD-related glycoproteins. In addition to the above five families, there are several oligosaccharide antigen molecules whose functions are not fully understood. It is expected that CD65, CD75 and CD75s participate in cell-to-cell interaction. CDw15, CD60a and CD77 are glycosphingolipids, located in rafts, and adjust the functions of glycoprotein-type receptors. Although not included in Table 1, IgSF and integrin families are glycoproteins. Research on the function of the oligosaccharide portions of IgSFs and integrins in cell adhesion is currently underway. For many C-type lectins and Siglec(s) (CD22, CD33, CD169, CD170, CDw327, CDw328, CDw329) in the subfamily of IgSF, their functions are not yet fully elucidated.

Applications in Blood Cell Glycobiology

The oligosaccharide-related antigens serve as markers of blood cell differentiation lineages and cancer cells, and some are already in use as diagnostic reagents. It is strongly expected that elucidating the site of action in blood cell glycobiology-related cell-to-cell or molecule-to-molecule interactions will lead to the discovery of new drugs in the field of immunity, oncology and so on. Blood cells shall continue to be important and versatile clinical samples. Blood continues to be a treasure box of clinical information. Areas where blood cell glycobiology can contribute are broad.

In Vivo Assay for Extramedullar Infiltration of Human Leukemia Cell Lines

1. Prepare human leukemic cell sublines in which a gene of interest is knocked down.
2. Stain the competitor and test cells using carboxyfluorescein diacetate succinimidyl ester (CFSE) or tetramethylrhodamine-5-isothiocyanate (TRITC).
3. Mix the competitor and test cells at a ratio of 1:1 and inject them intravenously via the tail of γ-irradiated immunodeficient mice.
4. Kill the mice after a certain period of time and prepare single cell suspensions from the extramedullar organs.
5. Analyze the cells that migrated to the organs and count the numbers of cells labeled with fluoroprobes using a flow cytometer.

References

Mason D, André P, Bensussan A et al (2001) CD antigens 2001: aims and results of HLDA Workshops. Stem Cells 19:556–562

Mason D, André P, Bensussan A et al (2002) Reference: CD Antigens 2002. J Immunol 168: 2083–2086

Zola H, Swart B, Nicholson I et al (2005) CD molecules 2005: human cell differentiation molecules. Blood 106:3123–3126

Part 3
Glycosyltransferase Gene KO Mice

Section XX
Glycosyltransferase Gene KO Mice

Fucosyltransferase-9 Knockout Mouse

Takashi Kudo[1], Hisashi Narimatsu[2]

Introduction

Stage-specific embryonic antigen-1 (SSEA-1), an antigenic epitope of which was defined as a Lewis x [Le^x: Galβ1–4(Fucα1–3)GlcNAc–] carbohydrate structure, is widely expressed on the surface of mammalian cells, and is considered to be involved in cell–cell interactions during embryogenesis, differentiation, and neurodevelopmental processes. FUT3, FUT4, FUT5, FUT6 and FUT9 can synthesize the Le^x structure, while FUT7 cannot (Kudo et al. 1998). Detailed analysis of the substrate specificity of polylactosamine acceptor has revealed FUT9 to have more-efficient activity for synthesis of the Le^x structure than other α1,3FUTs (Nishihara et al. 1999).

For many years, SSEA-1 has been implicated in the development of mouse embryo as a functional carbohydrate epitop in cell–cell interaction during morula compaction. However, Fut9 knockout (Fut9$^{-/-}$) mice develop normally, with no gross phenotypic abnormalities, and are fertile, although expression of the SSEA-1 epitopes was completely absent in early embryos and in primordial germ cells of Fut9$^{-/-}$ mice despite the expression of *Fut4* gene (Kudo et al. 2004). Immunohistochemical results using anti-Le^x and anti-Fut9 mAbs strongly indicated that Fut9 is the enzyme most responsible for the synthesis of Le^x in the CNS (Nishihara et al. 2003). No obvious differences in the architecture of the cerebrum and cerebellum were observed in Fut9$^{-/-}$ mice, but an increase in the anxiety-like behavior and a decrease in the number of calbindin-immunoreactive cells in the amygdalar subdivision was observed (Kudo et al. 2007). We conclude that expression of the SSEA-1 epitope in the developing mouse embryo is not essential for embryogenesis and that in the amygdalar subdivisions play critical roles in functional regulations of interneurons.

Immuno-Thin Layer Chromatography with SSEA-1 Antibody

1. Neural glycolipids were extracted with a mixture of chloroform–methanol–water (30:60:8, v/v).
2. Apply the neural glycolipids to an HPTLC (Kieselgel 60) plate with a microsyringe.
3. Develop the HPTLC plate in a TLC chamber with a mixture of chloroform–methanol–water (60:35:8, v/v).
4. Coat the plate with poly-isobutyl-methacrylate beads in hexan.
5. Block with 1% BSA/PBS.

---

[1] Graduate School of Comprehensive Human Sciences, University of Tsukuba, 1-1-1 Tennodai, Tsukuba, Ibaraki 305-8575, Japan
E-mail: t-kudo@md.tsukuba.ac.jp

[2] Research Center for Medical Glycosciences, National Institute of Advanced Industrial Science and Technology (AIST), 1-1-1 Umezono, Tsukuba, Ibaraki 305-8568, Japan
E-mail: h.narimatsu@aist.go.jp

6. Diluted first antibody (SSEA-1) was overlaid on the plate in the humidified chamber.
7. The TLC plate was washed with 1% BSA/PBS and then incubated with biotinylated second antibody (anti-mouse IgM) for 2 h at room temperature.
8. The TLC plate was washed with 1% BSA/PBS and visualized using Vectastain Elite ABC kit and ECL detection kit (Kudo et al. 2007).

References

Kudo T, Ikehara Y et al (1998) Expression cloning and characterization of a novel murine α1,3-fucosyltransferase, mFuc-TIX, that synthesizes the Lewis x (CD15) epitope in brain and kidney. J Biol Chem 273:26729–26738

Kudo T, Kaneko M et al (2004) Normal embryonic and germ cell development in mice lacking α1,3-fucosyltransferase IX (Fut9) which show disappearance of stage-specific embryonic antigen 1. Mol Cell Biol 24:4221–4228

Kudo T, Fujii T et al (2007) Mice lacking α1,3-fucosyltransferase IX demonstrate disappearance of Lewis x structure in brain and increased anxiety-like behaviors. Glycobiology 17:1–9

Nishihara S, Iwasaki H et al (1999) α1,3-Fucosyltransferase 9 (FUT9; Fuc-TIX) preferentially fucosylates the distal GlcNAc residue of polylactosamine chain while the other four α1,3FUT members preferentially fucosylate the inner GlcNAc residue. FEBS Lett 462:289–294

Nishihara S, Iwasaki H et al (2003) α1,3-Fucosyltransferase IX (Fut9) determines Lewis X expression in brain. Glycobiology 13:445–455

Knockout Mice of α1,6 Fucosyltransferase (Fut 8)

Naoyuki Taniguchi[1,2]

Introduction

The core fucosylation (α1,6-fucosylation) of glycoproteins is widely distributed in mammalian tissues, and is altered under pathological conditions. To investigate the physiological functions of the core fucose, we generated α1,6-fucosyltransferase (*Fut8*)-null mice and found that disruption of *Fut8* induces severe growth retardation and death during postnatal development. Histopathological analysis revealed that $Fut8^{-/-}$ mice showed emphysema-like changes in the lung, verified by a physiological compliance analysis.

Results

Fut8 induces severe growth retardation and death during postnatal development. Histopathological analysis revealed that Fut8(−/−) mice showed emphysema-like changes in the lung, verified by a physiological compliance analysis. Biochemical studies indicated that lungs from Fut8(−/−) mice exhibit a marked overexpression of matrix metalloproteinases (MMPs), such as MMP-12 and MMP-13, highly associated with lung-destructive phenotypes, and a down-regulation of extracellular matrix (ECM) proteins such as elastin, as well as retarded alveolar epithelia cell differentiation. These changes should be consistent with a deficiency in TGF-β1 signaling, a pleiotropic factor that controls ECM homeostasis by down-regulating MMP expression and inducing ECM protein components. In fact, Fut8(−/−) mice have a marked dysregulation of TGF-β1 receptor activation and signaling, as assessed by TGF-β1 binding assays and Smad2 phosphorylation analysis. We also show that these TGF-β1 receptor defects found in Fut8(−/−) cells can be rescued by reintroducing Fut8 into Fut8(−/−) cells. Furthermore, exogenous TGF-β1 potentially rescued emphysema-like phenotype and concomitantly reduced MMP expression in Fut8(−/−) lung. The other phenotypic changes were also observed. The loss of core fucosylation also impairs the function of low-density lipoprotein (LDL) receptor-related protein-1 (LRP-1), a multifunctional scavenger and signaling receptor, resulting in a reduction in the endocytosis of insulin-like growth factor (IGF)-binding protein-3 (IGFBP-3) in the cells derived from Fut8-null (Fut8−/−) mice. These data clearly indicate that core fucosylation is crucial for the scavenging activity of LRP-1 in vivo. A group of genes, including trypsinogens 4, 7, 8, 11, 16, and 20, were down-regulated in Fut8−/− embryos as judged by DNA microarray analysis. The expression of trypsinogen

---

[1]Department of Disease Glycomics (Seikagaku Corporation), Institute for Microbial Diseases, Osaka University, 4th Floor Advance Innovation Center, Taniguchi Research Group, 2-1 Yamadaoka Suita, Osaka 565-0871, Japan
Phone: +81-6-6879-3427, Fax: +81-6-6879-3427
E-mail: tani52@wd5.so-net.ne.jp

[2]Disease Glycomics Team, Systems Glycobiology Research Group, Chemical Biology Department, Advanced Science Institute, RIKEN, 2-1 Hirosawa, Wako 351-0198, Japan
Phone: +81-48-467-9613, Fax: +81-48-462-4692

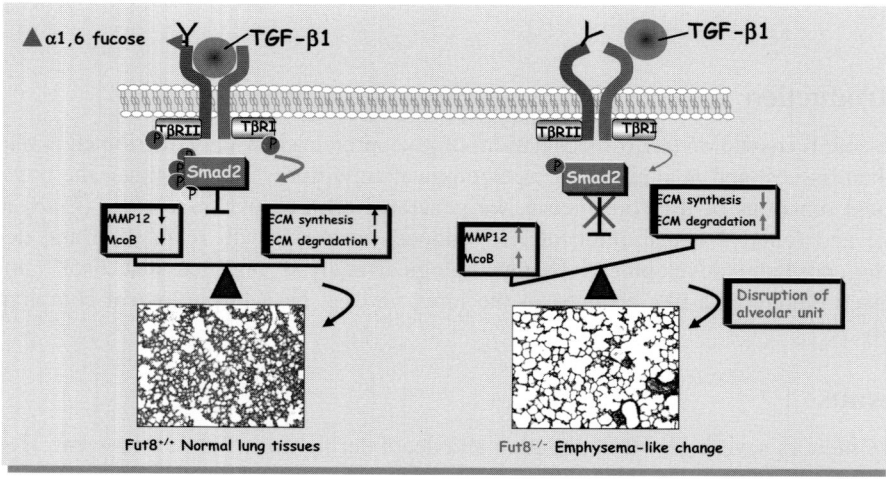

Fig. 1 Dysregulation of TGF-β1 receptor activation leads to abnormal lung development and emphysema-like phenotype in core fucose-deficient mice

proteins was also found to be lower in Fut8−/− mice in the duodenum, small intestine, and pancreas. Trypsin, an active form of trypsinogen, regulates cell growth through a G-protein-coupled receptor, the proteinase-activated receptor 2 (PAR-2). In a cell culture system, a Fut8 knockdown mouse pancreatic acinar cell carcinoma, TGP49-Fut8-KDs, showed decreased growth rate, similar to that seen in Fut8−/− mice. These data suggest that a correlation between Fut8 and regulation of EGF receptor (EGFR)-trypsin-PAR-2 pathway may exist in controlled cell growth and that the EGFR-trypsin-PAR-2 pathway is suppressed in TGP49-Fut8-KDs as well as in Fut8−/− mice (Fig. 1).

References

Lee SH, Takahashi M, Honke K, Miyoshi E, Osumi D, Sakiyama H, Ekuni A, Wang X, Inoue S, Gu J, Kadomatsu K, Taniguchi N (2006) Loss of core fucosylation of low-density lipoprotein receptor-related protein-1 impairs its function, leading to the upregulation of serum levels of insulin-like growth factor-binding protein 3 in Fut8−/− mice. J Biochem (Tokyo) 39:391–398

Li W, Nakagawa T, Koyama N, Wang X, Jin J, Mizuno-Horikawa Y, Gu J, Miyoshi E, Kato I, Honke K, Taniguchi N, Kondo A (2006) Down-regulation of trypsinogen expression is associated with growth retardation in alpha1,6-fucosyltransferase-deficient mice: attenuation of proteinase-activated receptor 2 activity. Glycobiology 16:1007–1019

Taniguchi N, Miyoshi E, Jianguo G, Honke K, Matsumoto A (2006) Decoding sugar functions by identifying target glycoproteins. Curr Opin Struct Biol 16:561–566

Wang X et al (2005) Dysregulation of TGF-β1 receptor activation leads to abnormal lung development and emphysema-like phenotype in core fucose-deficient mice. Proc Natl Acad Sci USA 102: 15791–15796

Glucuronyltransferase (GlcAT-P) Gene-Deficient Mice

Ippei Morita[1], Shogo Oka[2]

Introduction

The HNK-1 carbohydrate has a very unique structural feature comprising a sulfated trisaccharide (HSO_3-3GlcAβ1-3Galβ1-4GlcNAc-), whose biosynthesis is regulated by two glucuronyltranferases (GlcAT-P and GlcAT-S). The carbohydrate is expressed on a limited number of glycoproteins, such as neural cell adhesion molecule (NCAM), L1, phosphacan and so on, in the nervous system. To elucidate the role of the HNK-1 carbohydrate, we generated and analyzed mice with a targeted deletion of the GlcAT-P gene, which is mainly expressed in the nervous system.

Results

We constructed a targeting vector using a strategy in which most of the catalytic region of GlcAT-P (exons 4 and 5) is replaced by the neomycin resistance gene. The targeting vector was transfected into ES cells (E14-1), and homologous recombination was confirmed by Southern blotting. To generate chimeric mice, ES cells were transferred into the uteri of pseudopregnant mice. Mice heterozygous for the mutation were obtained by cross-breeding of the chimeras with C57BL/6 mice.

The GlcAT-P(–/–) mice exhibited normal birth and growth, and there was no significant difference in their appearance or body weight, although the mice had almost completely lost the HNK-1 carbohydrate. To examine the effect of the HNK-1 carbohydrate deficiency on synaptic plasticity, we analyzed LTP (long-term potentiation) in the hippocampal CA1 region of GlcAT-P(–/–) and wild-type (+/+) mice. We monitored excitatory postsynaptic potentials (EPSPs), which were evoked by stimulating afferent fibers in the stratum radiatum of the CA1 region, using the extracellular field potential recording technique. High-frequency stimulation of afferent fibers (100 Hz for 1 s) gave rise to LTP of excitatory synaptic transmission in the wild-type mice, while the magnitude of LTP in the GlcAT-P(–/–) mice was significantly lower than that in the wild-type ones (Yamamoto et al. 2002). Furthermore, considering the reduced LTP, two types of behavioral tests (Morris water maze test and water-filled multiple T-maze test) were carried out to evaluate spatial navigation and hippocampus-dependent memory formation. In the Morris water maze test, the time taken to reach the hidden platform (escape latency) was significantly longer for GlcAT-P(–/–) mice than wild-type mice during 4 days of training. In the water-filled multiple T-maze test, GlcAT-P(–/–) mice showed increased escape

---

[1] Department of Biological Chemistry, Graduate School of Pharmaceutical Sciences, Kyoto University, Kyoto 606-8501, Japan

[2] Department of Biological Chemistry, School of Health Sciences, Faculty of Medicine, Kyoto University, Kyoto 606-8507, Japan
Phone: +81-75-751-3959, Fax: +81-75-751-3959
E-mail: shogo@hs.med.kyoto-u.ac.jp

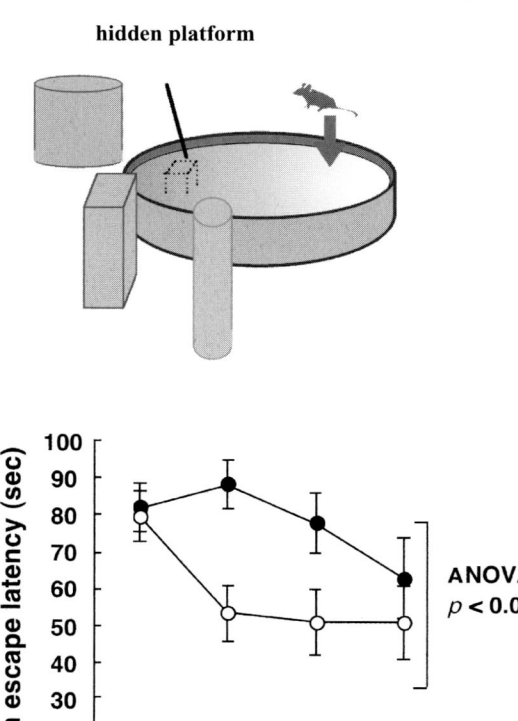

Fig. 1 **A** Schematic diagram of the Morris water maze test. **B** Mice were trained in a Morris water maze for 4 consecutive days. Learning performance is expressed as the mean escape latency for four trials per day. GlcAT-P(−/−) mice showed significantly greater escape latency than GlcAT-P(+/+) mice ($P < 0.05$, ANOVA). (Yamamoto et al. 2002)

latencies in the goal arm compared to wild-type mice. These lines of evidence indicate the involvement of the HNK-1 carbohydrate in higher order brain functions such as learning, memory and synaptic plasticity. In the near future, it is necessary to shed light on the molecular mechanisms by which the HNK-1 carbohydrate controls synaptic plasticity.

References

Yamamoto S, Oka S, Inoue M, Shimuta M, Manabe T, Takahashi H, Miyamoto M, Asano M, Sakagami J, Sudo K, Iwakura Y, Ono K, Kawasaki T (2002) Mice deficient in nervous system-specific carbohydrate epitope HNK-1 exhibit impaired synaptic plasticity and spatial learning. J Biol Chem 277:27227–27231

KO Mice of β1,4-galactosyltransferase-I

Masahide Asano

Introduction

β1,4-galactosyltransferases (β4GalTs) transfer galactose (Gal) from UDP-Gal to terminal GlcNAc of *N*- and *O*-glycans in a β1,4-linkage to synthesize the Galβ1–4GlcNAc structure. So far, seven β4GalT genes (β4GalT-I to VII) have been isolated and six of them are shown to be involved in the biosynthesis of the Galβ1–4GlcNAc structure. β4GalT-I is expressed ubiquitously and strongly in almost all tissues except neural tissues, suggesting that β4GalT-I is engaged in β4-galactosylation of many glycoproteins. β4-galactosylation of glycoproteins is widely distributed in mammalian tissues and involved in various physiological functions through interactions with selectins, galectins, asialoglycoprotein receptor (ASGPR) and so on. We, as well as Shur's group, generated mice lacking in β4GalT-I to elucidate the role of β4-galactosylation in vivo (Asano et al. 1997; Lu et al. 1997).

Results

The carbohydrate structures were analyzed in various tissues of β4GalT-I$^{-/-}$ mice. Not only type 2 *N*-glycans, but also core 2 *O*-glycans were largely reduced, indicating that β4GalT-I was responsible for the biosynthesis of these carbohydrate structures. Furthermore, the shift of the backbone structure from type 2 chain (Galβ1–4GlcNAc) to type 1 chain (Galβ1–3GlcNAc) was observed in β4GalT-I$^{-/-}$ mice. In relation to this change, sialyl linkage in *N*-glycans was also shifted from Siaα2–6Gal to Siaα2–3Gal. Thus, β4GalT-I deficiency was primarily compensated for by β1,3-galactosyltransferases, which resulted in altered backbone structures and different sialyl linkages (Kotani et al. 2001).

β4GalT-I$^{-/-}$ mice were born normally and were fertile, but showed growth retardation and semi-lethality before weaning. Epithelial cell proliferation of the skin and small intestine was enhanced, and cell differentiation in the intestinal villi was abnormal (Asano et al. 1997). The β4GalT-I$^{-/-}$ mice that survived exhibited blood leukocytosis, but normal lymphocyte homing to peripheral lymph nodes. Acute and chronic inflammatory responses including contact hypersensitivity (CHS) and delayed-type hypersensitivity (DTH) responses were significantly suppressed because of the defect in selectin-ligand biosynthesis mainly expressed on the core 2 *O*-glycans of neutrophils and macrophages (Fig. 1) (Asano et al. 2003). β4GalT-I$^{-/-}$ mice also showed significantly delayed skin wound healing with reduced re-epithelialization, collagen synthesis and angiogenesis. Neutrophil and macrophage recruitment at wound sites was impaired in these mice probably because of selectin-ligand deficiency (Mori et al. 2004). Recently, we have found that β4GalT-I$^{-/-}$ mice spontaneously developed human IgAN-like disease with IgA

Advanced Science Research Center, Kanazawa University, 13-1 Takara-machi, Kanazawa 920-8640, Japan
Phone: +81-76-265-2460, Fax: +81-76-234-4240
E-mail: asano@kiea.m.kanazawa-u.ac.jp

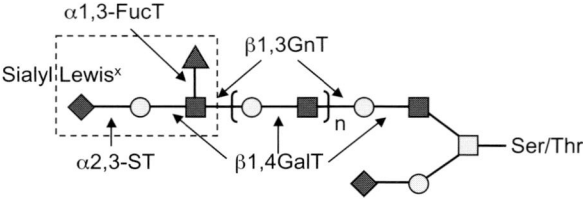

Fig. 1 Sialyl Lewis^x at the terminus of N-acetyllactosamine repeats on the core 2 branch

deposition and expanded mesangial matrix (see Chapter by M. Asano, this volume).

Generation of β4GalT-I KO Mice

Homologous Recombination in ES (embryonic stem) cells

1. Construct a targeting vector for the *β4GalT-I* gene, in which the Neo cassette (a positive selection marker) was replaced with exon 1 of the *β4GalT-I* gene, and the DT-A cassette (a negative selection marker) was ligated at the end of the left arm.
2. Introduce the linearized targeting vector (20 μg/10^7 cells) into ES cells by electroporation (250 V, 500 μF).
3. Add 180 μg/ml G418 to the medium 24 h after electroporation, and select G418-resistant colonies for 7–10 days.
4. Pick up a single colony, scatter cells by trypsin-treatment, and plate them into 24-well plate.
5. Screen homologous recombinant ES clones by PCR (polymerase chain reaction) using cell lysates of each clone.
6. Cryopreserve PCR-positive ES clones.
7. Confirm the homologous recombination by Southern blot and check the karyotype of positive clones.

Generation of Chimera Mice by the Aggregation Method from the Homologous Recombinant ES Clones

1. Prepare an aggregation plate by creating deep depressions in a plastic dish.
2. Collect 8-cell-stage embryos from super-ovulated female mice.
3. Remove the zona pellucida of the embryos by acid tyrode treatment.
4. Place zona-free 8-cell-stage embryos individually into each of the depressions in the aggregation plate.
5. Place a clump of ES cells into each of the depressions in the aggregation plate containing the embryo.
6. Culture the aggregates between ES cells and embryos overnight to form a single embryo at the late-morula or early-blastocyst stage.
7. Transfer the developed embryos into the uterus of a pseudopregnant female mouse.

Comments

There are two obstacles to overcome in generating KO mice. One is to obtain desired homologous recombinant ES clones and the other is to obtain germ-line transmission. The efficiency of homologous recombination is dependent on the locus of the targeted gene and the design of the targeting vector. It is difficult to expect efficiency—sometimes more than 10%, sometimes less than 1%. ES cells tend to differentiate and lose normal chromosome number very easily. Differentiated ES cells and ES cells with abnormal chromosomes are unable to contribute to germ cells in chimera mice. It is important to maintain the pluripotency of ES cells during screening.

References

Asano M, Furukawa K, Kido M, Matsumoto S, Umesaki Y, Kochibe N, Iwakura Y (1997) Growth retardation and early death of β-1,4-galactosyltransferase knockout mice with augmented proliferation and abnormal differentiation of epithelial cells. EMBO J 16:1850–1857

Asano M, Nakae S, Kotani N, Shirafuji N, Nambu A, Hashimoto N, Kawashima H, Hirose M, Miyasaka M, Takasaki S, Iwakura Y (2003) Impaired selectin-ligand biosynthesis and reduced inflammatory responses in β-1,4-galactosyltransferase-I-deficient mice. Blood 102:1678–1685

Kotani N, Asano M, Iwakura Y and Takasaki S (2001) Knockout of mouse β1,4-galactosyltransferase-1 gene results in a dramatic shift of outer chain moieties of N-glycans from type 2 to type 1 chains in hepatic membrane and plasma glycoproteins. Biochem J 357:827–834

Lu Q, Hasty P, Shur BD (1997) Targeted mutation in β1,4-galactosyltransferase leads to pituitary insufficiency and neonatal lethality. Dev Biol 181:257–267

Mori R, Kondo T, Nishie T, Ohshima T, Asano M. (2004) Impairment of skin wound healing in β-1,4-galactosyltransferase-deficient mice with reduced leukocyte recruitment. Am J Pathol 164:1303–1314

Sulfotransferases

Takashi Muramatsu[1], Kenji Uchimura[2]

Introduction

Molecular cloning has been accomplished for a large number of carbohydrate sulfotransferases. The majority of them are involved in the synthesis of glycosaminoglycans such as heparan sulfate, chondroitin sulfate, and keratan sulfate (Habuchi et al. 2006), whereas some of them play roles in formation of glycans in glycoproteins or glycolipids. Increasing numbers of carbohydrate sulfotransferases have been knocked out in mice to understand the role of sulfate groups more precisely (Table 1). Interestingly, all knockout mice exhibit phenotypes, underlining the physiological importance of carbohydrate sulfation.

Comments

The first carbohydrate sulfotransferase knocked out was heparan sulfate 2-O-sulfotransferase. Inactivation of the gene by Gene trap leads to neonatal death in mice due to failure of kidney formation (Bullock et al. 1998). The striking phenotype gave convincing evidence that heparan sulfate plays important roles in mammalian development.

Heparan sulfate/heparin *N*-deacetylase/*N*-sulfotransferase (NDST) catalyzes both the deacetylation and sulfation of an amino group in *N*-acetylglucosamine. NDST1 is involved in the synthesis of heparan sulfate, and NDST-2 in the synthesis of heparin. The action of NDST usually precedes the action of other sulfotransferases. However, in embryonic stem cells derived from NDST1 and NDST2 double knockout mice, which lack *N*-sulfated glucosamine, heparan sulfate with 6-sulfated *N*-acetylglucosamine was detected, indicating that the 6-sulfation step, which does not require *N*-sulfation, is present (Holmborn et al. 2004). Reflecting the physiological importance of heparan sulfate, NDST1 knockout mice die at the neonatal stage (Ringvall et al. 2000). Conditional knockout at a restricted tissue is required to understand the precise role of heparan sulfate chain in a physiological process (Wang et al. 2005).

N-acetylglucosamine-6-*O*-sulfotransferases (GlcNAc6STs) are involved in the synthesis of both L-selectin ligands and keratan sulfate. Decreased lymphocyte homing in GlcNAc6ST deficient mice is due to the impaired synthesis of L-selectin ligand and other phenotypes are due to failure in the synthesis of highly-sulfated keratan sulfate. In vivo lymphocyte homing assay is shown as a protocol. Further information concerning the

---

[1] Department of Health Science, Faculty of Psychological and Physical Science,
Aichi Gakuin University, 12 Araike, Iwasaki-cho, Nisshin, Aichi 470-0195, Japan
Phone: +81-561-73-1111, Fax: +81-561-73-1142
E-mail: tmurama@dpc.aichi-gakuin.ac.jp

[2] Section of Pathophysiology and Neurobiology, Department of Alzheimer's Disease Research, National Institute for Longevity Sciences (NILS), 36-3 Gengo, Morioka, Obu, Aichi 474-8522, Japan
Phone: +81-562-46-2311, Fax: +81-562-46-3157
E-mail: arumihcu@nils.go.jp

Table 1 List of knockout of carbohydrate sulfotransferases

Name of enzymes	Methods	Phenotypes	PubMed ID No. of key references
Heparan sulfate/ heparin N-deacetylase/ N-sulfotransferase-1	KO, conditional KO	Neonatal death; disturbed Calcium ion kinetics in myotube; cerebral hypoplasia and craniofacial defects; impaired neutrophil infiltration	10852901, 12692154, 15292174, 15319440, 16020517, 16056228
Heparan sulfate/ heparin N-deacetylase/ N-sulfotransferase-2	KO	Abnormal mast cells	10466726, 10466727
Heparan sulfate 2-O-sulfotransferase	Gene trap	Neonatal death due to renal agenesis; impaired retinal axon guidance	9637690, 16807321
Heparan sulfate 3-O-sulfotransferase-1	KO	Normal hemostasis; genetic background-specific lethality and intrauterine growth retardation	12671048
Heparan sulfate 6-O-sulfotransferase-1	Gene trap, KO	Impaired retinal axon guidance; Abnormalities in fetal microvessels	16807321, 17405882
Chondroitin 4-sulfotransferase	Gene trap	Severe chondrodysplasia	16079159
Chondroitin 6-sulfotransferase	KO	Decreased naïve T lymphocytes in the spleen of young mice	11696535
N-Acetylglucosamine-6-O-sulfotransferase-1	KO	Decreased lymphocyte homing; suppression of glial scar formation in the injured brain	15175329, 16227985, 16227986, 16624895
N-Acetylglucosamine-6-O-sulfotransferase-2	KO	Decreased lymphocyte homing	11520459, 14597732, 16227985, 16227986
N-Acetylglucosamine-6-O-sulfotransferase-3	KO	Thin cornea	16938851
HNK-1 sulfotransferase	KO	Alterations in synaptic efficacy, spatial learning and memory; reduced extracellular space in the brain	12213450, 12358771, 16262627
Cerebroside sulfotransferase	KO	Abnormalities in paranodal junctions; male sterility; impaired neutrophil infiltration	11917099, 12151530, 14583626, 15659616

phenotype of GlcNAc5ST knockout mice is provided in the section on GlcNAc6STs. Among five isozymes of GlcNAc6STs, three enzymes have been knocked out. Among isozymes of heparan sulfate 3-O-sulfotransferases, so far, knockout of only one enzyme has been published. This is also true for heparan sulfate 6-O-sulotransferases.

Cerebroside sulfotransferase is involved in the formation of both sulfatide (HSO3-3-galactosyl ceramide) and seminolipid (HSO3-3-monogalactosylacylglycerol). The neurological phenotype observed in cerebroside sulfotransferase knockout mice is ascribed to the lack of sulfatide, and male sterility in these mice is due to the impaired synthesis of seminolipid.

Protocol: In Vivo Lymphocyte Homing Assay

Lymphocytes circulate between the blood and the lymphoid organs. Lymphocyte homing to lymph nodes is initiated by the binding of L-selectin to its ligands. A major class of L-selectin ligands in lymphocyte homing is sialy 6-sulfo Lewis X. GlcNAc-6-O-sulfation is essential for the ligand activity. To assess the activity of homing ligand, an in vivo assay of lymphocyte homing is utilized. Lymphocyte donors and recipients in the assay described here are adult mice. Flow cytometry is used to quantify homing of lymphocytes to lymph nodes.

1. Dissect out mesenteric lymph nodes of 5- to 10-week-old CD-1 mice and place them in phosphate buffered saline (PBS).
2. Mush the nodes with two slide glasses to have single-cell suspensions in PBS.
3. Filter suspensions through a nylon cell strainer (40 µm pore size).
4. Metabolically label filtered cells with 5 µM 5-chloromethylfluorescein diacetate (CMFDA) at 37°C for 30 min.
5. Inject the cells (1.7×10^7 in 100 µl PBS/mouse) into tail veins of 8- to 10-week-old mice.
6. One hour after injection, kill the mice and dissect out peripheral lymph nodes, mesenteric lymph nodes, Peyer's patches, and spleen.
7. Create single-cell suspensions of these organs in PBS by teasing with 23-gauge needles.
8. Determine the fraction of CMFDA-labeled cells in suspensions by flow cytometry (5×10^5 cells per organ).
9. Acquire and analyze data with CellQuest software.

References

Bullock SL, Fletcher JM, Beddington RS, Wilson VA (1998) Renal agenesis in mice homozygous for a gene trap mutation in the gene encoding heparan sulfate 2-sulfotransferase. Genes Dev 12:1894–1906

Habuchi H, Habuchi O, Uchimura K, Kimata K, Muramatsu T (2006) Determination of substrate specificity of sulfotransferases and glycosyltransferases (proteoglycans). Methods Enzymol 416:225–243

Holmborn K, Ledin J, Smeds E, Eriksson I, Kusche-Gullberg M, Kjellen L (2004) Heparan sulfate synthesized by mouse embryonic stem cells deficient in NDST1 and NDST2 is 6-O-sulfated but contains no N-sulfate groups. J Biol Chem 279:42355–42358

Ringvall M, Ledin J, Holmborn K, van Kuppevelt T, Ellin F, Eriksson I, Olofsson AM, Kjellen L, Forsberg E (2000) Defective heparan sulfate biosynthesis and neonatal lethality in mice lacking N-deacetylase/N-sulfotransferase-1. J Biol Chem 275:25926–25930

Wang L, Fuster M, Sriramarao P, Esko JD (2005) Endothelial heparan sulfate deficiency impairs L-selectin- and chemokine-mediated neutrophil trafficking during inflammatory responses. Nat Immunol 6:902–910

KO Mice of Cerebroside Sulfotransferase

Koichi Honke[1,2]

Introduction

Sulfatide synthase catalyzes the transfer of sulfonate group to the C3 position of the nonreducing terminal galactose of glycolipid oligosaccharides, and is traditionally called cerebroside sulfotransferase (CST). We purified CST homogeneously from human renal cancer cells (Honke et al. 1996) and subsequently isolated a cDNA clone of *CST* from a cDNA library of human renal cancer cells on the basis of the partial amino acid sequences of the purified enzyme (Honke et al. 1997). Furthermore, we cloned mouse cDNA and genomic DNA (Hirahara et al. 2000). In order to know the biological roles of the sulfation of glycolipids and to determine whether a single enzyme is responsible for the biosynthesis of sulfatide and seminolipid which are different in their lipid moiety, CST-deficient mice were created by gene targeting (Honke et al. 2002).

Results

CST activity disappears in the whole body of $Cst^{-/-}$ mice (Honke et al. 2002). The mutant mice show a complete loss of sulfatide in brain and seminolipid in testis, proving that a single gene copy is responsible for the biosynthesis of sulfatide and seminolipid. $Cst^{-/-}$ mice are born healthy, but begin to display hindlimb weakness by 6 weeks of age and subsequently show a pronounced tremor and progressive ataxia. Although compact myelin is preserved, $Cst^{-/-}$ mice display abnormalities in paranodal junctions (Fig. 1). Furthermore, clustering of Na^+ and K^+ channels at the node is deteriorated in CST-null mice (Ishibashi et al. 2002). Myelin of central nervous system is produced by oligodendrocytes. Terminal differentiation of oligodendrocytes is enhanced in CST-KO mice (Hirahara et al. 2004), indicating that sulfatide plays a critical role in the regulation of oligodendrocyte terminal differentiation, in addition to their eventual roles as structural components of mature myelin. Despite the significant neurological disorders, *Cst*-null mice are able to survive more than 1 year of age. On the other hand, $Cst^{-/-}$ males are sterile because of an arrest in spermatogenesis before the first meiotic division (Fig. 1), whereas females were able to breed (Honke et al. 2002). Introduction of wild-type spermatogogenic cells into *Cst*-null testis results in the restoration of spermatogenesis, proving that seminolipid is essential for germ cell function in spermatogenesis (Zhang et al. 2005). Moreover, CST deficiency ameliorates L-selectin-dependent monocyte infiltration in the kidney after ureteral obstruction, an experimental model of renal interstitial inflammation, indicating that sulfatide is an endogenous ligand of L-selectin (Ogawa et al. 2004). Studies on the molecular mechanisms by which sulfoglycolipids participate in these biological processes are ongoing.

---

[1] Department of Biochemistry, Kochi University Medical School, Kohasu, Oko-cho, Nankoku, Kochi 783-8505, Japan
Phone: +81-88-880-2313, Fax: +81-88-880-2314
E-mail: khonke@kochi-u.ac.jp

[2] CREST, Japan Science and Technology Agency, Japan

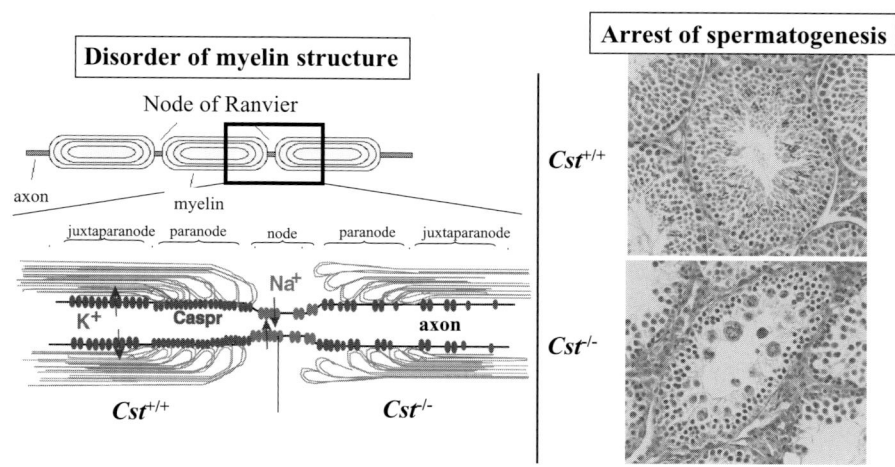

Fig. 1 Main phenotypes of CST-knockout mouse

References

Hirahara Y, Tsuda M, Wada Y, Honke K (2000) cDNA cloning, genomic cloning, and tissue-specific regulation of mouse cerebroside sulfotransferase. Eur J Biochem 267:1909–1917

Hirahara Y, Bansal R, Honke K, Ikenaka K, Wada Y (2004) Sulfatide is a negative regulator of oligodendrocyte differentiation: development in sulfatide-null mice. Glia 45:269–277

Honke K, Yamane M, Ishii A, Kobayashi T, Makita A (1996) Purification and characterization of 3′-phosphoadenosine-5′-phosphosulfate:GalCer sulfotransferase from human renal cancer cells. J Biochem (Tokyo) 119:421–427

Honke K, Tsuda M, Hirahara Y, Ishii A, Makita A, Wada Y (1997) Molecular cloning and expression of cDNA encoding human 3′-phosphoadenylylsulfate:galactosylceramide 3′-sulfotransferase. J Biol Chem 272:4864–4868

Honke K, Hirahara Y, Dupree J, Suzuki K, Popko B, Fukushima K, Fukushima J, Nagasawa T, Yoshida N, Wada Y, Taniguchi N (2002) Paranodal junction formation and spermatogenesis require sulfoglycolipids. Proc Natl Acad Sci USA 99:4227–4232

Ishibashi T, Dupree JL, Ikenaka K, Hirahara Y, Honke K, Peles E, Popko B, Suzuki K, Nishino H, Baba H (2002) A myelin galactolipid, sulfatide, is essential for maintenance of ion channels on myelinated axon but not essential for initial cluster formation. J Neurosci 22:6507–6514

Ogawa D, Shikata K, Honke K, Sato S, Matsuda M, Nagase R, Tone A, Okada S, Usui H, Wada J, Miyasaka M, Kawashima H, Suzuki Y, Suzuki T, Taniguchi N, Hirahara Y, Tadano-Aritomi K, Ishizuka I, Tedder TF, Makino H (2004) Cerebroside sulfotransferase deficiency ameliorates L-selectin-dependent monocyte infiltration in the kidney after ureteral obstruction. J Biol Chem 279:2085–2090

Zhang Y, Hayashi Y, Cheng X, Watanabe T, Wang X, Taniguchi N, Honke K (2005) Testis-specific sulfoglycolipid, seminolipid is essential for germ cell function in spermatogenesis. Glycobiology 15:649–654

KO Mice of β1,4-*N*-acetylgalactosaminyltransferase (GM2/GD2 Synthase)

Koichi Furukawa, Kogo Takamiya, Keiko Furukawa

Introduction

Complex gangliosides have been considered to play important roles in the development and differentiation of nervous systems in vertebrates. In order to directly address this issue, we generated knockout mice of beta 1,4-*N*-acetylgalactosaminyltransferase gene (Nagata et al. 1992) that is responsible for the synthesis of GM2 and GD2 (and GA2), and clearly showed that this enzyme is really critical for the synthesis of all complex gangliosides (Takamiya et al. 1996). Although the null mutant mice showed minor neurological disorders, their brain and nervous tissues were almost normally generated. In contrast, they showed aspermatogenesis (Takamiya et al. 1998) and immuno-dysfunctions. However, the null mutants demonstrated neurodegeneration with aging suggesting roles of complex gangliosides in the maintenance of integrity of nervous systems (Sugiura et al. 2005).

Procedure

Targeting vector was constructed using 7.5 kbp of genome fragment and a BlueScript-derived vector with DT-A by inserting neo-resistant gene into exon 4, and transfected into TT2 ES cell line derived from (B6xCBA) F1 (Takamiya et al. 1996). Homologous recombinants were obtained with PCR and then injected into blastcysts, resulting in the chimeric mice. Homozygotes were obtained by mating heterozygotes at about 25% ratio (Fig. 1).

Results

All complex gangliosides completely disappeared, and GM3 and GD3 accumulated at an extremely high level. The null mutant mice showed almost normal appearance and behavior. Pathological examination also revealed almost normal architecture of brains. Male mice turned out to be infertile due to aspermatogenesis. The reason was later clarified to be disordered testosterone transport. Reduced reaction of T lymphocytes to IL-2 was also reported by us (Zhao et al. 1997). Long-term observation revealed that the null mutant mice undergo neuronal degeneration in peripheral nerves such as sciatic nerve and spinal cords. Cerebellum also showed atrophic changes and deletion of Purkinje neuron, suggesting complex gangliosides are critical to maintain the integrity of nervous tissues.

Department of Biochemistry II, Nagoya University Graduate School of Medicine, 65 Tsurumai, Showa-ku, Nagoya 466-0065, Japan
Phone: +81-52-744-2070, Fax: +81-52-744-2069
E-mail: koichi@med.nagoya-u.ac.jp

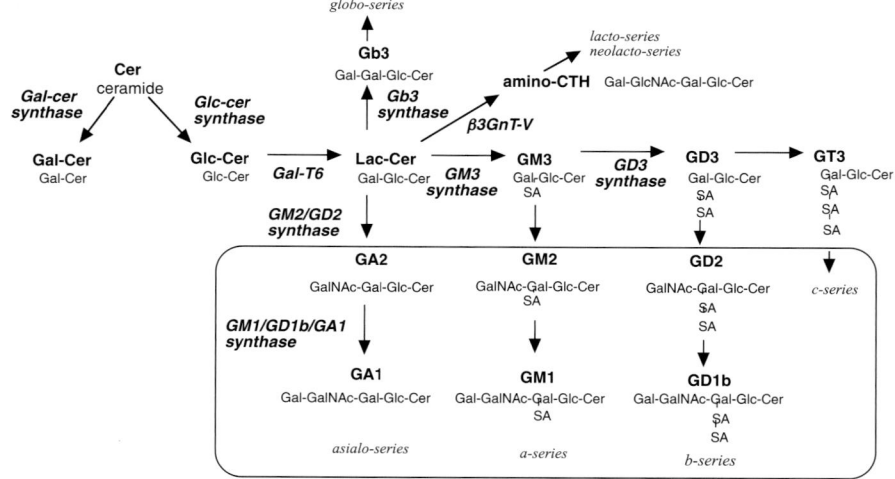

Fig. 1 Synthetic pathway of gangliosides showing that the disruption of GM2/GD2 synthase resulted in the complete loss of complex gangliosides

Protocols for Gene Targeting

1. Mouse β1,4GalNAc-T genome was cloned from BALB/c mouse genomic library using a 2.1-kb *Xba*-I fragment of mouse cDNA clone pTm3–5 as a probe.
2. A targeting plasmid was constructed containing a neomycin-resistant gene inserted into the exon.
3. The targeting vector (24 nM) was linearized with *Not*-I and was mixed with ES cell suspension (1 × 10^7), then electroporated at 0.25 kV, 960 μF, using a Bio-Rad Genepulser.
4. Forty-eight hours after electroporation, G418 was added to the medium at the concentration of 150 μg/ml.
5. After 7–8 days, G418-resistant clones were isolated and subjected to screening for homologous recombination by PCR.
6. The sense primer was 5′-TCGTGCTTTACGGTATCGCCGCTCCCGATT-3′ in 3′ terminus of PGK neo, and the antisense primer was 5′-GGGTGTGGCGGCATACATCT-3′ in the intron of the β1,4GalNAc-T gene.
7. The reaction was started one cycle of 95°C (2 min), 55°C (1 min), 74°C (5 min), thereafter 35 cycles of 94°C (1 min), 60°C (30 s), 74°C (1.5 min) were used. Homologous recombinant clones gave a 1.1-kb fragment.
8. Homologous recombinant ES lines were injected into 8-cells germ or blastcysts to obtain chimeric mice.

Comments

PCR reaction for the screening should be confirmed for the minimum limit of DNA amount to be analyzed. Sites for the insertion of neoresistant gene should be determined by considering the orientation of the enzyme product to completely disrupt the activity.

References

Nagata Y, Yamashiro S, Yodoi J, Lloyd KO, Shiku H, Furukawa K (1992) Expression cloning of β1,4 N-acetyl-galacto-saminyltransferase cDNAs that determine the expression of GM2 and GD2 gangliosides. J Biol Chem 267:12082–12089

Sugiura Y, Furukawa K, Tajima O, Mii T, Honda T, and Furukawa K (2005) Sensory nerve-dominant nerve degeneration and remodeling in the mutant mice lacking complex gangliosides. Neuroscience 135:1167–1178

Takamiya K, Yamamoto A, Furukawa K, Yamashiro S, Shin M, Okada M, Fukumoto S, Haraguchi M, Takeda N, Fujimura K, Sakae M, Kishikawa M, Shiku H, Furukawa K, Aizawa S (1996) Mice with disrupted GM2/GD2 synthase gene lack complex gangliosides, but exhibit only subtle defects in their nervous system. Proc Natl Acad Sci USA 93:10662–10667

Takamiya K, Yamamoto A, Zhao J, Furukawa K, Yamashiro S, Okada M, Haraguchi M, Shin M, Takeda N, Kishikawa M, Shiku H, Aizawa S, Furukawa K (1998) Complex gangliosides are essential in spermatogenesis of mice: possible roles in the transport of testosterone. Proc Natl Acad Sci USA 95:12147–12152

Zhao J, Furukawa K, Fukumoto S, Okada M, Miyazaki H, Shiku H, Aizawa S, Matsuyama M, Furukawa (1999) Attenuation of the interleukin 2 signals in complex ganglioside-lacking mice. J Biol Chem 274:13744–13747

KO Mice of α-2,8-sialyltransferase (GD3 Synthase)

Koichi Furukawa, Masahiko Okada, Keiko Furukawa

Introduction

Gangliosides have been considered to play important roles in the development and differentiation of nervous systems in vertebrates. In particular, b-series gangliosides were reported to be effective as a neurotrophic factor, e.g. as inducer of neurite extension. In order to directly address these functions, we generated knockout mice of α-2,8-sialyltransferase gene (Haraguchi et al. 1994) that is responsible for the synthesis of GD3 (and GT3), and clearly showed that this enzyme is really critical for the synthesis of b-series gangliosides (Okada et al. 2002). The null mutant mice showed no apparent neurological disorders, and their brain and nervous tissues were almost normally generated. They showed no defects in spermatogenesis and immunological examination. However, the null mutants demonstrated reduced neuroregeneration in the hypoglossal resection experiments, suggesting that b-series gangliosides are important in the repair of lesioned nerves (Okada et al. 2002).

Procedure

Targeting vector was constructed using a BlueScript-derived vector with DT-A by inserting neo-resistant gene into exon 1, and transfected into TT2 ES cell line derived from (B6xCBA) F1 (Okada et al. 2002). Homologous recombinants were obtained with PCR and then injected into blastcysts, resulting in the chimeric mice. Homozygotes were obtained by pairing heterozygotes at about 25% ratio (Fig. 1).

Results

All b-series gangliosides completely disappeared, and a-series gangliosides such as GM1 and GD1a accumulated at high levels. The null mutant mice showed almost normal appearance and behaviors. Pathological examination also revealed almost normal architecture of brains. Male mice are fertile. Apoptotic reactions via Fas/Fas-L were almost equivalent between the wild type and the null mutant mice, suggesting no reduced susceptibility to Fas-induced apoptosis despite the lack of GD3 synthesis (Okada et al. 2002). Even long-term observation revealed no neuronal degeneration in the null mutants. Therefore, the results of hypoglossal nerve resection/regeneration experiments suggested that strictly regulated ganglioside composition is needed for the correct regeneration of lesioned nerves. These results suggested that some carbohydrate groups of glycolipids differentially regulate various sensory nerves as demonstrated (Handa et al. 2005).

Department of Biochemistry II, Nagoya University Graduate School of Medicine, 65 Tsurumai, Showa-ku, Nagoya 466-0065, Japan
Phone: +81-52-744-2070, Fax: +81-52-744-2069
E-mail: koichi@med.nagoya-u.ac.jp

KO Mice of α-2,8-sialyltransferase (GD3 Synthase)

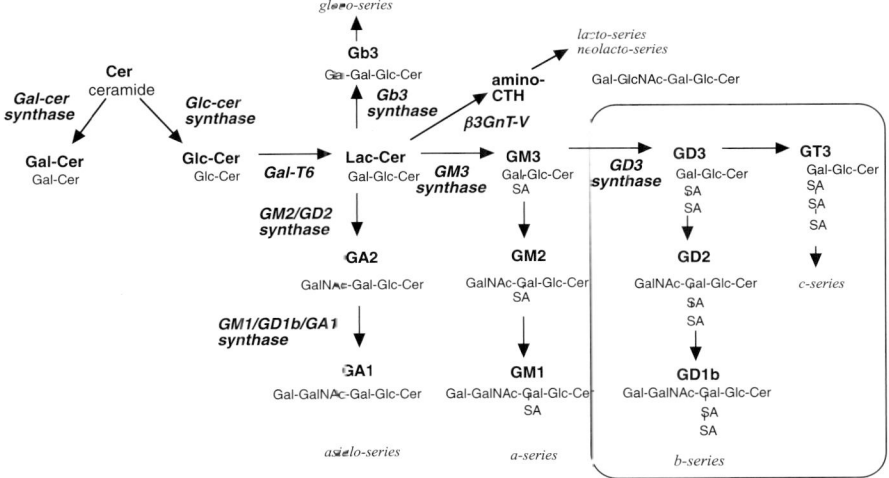

Fig. 1 Synthetic pathway of gangliosides showing that the disruption of GD3 synthase resulted in the complete loss of b-series gangliosides

Protocols for Gene Targeting

1. Chromosomal GD3 synthase gene was isolated from λgt11 phage library using GD3 synthase cDNA (pD3T-31) and mapped.
2. Neo^r gene was inserted between the *Bal*I and *Acc*I sites in exon 1 of the gene, and a 9.5-kb gene fragment was used as a targeting vector.
3. Diphtheria toxin A gene was attached to eliminate nonhomologous recombinants.
4. Homologous recombination was confirmed by Southern blotting, generating 4.5- and 2.8-kb fragments by *Bam*HI and 4.6- and 3.1-kb fragments by *Hin*dIII digestion in the wild-type and the recombinant allele, respectively.
5. Homologous recombinant ES cells were injected into 8-cell germ or blastcysts to generate chimeric mice.

Comments

PCR reaction for the screening should be confirmed from the minimum DNA to be analyzed. Sites for the insertion of neo-resistant gene should be determined by considering the orientation of the enzyme product to completely disrupt the activity.

References

Handa Y, Ozaki N, Honda T, Furukawa K, Tomita Y, Inoue M, Okada M, Furukawa K, Sugiura Y (2005) GD3 synthase gene knockout mice exhibit thermal hyperalgesia and mechanical allodynia but decreased response to formalin induced prolonged noxious stimulation. Pain 117:271–279

Haraguchi M, Yamashiro S, Yamamoto A, Furukawa K, Takamiya K, Lloyd KO, Shiku H, Furukawa K (1994) Isolation of GD3 synthase gene by expression cloning of GM3 alpha-2,8-sialyltransferase cDNA using anti-GD2 monoclonal antibody. Proc Natl Acad Sci USA 91:10455–10459

Okada M, Itoh M, Haraguchi M, Okajima T, Inoue M, Ohishi H, Matsuda Y, Iwamoto T, Kawano T, Fukumoto S, Miyazaki H, Furukawa K, Aizawa S, Furukawa K (2002) b-Series ganglioside deficiency exhibits no definite changes in the neurogenesis and the sensitivity to Fas-mediated apoptosis, but impairs regeneration of the lesioned hypoglossal nerve. J Biol Chem 277:1633–1636

Double KO Mice of β-1,4-N-acetylgalactosaminyltransferase (GM2/GD2 Synthase) and α-2,8-sialyltransferase (GD3 Synthase)

Koichi Furukawa, Orie Tajima, Yusuke Ohmi, Keiko Furukawa

Introduction

Gangliosides have been considered to play important roles in the development and differentiation of nervous systems in vertebrates. They have been also considered to have neurotrophic activity. In order to directly address these biological functions, we generated knockout mice of β-1,4-N-acetylgalactosaminyltransferase gene (Takamiya et al. 1996) that is responsible for the synthesis of GM2 and GD2 (and GA2), and those of α-2,8-sialyltransferase gene (Okada et al. 2000) that is responsible for the synthesis of GD3. These KO mice showed relatively mild abnormal phenotypes than expected. This seemed due to the compensatory effects of the remaining glycolipids in the individual KO mice (Furukawa et al. 2004). Therefore, we mated these two KO mice in order to generate double KO mice in which only GM3 should remain. Now we can largely eliminate the compensatory effects of remaining glycolipids and correctly observe the effects of ganglioside deficiency. DKO mutants were born with almost normal appearance, and grew up for a while after birth. However, they demonstrated marked neurodegeneration from early stages of life, and various abnormal phenotypes probably caused by neurodegeneration (Inoue et al. 2002).

Procedure

Targeting vectors for the individual single KO were described in their items (Takamiya et al. 1996; Okada et al. 2002). Screening of the homologous recombinants was also performed with PCR with the same procedure. Mating of heterozygotes of two KO mice resulted in the generation of DKO mice at a low ratio (theoretically 1:16). Later, DKO mice were generated by mating homozygous KO female of alpha-2,8-sialyltransferase gene and heterozygous male of β-1,4-N-acetylgalactosaminyltransferase gene to raise the chance to obtain DKO (Fig. 1).

Results

All gangliosides except for GM3 completely disappeared in the null mutant mice. Expression levels of GM3 did not change during the observation time in both brain and skin tissues. After 12 weeks of birth, the null mutants suddenly die gradually. They showed

---

Department of Biochemistry II, Nagoya University Graduate School of Medicine, 65 Tsurumai, Showa-ku, Nagoya 466-0065, Japan
Phone: +81-52-744-2070, Fax: +81-52-744-2069
E-mail: koichi@med.nagoya-u.ac.jp

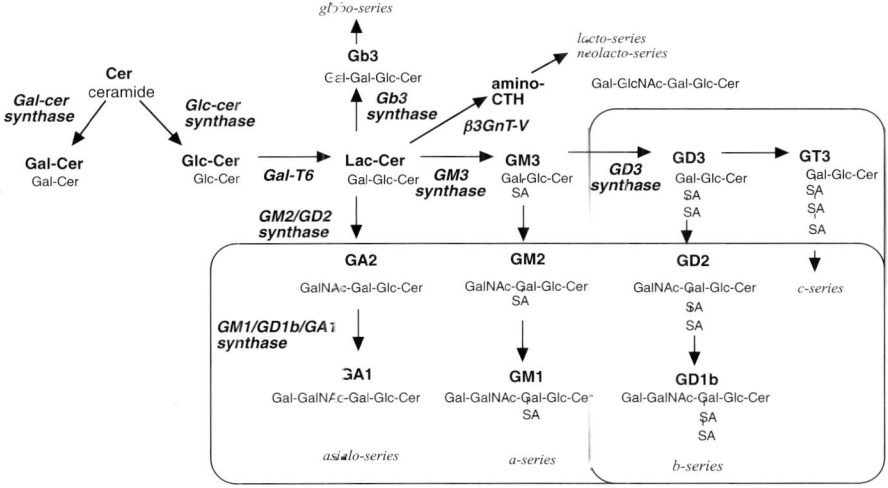

Fig. 1 Synthetic pathway of gangliosides showing that the disruption of GM2/GD2 synthase and GD3 synthase resulted in the complete loss of all gangliosides except GM3

skin lesions at face and neck. They appeared first around eyes and gradually expanded. Frequent scratching by themselves seemed to exacerbate the lesions. Neurodegeneration was found in sciatic nerves and trigeminal nerves at even 4-week-old mice, and might cause reduced sensory function, leading to frequent scratching. The null mutants showed reduced memory and learning, suggesting intrinsic circumstances were not well maintained in the absence of gangliosides. Presently, mechanisms for the regulation of nervous systems with gangliosides in vivo were under investigation.

Protocols

1. Generation of the DKO mice was primarily performed by mating heterozygotes of GM2/GD2 synthase and those of GD3 synthase. To efficiently obtain DKO, heterozygotic male mice of GM2/GD2 synthase and homozygotic female of GD3 synthase were mated.
2. Isolation of mice in metabolic cages to induce wounds: seven each of Wd/Wd-type and Ho/Ho-type 25-week-old male mice were housed in special cages usually used for metabolic study and were fed regular chow.
3. The mice were daily observed for 2 months and checked for body weight, generation of wounds, and frequency of skin scratching. Scratching measurements were performed for 5 min twice a week.

Comments

The same kind of DKO generated in NIH showed auditory seizure, which was not at all observed in our mutant mice. This difference in the phenotype of the same DKO might be due to different genetic back grounds.

References

Furukawa K, Tokuda N, Okuda T, Tajima O, Furukawa K (2004) Glycosphingolipids in engineered mice: insights into function. Semin Cell Dev Biol 15:389–396

Inoue M, Fujii Y, Furukawa K, Okada M, Okumura K, Hayakawa T, Furukawa K, Sugiura Y (2002) Refractory skin injury in the complex knock-out mice expressing only GM3 ganglioside. J Biol Chem 277:29881–29888

Okada M, Itoh M, Haraguchi M, Okajima T, Inoue M, Ohishi H, Matsuda Y, Iwamoto T, Kawano T, Fukumoto S, Miyazaki H, Furukawa K, Aizawa S, Furukawa K (2002) b-Series ganglioside deficiency exhibits no definite changes in the neurogenesis and the sensitivity to Fas-mediated apotosis, but impairs regeneration of the lesioned hypoglossal nerve. J Biol Chem 277:1633–1636

Takamiya K, Yamamoto A, Furukawa K, Yamashiro S, Shin M, Okada M, Fukumoto S, Haraguchi M, Takeda N, Fujimura K, Sakae M, Kishikawa M, Shiku H, Furukawa K and Aizawa S (1996) Mice with disrupted GM2/GD2 synthase gene lack complex gangliosides, but exhibit only subtle defects in their nervous system. Proc Natl Acad Sci USA 93:10662–10667

Part 4
Infrastructures and Research Resources

Section XXI
Glycosyltransferases and Related Genes, and Useful Cell Lines

Cell Lines for Glycobiology

Mitsuru Nakamura

Introduction

Cell culture, one of the most important technologies in glycobiological research, is discussed in this section with particular focus on mammalian cells.

Meaning of Cell Culture

In cell biology, one tries to understand life process in scientific ways. Scientific approaches involve analyzing the life process in a simple and controlled system in vitro. One of the most important technical accomplishments in biology is the cell culture technology; taking cells out of a complicated multi-cellular living body and growing them in vitro. A specific cell is separated from the influences of the body, the multi-cellular system. The first example of success was in 1907, when R. Harrison observed outgrowth of axons in tissue culture experiments. With advances in cell culture technology, one can obtain cells of one kind in large quantities, research the function of various cells, induce cells to differentiate into various lineages, and have cells respond to bioactive substances.

Glycobiology is also developing greatly in response to the benefits of cell culture technology. Now, with advances in cell culture technology, the function of a specific oligosaccharide can be analyzed at the cellular level. Notably, progress in the molecular cloning of glycosyltransferase genes or lectin genes enables one to artificially control the expression level of a sugar chain-related gene at the cellular level. The time is now right for researching the functions of a specific sugar chain.

Cell Culture Technologies

There are two types of cell culture: "primary cultures" using cells obtained from the living body, and "secondary cultures" using cells that were once cultured and then frozen. In primary cultures, cells are separated from each other by digesting, if required, cell adhesion-related proteins using proteases and suspended in a culture medium containing EDTA as a single cell suspension. Cells in the suspension are sometimes separated using a fluorescence-assisted cell sorter (FACS) and a centrifugal separation and fractionation (counter-current centrifugal elutriation) method. There are two primary culture systems; "mass cultures" and "clonal cell cultures". Colonies (clones) originating from single cells are formed with the latter method. Normal cells usually become impossible to culture

---

Cell Regulation Analysis Team, Research Center for Medical Glycoscience, National Institute of Advanced Industrial Science and Technology;
Function of Biomolecule. Medical Science for Control of Pathological Processes, Tsukuba University, Central-2, 1-1-1 Umezono, Tsukuba, Ibaraki 305-8568, Japan
Phone: +81-29-861-2745. Fax: +81-29-861-2744
E-mail: owl.nakamura@aist.go.jp

after several rounds of cell division and aging and are not necessarily suitable for research. We therefore generally use "cell lines" that have been transformed and can be divided and grown in vivo repeatedly without any restrictions. Cell lines are established from various kinds of animals.

Secondary cultured cells and cell lines are supplied by organized distributors or personally provided by other scientists who have established such cells. Distributors provide the cells on a commercial or non-commercial basis and include cell banks managed by local governmental, non-governmental, and non-profit organizations [Cell Bank and Gene Bank, The Japanese Tissue Culture Association, ISBN4-320-05488-1/ISBN978-4-320-05488-2; Riken Bioresource Catalogue, Riken Bioresource Center, JCRB/HSRRB Research Resource Catalogue, Human Science Research Resource Bank, http://cellbank.nibio.go.jp/; ATCC Cell Biology Catalogue, American Type Culture Collection (ATCC); ECACC Cell Lines Catalogue, European Collection of Animal Cell Culture (ECACC); DSMZ Online Catalogue, Deutsche Sammlung von Mikroorganismen und Zellkulturen GmbH (DSMZ)].

It is necessary to culture cells with additives that facilitate growth. Usually these additives are sera from mammals including humans, cows, sheep, and horses. The most common additive is fetal calf serum (FCS). On the other hand, in a "serum-free culture", cells are grown in the presence of only well-defined growth factors. Serum-free culture is a mandatory technology to investigate the function of a molecule. By adding the molecule to the cell culture medium, one tests the possible effects of the molecule on cell growth, apoptosis, cell motility, membrane trafficking, and so on. The culture media used in serum-free culture are called "serum-free media" (Barnes and Sato 1980). The most basic and essential supplements in serum-free media are insulin, transferrin, and selenium (ITS).

In glycobiological studies, one of the most useful cell culture-related technologies is a hybridoma that produces a monoclonal antibody (mAb) against the antigen epitope of a cell surface oligosaccharide. Most mAbs are commercially available and one does not necessarily have to obtain and culture the hybridoma. Sometimes, depending on our experiments, we may need to culture or prepare hybridoma with our own hands and produce mAb "in house" or we can alternatively ask services outside to prepare and order hybridomas.

Recently, genes encoding glycosyltransferases and lectins have been isolated. To investigate the functions of such glycogenes, a standard approach is to re-model artificially the machinery of sugar chain biosynthesis or sugar chain recognition. Using the over-expression technique, we can now knockdown glycogenes by RNA interference technology. The over-expression and knocking down can together be referred to as "sugar chain re-modeling" in the field of cell glycobiology. For introducing genes of interest, there are transfection methods using classical calcium phosphate, electroporation, and polycationic reagents including lipofectamine. Alternatively, methods using viral vector systems are preferable for hematopoietic cells, blood cells, and primary cultured cells in which gene transfection is known to be difficult. The most commonly used viral vectors are adenoviruses, retroviruses, and lentiviruses. The procedures of gene over-expression and knockdown using viral vectors are relatively complex. However, a powerful virus-based transfection technology is used in the fields of hematology and immunology, as the transfection efficiency is high and it is relatively easy to obtain transfected cells using selection markers including enhanced green fluorescence protein encoded by the

Cell Lines for Glycobiology

gene fused to a dependent promoter or IRES sequence of the vector. A protocol in which we are currently using lentiviral short hairpin RNA (shRNA) is presented in this section. In the very competitive non-glycobiological fields, the use of a dominant-negative mutant of the gene of interest is one of the most common techniques. As its effect on suppression of the gene function is clearer than that of gene knockdown, one can expect that a search for dominant-negative mutants shall start with important glycosyltransferase genes.

Cell culture and gene transfection-related technologies specific for glycobiological studies may not exist. However, a number of cell lines indispensable in glycobiology have been established. Typical examples are the Lec cells established by Dr. P. Stanley, 3LL cells of Dr. J. Inokuchi, and glycolipid-deficient cells of Dr. Y. Hirabayashi. They have been used for the isolation of glycogenes. Glycobiological studies will progress steadily with the establishment of cell lines deficient in or over-expressing glycogenes. Attempts to construct and maintain a glycobiology-related cell bank and distribute cell lines to researchers will become necessary in the future.

Preparation and Packaging of a Lentiviral shRNA Vector

1. Digest with an appropriate restriction enzyme and dephosphorylate a shRNA-expression lentiviral vector that contains an MCS connected with the downstream of the U6 promoter and an EGFP-coding sequence driven by the CMV promoter.
2. Anneal chemically synthesized forward and reverse oligonucleotides containing sequences corresponding to siRNA, phosphorylate both termini of the annealed oligonucleotide, and insert the oligonucleotide into the vector prepared above.
3. Inoculate 293FT cells in 20 mL of D'MEM supplemented with 10% FCS at a density of 1×10^5 cells/mL in a 75 cm^2 flask. ⟨day 0⟩
4. Transfection ⟨day 3⟩
    (1) Add 80 µL of lipofectamine 2000 to 3 mL of opti-MEM and leave at RT for 5 min.
    (2) Mix the lentiviral shRNA vector, packaging vectors (gag/pol, rev, and vsv-g), and opti-MEM (3 mL).
    (3) Mix gently (1) and (2) and leave at RT for 20 min.
    (4) Replace the medium of 293FT cells with 10 mL of fresh opti-MEM supplemented with 10% FCS.
    (5) Drop (3) into (4) and mix.
    (6) Culture the cells for 24 h in a CO_2-incubator.
5. Replace 10 mL of medium with 10 mL of fresh D'MEM supplemented with 10% FCS and culture the cells for 3 days. ⟨day 4⟩
6. Harvesting the lentivirus ⟨day 5⟩
    (1) Harvest the culture supernatant and clean it up using a centrifuge and 0.8 µm filter.
    (2) Spin down lentivirus particles using an ultracentrifuge at 68,000 × g for 90 min at 4°C.
    (3) Resuspend the particles in 25 µL of PBS(−) and clear the solution using a centrifuge (e.g. KUBOTA 3700) at 12,000 × g for 1 min.
    (4) Take out the supernatant and use it immediately for infection of the target cells.

7. Examine the cells for EGFP-positivity and purify the positive cells using a fluorescence-assisted cell sorter (FACS).
8. Characterize the cells or conduct functional assays.

Reference

Barnes D, Sato G (1980) Methods for growth of cultured cells in serum-free medium. Anal Biochem 102:258

Sugar Chain-Related Genes: Glycosyltransferases Expression System

Takashi Sato, Hisashi Narimatsu

Introduction

In the synthesis of glycans, it is best to apply organic chemistry if we require a large amount of single structure, while an enzymatic synthesis using various glycosyltransferases with different substrate specificities is a suitable method when we need a variety of glycans in small quantity. However, commercially available glycosyltransferases are limited in variety which then restricts the types of synthesizable glycans. The structural and functional analyses of glycans involve the syntheses of various glycan structures, hence it is vital to have a variety of glycosyltransferases.

As part of NEDO project, so-called "GlycoGene Project", we comprehensively produced and analyzed human novel glycogenes using bioinformatics technology. A glycogene mentioned here represents glycosyltransferase, molecular chaperon associated with glycan synthesis, sulfotransferase or nucleotide–sugar transporter genes. Among these genes, we mainly cloned glycosyltransferases and sulfotransferases of type-II membrane protein with transmembrane domain at N-terminus into GATEWAY®-entry vectors developed by Invitrogen. More than 90% of glycosyltransferases are the type-II membrane protein and are present at Golgi apparatus. We therefore removed transmembrane domain and obtained only catalytic domain using PCR, and cloned them into an entry vector in order to express them as a soluble enzyme. We call these resultant constructs Entry clones. GATEWAY cloning technology, which uses λ Phage-mediated site-specific recombination, enables us to easily transfer the gene of interest from Entry clone to Destination Vector. Glycosyltransferases can be easily expressed in various expression systems including *E. coli*, yeast, insect cells (Sf21, Hi5) and mammalian cells (293T, CHO, COS) if we prepare GATEWAY®-compatible destination vectors in advance. Besides the above system, tagged glycosyltransferases can also be expressed if we prepare tagged destination vectors such as FLAG and HIS in advance. Using these systems, we tried to express human glycosyltransferases in various expression systems and compared these enzymes by measuring the quantity of expressed proteins and enzyme activities.

In most glycosyltransferases, they showed the highest activity per given amount of protein when expressed in 293T cells derived from human embryonic kidney cell. It is not surprising that human enzymes work better when expressed in human cells. Not all enzymes, however, follow the above expression pattern. Even if expressed in insect cells and yeast, there are cases where the enzymes have similar enzymatic activities as those expressed in 293T cells. Therefore, it is important for us to evaluate the most suitable expression system for each enzyme. Furthermore, the expression systems of insect cells

---

Research Center for Medical Glycoscience, National Institute of Advanced Industrial Science and Technology (AIST), 1-1-1 Umezono, Tsukuba, Ibaraki 305-8568, Japan
E-mail: takashi-sato@aist.go.jp

and yeast can be expressed in large quantities and produced at a relatively low price even though their enzymatic activities are low. Therefore, it is also important to select the suitable expression system by balancing the cost performance of interest. On the contrary, in the case of mammalian cells, improvements such as the selection of high-expression clone and the adaptation for serum-free medium would be possible. Using the above systems, we succeeded in transferring more than 100 types of human glycosyltransferases into libraries in the form of active enzymes. We have been synthesizing human glycans using these enzymes.

We are also planning to increase the number of the available entry clones. The list of entry clones of glycosyltransferase genes as well as sequences of cloned genes and used primers will be managed and updated in GlycoGene DataBase(GGDB).

Procedures

Construction of Glycosyltransferase Expression Vector

Materials

Entry clone of glycosyltransferase (we used pENTR/D-TOPO (Invitrogen) vector containing catalytic domain of glycosyltransferase.)

Destination Vector for Expression

In this section, we used pFLAG-CMV3-DEST vector derived from pFLAG-CMV3 (Sigma-Aldrich). Destination vector can be introduced from your favorite vectors using Gateway® vector conversion system (Invitrogen).

Gateway® LR Clonase® Enzyme Mix (Invitrogen).

Expression of Recombinant Glycosytransferase

Materials

293T cells (available from ATCC), maintained in DMEM containing 10% FCS, penicillin, and streptomycin.
LipofectAMINE 2000 (Invitrogen)
Opti-MEM® I (Invitrogen)
Anti-FLAG M2 Agarose Affinity Gel (Sigma-Aldrich)
50 mM Tris-buffered saline (50 mM Tris–HCl, pH 7.4, and 150 mM NaCl)

Methods

1. One day before transfection, seed 2.0×10^6 of 293T cells in 10 ml of DMEM containing 10% FCS without antibiotics on 10 cm dish. Cells will be near confluent the next day.
2. Dilute 30 μg of DNA in 1.5 ml of Opti-MEM®I and mix gently.
3. Dilute 30 μl of Lipofectamine 2000 in 1.5 ml of Opti-MEM®I and mix gently. Incubate the mixture for 5 min at room temperature.
4. Mix the diluted DNA solution with diluted Lipofectamine 2000 solution (total volume = 3 ml). Mix gently and incubate for 20 min at room temperature.

5. Drop the mixed solution on confluent 293T cells. Mix gently and incubate for 48–72 ho at 37°C in a CO_2 incubator.
6. Recover the culture media and remove the cell debris by centrifugation. Mix with anti-FLAG M2 agarose affinity gel and incubate for O/N at 4°C with gentle agitation by rotary shaker.
7. Wash the enzyme–agarose gel mixtures three times with 50 mM Tris-buffered saline.
8. The enzyme–agarose gel mixtures can be suspended for enzyme reaction buffer and used for enzymatic glycan synthesis as enzyme sources.

Section XXII
Sugar Library

GALAXY Database and Pyridylaminated Oligosaccharide Library

Koichi Kato, Noriko Takahashi

Introduction

A wide array of standard oligosaccharides is a valuable glycomics tool for systematic and detailed analyses of structures and functions of sugar chains. The multi-dimensional HPLC mapping method developed by Takahashi and co-workers enables us to isolate N-linked oligosaccharides as pyridylaminated (PA) derivatives, discriminating isomeric structures. Therefore it becomes extremely useful not only for identification and profiling of N-glycans but also for purification and library formation of PA-oligosaccharides (Tomiya et al. 1998; Takahashi et al. 2007). Based on the accumulated 2-D/3-D HPLC data for more than 500 different structures of N-glycans, we developed a web application, GALAXY (*G*lyco*a*na*l*ysis by the three *a*xes of MS and chromatograph*y*), for effective utilization of the multi-dimensional HPLC map (Takahashi and Kato 2003).

Concept

In the 2-D/3-D HPLC mapping method, N-glycans are typically released by use of glycoamidase A from the protein portion of a natural glycoprotein (typically ca. 1 mg) and fluorescently labeled with 2-aminopyridine at their reducing ends. These PA-oligosaccharides are sequentially separated by HPLC using three different columns, viz., anion-exchange (DEAE), reversed-phase (ODS), and normal phase (amide-silica) columns (Fig. 1), calibrated individually by use of PA-derivatized isomalto-oligosaccharide mixture. On the basis of the elution positions on the HPLC columns, which are expressed in glucose unit (GU) values for standardization, the structures of PA glycans are estimated in comparison with reference to PA-oligosaccharides. The sample PA-oligosaccharide and one of the candidate references are co-injected into HPLC columns to confirm their identity by observing a single peak. Furthermore, trajectories of elution positions on the map are followed upon glycosidase and/or glycosyltransferase treatments of the sample PA-glycans for further confirmation. Since detection is based on fluorescence intensity from the PA moieties, this method offers quantitative glycosylation profiling at molecular, cellular, and tissue levels, which can be used, e.g., for diagnosis based on glyco-biomarkers.

The HPLC data for approximately 500 different N-glycan structures are available at the GALAXY web site (http://www.glycoanalysis.info/ENG/index.html), which contains information about molecular weight, sources, and references for each glycan (Fig. 2). In this database, the N-glycans so far documented are expressed by a diagram, in which

---

Graduate School of Pharmaceutical Sciences, Nagoya City University, Tanabe-dori 3-1, Mizuho-ku, Nagoya 467-8603, Japan
Phone: +81-52-836-3447, Fax: +81-52-836-3447
E-mail:kkato@phar.nagoya-cu.ac.jp

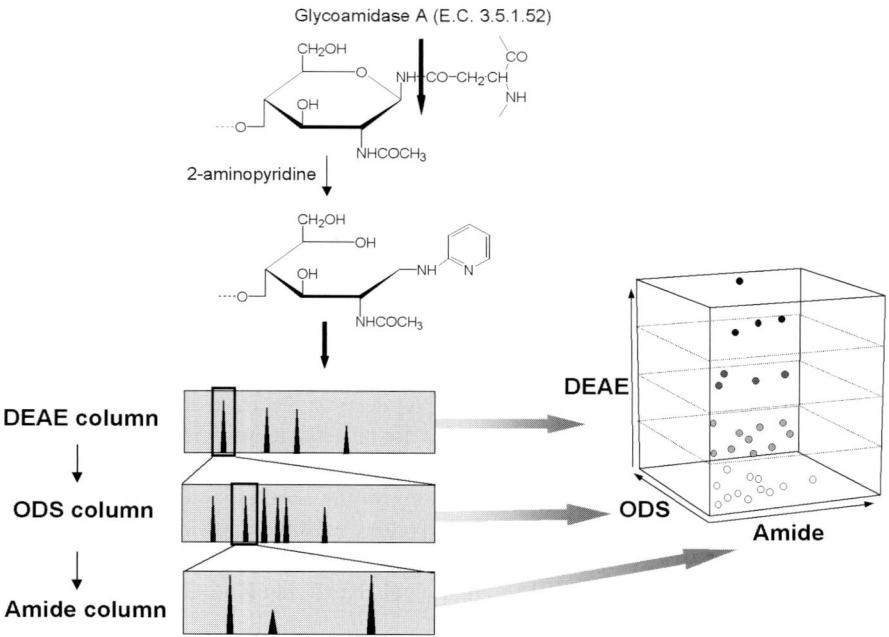

Fig. 1 Scheme of 2-D/3-D HPLC mapping method

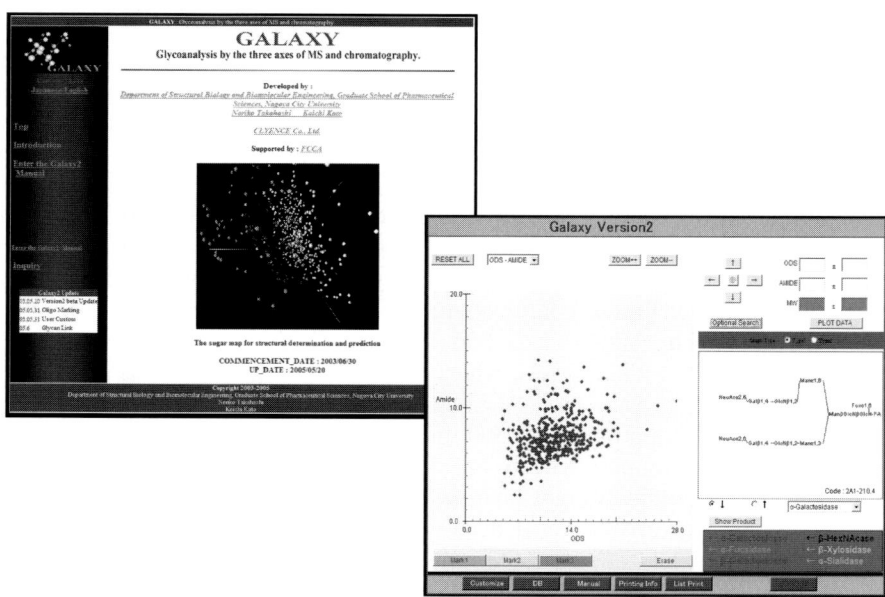

Fig. 2 A web application, GALAXY. User manual and list of PA-oligosaccharides are available at the GALAXY web site

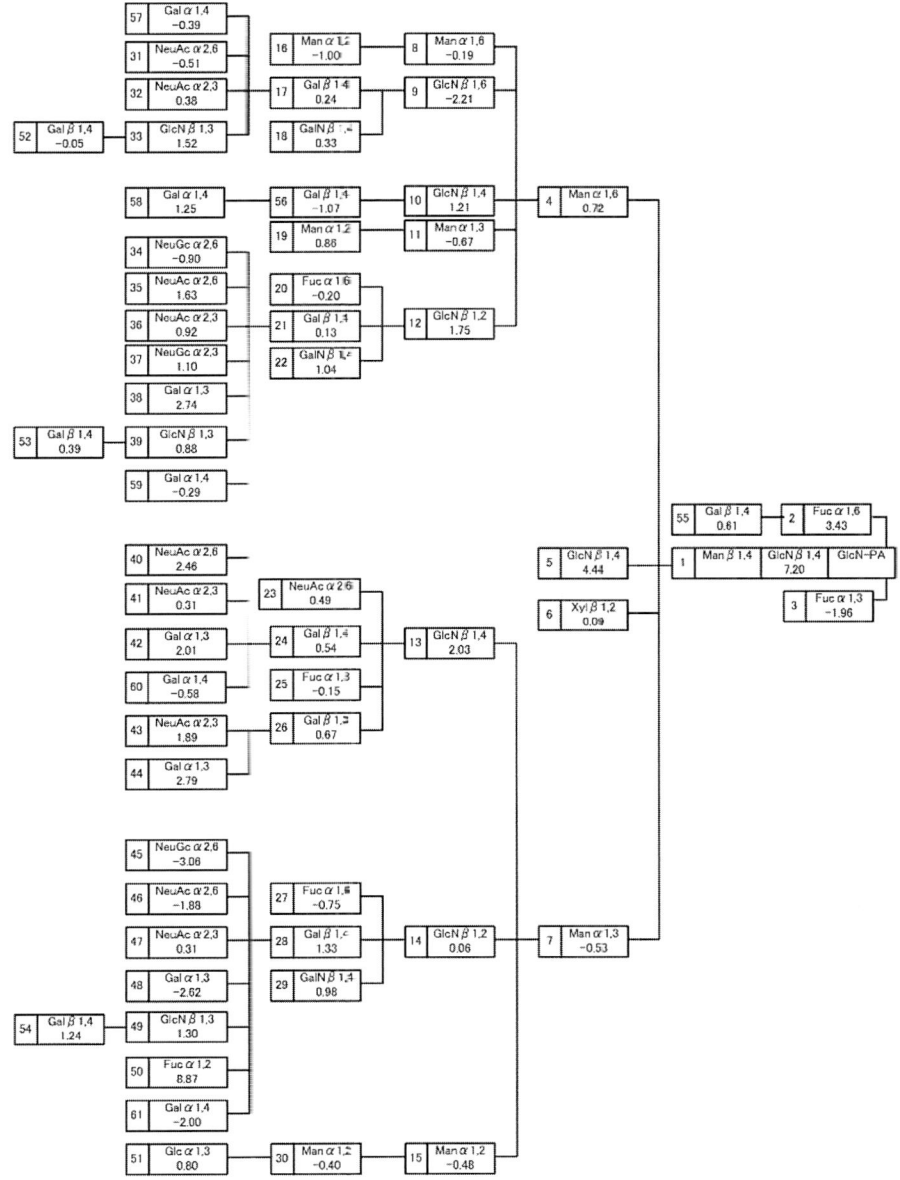

Fig. 3 Glycotree diagram with the UC values for GU on the ODS column

individual *N*-glycans are parts of the 'Glycotree' joining more than 60 different units of sugar residues (Fig. 3). Hence each PA-oligosaccharide is operationally connected to its precursors or the final product of a glycosidase treatment, which can be graphically indicated. The contribution of each component monosaccharide unit (unit contribution = UC) can be computed by linear multiple regression analysis on the assumption that the elution position of a given PA-glycan is represented by the sum of UC. Using the

Fig. 4 An example of code numbers for notation of N-linked oligosaccharides registered in the GALAXY database

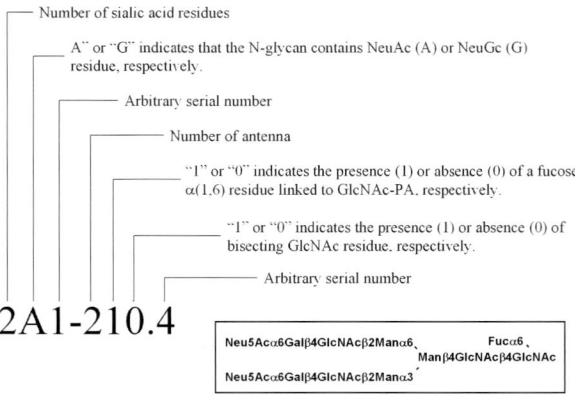

calculated UC data, GALAXY allows us to predict coordinates on the map even for putative PA-oligosaccharides resulting from glycosidase treatments.

In GALAXY, individual N-glycans are coded with alphanumeric characters (code numbers) such as M9.2 and 2A1-210.4 (Fig. 4). Using the GALAXY database as a guide, those PA-oligosacharides can be strategically purified from natural glycoprotein sources, and further derivatized by glycosidase and glycosyltransferase treatments. The sugar library thus constructed can be usefully applied to structural analyses by NMR and MS andthe characterization of sugar–protein interactions (Kamiya et al. 2005). The multi-dimensional HPLC data are still expanding. For example, an HPLC map of sulfated N-glycans has been reported (Yagi et al. 2005).

References

Kamiya Y, Yamaguchi Y, Takahashi N, Arata Y, Kasai K, Ihara Y, Matsuo I, Ito Y, Yamamoto K, Kato K (2005) Sugar-binding properties of VIP36, an intracellular animal lectin operating as a cargo receptor. J Biol Chem 280:37178–37182

Takahashi N, Kato K (2003) GALAXY (Glycoanalysis by the three axes of MS and chromatography): a web application that assists structural analyses of N-glycans. Trends Glycosci Glycotech 15:235–251

Takahashi N, Yagi H, Kato K (2007) The two/three dimensional HPLC mapping method for the identification of N-glycan structures. In: Kamerling H (ed) Comprehensive glycoscience. Elsevier, Amsterdam

Tomiya N, Awaya J, Kurono M, Endo S, Arata Y, Takahashi N (1988) Analyses of N-linked oligosaccharides using a two-dimensional mapping technique. Anal Biochem 171:73–90

Yagi H, Takahashi N, Yamaguchi Y, Kimura N, Uchimura K, Kannagi R, Kato K (2005) Development of structural analysis of sulfated N-glycans by multidimensional high performance liquid chromatography mapping methods. Glycobiology 15:1051–1060

Mass Production of Sugar Intermediates for the Synthesis of Functional Oligosaccharides

Yuji Matsuzaki

Introduction

To elucidate the functions of oligosaccharides, the chemical syntheses of many complicated oligosaccharides have already been undertaken. However, the synthesis of oligosaccharides takes a relatively long time, because of the many processes required to prepare the glycosyl acceptors or the glycosyl donors that are used for the glycosylation reactions. Therefore, we have established a system for supplying the large quantities of monosaccharide intermediates that are used in the chemical synthesis of oligosaccharides, and we expect that this will allow accelerated elucidation of the functions of oligosaccharides in the future.

Choice of the Anomeric Protecting Group for the Sugar Intermediates

The use of versatile intermediates is very important for various oligosaccharide syntheses. In addition, specific combinations of protecting groups are important to control regio- and stereo-selectivity in glycosylations, and an appropriate choice of protecting group at the anomeric center is particularly important for the successful outcome of oligosaccharide syntheses. Therefore, we have selected p-methoxyphenyl (pMP) glycosides, which can be prepared by Ogawa's method (Nakano et al. 1990; Matsuzaki et al. 1993), and which can be converted into the corresponding glycosyl donors by using Magnusson's procedure (Zhang and Magnusson 1996; Fig. 1).

Mass Production of Sugar Intermediates

We have recently succeeded in the mass production of several monosaccharide derivatives on the 5–100 kg scale by developing several manufacturing processes that use reaction facilities with capacities between 2 and 5 tons. Successful realization of the mass production of these raw materials has been possible due to our development of: (1) an efficient purification method that can be applied after massive-scale reactions, and which is compatible with actual production conditions, and (2) a method for the purification of the products without the need for column chromatography. Control of the crystallization process is the key technology that makes mass production possible (Fig. 2).

Glyco Synthetic Lab., Tokyo Chemical Industry Co., Ltd., 6-15-9 Toshima, Kita-ku, Tokyo 114-0003, Japan
Phone: +81-3-3919-5131, Fax: +81-3-5390-8387
E-mail: y-matsuzaki@tokyokasei.co.jp

Fig. 1 Conversion of *p*-methoxyphenyl glycosides into the corresponding glycosyl donors

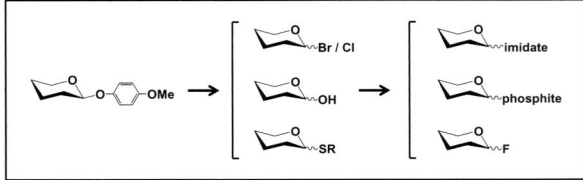

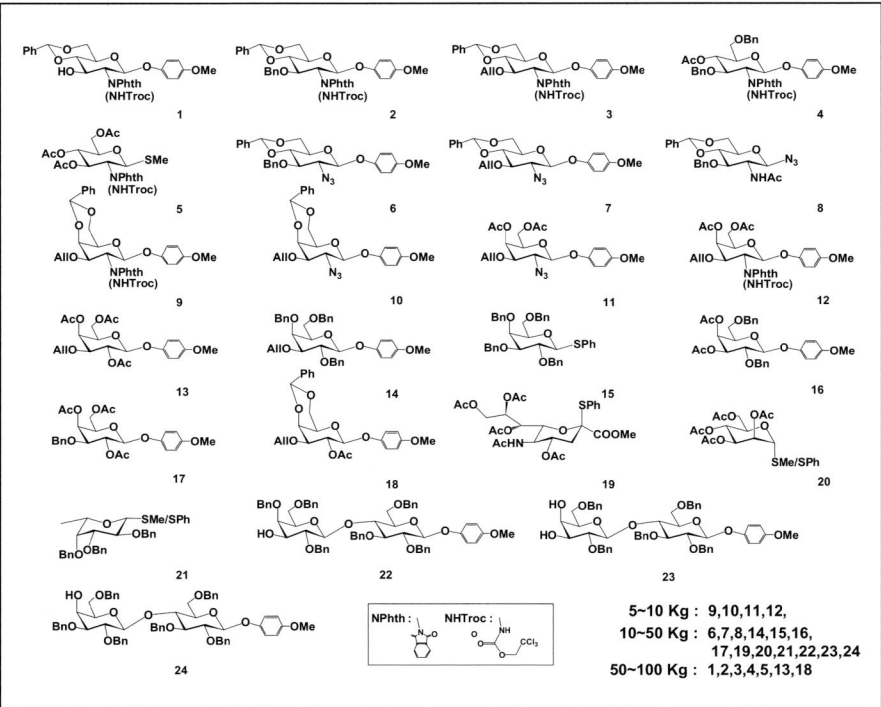

Fig. 2 Structures and quantities of mass-produced sugar intermediates. You can view the NMR data for these compounds on the Web site http://www.tci-asiapacific.com/product/bio-chem/glyco-chem/index.shtml

Disaccharide Building Blocks for the Syntheses of Oligosaccharides (Towards the Future)

Mass-produced sugar intermediates will be able to contribute to the mass production of the disaccharide building blocks that are found in many natural oligosaccharides, such as Gal-GlcN(β1–4, β1–3) and GlcN-Man(β1–2), or to the production of compounds that are difficult to construct stereo-selectively, such as Neu5Ac-Gal(α2–3, α2–6) and Man-GlcN(β1–4). Since it is believed that a variety of oligosaccharides can be synthesized by using these monosaccharide intermediates or disaccharide building blocks to complete sugar libraries, these building blocks should enable researchers to save a lot of time and labor (Fig. 3).

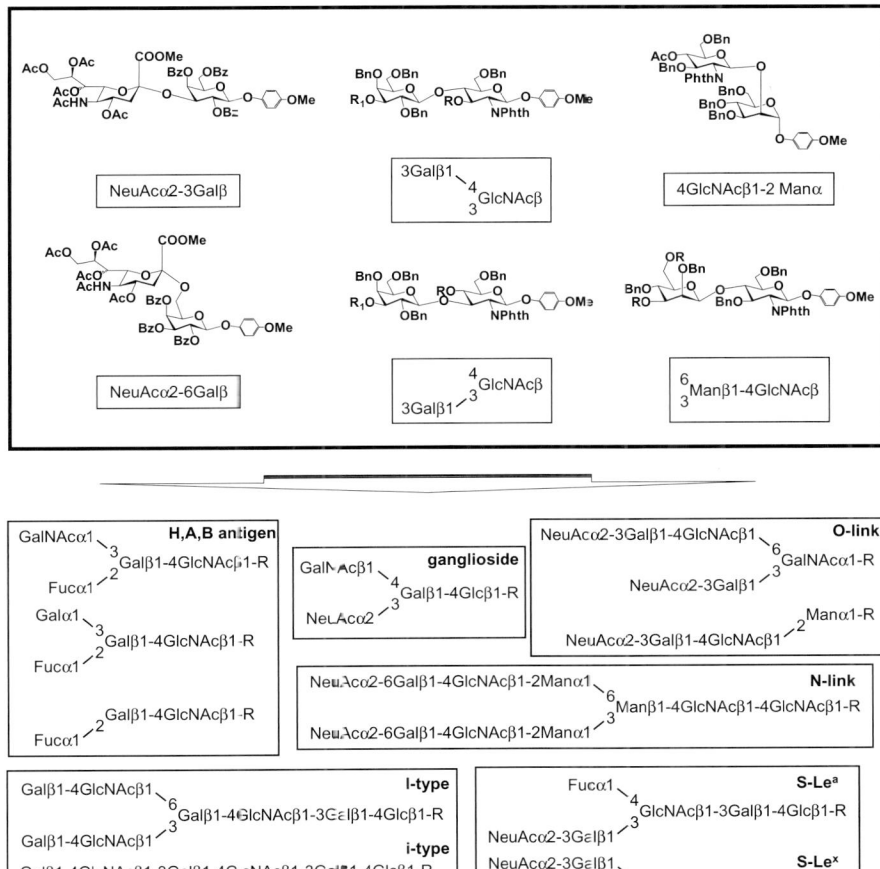

Fig. 3 Disaccharide building blocks for the syntheses of various glycoconjugates

Furthermore, manufacturing costs can be reduced by the mass production of these raw building blocks due to the merits of increased scale, and hence the price of chemically synthesized oligosaccharides should be reduced in the future.

This work is supported by the New Energy and Industrial Technology Development Organization (NEDO).

References

Matsuzaki Y, Ito Y, Nakahara Y, Ogawa T (1993) Synthesis of branched poly-N-acetyl-lactosamine type pentaantennary pentacosasaccharide: glycan part of a glycosyl ceramide from rabbit erythrocyte membrane. Tetrahedron Lett 34:1061–1064

Nakano T, Ito Y, Ogawa T (1990) Total synthesis of a sulfated glucuronyl glycosphingolipid,IV^{3}GlcA (3-SO$_3$)nLcOse$_4$Cer, a carbohydrate epitope of neural cell adhesion molecules. Tetrahedron Lett 31:1597–1600

Zhang Z, Magnusson G (1995) Conversion of p-methoxyphenyl glycosides into the corresponding glycosyl chlorides and bromides, and into thiophenyl glycosides. Carbohydr Res 295:41–55

Section XXIII
Database

A Database System for Glycogenes (GGDB)

Akira Togayachi, Kwon-Yeon Dae, Toshihide Shikanai, Hisashi Narimatsu

Significance of the Database

Glycogenes include genes of glycosyltransferase, sugar-nucleotide synthase, sugar-nucleotide transporter, sulfotransferase, etc. A total of 184 glycogenes are known to exist, and the authors collected the data of human glycogenes including the newly found ones in the "Construction of Glycogene Library" project (April 2001–March 2004) (Narimatsu 2004). There has been no database which comprehensively stores the information on human glycogenes. At present, over 184 genes of human glycosyltransferases and sulfotransferases are identified, cloned and expressed in various expression systems to analyze the activity for carbohydrate synthesis and its biological function. Therefore, in order to enable one-stop search of information necessary for the analysis of glycogenes, we constructed GlycoGene Database (GGDB) (Kwon 2004) which comprehensively includes all information of glycogenes obtained to date, and equips functions for analysis of homology and other purposes.

Current State of Public Databases in Glycobiology

Major public databases related to carbohydrates and glycogenes are CAZy Database (http://www.cazy.org/), KEGG Pathway Database (please see section 159 in this part) (KEGG, Kyoto Encyclopedia of Genes and Genomes, http://www.genome.jp/kegg/) Consortium for Functional Glycomics Database (http://www.functionalglycomics.org/), and others (please see other section in this part). CAZy database classifies carbohydrate genes of various living organisms into families according to resemblance of their sequence, and provides the EC number and accession number for each gene. Since KEGG pathway database provides pathways of biosynthesis and metabolism of glycans, the users can retrieve the relationship between glycogenes and structures of carbohydrate chains. Glycosylation pathways can be retrieved with Consortium for Functional Glycomics Database, and structures of carbohydrate chains can be obtained.

Present Status of GGDB

The objective of Glycogene Database (GGDB, http://ricdb.ibase.aist.go.jp/rcmg/ggdb/) (Fig. 1) is to create a database where entire information on glycogenes can be easily retrieved. In the GGDB, the following properties information of each glycogene are described and stored in a XML format: gene name, enzyme name, DNA sequence, tissue expression, substrate specificity, homologous gene, EC number, and external link to various databases. It graphically displays the information on substrate specificities, etc.

Research Center for Medical Glycoscience (RCMG), National Institute of Advanced Industrial Science and Technology (AIST), OSL Central 2, 1-1-1 Umezono, Tsukuba, Ibaraki 305-8568, Japan
E-mail: a.togayachi@aist.go.jp

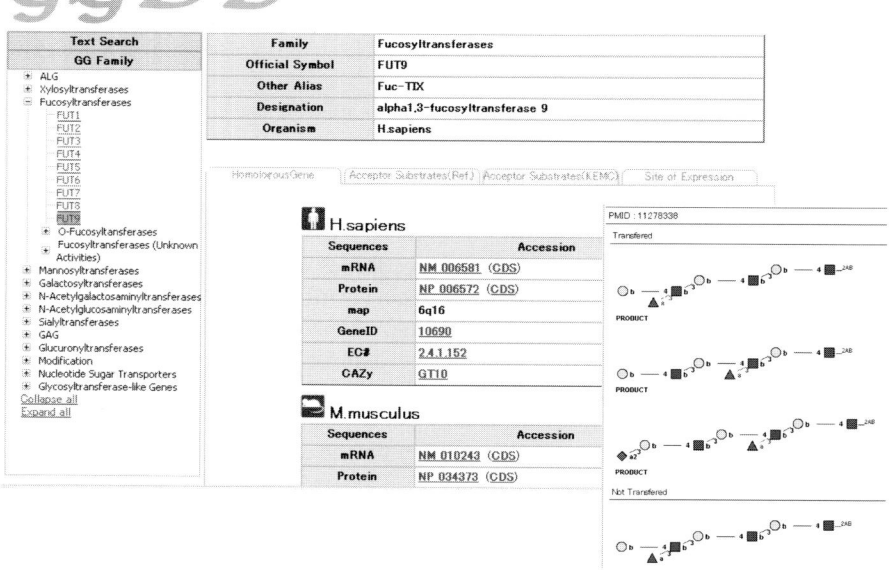

Fig. 1 Website of glycogene database (GGDB)

Future Perspectives

It is expected that the database greatly contributes to improve the efficiency of analysis required for progress in research of glycogenes in the future. With the development in the research on structural analysis of carbohydrate chains, research of synthesis of carbohydrate chains is currently making progress. In addition, GGDB will be updated to provide more integrated functions, for example, to display the information of enzymatic reaction and to link to mass spectrometry database of glycans. We are planning to add other property information to GGDB based on feedback from users, and to expand the interface in the future. In addition, we are undertaking the research on the data stored in GGDB, including the analysis of rules existing glycogenes, using data mining technique.

Protocols

In Silico Searching of Glycosyltransferase Genes for GGDB

We had developed a bioinformatics system for the identification and in silico cloning of human glycogenes (Narimatsu 2004). Candidates for glycosyltransferases are selected based on several parameters. For example, our system uses profile hidden markov model (HMM) method for clustering glycosyltransferase families. The parameters for glycogenes are acquired as follows;

1. For example, to annotate open reading frames (ORFs), EST sequences are assembled with Phrap, and a gene region (ORF) is predicted with GENSCAN software.

2. Motif search is executed to identify candidate sequences which possess the motifs of each glycosyltransferase subfamily. These motifs are determined using multiple EM software for motif elicitation. The motif information is also acquired from other databases such as Pfam (http://pfam.sanger.ac.uk/), PROSITE (http://www.expasy.ch/prosite/), MOTIF Search (http://motif.genome.jp/), etc.

3. The transmembrane domain which is localized at the N-terminal end, is identified as a region which contains 18–22 hydrophobic amino acid residues.

4. DXD (DXH) motif, essential for divalent cation binding, is identified. The DXD motif interacts with phosphate groups of sugar–nucleotide donor substrate through coordination of a divalent cation such as Mn^{2+}. However, some of glycosyltransferases which require no cation for their activities have no DXD (DXH) motifs.

5. The stem region, which is located between a transmembrane domain and a catalytic domain, is identified. The stem region has proline/serine/threonine-rich sequences.

6. The localization of cysteine residues in the sequence is also an important factor.

The information of cloned genes was stored in GGDB.

How to Use GGDB

Browsing Information in GGDB

All the present information on human glycogenes is summarized in a recently redeveloped GGDB. Researchers can access GGDB (http://riodb.ibase.aist.go.jp/rcmg/ggdb/) with their own web browser (i.e. Internet Explorer, FireFox, Safari, etc.), and browse GGDB to easily find information on glycogenes. For accessing each glycogene, users click checkbox of subfamily (category) and then click each gene symbol.

Text Search

1. Users can search on gene symbol or designation in GGDB by keyword.
2. Users click "Text Search" on the left of GGDB website.
3. Select appropriate information (gene symbol or designation) and then enter a search term(s) and click "SEARCH". Users can enter 1 or 2 keywords (two keywords are concatenated by AND).

References

Kwon Y-D GGDB (2004) A database system for glycogenes. The second symposium of Japanese consortium for glycobiology and glycotechnology, pp 42–43

Narimatsu H (2004) Construction of a human glycogene library and comprehensive functional analysis. Glycoconj J 21:17–24

Website of JCGGDB (also link to GGDB from here) http://jcggdb.jp

Website of Research Center for Glycoscience, AIST (also link to GGDB from here), http://unit.aist. go.jp/rcmg/ci/index.html

CabosDB: Carbohydrate Sequencing Database

Norihiro Kikuchi

Introduction

High-throughput methods for sequencing biomolecules such as DNA, RNA and proteins, and for profiling their patterns of expression have accelerated genome-wide analyses in living organisms. Further, they have enabled molecular biologists and biochemists to investigate biological functions and disorders, the so-called -omics studies. Glycomics, which has become one of the most important fields in the *post-genomic* era, has lagged behind genomics and proteomics, mainly because of the inherent difficulties in the analysis of glycan structures and functions. However, developments in technologies and strategies for analyzing glycan structures should overcome many of these difficulties. In April 2003, the NEDO structural glycomics project (SG project) was initiated under the NEDO (New Energy and Industrial Technology Organization) framework supported by METI (the Ministry of Economy, Trade, and Industry), Japan; the project ran for 3 years. The purpose of this initiative was to develop novel methodologies for structural glycomics. In this project, we used a glycoproteomics method based on liquid chromatography coupled with mass spectrometry (LC–MS) and multistage tandem mass spectrometry, and a lectin profiling method using frontal affinity chromatography (FAC) and lectin arrays. These methods were applied to analyze glycosylation sites, the glycan sequences attached to glycoproteins, and to profile complex features of glycans expressed on cell surfaces. In addition to these methods, database and bioinformatics tools were developed to integrate data from the experiments, to support data analysis, and to provide glycobiologists with these data and with bioinformatics tools for structural glycomic studies. Here, we describe a database named CabosDB (CArBOhydrate Sequencing DataBase).

Database Contents

CabosDB consists of an oligosaccharide database, a mass spectra database, a lectin affinity dataBase (LADB) and a glycoprotein database. The oligosaccharide database contains oligosaccharide structures used in MS and FAC experiments. These structures were described using carbohydrate sequence markup language (CabosML, see Chapter by N. Kikuchi in *Experimental Glycoscience: Glycochemistry*). The mass spectra database stores mass spectra (Fig. 1a) and experimental conditions. The mass spectra were automatically interpreted by comparisons to theoretical values of fragment ion structures. The oligosaccharide database provides a web service for the rapid identification of oligosaccharide structures from mass spectra data. The user can determine the structure of an analyte using interface software that connects Shimadzu/Kratos AXIMA QIT with our web service (Kameyama et al. 2005; see Chapter by A. Kameyama in *Experimental Glycoscience: Glycochemistry*). LADB contains 220 lectin molecules and a lectin map. The

---

Hitotsubashi SI bldg., 3-26 Kandanishiki-cho, Chiyoda-ku, Tokyo 101-0054, Japan
Phone: +81-3-5259-6570, Fax: +81-3-5280-2737
E-mail: kikuchi-norihiro@mki.co.jp

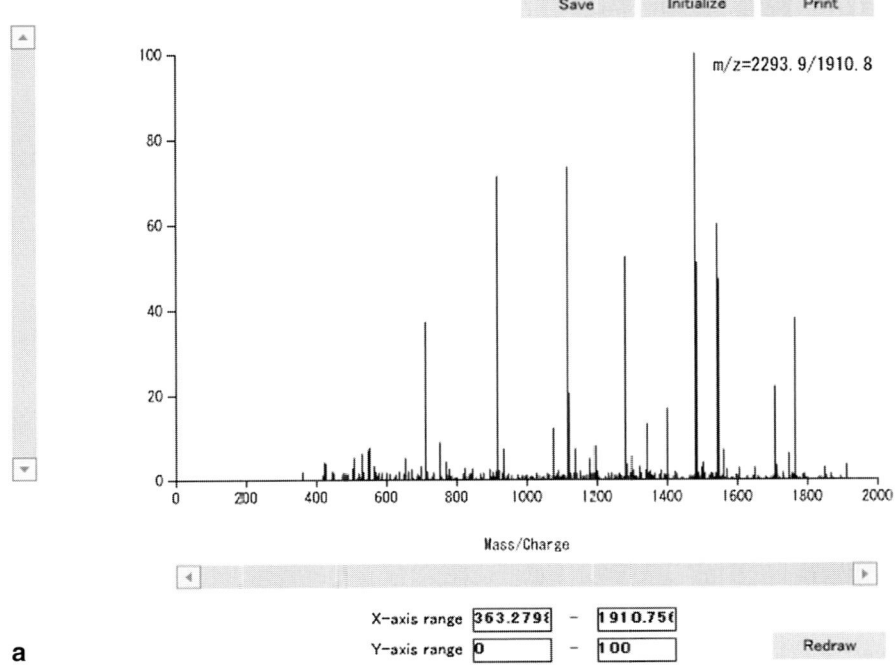

Fig. 1a Screen shots of CabosDB **a** An example of a mass spectrum analysis. The *x* axis represents *m/z* values and the *y* axis indicates the intensities of the peaks. **b** An example of a glycoprotein detected using the IGOT method. The *horizontal bar* represents the protein sequence. Glycosylation sites, detected using the lectins ConA and AAL, are represented by *circular symbols*, and the consensus sites of N-glycosylation are also indicated on the sequence. The membrane spanning region is represented by the *line* under each sequence

lectin molecules are organized by family, carbohydrate recognition domain (CRD), amino acid sequences, function, tertiary structures and binding sites (see Chapter by Y. Takahashi and J. Hirabayashi, this volume). The lectin map includes affinity constants between more than 100 lectins and more than 100 oligosaccharides measured by FAC (Nakamura et al. 2005). The glycoprotein database contains more than 1,000 glycoprotein sequences and glycan binding sites identified by the isotope coded glycosylation site specific tagging (IGOT) method (Fig. 1b) (Kaji et al. 2003; see Chapter by H. Kaji and T. Isobe in *Experimental Glycoscience: Glycochemistry*). CabosDB has a web interface to access information, including graphical representation of data (glycan binding sites, affinity constants, etc.).

Comments

System requirements

- OS: Windows 2000 or XP.
- Web Browser: Internet Explorer *version* 6 or later.

We are planning to update CabosDB to be open to the access as a part of a database for Japan Consortium for Glycobiology and Glycotechnology (JCGGDB, http://jcggdb.jp) in 2008.

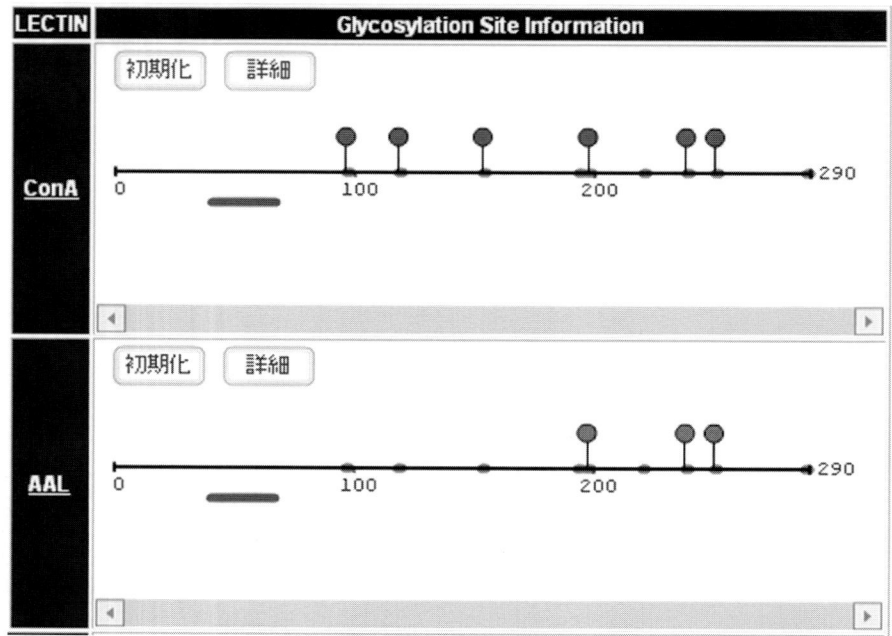

Fig. 1b

Acknowledgments We are grateful to A. Kameyama, S. Nakaya, S. Nakamura, T. Shikanai, H. Kaji, J. Hirabayashi and H. Narimatsu for fruitful discussions and for providing the data sources, to Y. Takahashi (MKI) for useful advice and support throughout the work, and to T. Kobayashi, M. Matsumoto and H. Yazawa for developing the database system. This work was supported by grants from the New Energy and Industrial Technology Development Organization (NEDO) of Japan.

References

Kaji H, Saito H, Yamauchi Y, Shinkawa T, Taoka M, Hirabayashi J, Kasai K, Takahashi N, Isobe T (2003) Lectin affinity capture, isotope-coded tagging and mass spectrometry to identify N-linked glycoproteins. Nat Biotechnol 21(6):667–672

Kameyama A, Kikuchi N, Nakaya S, Ito H, Sato T, Shikanai T, Takahashi Y, Takahashi K, Narimatsu H (2005) A strategy for identification of oligosaccharide structures using observational multistage mass spectral library. Anal Chem 77(15):4719–4725

Kikuchi N, Kameyama A, Nakaya S, Ito H, Sato T, Shikanai T, Takahashi Y, Narimatsu H (2005) The carbohydrate sequence markup language (CabosML): an XML description of carbohydrate structures. Bioinformatics 21(8):1717–1718

Nakamura S, Yagi F, Totani K, Ito Y, Hirabayashi J (2005) Comparative analysis of carbohydrate-binding properties of two tandem repeat-type Jacalin-related lectins, Castanea crenata agglutinin and Cycas revoluta leaf lectin. FEBS J 272(11):2784–2799

GlycoEpitope: A Database of Carbohydrate Epitopes and Antibodies

Toshisuke Kawasaki[1], Hiromi Nakao[2], Tomoko Tominaga[3]

Introduction

Recently, the involvement of carbohydrate chains in life sciences has been extended to diverse functions as cell to cell recognition, communication in neuronal tissues and immune systems, pathogen recognition, sperm-egg recognition, fertilization, regulation of hormonal half-lives in the blood, direction of embryonic development and differentiation, and direction of the distribution of various cells and proteins throughout the body. A large number of polyclonal and monoclonal antibodies have been used as very important tools for analyzing the expression of various carbohydrate chains and their functions. However, a large amount of important information on carbohydrate-recognizing antibodies is spread throughout a wide range of literature. In this database, useful information on such carbohydrate antigens, i.e., glyco-epitopes and antibodies has been assembled as a compact encyclopedia (Kawasaki et al. 2006). It has been developed with the cooperation of foremost researchers in the field of glycobiology and is maintained by the Ritsumeikan University Research Center for Glycobiotechnology. GlycoEpitope provides a wealth of information including lists of glycoproteins that express carbohydrate antigens, glycolipids of which part of the structure is a carbohydrate antigen, enzymes that take part in the synthesis and degradation of epitopes, the times and sites of expression of carbohydrate antigens, diseases to which carbohydrate epitopes are related, and suppliers from which carbohydrate-recognizing antibodies can be obtained. This database is useful for not only glycobiologists but also a wide range of life science researchers. The search criteria are very flexible, so, a user can easily find the information he needs. Here, we would like to give a general outline of the database and how to use it.

Concept

GlycoEpitope consists of five databases, i.e., general data on epitopes (General), antibodies recognizing these epitopes (Antibody), glycoproteins (Glycoprotein) and glycolipids (Glycolipid) expressing epitopes and enzymes that take part in the synthesis and degradation of epitopes (Enzyme). Additionally, all references are listed in the "References" (Fig. 1).

(1) General

"General"consists of general information on epitopes: epitope ID specific for this database, name, sequence (first to third if available) of the carbohydrate chain

---

[1-3] Research Center for Glycobiotechnology, Ritsumeikan University, Nojihigashi 1-1-1, Kusatsu, Shiga 525-8577, Japan
Phone: +81-77-561-3444, Fax: +81-77-561-3496
E-mail: [1] tkawasak@fc.ritsumei.ac.jp [2] nnv25087@fc.ritsumei.ac.jp,
[3] ttv25139@fc.ritsumei.ac.jp

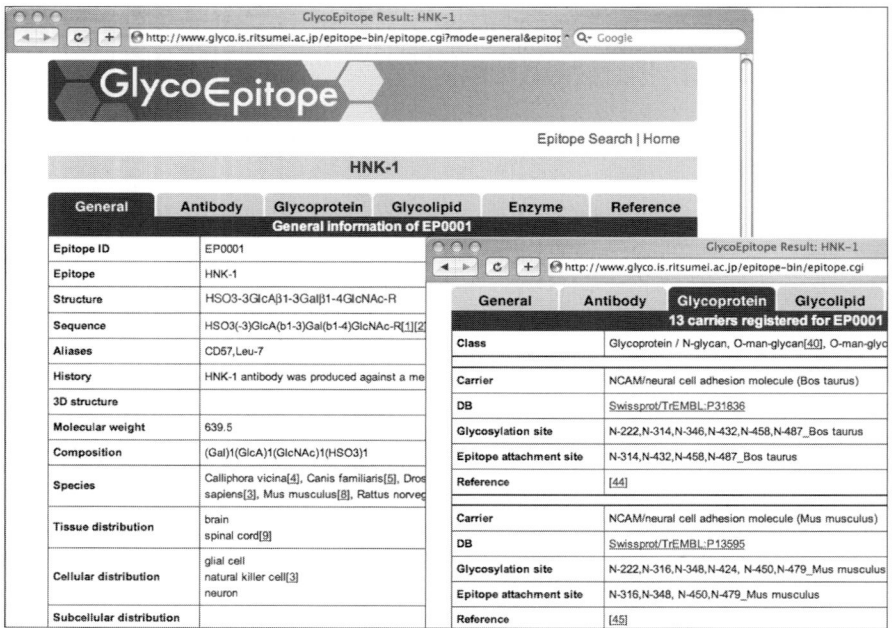

Fig. 1 In-depth data on an epitope in the database

structure, aliases, history of detection, molecular weight, monosaccharide components, species in which expression of the epitope is known, tissues and cells where the epitope is expressed, subcellular distribution, developmental changes of the expression, cell lines, receptors, functions, related diseases, and examples of epitopes and antibodies which are practically used.

(2) Antibody

In "Antibody", the following information on antibodies is given: name, species that produce the antibodies, isotype, whether monoclonal or polyclonal, recognition region, immunoprecipitation, immunoblotting, histochemistry, and availability. Further information on each antibody is available by clicking the name of the antibody.

(3) Glycoprotein

In "Glycoprotein", whether an epitope is found in an N-glycan or O-glycan is shown. The glycosylation site and epitope attachment site on its carrier protein are also shown. For gene sequences, amino-acid sequences, and 3D protein structures, there are links to external database websites.

(4) Glycolipid

In "Glycolipid", the name, alias, classification, glycolipid sequences, 3D structure, molecular weight, and monosaccharide components of a glycolipid are given. There are also links to external database websites.

(5) Enzyme

In "Enzyme", a user can find information on enzymes that take part in the synthesis and degradation of epitopes, including the name, catalyzed reaction, description, and EC number. Information regarding the availability of gene-deficient organisms is

also given. For the gene sequences, amino-acid sequences and 3D structures of these enzymes, there are links to external database websites.
(6) Reference

In "Reference", the titles, author's names, and PubMed ID of all references cited in the above databases are listed. There are links to the PubMed webpage.

How to Operate the Database

The database is available to the public through the Internet (URL http://www.glyco.is.ritsumei.ac.jp/epitope/). By clicking "List all epitopes" or "List all antibodies", lists of all epitopes or antibodies in the database will be shown. User guides in English and Japanese are available at the bottom of the top page. A user can search for epitopes or antibodies by using search function "Epitope Search" or "Antibody Search".

(1) Epitope Search

"Epitope Search" is useful for finding information on epitopes. By either entering or selecting keywords for an item, which a user wants to search for, and pressing the search button, the search will be initiated. By clicking "Epitope ID", the "General" data will be shown. A user can also easily access to a database other than "General", i.e., Antibody, Glycoprotein, Glycolipid, Enzyme or Reference, by clicking a tab at the top.

(2) Antibody Search

"Antibody Search" is useful for finding information on antibodies. By either entering keywords in "All-fields" or "Antibody name", and by pressing the search button, the search will be initiated. For "Antibody name", a user can alternatively choose a keyword from a pull-down menu. A list of Antibody IDs, Epitope IDs, names and recognition regions will be given as a search result. By clicking "Antibody ID" and "Epitope ID", the general data will be shown.

As of March 2007, "GlycoEpitope" includes 196 epitopes and 358 antibodies. We welcome entries from users to this database. An entry page can be obtained by clicking "Feedback form" in the footnote on the front page. Please use the entry page to inform us of any questions and comments. We hope "GlycoEpitope" will be useful for your research.

Reference

Kawasaki T, Nakao H, Takahashi E, Tominaga T (2006) GlycoEpitope: the integrated database of carbohydrate antigens and antibodies. Trends Glycosci Glycotechnol 18:267–272

A Novel Lectin-Affinity Database for Structural Glycomics

Yoriko Takahashi[1], Jun Hirabayashi[2]

Introduction

Lectins, non-enzymatic glycan-binding proteins, are ubiquitous in nature and act as recognition molecules in many biological processes including cell–cell and cell–molecule interactions, innate immunity, morphogenesis and fertilization. Over the past 120 years numerous lectins have been isolated from plants, microorganisms, fungi, animals and viruses, resulting in a wealth of data. Lectins are also used as invaluable tools for the detection, isolation and characterization of various glycoproteins, or for histopathological examination of cells and tissues and for clinical diagnosis of carcinoma and leukemia.

In the post-genome era, glycomics has become one of the most important fields of life sciences. It is now imperative to perform the structural analysis of glycosylation at the cell surface in a rapid, sensitive and high-throughput manner, and novel methodologies are being developed to achieve this goal. A strategy using sets of lectins (Hirabayashi 2004), as represented by lectin array (Kuno et al. 2005), is a powerful tool for such analysis along with that using mass spectrometry (Kameyama et al. 2005). In particular, a lectin-based approach is ideal for profiling complex features of glycans expressed at the cell surface because lectins can discriminate subtle differences in branching, linkage, and terminal modifications on individual cells. However, no database was available to integrate knowledge concerning the properties of various lectins and their specificities for a wide variety of glycans.

This chapter describes a novel structural glycomics-oriented database, designated Lectin Affinity DataBase (LADB). LADB provides comprehensive affinty information about lectins and glycans, together with useful search and browse functions over the Web. This database system is integrated into CabosDB (Carbohydrate sequence DataBase, see Chapter by N. Kikuchi in this volume).

Database Contents

LADB provides molecular information about 220 lectins derived from plants, fungi and animals including the common name, formal nomenclature, source organism and tissue, lectin family, number of CRDs (carbohydrate recognition domains), amino acid sequence, polypeptide fold, tertiary structure, references and links to other public databases (e.g.,

---

[1] Bioscience Group, Mitsui Knowledge Industry Co., Ltd, Hitotsubashi SI bldg., 3-26, Kandanishiki-cho Chiyoda-ku, Tokyo 101-0054, Japan
Phone: +81-3-5259-6570, Fax: +81-3-5280-2737
E-mail: takahashi-yoriko@mki.co.jp

[2] Research Center for Medical Glycoscience, AIST, AIST Tsukuba Central 2, Tsukuba, Ibaraki 305-8568 Japan
Phone: +81-29-861-3187, Fax: +81-29-861-3125
E-mail: jun-hirabayashi@aist.go.jp

Fig. 1 A snapshot of molecular information of a lectin

GenBank, Protein Data Bank, Pfam) (Fig. 1). In addition, LADB gives comprehensive, experimental data on the affinities of about >100 lectins and >100 structurally defined glycans determined by high-throughput frontal affinity chromatography (Nakamura et al. 2005) in the course of the "hect-by-hect project" (Hirabayashi 2004). Glycan structures used in these experiments are described in CabosML (Kikuchi et al. 2005).

LADB also provides sophisticated search and browse functions. The entries are retrieved from the database using a variety of search options: they include keywords, molecular properties of lectin, structural features of glycan, and affinity level of lectin–glycan combination. The outcomes are displayed either in the form of a graph or table (Fig. 2). Users can then readily browse affinity patterns, characteristics of lectins and related structures of glycans of interest.

Comments

LADB is a novel, comprehensive knowledge base for lectin-based structural glycomics on the web. The most significant feature of LADB is to unify comprehensive data on affinities, which were experimentally obtained for a number of lectins and glycans, and those on glycan structures described in the context of bioinformatics. LADB allows researchers to compare the characteristics of lectins of interest with structural features of their target glycans in a simultaneous manner, thereby giving fresh scientific insights into these molecular interactions. Thus, LADB provides comprehensive, valuable information that should benefit researchers in glycobiology as well as related fields such as cell biology, histology, immunology, and clinical diagnosis.

We are planning to update LADB to be open to the public as a part of a database for Japan Consortium for Glycobiology and Glycotechnology (JCGGDB, http://jcggdb.jp) in 2008.

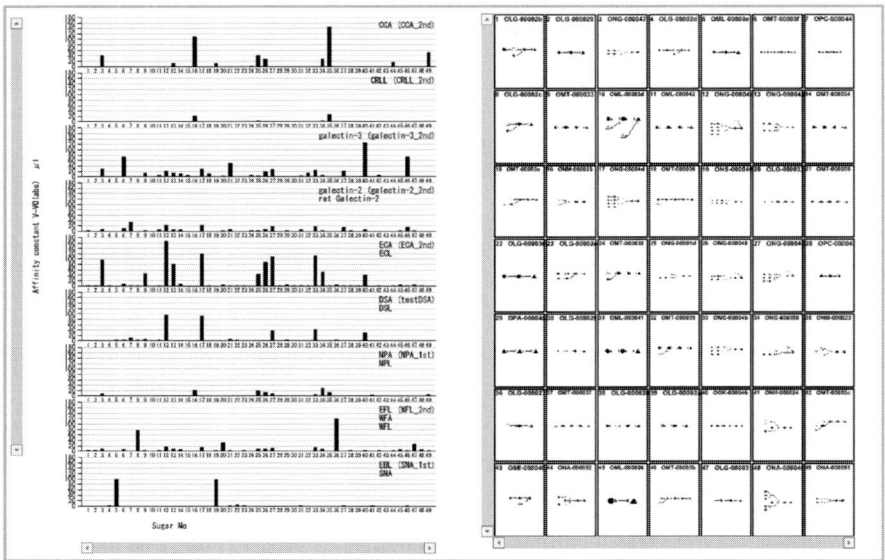

Fig. 2 A snapshot of affinity data between lectins and glycans. *Bar graphs* show affinity patterns of targeted lectins toward diverse glycans (*left* panel), where y-axis is affinity strength and the x-axis is glycan IDs. Glycan structures are displayed in a user-specified format (CFG format here) (*right* panel)

System Requirements

1. OS: Windows 2000 or XP
2. Web Browser: Internet Explorer *version* 6 or later.

Acknowledgements We are grateful to S. Nakamura for fruitful discussions and providing data source, to N. Kikuchi for useful advice and support throughout the work, and to T. Kobayashi, M. Matsumoto and H. Yazawa for developing the database system. This work was supported by grants from the New Energy and Industrial Technology Development Organization (NEDO) of Japan.

References

Hirabayashi J (2004) Lectin-based structural glycomics: glycoproteomics and glycan profiling. Glycoconj J 21:35–40

Kameyama A, Kikuchi N, Nakaya S, Ito H, Sato T, Shikanai T, Takahashi Y, Takahashi K, Narimatsu H (2005) A strategy for identification of oligosaccharide structures using observational multistage mass spectral library. Anal Chem 77:4719–4725

Kikuchi N, Kameyama A, Nakaya S, Ito H, Sato T, Shikanai T, Takahashi Y, Narimatsu, H (2005) The carbohydrate sequence markup language (CabosML): an XML description of carbohydrate structures. Bioinformatics 21:1717–1718

Kuno A, Uchiyama N, Koseki-Kuno S, Ebe Y, Takashima S, Yamada M, Hirabayashi J (2005) Evanescent-filed fluorescence-assisted lectin microarray: a new strategy for glycan profiling. Nat Methods 2:851–856

Nakamura S, Yagi F, Totani K, Ito Y, Hirabayashi J (2005) Comparative analysis of carbohydrate-binding properties of two tandem repeat-type Jacalin-related lectins, *Castanea crenata* agglutinin and *Cycas revoluta* leaf lectin. FEBS 272:2784–2799

Glycoconjugate Data Bank

Nobuaki Miura[1], Shin-Ichiro Nishimura[2,3]

Introduction

Glycomics and glycoproteomics are emerging fields with respect to post-translational modification. Exceptional development of modern experimental techniques, such as mass spectrometry, nuclear magnetic resonance, and knockout mice enables one to accumulate a great deal of information related to carbohydrate structure and functions. On the other hand, bioinformatics approaches to carbohydrate research have been significantly lagging. One of the reasons has been the lack of data, due to their structural complexity. In recent years, construction of a database with respect to carbohydrates and development of related bioinformatics tools have attracted increasing attention in various biological research areas such as glycobiology and drug discovery.

There are three databases which have been aggressively developed, the KEGG-glycan (Hashimoto et al. 2006), GLYCOSCIENCES.de (Lütteke et al. 2006), and the US Consortium for Functional Glycomics (CFG) (Raman et al. 2006). These are based on the Complex Carbohydrate Structure Database (CCSD; Doubet et al. 1989; Doubet and Albersheim 1992), which was developed by the CarbBank Project in the 1990s. All these databases have various powerfully developed bioinformatics tools with respect to carbohydrates and have supported studies in the field with respect to post-translational modification. However, from the viewpoint of drug discovery using carbohydrates as candidates of the drug seeds and of functional networks among carbohydrates and related glycosyltransferases, the linkage information among carbohydrates, glycosyltransferases, its inhibitors, and various diseases is still poor in these databases. CFG holds a variety of aminosugars for therapy with carbohydrate chips. Although the biosynthesis pathway is implemented in the KEGG-glycan, information about the inhibitor or related chemicals of the glycosyltransferase is lacking. In addition, there is no database that shows the step-by-step functional network among carbohydrates, glycosyltransferase, and various diseases. The three-dimensional (3D) structures of carbohydrates are also essential information. GLYCOSCIENCE.de is addressing this problem.

We are developing a complementary database with information that is lacking in other databases. Here, we will introduce the current status of a database named "Glycoconjugate Data Bank" (GDB) (http://www.glycoconjugate.jp/), which we are developing and

---

[1] Sun Microsystems Laboratory for Computational Molecular Life Science, Graduate School of Advanced Life Science, Hokkaido University, Sapporo 001-0021, Japan

[2] Division of Advanced Chemical Biology, Graduate School of Advanced Life Science Frontier Research Center for Post-Genomic Science and Technology, Hokkaido University, Sapporo 001-0021, Japan

[3] Drug-Seeds Discovery Research Laboratory, Hokkaido Center, National Institute of Advanced Industrial Science and Technology (AIST), Sapporo 062-8517, Japan
Phone: +81-11-706-9043, Fax: +81-11-706-9042
E-mail: shin@glyco.sci.hokudai.ac.jp

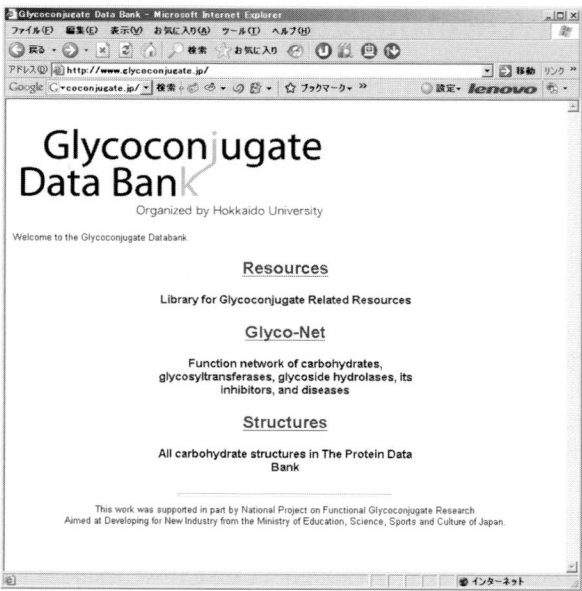

Fig. 1 Top page of the Glycoconjugate Data Bank

relating to the information of chemical resources, 3D structures, and functional networks of glycoconjugates (Fig. 1).

Glycoconjugate Data Bank

The GDB has been available since December 2003 and is composed of three databases called "Resources", "Structures", and "Glyco-Net". The URLs for these databases are shown in Table 1. We focus on the carbohydrate-related chemical resources as seeds of the drug candidates (Resources), 3D structures of carbohydrates extracted from the Protein Data Bank (http://www.rcsb.org/pdb/) (Structures), and functional networks among carbohydrates, glycosyltransferases, glycoside hydrolases, its inhibitor, phenotype and disease (Glyco-Net).

Resources

"Resources" holds information on the chemical resources mainly synthesized in our laboratory containing some candidates for the inhibitor of the glycosyltransferases or glycoside hydrolases. "Resources" has about 1,000 compounds produced by organic synthesis and 110 compounds extracted from bio-resources. Figure 2 shows an example of such compounds. It includes molecular name, molecular weight, formula, NMR spectra, synthetic pathway, reference information, and contact information. We will implement the 3D structure in the MDL mol file format in the near future.

Structures

"Structures" is a database of 3D structures of carbohydrates as ligands, N-glycan, or O-glycan extracted from PDB. "getCARBO" program code was developed and implemented in this database, and the 3D structures of carbohydrates extracted from PDB.

Glycoconjugate Data Bank

Table 1 Databases in the Glycoconjugate Data Bank

Description	URL
GDB home	http://www.glycoconjugate.jp/
Resources	http://www.glycoconjugate.jp/resources/
Structures	http://www.glycostructures.jp//
Glyco-Net	http://www.glycoconjugate.jp/functions/

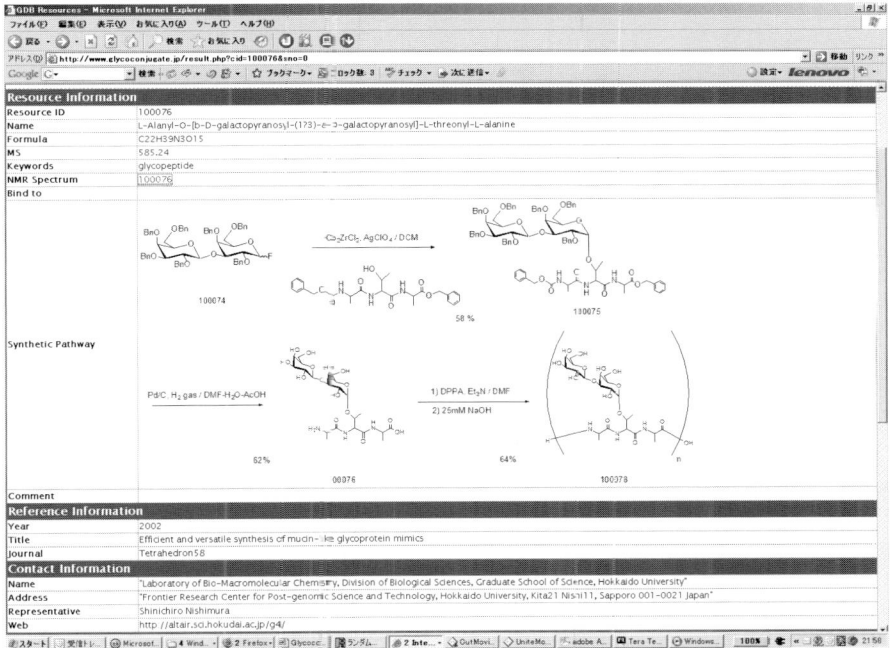

Fig. 2 Overview of entries of Resources

"Structures" contains many unnatural glycans that are likely to be artifacts. For all N-glycans in this database, a doubtful (sub)structure is marked with a question mark. In addition, the same glycans were simultaneously searched from KEGG-glycan for comparison (Fig. 3). Further, a program code to judge the structure of N-glycans in PDB, whether the glycan structures match typical N-glycan or not, is also implemented.

Glyco-Net

"Glyco-Net" can show functional networks among glycoconjugates, their related enzymes, phenotypes, and various diseases. The concept is accumulation of binary relations for carbohydrate-related information, as shown in Fig. 4. The first attempt to represent molecular functions in binary relations was the KEGG BRITE database by Prof. Kanehisa, Kyoto University (http://www.genome.jp/kegg/brite.html). Now, we obtain data from the research article manually. In one research article, some experimental facts

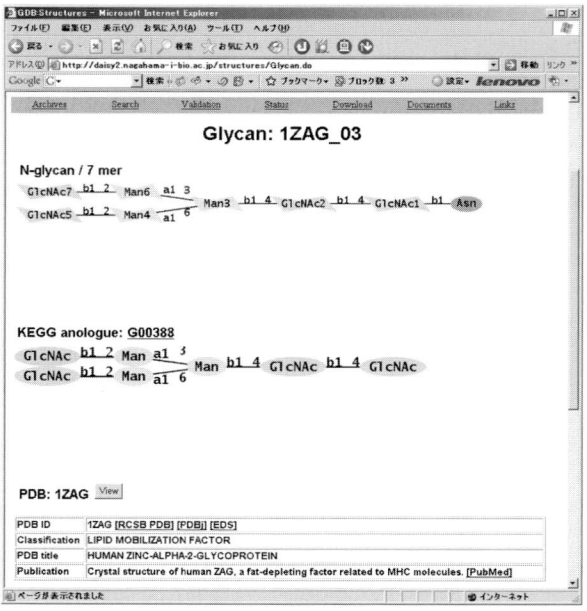

Fig. 3 An example of a search for a glycan structure

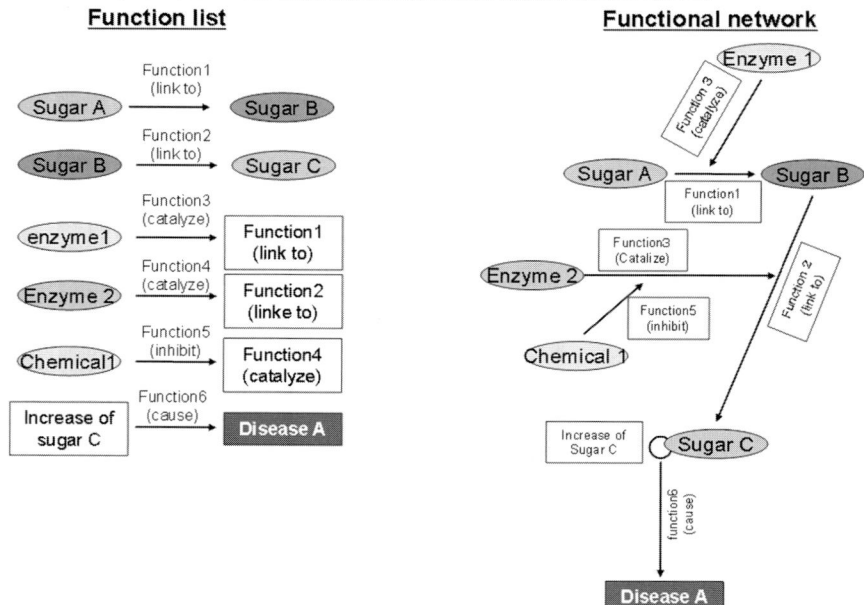

Fig. 4 Binary relations in Glyco-Net

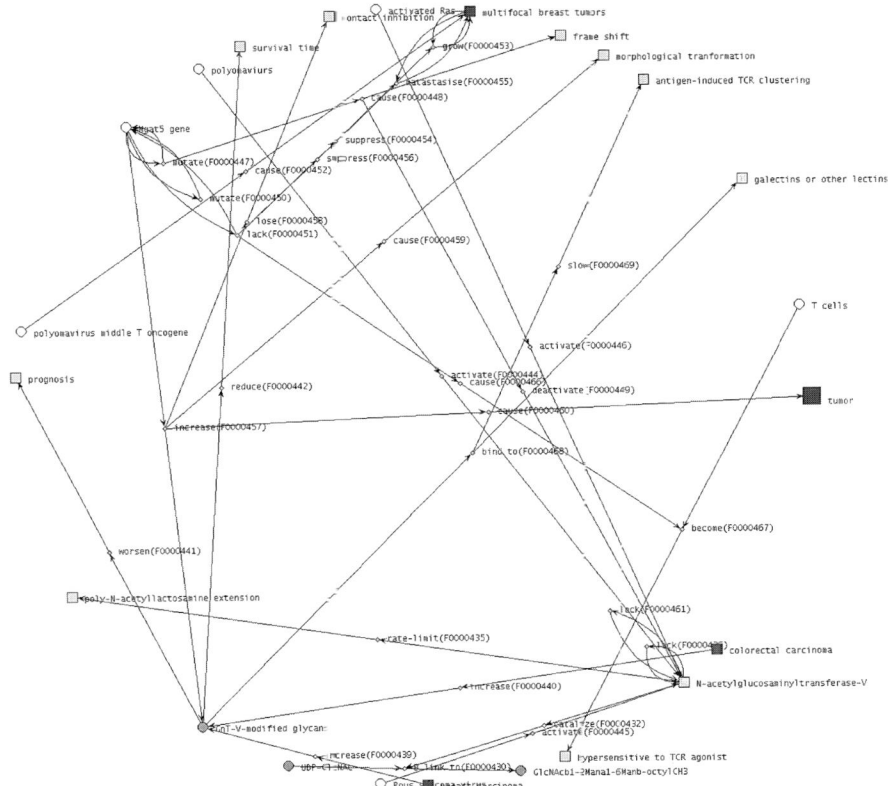

Fig. 5 A search result in Glyco-Net

are usually described via a certain assay method. Binary relations with respect to biosynthetic pathways, inhibition of enzymes, phenotypes, and diseases, which correlate to the functions of glycoconjugates have been extracted. In addition, we are constructing a kind of dictionary in the field of glycoconjugate research in order to automatically gather the above binary relations from web articles such as PubMed (http://www.pubmed.org). Currently, "Glyco-Net" has about 1,000 objects with respect to glycosyltransferase, about 2,500 of its genes, 2,500 data of functions, 1,200 pieces of data related to the assay that verifies the functions of glycoconjugates, and 95 pieces of data with respect to disease that are caused by carbohydrate abnormalities. These data were obtained from the *Handbook of Glycosyltransferases and Related Genes* (Taniguchi et al. 2003). Figure 5 shows an example of the search results of Glycol-Net.

References

Doubet S, Albersheim P (1992) CarbBank. Glycobiology 2:505
Doubet S, Bock K, Smith D, Darvill A, Albersheim P (1989) The complex carbohydrate structure database. Trends Biochem Sci 14:475–477
Hashimoto K, Goto S, Kawano S, Aoki-Kohinata KF, Ueda N, Hamajima M, Kawasaki T, Kanehisa M (2006) KEGG as a glycome informatics resources. Glycobiology 16:63R–70R

Lütteke T, Bohne-Lang A, Loss A, Goetz T, Frank M, von der Lieth C-W (2006) GLYCOSCIENCES.de: and internet portal to support glycomics and glycobiology research. Glycobiology 16:71R–81R

Raman R, Venkataraman M, Ramakrishnan S, Lang W, Raguram S, Sasisekharan R (2006) Advancing glycomics: Implementation strategies at the consortium for functional glycomics. Glycobiology 16:82R–90R

Taniguchi N, Honke K, Fukuda M (eds) (2003) Handbook of glycosyltransferases and related genes. Handbook of glycosyltransferases and related genes. Springer, Tokyo

KEGG GLYCAN for Integrated Analysis of Pathways, Genes, and Structures

Kosuke Hashimoto, Minoru Kanehisa

Introduction

While much knowledge about glycans has accumulated, most of the primary data for both experimental and bioinformatics analyses are information on glycan structures and glycan-related genes. Therefore, it is necessary to aggregate all these chemical and genomic information together in an effort to predict complex cellular processes and to understand organism behaviors at a higher level. Following this perspective, we have been developing KEGG (Kyoto Encyclopedia of Genes and Genomes) (Kanehisa et al. 2008) to include three basic glycan resources: (1) GLYCAN, a database of glycan structures; (2) the glycosyltransferase and the glycosidase reaction library; and (3) glycan-related pathways (Hashimoto et al. 2006). KEGG also includes three useful tools: (4) the Composite Structure Map (CSM), a map illustrating all possible variations in glycan structures within organisms; (5) KegDraw, an intuitive drawing tool for chemical structures; and (6) KCaM, a glycan search and alignment tool (Aoki et al. 2004). All resources are available at KEGG (http://www.genome.jp/kegg/glycan/).

KEGG GLYCAN

GLYCAN, which contains about 11,000 unique glycan structures, has a central role in the glycan resources of KEGG. Each entry is annotated with information on the structure, including composition, class, and mass, as well as a separate GIF image file for display. Additionally, entries may contain links to other databases in KEGG, such as COMPOUND, REACTION, and PATHWAY, summarized in Table 1. All data are accessible via both the web site and over the Internet through an Application Programming Interface (API) accessible from software.

Glycosyltransferase and Glycosidase Reaction Library

This library provides genomic information including glycosyltransferases and glycosidases. Ortholog groups of these enzymes are related to the reactions that they catalyze. Reaction specificity is characterized by the following three features: acceptor monosaccharide residues, the donor monosaccharide residue, and the linkage between them. This

---

Bioinformatics Center, Institute for Chemical Research, Kyoto University, Gokasho, Uji, Kyoto 611-0011, Japan
Phone: +81-774-38-3270, Fax: +81-774-38-3269
E-mail: kanehisa@kuicr.kyoto-u.ac.jp

Table 1 Annotation of entries in GLYCAN

Attribute	Description
Name	Common name(s) for this entry, if any
Composition	Textual description of the monosaccharide composition
Mass	Molecular mass, calculated by summing the mass of the monosaccharides minus the number of bonds times the mass of water
Class	Classification of glycans, as in (glycoprotein; N-glycan)
Remark	Any comments, such as on lectins and other interacting molecules
Reaction	Corresponding REACTION database entry if this entry is a reactant
Pathway	Corresponding PATHWAY database entry for the biological process in which this entry is involved
Enzyme	Corresponding ENZYME database entry (EC number) for catalytic actions on this entry
Ortholog	Corresponding KO (KEGG Orthology) database entry to an ortholog group, which is an identifier for the orthologous genes from different organisms
Reference	Literature citation for this entry with links to PubMed
DBLINKS (other DBs)	Links to corresponding entries in other external databases, such as CarbBank

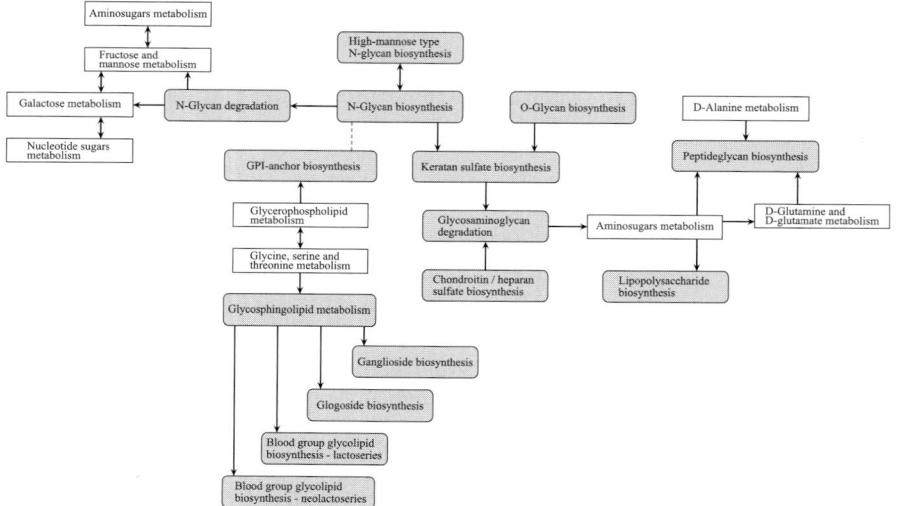

Fig. 1 The overall relationship of the 15 KEGG pathway maps for glycan biosynthesis and metabolism (*shaded*) and some other metabolic pathway maps

resource is applied to the prediction of glycan structures from microarray data (Kawano et al. 2005; Suga et al. 2007).

Glycan Biosynthetic and Metabolic Pathways

KEGG PATHWAY is a database that represents molecular interaction networks, including metabolic pathways, regulatory pathways, and molecular complexes. A unique feature of KEGG PATHWAY is its role in integrating the glycan structures, reactions, and enzymes related to glycans. For example, the pathway map of lactoseries glycolipid biosynthesis is composed of 13 glycosyltransferase reactions including 9 glycosyltransferases and 13 glycans. Figure 1 illustrates the overall relationship of the 15 pathway maps currently available for glycan biosynthesis and metabolism as well as other metabolic pathway maps. There are also additional maps called structure maps, which show glycan structures connected to its catalyzing enzymes, such as *N*-glycan, *O*-glycan, and glycosaminoglycan.

Composite Structure Map

KEGG PATHWAY represents the step-by-step process of biochemical reactions involving genes and glycan structures. On the other hand, the CSM is a static representation of all possible variations of glycan structures in a tree format using entire KEGG GLYCAN entries. Any node corresponds to a list of glycan structures that are located on the single path from the root to that node. Thus, each node is hyperlinked to its corresponding list. Because glycosyltransferases catalyze the biosynthesis of glycan structures by the addition of individual linkages, each edge in the CSM is hyperlinked to its corresponding glycosyltransferase-related information, if known. It is possible to color the edges corresponding to a specification of colors and genes stored in a local file. Thus, a variety of analyses can be performed using this tool. For example, microarray gene-expression data can be displayed in this tree by color coding the up- or down-regulation of genes involved in glycan synthesis, thereby linking gene expression to glycan structures.

KegDraw: Glycan Structure Drawing Tool

KegDraw is a software tool for drawing chemical structures, including glycans. It is a Java application, so it runs locally in a platform-independent manner. KegDraw consists of two drawing modes, 'Compound' and 'Glycan.' 'Compound' allows for the drawing of chemical compounds in a similar way as ChemDraw and ISIS/Draw, and 'Glycan' is for drawing glycans with monosaccharide units. In 'Glycan' mode, glycan structures can be drawn in a variety of ways. The simplest method is selecting monosaccharides and binding conformations one by one from popup menus. However, convenient functionalities such as cut-and-paste and pre-defined template structures are also available. Some other functions for easy drawing include swapping branches and arranging nodes along a straight line or at even intervals. KegDraw currently handles KCF (KEGG chemical function), so glycan structures can be entered by representing the data in KCF format. After drawing the glycan structure, it can be transferred into KCF as output data and/or it can be uploaded to the GLYCAN database for use as a structure search tool.

KCaM: Structure Search Tool

The structures in the GLYCAN database can be accessed by queries using keywords such as commonly used names or by using a structure search. The structure search tool employed by KEGG GLYCAN is called KCaM, which utilizes a dynamic programming technique and a theoretically proven efficient algorithm for finding the maximum common subtree between two trees (Aoki et al. 2004). This tool is available both at the KEGG web site and through KegDraw. KCaM consists of two main variations, namely, an approximate matching algorithm and an exact matching algorithm. The former aligns monosaccharides allowing gaps in the alignment, whereas the latter aligns linkages and disallows any gaps, resulting in a stricter criterion for alignment.

References

Aoki KF, Yamaguchi A, Ueda N, Akutsu T, Mamitsuka H, Goto S, Kanehisa M (2004) KCaM (KEGG Carbohydrate Matcher): a software tool for analyzing the structures of carbohydrate sugar chains. Nucleic Acids Res 32:W267–W272

Hashimoto K, Goto S, Kawano S, Aoki-Kinoshita KF, Ueda N, Hamajima M, Kawasaki T, Kanehisa M (2006) KEGG as a glycome informatics resource. Glycobiology 16:63R–70R

Kanehisa M, Araki M, Goto S, Hattori M, Hirakawa M, Itoh M, Katayama T, Kawashima S, Tokimatsu T, Yamanishi Y (2008) KEGG for linking genomes to life and the environment. Nucleic Acids Res 36:D480–D484

Kawano S, Hashimoto K, Miyama T, Goto S, Kanehisa M (2005) Prediction of glycan structures from gene expression data based on glycosyltransferase reactions. Bioinformatics 21:3976–3982

Suga A, Yamanishi Y, Hashimoto K, Goto S, Kanehisa M (2007) An improved scoring scheme for predicting glycan structures from gene expression data. Genome Informatics 18:237–246

Section XXIV
Microarray

DNA Microarray in Glycobiology

Harumi Yamamoto[1], Hiromu Takematsu[2,3], Yasunori Kozutsumi[1,2,3]

Introduction

DNA microarray analysis enables us to provide a considerable amount of information on glyco-chain expression (Brazma et al. 2001; Ide et al. 2006). DNA microarray experiments are developed to obtain overall transcriptome of cells using DNAs spotted on a slide glass as probes and fluorescently labeled DNAs derived from mRNA samples. DNA microarray experiments require high-quality RNA. Thus, it is advisable to prepare RNA as intact as possible. The quality of RNA is usually assessed by the intactness of the peaks and the lack of degradation of RNA by capillary electrophoresis. This step is essential for the reliable final readout of microarray experiments. Following is the normal procedure operated in our laboratory for the DNA microarray experiments using RIKEN human and mouse glyco-chain-related DNA microarrays version 2 (human version: GEO platform GPL 3465 http://www.ncbi.nlm.nih.gov/geo/), which were designed by us and spotted by Takara Bio Inc.

Poly(A)$^+$ RNA Preparation

As far as RNA degradation and ribosomal RNA contamination are limited, any protocol should work for the preparations. Usually an mTRAP Midi Kit (Active Motif #23024) is used for poly(A)$^+$ RNA and mTRAP Total (Active Motif #23012) for total RNA. The details are described in the manufacturer's instructions.

RNA Quality Check

The quality check and quantitative determination of purified RNA are usually carried out using the Bioanalyzer (Agilent Technologies) according to the manufacturer's instructions (200–500 ng of purified poly(A)$^+$ RNA is required per assay).

Fluorescence Labeling of Poly(A)$^+$ RNA

Unbiased labeling of RNA is very important, especially for the two-color comparison of DNA microarray. Poly(A)$^+$ RNA is usually labeled using the CyScribe Post-Labeling Kit (GE Healthcare RPN #5660) according to the manufacturer's instructions. This was the

---

[1] Supra-biomolecular System Research Group, RIKEN Frontier Research System, Wako, Saitama 351-0198, Japan

[2] Laboratory of Membrane Biochemistry and Biophysics, Graduate School of Biostudies, Kyoto University, Yoshida Shimoadachi-cho, Sakyo-ku, Kyoto 606-8501, Japan

[3] CREST, JST, Kawaguchi, Saitama, Japan
Phone: +81-75-753-7684, Fax: +81-75-753-7686
E-mail: yasu@pharm.kyoto-u.ac.jp

most effective among systems tested to minimize the bias caused by the difference in fluorescent functions between the two colors for our microarrays.

Microarray Hybridization and Signal Detection

Resuspend the combined Cy-dye labeled sample with the hybridization solution (20 µl) and heat at 95°C for 2 min.

Incubate DNA microarray slide(s) with the pre-hybridization solution (20 µl) for 2 h and wash with 2× SSC and 0.2× SSC.

Add the Cy-dye labeled sample onto the slide(s) and incubate 65°C for overnight.

Wash the slide(s) with 2× SSC-0.2% SDS at 55°C for 5 min twice and at 65°C once.

Rinse the slide(s) with 0.05× SSC.

Scan fluorescence signal on the slide(s) by the 428 Array Scanner (Affymetrix).

Quantify signal intensity of each spot by software ImaGene (BioDiscover).

Hybridization solution: $6 \times$ SSC, 0.2% SDS, $5 \times$ Denhart's solution, 0.2 mg/ml denatured salmon sperm DNA.

Pre-hybridization solution: $6 \times$ SSC, 0.2% SDS, $5 \times$ Denhart's solution, 2 mg/ml denatured salmon sperm DNA.

Bioinformatic Analysis of DNA Microarray Data

Massive quantity of the data is automatically supplied from the microarray experiments. Thus, handling of the data could be very different from the normal molecular biology experiments. Following background subtraction, the bias between the two colors should be normalized to determine whether genes of interest are up-regulated or down-regulated (Koike et al. 2004).

Cross-Sample Comparison

Cross-sample comparison is a method to compare gene expression among multiple samples and is effective to directly determine gene(s) related to phenotype(s) of interest (Naito et al. 2007). For this purpose, each RNA sample is compared to commercially available Universal Reference RNA to eliminate the aforementioned bias. Here is an example in which one of sialyltransferase genes was determined to be involved in the expression of GL7 epitope, a marker for B cell activation.

Measure the mean fluorescence intensity of the binding of GL7 monoclonal antibody to each cell sample using flow cytometry to obtain the relative expression profile of GL7 epitope among samples (Fig. 1A).

Label universal reference RNA (Clontech #636538) with Cy3 and poly(A)$^+$ RNA from each cell sample with Cy5, perform microarray analysis, and calculate Cy5/Cy3 value of each sample to obtain the relative gene-expression profile of each gene among samples (Fig. 1B).

Calculate Pearson's correlation coefficient of each gene between the relative gene-expression profile and the relative expression profile of GL7 epitope. In fact, one of sialyltransferase genes, *ST6GAL1*, with the highest correlation coefficient index was shown to be involved in the expression of GL7 epitope (Naito et al. 2007).

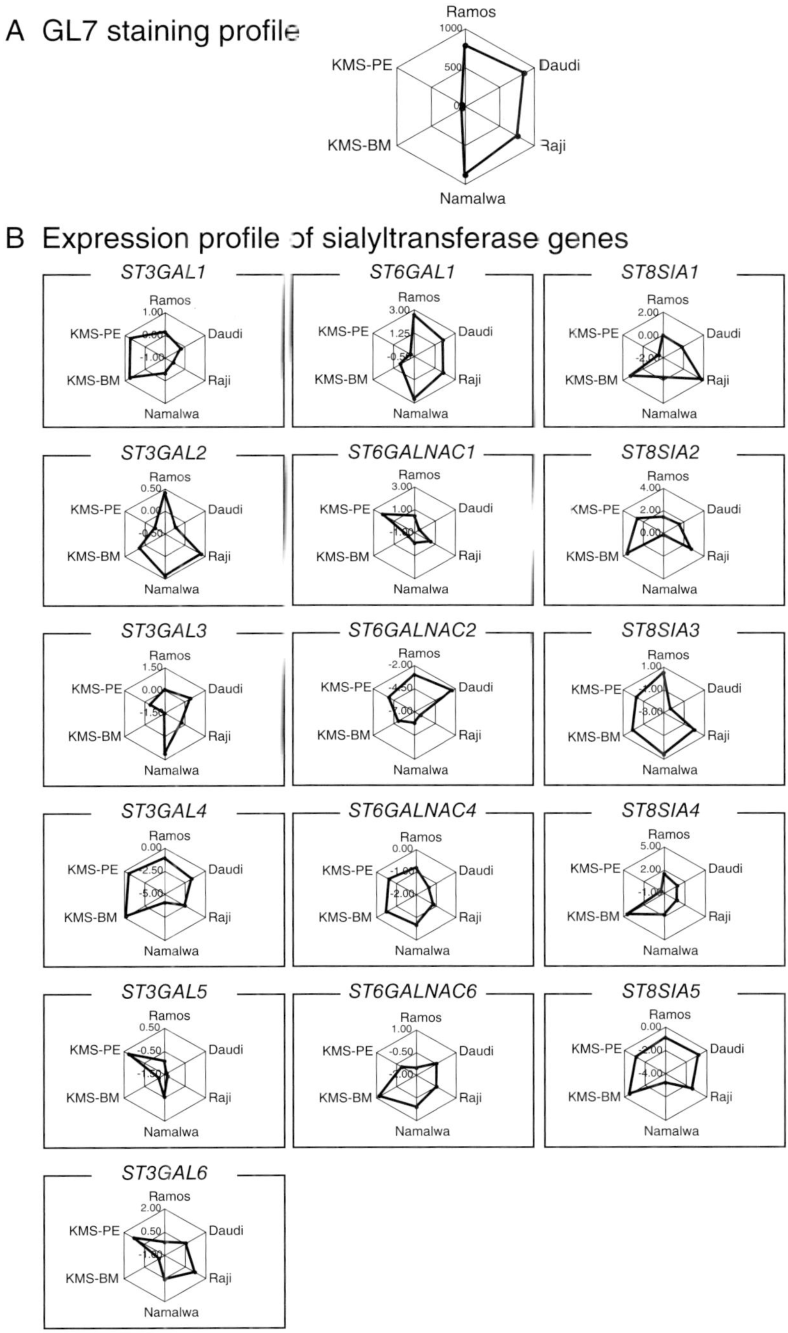

DNA Microarray in Glycobiology

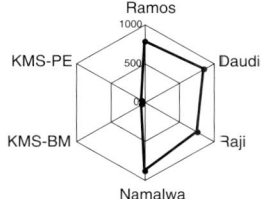

A GL7 staining profile

B Expression profile of sialyltransferase genes

Fig. 1 Relative expression profile of GL7 epitope (**A**) and relative gene-expression profiles of sialyltransferase genes (**B**) among human B cell lines

References

Brazma A, Hingamp P, Quackenbush J, Sherlock G, Spellman P, Stoeckert C, Aach J, Ansorge W, Ball CA, Causton HC, Gaasterland T, Glenisson P, Holstege FC, Kim IF, Markowitz V, Matese JC, Parkinson H, Robinson A, Sarkans U, Schulze-Kremer S, Stewart J, Taylor R, Vilo J, Vingron M (2001) Minimum information about a microarray experiment (MIAME)-toward standards for microarray data. Nat Genet 29:365–371

Ide Y, Miyoshi E, Nakagawa T, Gu J, Tanemura M, Nishida T, Ito T, Yamamoto H, Kozutsumi Y, Taniguchi N (2006) Aberrant expression of N-acetylglucosaminyltransferase-IVa and IVb (GnT-IVa and b) in pancreatic cancer. Biochem Biophys Res Commun 341:478–482

Koike T, Kimura N, Miyazaki K, Yabuta T, Kumamoto K, Takenoshita S, Chen J, Kobayashi M, Hosokawa M, Taniguchi A, Kojima T, Ishida N, Kawakita M, Yamamoto H, Takematsu H, Suzuki A, Kozutsumi Y, Kannagi R (2004) Hypoxia induces adhesion molecules on cancer cells: a missing link between Warburg effect and induction of selectin-ligand carbohydrates. Proc Natl Acad Sci USA 101:8132–8137

Naito Y, Takematsu H, Koyama S, Miyake S, Yamamoto H, Fujinawa R, Sugai M, Okuno Y, Tsujimoto G, Yamaji T, Hashimoto Y, Itohara S, Kawasaki T, Suzuki A, Kozutsumi Y (2007) Germinal center marker GL7 probes activation-dependent repression of N-glycolylneuraminic acid, a sialic acid species involved in the negative modulation of B cell activation. Mol Cell Biol 23:3008–3022

Lectin Microarray

Atsushi Kuno[1], Jun Hirabayashi[2]

Introduction

Lectin-based glycan profiling is an emerging technique to estimate an entire feature of glycosylation of glycoproteins and cells through biochemical interaction analysis between the analyte glycan and multiple lectins. Hereby, the strategy is basically distinct from other conventional methods represented by MS and HPLC (Hirabayashi 2004). Lectin microarray described in this chapter is one of the most sensitive, rapid and high-throughput profiling methods, which enables analysis of over a 100 of lectin–glycan interactions in a simultaneous manner. However, the biggest charm of the lectin microarray is its simple manipulation in order to analyze glycan structures in both pure and crude forms of glycoproteins, because analyte glycans need not be released from protein moiety nor highly purified as in the cases of HPLC and MS analyses. In fact, the lectin microarray has been recognized as a unique method to analyze glycosylation feature of diverse glycoproteins (Kuno et al. 2005; Rosenfeld et al. 2007), which include those of crude cell lysates, sera, and bacteria (Zheng et al. 2005; Ebe et al. 2006; Hsu et al. 2006; Pilobello et al. 2007). Through comparative microarray analysis (i.e., differential profiling), structural differences can be detected as the changes in signal patterns of the lectin-binding intensities. Therefore, this system is a good means for quality control of various glycoprotein products (e.g., antibody drugs), differential analysis of appropriate clinical samples for investigating useful glyco-biomarkers, and so on.

Because the interaction between lectins and glycans is relatively week compared with those between antigen and antibodies, a microarray scanner without washing processes should be used for the lectin microarray analysis. In fact, any washing process substantially wipes out relatively weak but significant binding-signals as a result of enhanced dissociation (Uchiyama et al. 2006). In this chapter, we describe below a standard protocol for glycan profiling using a developed lectin microarray system (Kuno et al. 2005) with the aid of an evanescent-field fluorescence-assisted array scanner recently commercialized by Moritex Co.

Protocol

We describe here one of the most frequently used protocols for lectin microarray analysis, i.e., a direct fluorescent labeling method prior to detection. In this protocol, analyte glycoproteins or glycopeptides are labeled with an appropriate fluorescence reagent. Excess

---

[1,2] Research Center for Medical Glycoscience (RCMG), National Institute of Advanced Industrial Science and Technology (AIST), Tuskuba Central 2, 1-1-1 Umezono, Tsukuba, Ibaraki 305-8568, Japan
Phone: +81-29-861-3187, Fax: −31-29-861-3125
E-mail: [1] atsu-kuno@aist.go.jp, [2] un-hirabyashi@aist.go.jp

reagents, which should generate high background noise, must be removed before probing on the lectin microarray.

Fluorescent Labeling of Glycoprotein

1. Prepare a glycoprotein solution in an appropriate reaction buffer free from primary amine. The concentration and amount of the glycoprotein solution should be at least 50 μg/ml and 1 μg, respectively.
2. Add an equal amount of Cy3-dye (GE Healthcare Bioscience) by weight to the glycoprotein solution.
3. Mix and incubate the solution at room temperature for 1 h under darkness so that the labeling reaction proceeds.
4. Fill up the reaction solution to 70 μl with the buffer.
5. Apply the resultant solution to the Zeba™ Desalt Spin Column (PIERCE) to remove the free Cy3-dye.
6. Centrifuge at $1,500 \times g$ for 2 min to collect a reagent-free sample.
7. Adjust the volume of the purified sample to 100 μl. The solution thus adjusted is designated fluorescent-labeled glycoprotein solution.

In case a relatively high-background scan image is obtained, the ratio of Cy3-dye to glycoprotein should be made 10 times higher than that of the above standard protocol for signal-to-noise improvement.

Lectin–Glycoprotein Interaction Reaction

1. Prepare 100 μl of the probing solution described above by dilution of the fluorescent-labeled glycoprotein solution to an appropriate concentration with PBS containing 1% Triton X-100 (PBSTx).
2. Discard the keeping buffer from a glass slide of Lectin Microarray, version 6.0 (Moritex, Co.), which consists of seven divided reaction wells, each comprising triplicate spots of 43 lectins.
3. To avoid drying, immediately after discarding the keeping buffer, apply 60 μl of the probing solution to each reaction well of the glass slide.
4. Place the glass slide in an appropriate humid chamber box and incubate at 20°C for over 3 h with shaking (the following steps are optional: if the scan data are expected to show high background noise due to an excess amount of unbound fluorescent compounds, the reaction solution should be replaced with an appropriate buffer).
5. Remove the reaction solution from each and rinse it with PBSTx three times.
6. Keep the glass slide on the wet condition by applying 60 μl of PBSTx upon scanning.

If you want to get stronger signals, the incubation time should be extended up to overnight.

Scanning and Data Analysis

1. Turn on a scanner, SC-profiler (Moritex, Co.).
2. Double click the "SC-Profiler" icon to start operation software.
3. Wait for at least 1 h to make the halogen lamp power and CCD camera stable.
4. Set a glass slide on the stage and push the "insert" button.

Fig. 1 Optimization of scan conditions for comparative analysis (differential profiling). Three sets of bar graphs show apparently different signal patterns, which are calculated from scan images of the same sample with different scan conditions (i.e., sample concentration and gain setting). **a** In case of a low sample concentration or low gain setting. **b** In case of an optimum condition. **c** In case of a high sample concentration or high gain setting

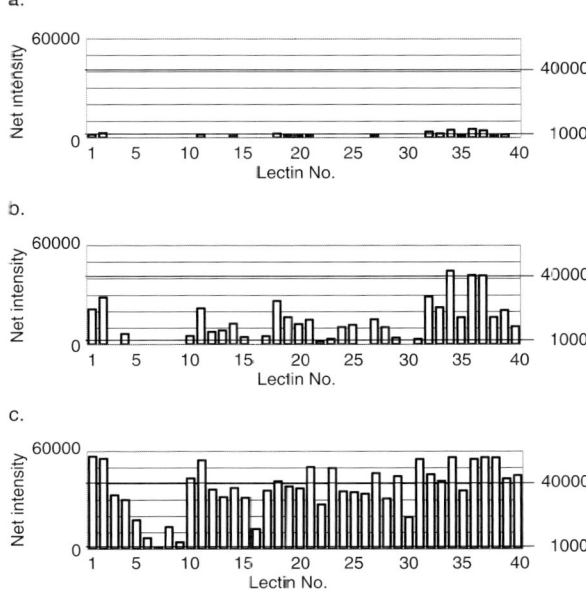

5. Select a scan area and integration times.
6. Set the gain, exposure time and file name.
7. Push the "start" button.
8. After scanning, make sure that the scan image has been saved as both TIFF and JPEG image files.
9. Eject the glass slide.
10. Quit the operation software and computer.
11. Turn off the SC-Profiler.
12. Analyze the obtained scan image saved as TIFF file by the Array Pro Analyzer version 4.5 (Media Cybernetics, Inc.). The net intensity value for each spot is calculated by subtracting the background value, and then three identical spots are averaged.
13. Standardize each set of signal data as relative value compared with that showing maximal fluorescent intensity observed on each lectin array.
14. Express the above processed data as a bar graph in an EXCEL format.

Due to the specificity of the CCD camera, the gain and exposure time should be set so that the observed fluorescent intensities of almost all positive spots on the glass slide fall within the range 1,000–40,000, which shows significant linearity (see Fig. 1).

References

Ebe Y, Kuno A, Uchiyama N, Koseki-Kuno S, Yamada M, Sato T, Narimatsu H, Hirabayashi J (2006) Application of lectin microarray to crude samples: differential glycan profiling of Lec mutants. J Biochem 139:323–327

Hirabayashi J (2004) Lectin-based structural glycomics: glycoproteomics and glycan profiling. Glycocoj J 21:35–40

Hsu KL, Pilobello KT, Mahal LK (2006) Analyzing the dynamic bacterial glycome with a lectin microarray approach. Nat Chem Biol 2:153–157

Kuno A, Uchiyama N, Koseki-Kuno S, Ebe Y, Takashima S, Yamada Y, Hirabayashi J (2005) Evanescent-field fluorescence-assisted lectin microarray: a novel strategy for glycan profiling. Nat Methods 2:851–856

Pilobello KT, Slawek DE, Mahal LK (2007) A ratiometric lectin microarray approach to analysis of the dynamic mammalian glycome. Proc Natl Acad Sci USA 104:11534–11539

Rosenfeld R, Bangio H, Gerwig GJ, Rosenberg R, Aloni R, Cohen Y, Amor Y, Plaschkes I, Kamerling JP, Maya RB (2007) A lectin array-based methodology for the analysis of protein glycosylation. J Biochem Biophys Methods 70:415–426

Uchiyama N, Kuno A, Koseki-Kuno S, Ebe Y, Horio K, Yamada M, Hirabayashi J (2006) Development of a lectin microarray based on an evanescent-field fluorescence principle. Methods Enzymol 415:341–351

Zheng T, Peelen D, Smith LM (2005) Lectin arrays for profiling cell surface carbohydrate expression. J Am Chem Soc 127:9982–9983

Carbohydrate Microarray for Deciphering the Information Embedded in Oligosaccharide Structures

Shigeyuki Fukui

Introduction

The neoglycolipid (NGL) technology for generating lipid-linked oligosaccharide probes has been shown to be an extremely useful tool for detection and characterization of bioactive oligosaccharides that are extremely minor components (Feizi and Child 1994). As the affinities of most carbohydrate–protein interactions are thought to be very low, and the oligosaccharide portions that play a role in biological function are generally believed to be of a small region ranging from monosaccharide to decasaccharide, the neoglycolipid technology is eminently adaptable for microarray design for high-throughput detection and specificity assignments of carbohydrate–protein interactions.

The neoglycolipid technology is also a useful tool to purify the sequence-defined oligosaccharides from the mixtures of oligosaccharides and glycolipids which have been obtained from biological sources such as cells and whole organs, because the amounts of individual oligosaccharides that can be isolated from natural sources are usually limited.

Fukui et al. (2002) have explored the carbohydrate microarray system applicable both to structurally defined oligosaccharides and to those derived from biological sources, glycoproteins, proteoglycans, cells and even a whole organ, and found it to be a useful tool for high-throughput detection of ligands for carbohydrate-binding proteins even though 1 pmol of oligosaccharides per spot was applied on the nitrocellulose membrane. One picomole of NGLs prepared from the fragments of chondroitin sulfates was recognized by specific antibodies on the nitrocellulose membranes as an example (Yamaguchi et al. 2006).

Recently, cytokines, growth factors, cytokines and chemokines are shown to bind to a range of sulfated or sialylated probes. Thus, the carbohydrate microarray technology brings us one step closer to surveying the entire glycome.

Concept

Carbohydrate microarray on which various kinds of NGLs were immobilized has become a more sensitive and convenient assay system than that of affinity chromatography and ordinary ELISA methods, and also enabled us to get high-throughput detection of ligand carbohydrate even in heterogeneous glycan populations. Although this NGL-based microarray, as shown in Fig. 1, does not give us any kinetic parameters for the

---

Department of Biotechnology, Faculty of Engineering, Kyoto Sangyo University, Motoyama, Kita-ku, Kyoto 603-8555, Japan
Phone: +81-75-705-1892, Fax: +81-75-705-1914
E-mail: shifukui@cc.kyoto-su.ac.jp

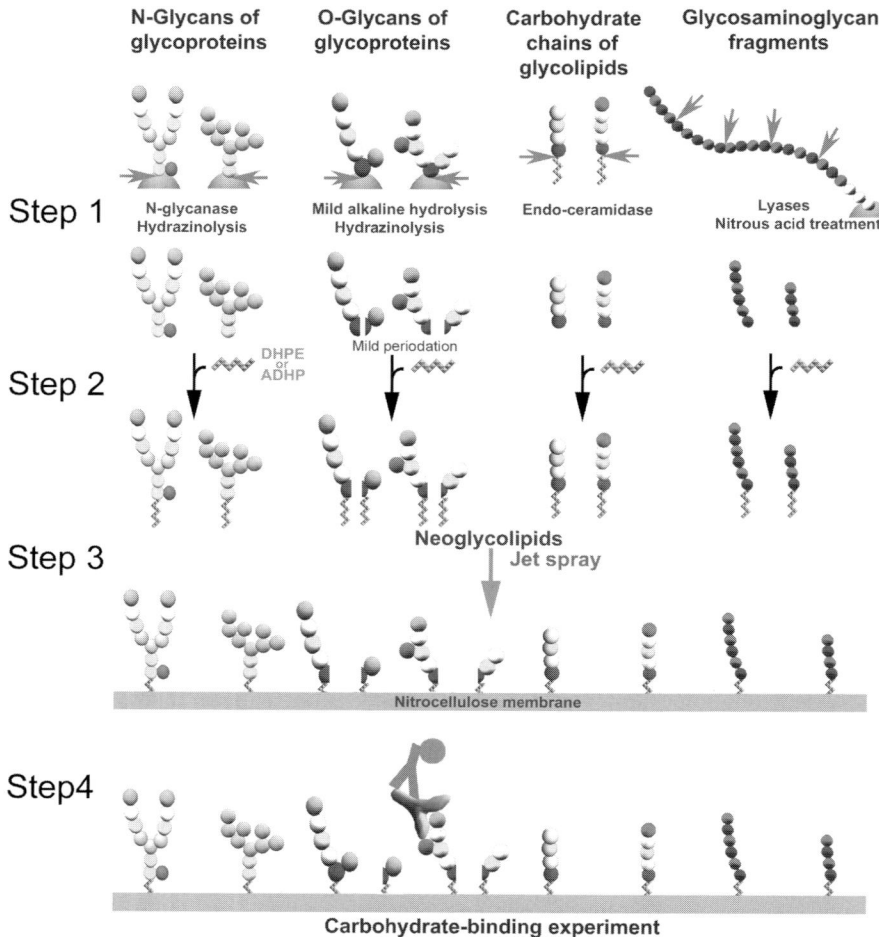

Fig. 1 Schematic representation of NGL-based carbohydrate microarray technology.
Reducing oligosaccharides are prepared from animal milk that can be isolated or that can be chemically synthesized, or that released by either chemical degradation or enzymic treatment of native glycoproteins and natural glycolipids and proteoglycans, as well as from reduced oligosaccharides alditols that are released by reductive alkaline elimination from O-linked mucin-type oligosaccharides followed by the treatment of mild periodation (Step 1). The oligosaccharides isolated are conjugated with DHPE or ADHP by a reductive-amination procedure to an aminolipid (neoglycolipid, NGL) (Step 2). Jet spray allows the NGLs to be immobilized on the surface of nitrocellulose membranes (Step 3). Immobilized NGLs are then probed for binding. After treatment of HRP-conjugated second antibody, the binding signals of carbohydrate-binding proteins were visualized by using diaminobenzidine reagent (Step 4)

association and dissociation constants between ligand oligosaccharides and carbohydrate-binding proteins, this technology has an advantage of easier handling and higher sensitivity than several other published carbohydrate microarray systems that involve either the non-covalent immobilization of carbohydrates and their derivatives on solid matrices or the covalent attachment of the carbohydrates to solid matrices by the chemical treatment such as Diels-Alder reaction (Houseman and Mrksich 2002; Schwarz et al.

2003). However, another advantage of this NGL-based microarray is that the visual-binding signals is well correlated with the strength of the binding affinity between the ligand oligosaccharide and the protein due to the necessity of another step to overlay second antibody, and following several washing steps before visualization of the binding signals.

Recently immobilized carbohydrate arrays with glycosaminoglycan-derived oligosaccharide fragments were applied to screen their binding activities to hepatocyte growth factor (HGF) and fibroblast growth factor (FGF-7), and it was demonstrated that the minimum size of dermatan sulfate chains required for binding to both factors are 8- or 10-mers, and almost all of the internal uronic acid residues are iduronic acid (Yamaguchi et al. 2006).

Preparation of Neoglycolipids for Carbohydrate Microarray According to the Method of Feizi et al. (1994)

1. To the dry oligosaccharide(s) (10–50 nmol) in the glass microvials (fitted with Teflon-lined caps), add either 5 mg/ml 1,2-dihexadecyl-sn-glycero-3-phosphoethanol-amine (DHPE) in chloroform–methanol (1:1, v/v) or 2 mM N-aminoacetyl-N-(9-anthracenylmethyl)-1,2-dihexadecyl-sn-glycerol-3-phosphoethanolamine (ADHP) in chloroform–methanol–water (10:10:1, v/v) to give 8 mol ratio of DHPE or 3 mol ratio of ADHP, respectively.

2. Sonicate for 5 min and heat at 60°C for about 1 h

3. In case of synthesizing DHPE-neoglycolipid, add freshly prepared sodium cyanoborohydride, 10 mg/ml in methanol, to give a 1:2 mole ratio of sodium cyanoborohydride to DHPE. For ADHP-neoglycolipid, add freshly prepared tetrabutylammonium cyanoborohydride, 28.2 mg/ml in methanol, to give a 1:10 mole ratio of tetrabutylammonium cyanoborohydride to ADHP.

4. To purify the neoglycolipids, thin layer chromatography is carried out. HPTLC plates, glass-backed or aluminum-backed, are typically cut to 10 cm × 10 cm.

5. Reaction mixture is applied 15 mm from the bottom edge as a 70 mm band with 15 mm free spaces at both sides. Chromatography is performed at ambient temperature (15°–23°C) until the solvent front reaches 5 mm from the top edge.

6. After development and drying the HPTLC plates, cut both the edges with 16 mm width in case of DHPE-neoglycolipid. The position of DHPE-neoglycolipid bands can be identified under UV lamps at 365 nm after spraying with primulin reagent at both edge plates. The position of ADHP-neoglycolipid can be seen as a blue band under UV at 254 nm. The outline of bands is pencil-marked.

7. The silica gel within the pencil-mark is scraped from the HPTLC plate. Neoglycolipids are extracted several times with a solvent of chloroform–methanol–water, 25:25:8, v/v.

Comment

For neutral and acidic oligosaccharides up to octa- and tetrasaccharides, respectively, add water to the dry oligosaccharides to give up 5–10%(v/v) water in the final reaction volume.

References

Feizi T, Child RA (1994) Neoglycolipids: probes in structure/function assignments to oligosaccharides. Methods Enzymol 242:205–217

Feizi T et al (1994) Methods Enzymol 230:484–519

Fukui S, FeiziT, Galustian C, Lawson AM, Chai W (2002) Oligosaccharide microarrays for high-throughput detection and specificity assignments of carbohydrate–protein interactions. Nat Biotechnol 20:1011–1017

Houseman BT, Mrksich M (2002) Model systems for studying polyvalent carbohydrate binding interactions. Top Curr Chem 218:1–44

Schwarz M, Spector L, Gargir A, Shtevi A, Gortler M, Altstock RT, Dukler A, Dotan N (2003) A new kind of carbohydrate array, its use for profiling antiglycan antibodies, and the discovery of a novel human cellulose-binding antibody. Glycobiology 13:749–754

Yamaguchi K, Tamaki T, Fukui S (2006) Detection of oligosaccharide ligands for Hepatocyte growth factor/Scatter factor (HGF/SF), Keratinocyte growth factor (KGF/FGF-7), RANTES and Heparin cofactor II by neoglycolipid microarrays of glycosaminoglycan-derived oligosaccharide fragments. Glycoconjugate J 23(7–8):513–523

Glossary

8-aminopyrene-1,3,6-trisulfonate (APTS) It is a fluorescent-labeling group for reducing ends of glycans. This labeling helps to achieve fast and supersensitive detection of glycans by capillary electrophoresis with argon laser-induced fluorescence detection.

α-1,6-fucosyl transferase This enzyme is also called Fut8. It transfers Fucose to the GlcNAc residue at the most proximity of N-linked glycan chain by the linkage of α-1,6. It has been reported to have pivotal roles in the context of development or differentiation.

α-dystroglycanopathy α-dystroglycanopathy collectively means the muscular dystrophy syndromes resulted from glycan structure deficiency of α-dystroglycan. It is proposed that the malformation of the glycans in α-dystroglycan hinders the binding to ligands, such as laminin, and results in muscular dystrophy because of the degeneration of the intra- or intercellular connection.

α-galactose epitope Humans, primates, and old world monkeys do not have the glycan epitope Galα1-3Gal structure, because the α3 galactosyltransferase responsible for the structure became pseudo-gene in those mammals. Human serum contains approximately 1% of natural antibody IgG against Galα1-3Gal, which would give rise to a serious problem when porcine organs are to be subjected to xenograft into humans.

α-galactosyl ceramide See "NKT cells".

α-mannosidase α-mannosidase is one of the hydrolyzing enzymes, which cleaves off the α1-2, α1-3, and/or α1-6 mannose residue. It includes the processing enzymes localized in the endoplasmic reticulum or Golgi structure, and degrading enzymes that function on the glycoproteins located in the lysosome or cytosol.

α-N-acetylglucosamine residue It was discovered with a structure, GlcNAcα1-4Galβ, at the non-reducing end of O-linked glycan. Humans have such a structure, specifically in the mucus secreted by gland mucus cells dwelling in the middle to the lower layer of the gastric mucosa, the Brunner's gland in the duodenal mucosa, or pancreatic duct epithelial cells showing metaplasia into gastric pyloric gland tissue. This structure was also highly expressed in the gastric, pancreatic or biliary cancer. Thus, it is considered that the α-GlcNAc residue is the cancer-related glycoantigens in those cancers.

ABO blood group antigens The ABO antigens, major blood group antigens, were originally identified in erythrocytes, but they are also found on the mucosal epithelial cells or in the mucosa of various tissues. A type antigen was identified to have a glycoantigen structure, GalNAcα1,3Gal(Fucα1,2)-R, whereas B type has Galα1,3Gal(Fucα1,2)-R, and they are found in glycoproteins and glycolipids.

Highly homologous allelic genes at the ABO blood group locus on the chromosome 9 specify the specificity of transferases involved in the ABO antigen synthesis. H antigen (Fucα1,2Gal-R) is the precursor for those structures.

Adhesion molecule Adhesion molecule is involved in cell–cell and cell–matrix interactions. Epithelial cell shows three types of cell junctions: tight junction, adherence junction, and desmosome. A variety of adhesion molecules, claudine, ocludin, or cadherin, are known to participate in cell–cell interactions. Hemidesmosome is known as the adhesion machinery between cell and extracellular matrix, and integrin molecules are involved in the cell-matrix interaction.

Algorithm Algorithm is a type of effective method, which has a definite list of well-defined instructions for completing a task. Program is the instruction for a computer to carry out the task on the basis of the algorithm.

Alzheimer disease/Alzheimer's disease/AD Alzheimer disease (AD) is a neurodegenerative disease and is the most common cause of dementia. AD is characterized by loss of neurons and synapses, which results in gross atrophy of the affected brain regions. Typical pathological features of AD are deposition of β-amyloid peptide (Aβ) and neurofibrillary tangles in the brain. Deposition of Aβ in the brain is considered to be a major cause of AD, and therapeutic agents to inhibit Aβ production or to stimulate Aβ clearance are in a developmental stage.

Animal lectin Modern research on lectins, carbohydrate-binding protein, started from a study on plant lectins from legume seeds. Carbohydrate-binding proteins have been found in a wide variety of mammalian organs, such as a liver, and body fluids, and are called animal lectins. Animal lectins include C-type lectin, galectin, siglec, etc. Animal lectins are characterized by a cluster effect and carbohydrate recognition known as pattern recognition. The functions of especially host defense and ER quality control mechanisms are being clarified.

β-1,3-linkage glycosyltransferase motif There is a stretch of amino acid sequence, well conserved among the enzymes, which transfer sugars by β1,3-linkage (galactosyltransferase, N-acetylglucosaminyl transferase, N-acetylgalactosaminyl transferase, etc.). Each of this enzyme family has three rows of amino acid motif and each motif spans several to dozen amino acid residues. Meanwhile, the motifs are not very well conserved in core 1 Gal-T (C1GALT1,2) or iGnT.

β-1,3-N-acetylglucosaminyl transferase This is a generic name for the enzymes transfer GlcNAc by β1,3-linkage. The reported enzymes in this group are the following: β3GnT5, which is involved in glycolipid Lc3Cer; β3Gn-T6, which synthesizes O-linked glycan core3 structure; β3Gn-T2, which functions for the synthesis of keratan sulfate; T3, which is involved in the extension of core1, T4 and T8.

β-1,4-galactosyltransferase The enzymes in this group transfer galactose by β1,4 linkage. Seven human enzymes of this group have been reported so far and they were found to contain β1,4-linkage glycosyltransferase motif in the polypeptide.

β-1,4-N-acetyl galactosaminyl transferase This enzyme transfers N-acetylgalactosaminyl residues by β1,4-linkage. The enzymes in this group are classified into three groups according to the difference in the acceptor substrate: (1) enzymes, such as β4GalNAc-T1 and T2, whose acceptor is sialylated galactose to produce GM2 or Sda antigen, (2) enzymes, such as β4GalNAc-T3 and T4, whose substrate is N-acetylglucosamine and produce LacdiNAc structure, and (3) five enzymes involved in the synthesis of chondroitin sulfate from glucronic acid as substrate.

β-1,6-N-acetylglucosaminyl transferase The enzymes in this group transfer N-acetylglucosamine by β1,6-linkage. They include enzymes, which are categorized as follows: enzymes involved in N-linked glycan synthesis (MGAT5 and MGAT9), enzymes that participate in the synthesis of O-linked glycan-core structure, such as core2 Gn-T (C2GnT1 ... 3), IGnT, which produces I-type blood antigen, and others.

β-4 galactosyltransferase motif There is a stretch of amino acids well conserved among β-1,4 galactosyl transferase. This motif is also found in some β-1,4-linkage glycosyltransferases other than β-1,4 galactosyl transferases. The consensus sequence of this motif is WGXEDD/V/W.

β-amyloid peptide It is strongly believed that the deposition of β-amyloid peptide (Aβ) in the brain is implicated in the pathogenesis of Alzheimer disease. Aβ consists of 40–42 amino acids and is generated from β-amyloid precursor protein (APP) by the proteolytic β and γ cleavage.

β-galactose 3-O-sulfotransferase Sulfation of glycans at the position 3 of galactose residues has been found in glycolipids and glycoproteins in mammal, and occurs by the function of β-galactose 3-O-sulfotransferase (Gal3ST) family. This family constitutes four member enzymes, which were numbered according to the order of discovery. Only Gal3ST-1 (cerebroside sulfotransferase, CST) functions on glycolipid, whereas others work on glycoproteins.

β-galactoside β-galactoside is a β anomeric isomer of the galactose residue. Galectin was the lectin originally called β-galactoside-binding lectin, because it recognizes the β-galactoside structure specifically, until it was renamed Galectin in 1993.

β-galactosyltransferase β-galactosyl transferase transfers galactose residues by β-linkage. In humans, six β-1,3-galactosyltransferases, seven β-1,4-galactosyltransferases, and one galactosyl ceramide synthases are reported.

BACE1 BACE1 is a protease (β-secretase), which cleaves beta-Amyloid precursor protein (APP) at the β-site. BACE1 cleavage initiates the production of β-amyloid peptide (A β) to form amyloid plaque, which is considered to be a major cause of Alzheimer disease (AD). Therefore, BACE1 inhibitors are being developed as promising therapeutic agents for AD.

Bifidobacteria growth stimulator Milk and some other particular foods have components to stimulate the growth of bifidobacteria, which reside in the colon and have a protective role against infectious disease. Milk oligosaccharides are believed to be bifidobacteria growth stimulators. Recently, oligosaccharides having lacto-N-neo-biose I (e.g., lacto-N-neo-tetraose) have been hypothesized to be growth stimulators for *Bifidobacterium bifidus* and *Bifidobacterium longum*.

Biocombinatorial synthesis of glycan chains Biocombinatorial synthesis of glycan chains is a technique, which involves the synthesis of diverse oligosaccharide structures by combination of glycan chain primers and different kinds of cell types. By choosing appropriate glycan primers, considering cell types depending on their glycan biosynthetic pathway, this technique would be promising in synthesizing diverse kinds of glycan chains.

Bioinformatics Bioinformatics is a coinage designed in the 1990s from the combination of "biology" and "informatics". Bioinformatics and computational biology are used interchangeably. In a narrow sense, bioinformatics is concerned with biological information. In a wide sense, bioinformatics include the field of computational and statistics related to diverse biological phenomena (e.g., sequence alignment, protein structure alignment, and genome annotation).

Botulinum toxin/botulin Botulinum toxin is a neurotoxin protein produced by the bacterium *Clostridium botulinum*. Orally administered toxin is absorbed by the small intestinal epithelium and hematogenously reaches neurons, in which the toxin is endocytosized. The light chain of botulinum toxin is a protease that degrades SNAP-25, a membrane bound Q-SNARE protein. Because SNAP-25 is required for the release of neurotransmitters, such as acetylcholine, botulinum toxin suppresses the release of acetylcholine. Botulinum toxin is now used in the treatment of blepharospasm, urgency incontinence, and cosmetic surgery.

CabosDB CabosDB is an abbreviation for carbohydrate sequencing database and enables the researchers to find out information on glycan structural analysis. It includes information on oligosacharide structures and their mass analysis spectrum, lectins and their binding specificities to glycans, and glycoprotein and their glycosylated positions.

Caenorhabditis elegans (C. elegans) *Caenorhabditis elegans* is a model organism widely used in genomic and post-genomic studies. The mechanism of apotosis was elucidated from the study of *C. elegans*. RNAi and GFP technologies were developed using *C. elegans*. *C. elegans* consists of ca 1000 somatic cells and the length is about 1 mm. *C. elegans* could be cryopreserved and all the developmental processes and neural network system have been already elucidated. *C. elegans* is the only multicellular organism in which high-throughput gene disruption is possible.

Cancer metastasis Metastasis is the major factor that influences the prognosis of cancer. In cancer metastasis, cancer cells can break away from a primary tumor, enter lymphatic and blood vessels, circulate through the bloodstream, and settle down to grow within normal tissues elsewhere in the body. Metastasis can be roughly classified into hematogenous metastasis, lymphgenous metastasis, or peritoneal disseminated metastasis, depending on the metastatic path, though they are not mutually exclusive but can be related with each other, progressing simultaneously.

Capillary affinity electrophoresis Capillary affinity electrophoresis (CAE) is carried out in the electrolyte containing ligands or proteins that bind to ligands. It is a technique in which the migration patterns of interacting molecules in an electrical field are observed and used to quantify specific binding and estimate binding constant. When a carbohydrate in the mixture interacts with carbohydrate-binding protein, CAE allows determining subtle difference in affinities among the mixtures of carbohydrates. CAE also allows characterization of the structures of carbohydrates, on the basis of their affinities for an appropriate set of carbohydrate-binding proteins.

Capillary electrophoresis Capillary electrophoresis is a high-resolution separation technique that works on the basis of the different behavior of charge molecules in a high electric field (typically 10–30 kV). The technique uses a narrow (25–100 mm i.d.) capillary column that has borne silica, and its inner surface is coated with stationary phase. Thus, it provides highly efficient separations up to 1 000 000 theoretical plates. By connection with laser-induced fluorescent detector and with mass spectrometer, the technique also allows high-sensitive detection and structural characterization of samples.

Carbohydrate-deficient glycoprotein syndrome (congenital disorders of glycosylation, CDG) The standard name was changed from "carbohydrate-deficient glycoprotein syndrome" to "congenital disorders of glycosylation syndrome". In the narrow sense, CDGs are generic names of syndromes caused by genetic disorders of the enzyme related to N-glycosylation. CDGs can be distinguished into the disorder of biosynthesis of dolichol-linked oligosaccharide in the rough-surfaced endoplasmic reticulum (ER) and glycosylation of protein (CDG-I) and into that of the processing in the ER and Golgi apparatus of the N-linked glycans attached to proteins (CDG-II). The name of the disease is described by fixing alphabets on the basis of each responsible gene. The condition affects a wide variety of organs, especially the nervous system.

Carbohydrate-deficient phenomenon This is a concept proposed by Hakomori et al. as one of the patterns in the change of carbohydrates with malignant transformation. In normal cells, the structure of carbohydrates being synthesized is of the complex type. Malignant transformation causes trouble in a part of the carbohydrate synthesis, and in malignant cells, carbohydrates that have a simpler structure than in normal cells accumulate. Such a phenomenon is called carbohydrate-deficient phenomenon.

Carbohydrate library Carbohydrate library is a series of various carbohydrates or their derivatives constructed by organic synthesis, enzyme reaction, and extraction from natural resources. Besides being used as standard samples for structural analysis, carbohydrate library could be applied to a comprehensive analysis of interactions by combination with microarray technology, which would become a powerful tool for glycomics.

Carbohydrate ligands for selectins Selectin molecules mainly recognize carbohydrates as ligands, and each selectin binds to a specific ligand. E-selectin binds to sialyl Lewis x and sialyl Lewis a. P-selectin binds to sialyl Lewis x on the PSGL-1 molecule carrying sulfated tyrosine residue(s) in its N-terminal region. L-selectin binds to sialyl 6-sulfo Lewis x.

Carbohydrate microarray Carbohydrate microarray is prepared by immobilizing a variety of carbohydrates onto microarray slides such as glass plates. The microarray is typically used for the observation of the interactions between immobilized carbohydrates and fluorescent-labeled proteins. Sometimes, carbohydrate microarray is confused with carbohydrate chip. In general, the number of sugars on a carbohydrate microarray is larger than that on a sugar chip. Recently, a new type of carbohydrate microarray has been developed to observe the interaction with surface plasmon resonance (SPR) techniques, in which fluorescent labeling of protein is not necessary.

Carbohydrate receptor Carbohydrate receptor is a generic name of a receptor or a lectin binding to sugar chains in the broad sense. In enzyme replacement therapy for lysosomal disease (lysosomal storage disease), it is a receptor, such as a cation-independent mannose 6-phosphate receptor or a mannose receptor, on the target cell surface that recognizes and binds to a specific sugar chain structure and mediates internalizing the sugar chain via endocytosis and transportation into lysosome.

Carbohydrate recognition cytokine Cytokines that are small amounts of various bioactive substances bind to corresponding specific receptors and activate cells. In case cytokines bind to receptor subunits to form high-affinity complex, (cytokine)$_n$(receptor subunit)$_n$, carbohydrate recognition cytokines with lectin-like activity that trigger the recognition of receptors themselves, glycolipids close to receptors, GPI anchors, high-mannose-type sugar chains or heparan sulfate chains exist.

Carbohydrate xenoantigen Carbohydrate xenoantigens are glycan epitopes which are not found commonly in humans, but in other animals. For example, H–D antigen, Forssman antigen (GalNAcα1-3GalNAcβ1-4Galα1-4Galβ1-4Glcβ1-ceramide), Paul–Bunnell (P–B) antigen (erythrocyte glycoprotein antigen found in sheep, horse, or cow), or Galβ1-3Gal antigen are not found in humans, primates, and old world monkeys. These antigens are potential cause for immunological rejection against glycoprotein drugs prepared from non-human cells, xenotransplantation, or transplantation of human ES cells cultured under FBS (fetus bovine serum).

CD1d CD1d is an MHC class I-like molecule on antigen-presenting cells and immature thymus cells, which present the glycolipid with Galα1-linkage to T cell receptor (TCR) on NKT cells. Recently, its crystal structure was resolved and putative binding sites to glycolipids was proposed. Meanwhile, it is still disputed about which glycolipid is the in vivo ligand. CD1d is getting to be a hot topic in its relation with the activation of NKT cells.

Ceramidase Ceramidase is an enzyme, which hydrolyzes the amide bond between the sphingosine base and fatty acid in free ceramide. This enzyme is widely conserved from bacteria to humans.

It is classified into: (1) neutral ceramidase, (2) acid ceramidase, and (3) alkaline ceramidase according to the primary structure and optimum pH. The acid ceramidase exists in lysosome and the deletion of the enzyme results in Fabry's disease. The neutral ceramidase is mainly localized in the plasma membrane and modulates signal transduction through sphingolipid mediator.

Ceramide Ceramide is an *N*-acylsphingosine lipid that anchors the glycan of glycosphingolipid or phosphocholine of sphingomyelin onto the plasma membrane. Ceramides are produced by the degradation of sphingolipid or by a de novo biosynthetic pathway, and the amount of production is enhanced by the following stimuli: cytokine, heat, or ultraviolet. Ceramides are known to control a variety of signal transduction systems. Phosphorylated sphingosine is a ligand for G-protein coupled receptors (S1P, Edg).

Cerebroside sulfotransferase (CST) Cerebroside is a common name for galactosylceramide. Cerebroside sulfotransferase is a sulfatide synthase that transfers a sulfate onto the 3 position of galactose of galactosylceramide (cerebroside). This enzyme synthesizes not only sulfatide, but also lactosylceramide sulfate and seminolipid.

Chemoenzymatic synthesis To synthesize glycoconjugates, there are two major methods: chemical method and enzymatic method. Chemoenzymatic synthesis is the combination of these two methodologies, to utilize the advantages of the respective methods. For instance, the chemically synthesized glycopeptides with a monosaccharide can be utilized for synthesis of neoglycoconjugates through transglycosylation reaction of endoglycosidases.

Chondroitin sulfate Chondroitin sulfate is a glycosaminoglycan that consists of repeating disaccharide units of glucuronic acid and GalNAc. Various sulfation reactions give rise to structural diversity of this glycan. It occurs ubicuitously in most tissues, most abundantly in cartilage. Various bioactivities such as neurite outgrowth-promoting activity have been reported.

Chondroitin sulfate glucuronic acid transferase Chondroitin sulfate glucuronic acid transferase has a glucuronic acid transferase activity. The enzyme is involved in the synthesis of chondroitin (sulfate), having the repeating disaccharide unit, -3GalNAcβ1-4GlcNAcAβ1. It forms a gene family with chondroitin synthase, chondroitin polymerizing factor (ChPF) and chondroitin sulfate N-acetylgalactosamine transferase.

Chondroitin sulfate-N-acetylgalactosamine transferase Chondroitin sulfate-N-acetylgalactosamine transferase is involved in the synthesis of a repeating unit of chondroitin (sulfate), 3GalNAcβ1-4GlcAβ1. So far two distinct enzymes are reported, ChGn-1 and 2. It forms a gene family with chondoroitin synthetase, chondroitin-polymerizing factor, and chondroitin sulfate glucuronyltransferase.

Chondroitin sulfate synthase Chondroitin sulfate synthase is an enzyme having both N-acetylgalactosamine transferase and glucuronic acid transferase activities, participating in the synthesis of chondroitin (sulfate). Two kinds of enzymes, chondroitin synthase (ChSy)-1 and -2 were reported. Each ChSy can form a complex with chondroitin polymerizing factor (ChPF).

Choresteryl-α-D-glucopyranoside Choresteryl-α-D-glucopyranoside is a glycolipid where α1-3-linked glucose is attached to chresterol, often abbreviated as CGL. CGL, together with FAG (Chroresteryl-6-O-tetradecanoyl-α-D-glucopyranoside) or CPG (choresteryl-6-O-phosphatidyl-α-D-glucopyranoside), occurs in the cell wall of Helicobacter species including *H. pylori*, and is involved pivotally in growth, motility, and maintenance of morphology of these bacteria.

Chromosome mapping Chromosome mapping is to determine the position of genes on chromosomes. Fluorescently labeled DNA (e.g., FITC-DNA) is used as a probe for chromosome mapping. The position of the gene can be determined by the correlation between the chromosome band and the fluorescent signal.

C-linked glycosylation of protein Protein glycosylation was classified into two types: N-linked type (Asn-linked type) and O-linked type (Ser/Thr-linked type). Recently, tryptophan-linked mannose (C-mannosyl tryptophan) was discovered and the function is a topic of great interest.

COG complex COG is an abbreviation for conserved oligomeric Golgi complex. It is an octameric complex, which is considered to regulate the COP-I vesicular transport between Golgi stacks. Recently, this complex has received special attention after the discovery of a pathogenesis showing irregular glycosylation (CDG; congenital disorders of glycosylation) because of the deficiency in a part of the complex.

Collectin Collectin is a collective name of animal lectins bearing a conserved carbohydrate-recognition domain (CRD) as well as collagen-like domain (Gly-X-Y triplet), collagen-like lectin. It includes mannan-binding protein (MBP), conglutinin, CL-43, CL-46, surfactant protein A and D (SP-A and SP-D), CL-L1, CL-L2, and membrane type CL-P1. This protein is known to be implicated as modulators of the innate immune response, where it has a key role in the first line of defense against invading microorganisms.

Combinatorial chemistry Combinatorial chemistry is to access vast numbers of molecules efficiently in a combinatorial fashion. Combinatorial chemistry is exploited in the optimization of chemical reaction or generation of novel functional materials. Solid-phase or multi-component synthesis is a representative of combinatorial chemistry.

Complement-dependent lectin pathway The complement system is a host defense system designed to destroy pathogens. Once the complement system is activated, a chain of reactions involving proteolysis and assembly occurs, resulting in the destruction of the membranes of the pathogens. There are three categories of activation pathways: the classical pathway, alternative pathway, and lectin pathway. As compared with the classical pathway in which antibodies are required, the latter two pathways do not require antibodies. Complement-dependent lectin pathway is activated by the binding of mannan-binding protein (MBP, also called MBL) to carbohydrates on the surface of pathogens. Binding of MBP to pathogens triggers the activation of MASP (MBP-associated serine protease), which is bound to MBP, and C4, C2, and C3 subsequently undergo limited proteolysis to be activated. Ficolin, GlcNAc-binding lectin, has been recently identified as another recognition molecule of the lectin pathway.

Complex-type N-glycan Vertebrate N-glycans are classified into three types: high-mannose, hybrid, and complex. In case of the complex-type N-glycan, both $\alpha 3$- and $\alpha 6$-linked mannose residues are substituted with GlcNAc residues. Most secreted and cell surface glycoproteins have complex-type N-glycan.

Conditional knockout mice A knockout mouse is a genetically engineered mouse that has a particular gene inactivated by replacing or disrupting it with an artificial piece of DNA. Analysis of knockout mice will provide researchers with valuable clues about the gene function in vivo. *Conditional knockout strategy* works on the basis of a tissue-specific inactivation of the gene of interest, and it is a powerful technique especially when constitutive knockout mice are lethal. This can be achieved by means of Cre-Lox recombination system, which involves the targeting of a specific sequence of DNA having lox P sites (consisting of 34 bp) and splicing it with Cre protein, a site-specific DNA recombinase.

Congenital muscular dystrophy The term congenital muscular dystrophy is used to describe the muscular dystrophy present at birth. Muscular dystrophies are characterized by progressive skeletal weakness, defects in muscle proteins, and the death of muscle cells and tissues. Fukuyama congenital muscular dystrophy is a form of muscular dystrophy mainly described in Japan.

Core-3 glycans Core-3 glycans is one of the core structures of mucin type, O-linked glycans, GlcNAcβ1-3GalNAcα1-Ser/Thr. It is found in normal stomach and intestine, whereas the level of core 3 structure is reduced in cancer tissues.

Dendrimer Dendrimers are repeatedly branched molecules showing dendric architectures. Dendrimers are usually sphere or sector structures. The number of branching is called "generation". Dendrimers are composed of three elements: core, interior, and surface and have the potential to express a variety of functions by designing each element. With great attention, dendrimers are regarded as a new nanomaterial in glycotechnology and other material science fields.

Diabetes mellitus Diabetes mellitus is a metabolic syndrome characterized by chronic high blood sugar (hyperglycemia), which is related to genetic factor and environmental factors such as lifestyle. Diabetes mellitus is characterized by a deficiency in the effect of insulin, which is caused by absolute or relative insufficiency of insulin and/or decrements in sensitivity to insulin. Diabetes mellitus is roughly classified into type 1, caused by auto-immune mechanism, and type 2. It is reported that, in carbohydrates, the activation of the hexosamine pathway and errors in ganglioside metabolism are the factors of insulin resistance.

DNA microarray DNA microarray is a small chip made of glass or some other chemical materials on which various DNA fragments are spotted and arranged, and is used to analyze the expression of multiple genes simultaneously. Thus, it can tell the gene expression profile. Generally, it is used to compare the relative expression levels of those genes in a couple of samples by comparing the signal intensity after the hybridization of cDNA from both samples labeled with different fluorescent dyes.

Drosophila *Drosophila* (*Drosophila metanogaster*) is a small fly that is 2–4 mm long. Full-length genome of *Drosophila metanogaster* has been sequenced, and *Drosophila* is extensively used as a model organism in many scientific fields. Recently a lot of glycogenes have been identified and the functional analyses are now in progress.

Dystroglycan Dystroglycan is a component of dystrophin-glycoprotein complex formed in sleletal muscle. Dystroglycan is encoded by a single gene and the protein is cleaved into α and β subunits by posttranslational processing. α-Dystroglycan is an extracellular glycoprotein and binds to laminin. On the other side, β-Dystroglycan is a transmembrane glycoprotein that binds to dystrophin in the cytosol. Lack of dystroglycan in knockout mice results in embryonic lethality. Lack of dystroglycan in conditional knockout mice, specifically in the brain, peripheral nerve, or skeletal muscle, avoids embryonic lethality, resulting in abnormalities of each tissue.

EGF domain EGF is an abbreviation for epidermal growth factor. EGF domain is an amid acid motif homologous to epidermal growth factor (EGF) and recently it was shown that sometimes this domain is *O*-fucosylated and/or *O*-glucosylated.

EGF receptor EGF receptor is a type I transmembrane protein composed of 1186 amino acids and has 11 possible *N*-linked glycosylation sites. It transduces signals to the downstream molecules by dimerization and autophosphorylation on stimulation by the ligand (e.g., EGF) binding. PhospholipaseCγ/CaMK/PKC pathway and Ras/Erk cascade are some of the well-known downstream pathways of the EGF receptor.

Endo-α-*N*-acetylgalactosaminidase It is an enzyme that cleaves mucin-type glycans from glycoproteins. First found in the culture media of *Diplococcus pneumonia* in 1976, similar activity was observed in the culture media of *Clostridium* or *Alcaligens*. These enzymes have very strict substrate specificity, and only act on *O*-glycans with disaccharides, Galβ1-3GalNAc. On the other hand, new type of endo-α-*N*-acetylgalactosaminidase with broader glycan specificity has recently been isolated from *Streptomyces*, which has more potential to become a powerful tool reagent.

Endo-β-*N*-acetylglucosaminidase (ENGase) ENGase is a glycosidase that cleaves a glycoside bond in the proximal *N,N'*-diacetylchitobiose (GlcNAcβ1-4GlcNAc), releasing *N*-glycans from glycopeptides/glycoproteins. Various distinct enzymes with different glycan specificity have been identified. For instance, an enzyme called Endo H predominantly acts on high mannose glycans, whereas another enzyme called Endo D can also work on some complex-type glycans. On the basis of primary structures there are two types of ENGase: one is categorized as glycoside hydrolase family 18 (e.g., Endo H, Endo F), whereas the other is categorized as glycoside hydrolase family 85 (e.g., Endo D, Endo M and human cytosolic ENGase).

Endo-β-N-acetylglucosaminidase M (Endo M) ENGase isolated from *Mucor hiemalis*, Endo M, has relatively broad substrate specificity and can release N-glycans from all types of glycans (high-mannose; hybrid and complex type). This enzyme also has high transglycosylation activity, and is used as a tool reagent for chemoenzymatic synthesis of neoglycoconjugates.

Endocytosis There are two major pathways in eukaryotic cells for material transport. One of the pathways is called endocytosis, a process whereby cells absorb materials from the outside of cells by engulfing them with their cell membrane. Mainly the endocytosed materials are transported to lysosome via endosome. On the other hand, exocytosis is a process by which cells direct materials (intracellular vesicles) out of the cell membrane.

Endoglycoceramidase Endoglycoceramidase, also called ceramide glycanase, is an endoglycosidase, which cleaves glycoside bonds between glycans and ceramide of glycosphingolipids, thereby forming oligosaccharides and ceramides. It was originally identified in *Rhodococcus* sp., and later in leech, short-necked clam, and jellyfish. This enzyme can also catalyze transglycosylation reactions.

Endoglycosidase An endoglycosidase is an enzyme that releases oligosaccharides from glycoconjugates (glycoproteins, glycolipids, and proteoglycans). In a broader definition, it also includes enzymes that act on polysaccharides, such as chitinase or glucanase. These enzymes are used as tool reagents to study the structures and functions of glycans, because they can release glycans from aglycons under milder conditions compared with chemical methods. The transglycosylation activity of these enzymes is also utilized to form neoglycoconjugates.

Endosialidase Endosialidase specifically recognizes α2,8-linked oligo/polysialic acids, cleaving the glycoside bond between sialic acids. A soluble enzyme derived from bacteriophage K1F (Endo-*N*) catalyzes the following reaction: $(\rightarrow Neu5Acyl\alpha 2 \rightarrow)_n\text{-}X$ $(n > 5) \rightarrow (\rightarrow Neu5Acyl\alpha 2 \rightarrow)_{2-4} + (\rightarrow Neu5Acyl\alpha 2 \rightarrow)_2\text{-}X$. Endo-*N* is a useful reagent to confirm the occurrence of oligo/polysialic acids, and also to analyze their function.

Endosulfatase Endosulfatase is an enzyme that acts on polysaccharides, releasing sulfate. The best known enzyme is the one acting on 6-*O*-sulfates of heparan sulfate. This enzyme has been extensively studied in terms of the relationship with development and cancer formation, because the action of this enzyme results in upregulation of Wnt signaling concomitant with inhibition of FGF signaling pathway.

Enzyme cue synthesis strategy This is a strategy for the simultaneous production of glycans with various structures in a single tube by quitting sequel glycosylating reactions in the middle of each step. First, the glycan synthesis is aborted in the middle to have a mixture of products and unreacted materials. Then, the next glycosylation step is again aborted before completion to have a mixture of the four variants of the glycosylated materials, e.g., first products and unglycosylated materials in the first reaction and their reacted products.

Enzyme replacement therapy Absence or deficiency of enzyme results in accumulation of excess substrates or toxic metabolic intermediates. Enzyme replacement therapy is a medical treatment replacing an enzyme in patients in whom the particular enzyme is absent or deficient. Clinical application of enzyme replacement therapy is performed for lysosomal enzyme disorder and adenosine deaminase deficiency (a kind of severe combined immunodeficiency).

ERAD ERAD is an abbreviation for endoplasmic reticulum-associated (protein) degradation. If proteins newly synthesized in the ER cannot fold properly, such misfolded proteins are retrotranslocated from the ER into the cytosol, and finally they will be degraded by proteasome in the cytosol. This protein degradation cascade is called ERAD and is conserved from yeast to human.

ERQC ERQC is an abbreviation for endoplasmic reticulum quality control. Newly synthesized proteins are allowed to proceed to the vesicular transport system when they acquire proper folding, whereas misfolded and/or misassembled proteins are retained at the ER. ERAD pathway of misfolded proteins is also included in ERQC.

Evolution Diversity of oligosaccharide structures among various biological species is acquired through evolutionary process. The evolution of sugar chains was not well investigated because of the lack of structural data on oligosaccharides. However, a lot of glycosyltransferase genes have been cloned and the evolution of sugar chains is now well studied with the molecular evolution approach.

Expression cloning Expression cloning is a technique in DNA cloning that uses expression vectors to generate a library of clones, with each clone expressing one protein. This expression library is then screened for the property of interest, according to the antigen epitope or enzyme activitiy of the expressed proteins.

EXT gene family EXT gene family is composed of five genes: the EXT1 and EXT2, which are responsible for hereditary multiple exostoses (HME), and their homolog EXT-like genes (EXTL1, EXTL2 and EXTL3). It is not known if there is any relationship between HME and EXTL1 . . . L3.

All these five EXT genes encode glycosyltransferases for the synthesis of heparan sulfate.

Fluorescent labeling of glycans Reducing end of glycan chains can be labeled by organic compounds, forming chemically stable fluorescent derivatives. This derivative achieves highly sensitive detection through HPLC or mass spectrometry. Especially, this method is often used for structure analysis of glycans using reversed-phase HPLC.

Fluorescent polarization measurement Fluorescence polarization, excited with polarized light and emitted from the solution of the fluorescent-labeled molecules, is depolarized by the Brownian motion of the molecule itself. Because the Brownian motion can be correlated with the size of the molecule, measurement of the real-time interaction of fluorescent probes with other macromolecule is possible through the fluorescent polarization measurement. This method achieves real-time measurement of biomolecular interaction analysis without, not like other chip-based methods, a problem of non-specific interaction. See also "Single molecule fluorescent measurement (fluorescent correlation spectroscopy)".

Frontal affinity chromatography Frontal affinity chromatography (FAC) is a quantitative affinity chromatography, developed in 1974 by Kasai et al., which can accurately determine affinity constants between biomolecules such as enzymes–substrate analogues and lectin–oligosaccharides. FAC is more advantageous than other methods for the analysis of relatively weak interactions, such as sugar–protein interaction.

GALAXY (GlycoAnaLysis by the three AXes of MS and chromatographY) GALAXY is a web application for supporting structural analysis of N-linked glycans. Its database includes 2-D/3-D HPLC maps and molecular weight data for about 500 pyridylaminated oligosaccharides on the Intenet. The URL is http://www.glycoanalysis.info/

Galectin Galectins are defined as lectins having both β-galactoside binding and amino acid sequence similarity. It occurs not only in animals, but also in fungi (mushroom). This protein is known to be involved in quite diverse biological events, i.e., development, differentiation, apoptosis, growth, inflammation, and cancer metastasis.

Ganglioside Gangliosides are defined as sialic acid-containing glycoshingolipids. Most of the gangliosides are composed of ganglio-series, whereas it also includes sialic acid containing lacto-, neolacto- and globo series glycosphingolipids. It is most abundantly found in neuronal tissues, and it has been shown that gangliosides are involved in various biological events such as development and differentiation.

Gb3/CD77 Gb3 stands for globotriaosylceramide (Galα1,4Galβ1,4Glc-ceramide). It is often described as Gb3/CD77, and it was also named as CD77 antigen after being found on Burkitt lymphoma and immature B lymphocyte.

It is also known as a receptor for verotoxin, which is produced by pathogenic *Escherichia coli* O157. In humans, expressions are observed on the erythrocytes, kidney cells, splenocytes, and vascular endothelial cells. Its glycan structure is produced by α1,4 galactosyl transferase.

GD2 GD2 stands for GalNAcβ1,4 (NeuAcα2,8NeuAcα2,3) Galβ1,4Glc-ceramide. It is considered to be a cancer-related antigen, because it is often found in embryonic brain and also highly expressed in tumor derived from neuroectoderm, such as melanoma, neurblastoma or glycoma. It was also reported to be expressed in a subset of T cell lymphoma or small cell lung carcinoma.

The glycan moiety is synthesized by the action of β-1,4-N-acetylgalactosaminyltransferase.

GD3 GD3 stands for NeuAcα2,8NeuAsα2,3Galβ1,4Glc-ceramide. It is highly expressed on embryonic brain, melanoma, T cell lymphoma or activated T cells. This molecule is proposed to be the target for the antibody therapy for melanoma. Recently it was reported that this antigen causes the expression of cancerous phenotypes and also induces apoptosis.

This glycan structure is produced by ST8SiaI.

GDP-L Fucose synthase GDP-Fucose, which is a donor substrate of fucosyltransferase, is synthesized in cytosol and translocated into Golgi by the function of GDP-fucose transporter. There are a couple of GDP-Fuc synthesis pathways known: one is "salvage pathway", which includes Fuc kinase and GDP-Fuc pyrophosphorylase, and another is "de novo pathway", which includes GMD and FX.

Gene expression profiling Gene expression profiling is a whole picture of expression pattern of multiple genes. This can be obtained by methods such as SAGE (serial analysis of gene expression) or DNA microarray technique. Normally, these data are processed by clustering analysis, allowing us to envisage the character of cells and/or function of genes of interest. This method is also utilized to identify genes responsible for differentiation or malignant transformation of cells by comparing the expression profile between different cells.

Gene knockdown Gene knockdown is a method of downregulation of gene expression. In mammalian cells, short interfering RNA (siRNA) or short hairpin RNA (shRNA) are often used for this purpose, because such RNAi-based method will avoid interferon response.

Gene knockout Gene knockout is a method of complete inactivation of gene function. In the case of mouse, KO animal is generated through replacement of the target gene to, in most cases, a marker gene, which confers resistance to a certain toxic agent for selection. Tissue-specific KO mice can also be generated using Cre/loxP system.

Genetic polymorphisms Genetic polymorphisms arise from individual difference in DNA (genomic) sequence. In blood typing (ABO system, P system or Ii system), variable glycan antigens are formed because of single polymorphisms, resulting in a change in substrate specificity or reduced/loss of activity of responsible enzymes.

Glucosyl cereamide Glucosyl cereamide is a glycolipid in which a glucose residue is attached to ceramide. This glycolipid distributes ubiquitously in cells and serves as precursors of various glycolipids. It can also regulate the level of free ceramide in cells, controlling cellular signals. This lipid is synthesized on the cytosolic face of the endoplasmic reticulum by glucosyl ceramide synthetase.

Glycan database Recently, a variety of glycan-related databases have been constructed: (1) Database about glycan structures, glycan-related enzymes such as glycosyltransferases, anti-glycan antibodies, and lectins. (2) Web applications that support the analyses of two-dimensional HPLC mapping and mass spectrometry. (3) Integrated database of these databases.

Glycoamidase A Discovered in almond seeds in 1977, this enzyme is the first glycoamidase (peptide:N-glycanase) described in nature, having an optimal pH of 4. See also "peptide:N-glycanase".

Glycocluster effect The term, glycocluster effect, is used to represent the sugar-recognition mode of animal lectins. Although animal lectin has weak affinity for monosaccharide, apparent affinity of the lectins for monosaccharide-coated protein (neoglycoprotein) becomes high, depending on the number of the monosaccharides attached to a carrier protein. Animal lectins show high affinity toward the assembled non-reducing sugar residues.

Glycoconjugates Glycoconjugates represent carbohydrates, which are covalently attached to a nonsugar moiety (lipids or proteins). The major glycoconjugates are glycoproteins, glycolipids, and proteoglycans.

Glycoepitope (Glycotope) An epitope is the part of an antigen that is specifically recognized by antibodies. Generally, oligosaccharide-binding antibodies recognize several monosaccharides as an epitope (glycoepitope). Anti-carbohydrate antibodies can be produced because of the high antigenicity (immunogenicity) of the glycans. A variety of anti-carbohydrate monoclonal antibodies are now produced.

GlycoEpitope DB There have been tremendous numbers of mono- or polyclonal antibodies produced and used for glycan function analyses and medical examinations by tissue staining, immunoprecipitation, Western blotting, or ELISA.

Thus GlycoEpitope DB was innovated to store the Glycoepitopes or glycan structure-specific antibodies in an organized manner. Its Internet URL is http://www.glyco.is.ritsumei.ac.jp/epitope/

Glycogene Glycogene is a word that was coined by Narimatsu (AIST, Japan) several years ago. It is a family of genes that codes synthesizing, degrading, or binding of proteins to sugar chains. Glycogene includes genes associated with (1) glycosyltransferases, (2) sulfotransferases, which transfer a sulfate group to a sugar chain, (3) sugar–nucleotide transporters, (4) sugar–nucleotide syntheses, etc.

GlycoGene DB (GGDB) GGDB is an Internet database for the genes involved in glycosylation, such as glycosyltransferases, sulfotransferases or nucleotide-sugar transporters.

Glycolipid Glycolipids are a series of glycoconjugates synthesized by attaching glucose and/or galactose to a ceramid. The structures from the former are predominant. Sialic acid, GlcNAc, or Gal attach to a lactoceramide, and diverge into ganglio, lacto/neolacto, or globo series, respectively. They are mainly expressed on the cell surface and are thought to be especially involved in the tissue or the cell. Besides these, glyceroglycolipids also exist.

Glycomics The word "glycome" is a relatively new word, first defined at the end of the twentieth century as the entire complement of sugars, the third biological chains next to DNA and proteins, of an organism, just like genomes for DNA and proteomes for proteins. The research that studies glycomes is referred to as glycomics.

Glycopeptide A glycopeptide is a complex molecule consisting of carbohydrate and peptide, and is one of the glycoconjugates. A glycopeptide is derived from the digestion of glycoprotein by a protease.

Glycoprotein Glycoproteins, complex morecules of carbohydrates and proteins, are one of the glycoconjugates. Proteoglycans are often classified into different categories.

Glycosaminoglycan Glycosaminoglycans (GAG), previously known as mucopolysaccharides, are long unbranched polysaccharides consisting of repeating disaccharide units. Examples of GAGs include heparin/heparan sulfate, chondroitin sulfate/dermatan sulfate, keratan sulfate, and hyaluronan. Most of the GAGs, except for hyaluronan, are sulfated and are covalently linked to a protein to form proteoglycans.

Glycoside hydrase Glycoside hydrase is a generic term for enzymes involved in the hydrolysis of oligosaccharides and exists in lysosomes and cytosol. Lysosomes contain various glycoside hydrases and the gene deletion of glycoside hydrase results in lysosomal deficiency caused by accumulation of glycoconjugates.

Glycosphingolipid Glycosphingolipid is a generic name for glycolipid-containing sphingosine base. Glycosphingolipids usually form microdomains at the outer plasma membrane in a cholesterol-dependent or -independent manner. Glycosphingolipid-enriched mirodomain can regulate the functions of many membrane proteins, e.g., signal transduction across the membrane or cell–cell adhesion. Glycosphingolipid can function as a receptor for pathogenic bacteria, virus, and their toxins. Glycosphingolipid containing sialic acid is called ganglioside.

Glycosyl fluoride Glycosyl fluoride is an oligosaccharide replaced with fluoride. Glycosyl fluoride derivatives, in which C-1 hydroxyl group is substituted with fluoride, are useful synthetic intermediates for organic chemical- and enzymatic glycosylation.

Glycosyltransferase Glycosyltransferases are an enzyme family related to the biosynthesis of sugar chains located in the Golgi apparatus. There are approximately 300 glycosyltransferase genes, and two-thirds of them have been cloned to date. Glycosyltransferases exist in only very small amounts in the body and have specific expression patterns in a variety of organs. Moreover, the expression changes during development, differentiation, malignant transformation, etc.

Glycovirology Glycovirology is a research field covering both glycobiology and virology. The First International Meeting on Glycovirology was held in Sweden in June 2007. Sugar chain will surely be a very important target for research and drug discovery in the twenty-first century because of its close relationship to viral infection. Development in the field of glycovirology and fostering specialists in this research are derived.

GM2 GM2 stands for GalNAcβ1,4(NeuAcα2,3) Galβ1,4Glc-ceramide. It is expressed in embryonic brains, melanoma. glioma, and lung carcinoma; it is often observed to express epithelial carcinoma cells. This structure is synthesized by β-1,4-N-acetylgalactosaminyltransferase.

GM2/GD2 synthase This enzyme catalyzes the reaction of the synthesis from GM3, GD3 into GM2, GD2. It is also called β-1,4-N-acetylgalactosaminyltransferase (β4GalNAc-TI). It also can catalyze the GA2 synthesis reaction from lactosylceramide. The gene encoding this enzyme was cloned in 1992 for the first time among the glycolipid synthase. This enzyme functions on the synthesis of every single complex-type ganglioside.

GM3 GM3 stands for NeuAcα2,3Galβ1,4Glc-ceramide. It is located at the beginning of the synthesis pathways of sialylated glycolipids, and is synthesized by ST3SiaV. It is not highly expressed in the neuronal cells even with the high levels of its synthesis, because the product will be transformed into other glycolipids. It is observed in any of normal cancerous cells. Recently, its significance for the regulation of signaling pathways of EGF or insulin has been reported.

Golgi apparatus Golgi apparatus (also called Golgi body or Golgi complex) is an organelle composed of membrane-bound stacks called cisternae. It is found in most eukaryotic cells, bearing various glycosyl transferases and related proteins (e.g., sugar nucleotide transporters, etc.), and thus is involved importantly in glycosylation.

Growth factor Growth factor is a protein that promotes cellular proliferation and differentiation. Fibroblast growth factor (FGF) is one of the well-known growth factors.

The interaction with heparan sulfate is usually essential for the functional expression of the growth factor.

HCDM (human cell differentiation molecules) Originally known as HLDA (human leukocyte differentiation antigen), they are generally called the CD (cluster of differentiation) molecules. HLDA was previously restricted to being the cell surface molecules on leukocytes and its precursors, but currently the HCDM represents any differentiation antigens on other cell types, such as platelets, erythrocytes, endothelial cells, etc., and even includes the intracellular markers. Therefore, with the ever-changing field of studies surrounding HLDA, it was decided to change the name to HCDM.

H–D antigens Hanganatziu–Deicher antigen is named after the researchers who found the antigen. It has been known that patients who are treated with horse, sheep or goat serum often experience anaphylactic shock on the second treatment. The antigen responsible for this was found to be NeuGc-containing glycans, which is normally absent in human tissues.

Helicobacter pylori *Helicobacter pylori* is a Gram-negative, helical shaped bacteria having several flagella lives in the stomach. Urease produced from *H. pylori* metabolizes urea to carbon dioxide and ammonia (which neutralizes gastric acid). *H. pylori* infection is strongly related to not only chronic gastritis and peptic ulcer, but also stomach cancer and gastric lymphoma. Finding of *H. pylori* shed light on the Gastroenterology and infectious disease society. In 2005, Dr. Marschall and Dr. Warren, who found *H. pylori* were awarded the Novel prize in Medicine.

Heparan sulfate (HS) Heparan sulfate (HS), a member of glycosaminoglycan, consists of repeated disaccharide units of glucuronic acid (GlcA) and GlcNAc. HS is found on the cell surface and extracellular matrix of most animal cells. HS modification, in which the N-acetyl group of GlcNAc and various hydroxyl groups are partially sulfated, create structural heterogeneity and thereby enables the HS interaction with a variety of biologically active proteins such as growth factors, morphogenic ligands, and cyokine.

Heparan sulfate O-sulfoglycosyltransferase After the heparan sulfate (HS) chain is modified by N-deacetylase/N-sulfotransferase (NDST) and glucronyl C5 epimerase, which converts GlcA to iduronic acid (IdoA), HS O-sulfation occurs at various positions. Heparan sulfate O-sulfotransferases are cateogorized into three types: HS2ST(2-O-sulfation of GlcA/IdoA), HS6ST (6-O-sulfation of GlcN), and HS3ST(3-O-sulfation of GlcN).

Heparin Heparin is a highly sulfated glycosaminoglycan (GAG), which is composed of a 2-O-sulfated iduronic acid, and 6-O- and N-sulfated glucosamine, IdoA(2S)-GlcNS(6S). Heparin has strong anticoagulant activity, and thereby is used for anticoagulation in surgery. Heparin's exact physiological role is still unclear, because blood anti-coagulation is achieved mostly by endothelial cell-derived heparan sulfate proteoglycans. It has been proposed that, rather than anticoagulation, the main purpose of heparin is a defensive mechanism at sites of tissue injury against invading bacteria and other foreign materials.

Heparin-binding growth factor Heparin-binding growth factor represents a growth factor family having heparin-binding activity and contains basic FGF, HB-EGF, and midkine. Heparan sulfate functions as a co-receptor for the high affinity receptors of various growth factors.

Hidrazinolysis In a narrow sense, hidrazinolysis is a technique that releases glycan chains. Free N-linked glycan chains can be released by heating glycoproteins in hidorazin. Free amino groups caused by hidrazinolysis are usually acetylated. There is a report of a modified hidrazinolysis technique to obtain free O-linked glycan chains.

High endothelial cell High endothelial cell is expressed in high endothelial venues found in secondary lymphoid organs such as lymph nodes. High endothelial cells are characterized by their cuboudal form as opposed to squamous cells found in regular endothelial venues. Lymphocytes expressing L-selectin can move between high endothelial cells into the secondary lymphoid organs. L-selectin molecules weakly interact with specific oligosaccharides expressed on high endothelial cells.

High-mannose type glycan High-mannose type glycan is a kind of N-linked glycans that exist in eukaryotes. The glycan normally consists of 5–9 mannose and two N-acetylglucosamine residues. It is a biosynthetic precursor of hybrid-type and complex-type N-glycan. The high-mannose glycan seems to be related to the intercellular transport of glycoproteins.

HNK-1 antigen HNK-1 antigen is a carbohydrate antigen recognized by anti HNK-1 (human natural killer-1) antibody. HNK-1 antigen, which contains 3-O-sulfated glucronic acid residue at its non-reducing terminus, is mainly expressed in neuronal cells.

HNK-1 Glucronyltransferase This enzyme transfers glucuronic acid to a terminal galactose residue of N-acetyllactosamine structure in glycans, forming HNK-1 epitope structure.

HPLC (high-performance liquid chromatography) HPLC is a form of column chromatography used frequently in biochemistry and analytical chemistry. It is also referred to as high-pressure liquid chromatography. The use of small-particle resins makes it possible to be compatible for high pressure, giving the components less time to diffuse within the column and thus leading to improved resolution.

Hyaluronidase Hyaluronidases are a family of enzymes that degrade hyaluronic acid by cleaving β1,4-glycosidic linkage between GlcNAc and GlcUA. Bacteria hyaluronidase catalyzes the elimination reaction to cleave hyaluronic acid, and animal-type hyaluroniade catalyzes the hydrolysis reaction. Both types of enzymes have activities to cleave chondroitin sulfuric acids. In humans, seven hyaluronidase-like genes are identified, and some of the gene products may be involved in cancer progression.

Hyaluronic acid/hyaluronan/hyaluronate Hyaluronic acid is a non-sulfated glycosaminoglycan that has a repeated linear structure of disaccharide unit of (-4GlcUAβ1,3GlcNAcβ1-). Hyaluronic acid is synthesized by hyaluronan synthases. Hyaluronic acid generally consists of several thousand disaccharide repeats in length, and range in size from 10^3 to 10^6 Da depending on the distribution or the physiological condition. Hyaluronic acid is also one of the major components of the extracellular matrix, and contributes to tissue formation and regulation of diverse cellular functions.

Hybrid-type glycan Hybrid-type glycan is a type of *N*-linked glycans that exists in multicellular eukaryotes. The hybrid-type glycan contains common structural features of high-mannose type and complex-type glycans.

Hypoxia-inducible factor (HIF) Hypoxia-inducible factors (HIF) are transcription factors that respond to decrease in available oxygen in the cellular environment. HIFs have the ability to induce a series of genes to adapt to or resist hypoxia. In the normal oxygen condition, the α subunit of HIF is localized in the cytosol, and the degradation of proteasome occurs after rapid ubiquitination of the protein. In low-oxygen conditions, the degradation of the α subunit is stopped and the protein binds to the β subunit after nuclear translocation. The αβ complex induces the transcription of a series of genes with hypoxia-responsive element.

IgA nephropathy IgA nephropathy is primarily characterized by deposition of IgA in glomerulus, and the most frequently seen symptoms are proteinuria and hematuria. Berger and Hinglais, in 1968, were the first to describe IgA deposition in this form of glomerulonephritis. It is commonly seen in Japan, and as much as 40% of the patients end up with the introduction of dialysis because of chronic, progressive renal failure. It is estimated that 30% of patients newly introduced to dialysis treatment have this disease. Though the direct cause of this disease is still unclear, the involvement of IgA-containing protein complex or abnormality of glycans on IgA is predicted.

IgA1-binding protein IgA1-binding protein (IgA1-BP) represents serum proteins bound to asialo- and agalacto-IgA1. These proteins can be separated from serum through asialo-, agalacto-IgA1 affinity chromatography. As the IgA1 content in IgA1-BP is significantly higher in IgA nephropathy patients, the relationship between IgA nephropathy and glomerular deposited IgA is predicted.

Ii blood type antigens Allo antigen determined by the glycan structures on erythrocytes. The polylactosamine structure [(Galβ1-4GlcNAcβ1-3)n] is called i antigen, whereas the branched structure of GlcNAcβ1-3(GlcNAcβ1-6)Galβ1-4 is called I antigen. These antigens are expressed as blood type antigens on erythrocytes as well as other tissues.

Immunostimulating glycoconjugate Innate immunity recognizes the specific substances of microorganisms (pathogen-associated molecular pattern: PAMP) and activates the immune system. Most of the bacterial glycoconjugates are recognized as PAMP and possess immunostimulating activity. Although most of these glycoconjugates are high molecular weight compounds, a minimum unit for the activity is usually a part of them. Peptidoglycan, lipopolysaccharide of Gram-negative bacteria and lipoteichoic acid etc., are typical of *immunostimulating glycoconjugates*.

Influenza virus It is an RNA virus containing eight pieces of segmented negative-sense single-strand RNA coated with a lipid bilayer structure called envelope. There are three types of influenza virus: influenza virus A, B, and C, on the basis of the antigenicity of the proteins. Influenza virus A can be isolated from not only humans, but also from animals such as domestic fowls, aquatic birds, pigs, or horses. Influenza virus A is thus regarded as zoonosis, and causes an outbreak or gives rise to human influenza pandemics. Influenza virus A and B contain hemaglutinin (H), which recognizes sialoglycoconjugates, and neuraminidase (N) spike. In Influenza virus A, there are 16 subtypes for H (H1–16) and 9 subtypes of N (N1–9) so far known.

KEGG DB KEGG (Kyoto Encyclopedia of Genes and Genomes) is a database of biological systems, consisting of genetic building blocks of genes and proteins, chemical building blocks of both endogenous and exogenous substances, molecular wiring diagrams of interaction and reaction networks, and hierarchies and relationships of various biological objects.

Keratan sulfate Keratan sulfate is any of several glycosaminoglycans that have been found especially in the cornea, cartilage, and brain. Keratan sulfate is a glycan polymer that consists of repeating disaccharides, [Galβ1-4GlcNAc(6-O-SO$_3$H) β1-3], attaching to the core proteins via either N-linked or O-linked glycans. In the cornea, it has been shown that this glycan provides water retention capability.

Laboratory automation Laboratory automation represents automation of all the processes performed in a laboratory such as synthesis, purification, and analysis of diverse compounds, to increase productivity, reduce lab process cycle times, and elevate experimental data quality. The most widely known technology is laboratory robotics. More recently, the field of laboratory automation comprises many different automated laboratory instruments, devices, softwares, and algorithms.

Lectin Lectin is any of group of carbohydrate-binding proteins or glycoproteins except for enzymes, antibodies, and transporter, which specifically bind to carbohydrates. Orginally, lectin was restricted to multivalent proteins capable of agglutination. However, today lectin is used in a broad sense to denote all types of carbohydrate-binding proteins that do not catalyze reactions with their ligands.

Lectin affinity Affinity for lectin is often used to purify glycoproteins or determine the interactions with carbohydrates that have complex structures Today, lectin affinity is also applied to profile characteristic structures of carbohydrates.

Lectin database Web-database, one of the sub-database in CabosDB (Carbohydrate sequencing database), consists of information on the features of more than 200 lectins and on the binding constants of more than 100 standard carbohydrates to them. On the basis of the database, useful lectins for profiling carbohydrate structures have been retrieved and the interactions between carbohydrates and lectins have been examined from various directions.

Lewis blood group antigen Lewis blood group antigens are the molecules expressed on the surface of human red blood cells. However, they are not produced by the erythrocyte itself. Instead, they are components of body secretions, and are subsequently adsorbed onto the surface of erythrocytes. They are also expressed on many other tissues including digestive tract and tumor cells. There are two types of Lewis antigens, Lewis a and Lewis b.

Liquid chromatography-mass spectrometry (LC/MS) It is an analytical system to analyze compounds separated by liquid chromatography directly subjected to electrospray ionization-mass spectrometry.

Lymphocytes homing Lymphocytes homing indicate that native lymphocytes carried to the blood migrate into specific microenviroments within secondary lymphoid tissues (lympho node, Peter's patches, and spleen). Then the lymphocytes recirculate through these sites until they either die or encounter their specific antigen. Memory and effector T cells can efficiently extravasate in extralymphoid inflammatory sites, with subsets displaying targeted trafficking through, for example, inflamed skin, and intestinal mucosa. Therefore, this tissue-selective lymphocyte homing is also regarded as lymphocyte homing.

Lysosomal enzymes Lysosomal enzymes are hydrolases responsible for breaking down complex chemicals (biological substances or chemicals derived from microbes) in an organelle called lysosome.

Mannan-binding protein (MBP) Mannose or mannan-binding protein (MBP), also called mannose-binding lectin, belongs to the collectin family in the C-type lectin. MBP binds to terminal non-reducing sugar residues, mannose, N-acetylglucosamine, or fucose, in a calcium-dependent manner. MBP is produced by the liver and secreted into circulation. It is involved in innate immunity, and its binding to microorganisms results in activation of the complement system. MBP has an oligomeric structure, built of subunits that contain identical peptide chains of 32 kDa each.

Mass spectrometry Mass spectometry is an analytical technique used to obtain structural information on the target molecules. The main steps in measuring are (1) ionization of the sample, (2) separation of ions with different masses (m/z), and (3) detection of the number of ions in each mass.

Matrix-assisted laser desorption/ionization (MALDI) MALDI is a soft ionization technique used in mass spectrometry. Soft ionization technique affords molecular ions of biomolecules such as proteins, peptides, and carbohydrates, which tend to be fragile and quickly lose structure when ionized by conventional ionization methods. A matrix is used to protect the molecule from being destroyed by the direct laser beam. In the procedure, the sample solution is mixed with a large excess matrix solution, and the aliquots of the resulting mixture are spotted on a target plate for crystallization. Finally, the laser is fired at the crystals for ionization.

Methylation analysis Methylation analysis is a chemical approach for determining linkage position of the monosaccharide residues in an oligosaccharide. This method works on the basis of the acid stability of methyl ethers and the acid lability of glycosidic linkage. First, a stable methyl group is introduced on each free hydroxyl group of the oligosaccharide. The glycosidic linkages are then cleaved, producing individual monosaccharide residues with new free hydroxyl groups that appear at the positions that were previously involved in a linkage. The monosaccharides are reduced after N-acetylation to produce volatile compounds and are analyzed by gas–liquid chromatography coupled to a mass spectrometer.

MGL/CD301 MGL/CD301 is the macrophage galactose-type, C-type lectin. In humans, a single gene is located in chromosome 17, whereas mouse has two orthologous genes in chromosome 11. This type II transmembrane molecule is expressed in the bone marrow-derived macrophages or immature dendritic cells. C-terminal carbohydrate recognition domain binds to Gal and GalNAc as monosaccharides. This protein is involved in carbohydrate-dependent endocytosis of antigens and/or cellular trafficking. MGL-KO mice exhibit lack of antigen-dependent tissue regeneration.

Microdomain (Lipid raft) Lipid raft is a cholesterol-enriched microdomain in the cell membrane, and is also called "caveola" in case it contains caveolin. It can be isolated from cells as low-density, detergent-insoluble membrane fractions, which are rich in cholesterol, sphingolipid, glycosphingolipid, GPI-anchor protein, and a variety of signaling molecules. It has been reported that it is involved in various cellular events such as signal transduction, apotosis, endocytosis, infection, and membrane trafficking.

Midkine Midkine, a heparin-binding growth factor or cytokine, promotes cell survival and cell migration, and is deeply involved in cancer progression, the onset of inflammatory diseases, and the preservation and repair of injured tissues. Midkine also is involved importantly in development and is strongly expressed during midgestation. In the adult, midkine expression is strongly induced during oncogenesis, inflammation, and repair. Midkine binds to the trisulfated unit of heparan sulfate and chondroitin sulfate E unit, and the binding is essential for expressing its activity.

Milk oligosaccharides Mammalian milk usually contains, in addition to lactose as a dominant saccharide, a variety of other saccharides, called milk oligosaccharides, which commonly have a lactose unit at their reducing end. For example, human milk contains more than 100 milk oligosaccharides. The oligosaccharide content of mature human milk is from 12 g to 14 g per liter. Milk oligosaccharides act as anti-microbial defense factors against pathogenic bacteria. They appear to firstly act as prebiotics, stimulating the growth of beneficial organisms such as bifidobacteria, and secondly act as receptor analogues, competing with pathogenic bacteria for attachment to gastrointestinal receptor sites.

Molecular dynamics calculation Molecular dynamics calculation is a computer simulation where atoms and molecules are allowed to move under Newton's laws, giving a view of the motion of the atoms. The motion of atoms (coordinates, velocity, and acceleration) is calculated by solving Newton's laws iteratively (every one femto (10–15) seconds). The term, MD (molecular dynamics) simulation is also used for molecular dynamics calculation.

Monoclonal antibody Monoclonal antibodies are identical immunoglobulin molecules that are produced by cloned antibody-producing cells. Most monoclonal antibodies are produced by hybridomas, hybrid of B cells and myeloma cells. A lot of monoclonal antibodies against specific oligosaccharide antigens are produced and used for the analyses of oligosaccharides.

Mucin-type glycans O-linked glycans are usually attached to Ser/Thr residues of the core protein through the GalNAc residue at the reducing end. These are usually referred to as "mucin-type" glycans because of their predominant occurrence on the mucus glycoproteins.

Multi-dimensional HPLC mapping method Multi-dimensional HPLC mapping is a method to determine the structures of N-linked oligosaccharides by HPLC. N-linked oligosaccharides are released from proteins, and the reducing ends of released oligosaccharides are fluorescently labeled with 2-aminopyridine. Pyridylaminated oligosacchardes are first separated with their charge using the DEAE column. Next, each fraction is separated sequentially by HPLC using ODS and amide columns. The structure of PA-oligosaccharide can be deduced by comparing its elution position with that of the reference PA-oligosaccharide on the map.

Multistep tandem mass spectrometry Tandem mass spectrometry is performed by: (1) selection of a precursor ion, (2) collision of selected ions with rare gas, (3) separation of fragment ions, and (4) detection of fragment ions. Multistep tandem mass spectrometry is a method used to repeat the selection and collision of fragment ions. Detailed information on the structure of biomolecules (protein, oligosaccharides) can be obtained.

Myelin Myelin is an electrically insulating dielectric phospholipid layer that surrounds the axons of many neurons. The insulating properties of the compact myelin around the intermodal segment of the axon enable the fast propagation of electrical signals at the myelin-free node of Ranvier. It is a product of Schwann cells in the peripheral nervous system and of oligodendrocytes in the central nervous system. The high ratio of lipids to proteins is unique for myelin and differs from the ratio seen in other membranes. Specifically in myelin, glycosphingolipids are enriched. The myelin glycosphingolipids consist of two major components, galactosylceramide and its sulfated form.

N-acetylglucosamine 6-O-sulfotransferase N-acetylglucosamine 6-O-sulfotransferase, often abbreviated as GlcNAc6ST, is a sulfotransferase, which transfers sulfate onto GlcNAc residue, forming 6-O-sulfated GlcNAc structure. There are various molecular species of GlcNAc6STs, with different tissue distribution and substrate specificity. This enzyme is involved in the biosynthesis of L-selectin ligand as well as keratan sulfate. A sulfate group of L-selectin ligand in peripheral lymph nodes are synthesized by the action of both GlcNAc6ST-1 and GlcNAc6ST-2.

Neoglycolipid Neoglycolipid, a generic name of glycolipid with the structure being nonexistent or not found in nature, is synthesized by artificially replacing the sugar moiety of gangliosides or sphingoglycolipids. The one in which the structure of the lipid moiety was converted to the non-natural type is often included in this category.

Neoglycoprotein Neoglycoprotein, a generic name of glycoprotein with the structure being nonexistent or not found in nature, is derived by chemically inducing sugars to proteins. Most neoglycoproteins are generated by covalently bonding sugars to the lysine amino groups on the side chain of the protein.

Neoproteoglycan (probe) Neoproteoglycan, a kind of artificial glycoconjugates, is synthesized by covalently bonding polysaccharide chains such as glycosaminoglycans to the proteins. Neoproteoglycans are used for the search, purification, analyses of the interactions and functional substitutions, etc. of the substances with the affinity as the models of proteoglycans.

Neuroglycan C Neuroglycan C, which is abbreviated to NGC, is a transmembrane chondroitin sulfate proteoglycan that is exclusively expressed in the central nervous system. The human NGC gene is written as CSPG5, and is located on chromosome 3p21.3. It is known that gene expression changes are elicited by drug addiction and nerve injury, NGC activates epidermal growth factor (EGF) receptors, ErbB2 or ErbB3, and regulates neurite outgrowth. Studies in knockout mice have suggested that NGC regulates synapse formation.

NKT cells NKT cells are a subset of T cells, which express both TCR (T cell receptor) and NK cell receptor. They are involved in self-tolerance, allograft rejection, and cancer immunity. Galα1-containing glycolipids (Galα1-ceramide, Galα1-3Galβ1-4Glcβ1-ceramide) are presented by CD1d, an MHC class I-like molecule on antigen-presenting cells, and therefore cells recognize the complex with TCR, leading to their activation. Much attention has been paid to NKT cells with respect to their relationship of glycans with cancer immunity, and self-tolerance.

N-linked glycans N-linked glycans is one of the most common glycosylation types in eukaryotes, linked to the side chain of Asn residues in proteins. In eukaryotes, N-glycosylation only occurs at the consensus sequence called sequon, (Asn-X-Ser/Thr; X is any amino acids except Pro), whereas there is a rare occurrence of N-linked glycans at the Asn-X-Cys sequence.

Notch signal The Notch signaling pathway controls multiple cell differentiation processes during the developmental stages. Notch receptor interaction with a ligand (Delta-like, jagged in mammals) on neiboring cells induces proteolytic cleavage and release of the intracellular domain, which then moves to the nucleus to alter gene expression. The Notch extracellular domain has EGF-like repeats (e.g., Notch 1 has 36 of these repeats). Each EGF-like repeats can be multiply modified with O-fucose and O-glucose. O-Fucose can be elongated. Without O-fucosylation, Notch signal is not propery transduced.

Nucleotide sugar transporter Nucleotide sugar transporters are hydrophobic proteins in the Golgi apparatus or endoplasmic reticulum. They specifically antiport nucleotides sugars pooled in the cytosol into the lumen of Golgi or endoplasmic reticulum with the corresponding nucleoside monophosphates. With the molecular identities of nucleotide sugar transporters being unveiled by cloning in 1996, it has been reported that various nucleotide sugar transporters have different specificities in many eukaryotes. They present differently in different types of organisms; GDP-mannose transporter is present in plant, yeast, and protozoan, whereas it is absent in mammal.

O-Fuc glycans O-Fuc glycans is a type of O-glycans, where Fuc residues are attached to Ser/Thr residues in glycoproteins. There are two distinct types of O-Fuc modifications; one attached to EGF-like domain, and the other attached to thrombospondin type I repeat domain. The elongated structures of O-Fuc glycans are known to be different between these two types.

Oligosaccharide processing The term, oligosaccharide processing, implies the process of oligosaccharide biosynthesis. In general, oligosacchairdes are synthesized by a series of glycosyltransferases using sugar nucleotides as the donor substrates in the Golgi complex. Although the processing pathways of O-glycans, N-glycan, and proteoglycan are different, each glycan is in principle synthesized with specific glycosyltransferases.

Oligosaccharide profiling Oligosaccharide profiling is to characterize the oligosaccharide structures of the sample. The profiling is performed by identifying the oligosaccharides with liquid chromatography, electrophoresis, or lectins.

O-linked glycans See "O-linked glycopeptides".

O-linked glycopeptides Defined narrowly, O-linked glycopeptides stand for the ones containing mucin-type O-linked glycans (Ser/Thr-linked GalNAc). In a broad sense, they can include any glycopeptides containing O-linked glycans (not only O-GalNAc, but other type of O-linked glycans such as O-Fuc, O-Man, O-GlcNAc, O-Glc, etc.).

O-linked N-acetylglucosamine In this type of glycosylation, GlcNAc is linked to the hydroxyl group of Ser or Thr residues on proteins. O-linked N-acetylglucosamine is found in cytoplasm/nuclear proteins and an interplay between O-GlcNAc and phosphorylation has been suggested.

O-Man glycans O-Man glycans is a type of O-glycans, where Man residues are attached to Ser/Thr residues in glycoproteins. It is commonly observed in yeast, but similar modifications can be found in mammalian cells. The terminal structures of O-Man glycans are quite distinct between yeast and mammalian cells. This modification is essential for viability in yeast, whereas the deficiency of O-Man modifications results in congenital muscular dystrophy in humans.

One-pot glycosylation One-pot glycosylation method refers to one in which several glycosyl donors are allowed to react sequentially in the same flask, resulting in a single main oligosaccharide product. This method integrates several glycosylation steps into one synthetic operation to furnish target oligosaccharides in a short period of time without the need for protecting group manipulation and intermediate isolation. Therefore, this method is an attractive method in solution-phase methodology for the synthesis of oligosaccharide library.

Ozonolysis Ozonolysis is a chemical reaction, which selectively cleaves olefins under mild conditions. This method can be applied to release oligosaccharides from glycosphingolipids. After ozone gas is saturated into the glycoshingolipid solution, it is evaporated to dryness at room temperature. Alkaline treatment of the residue causes the elimination reaction to liberate intact oligosaccharides, whereas the ceramide portion is destroyed.

PAPS transporter It transports PAPS (3′-phosphoadenosine 5′-phosphosulfate) from the cytosol into the lumen of the Golgi apparatus. PAPS is synthesized by PAPS synthetase in the cytosol; therefore, the activity of PAPS transporter is essential for the sulfation of proteins/glycans in the Golgi. The gene was cloned from both humans and fruitfly in 2003.

Paroxysmal nocturnal hemoglobinuria Paroxysmal nocturnal hemoglobinuria is an acquired hemolytic anemia caused by somatic mutation of PIG-A gene in hematopoietic stem cells. PIG-A gene encodes a protein required for the synthesis of the GPI anchor. The gene that codes for PIG-A is inherited in an X-linked fashion, which means that only one active copy of the gene for PIG-A may exist. If a mutation occurs in the PIG-A gene in a bone marrow stem cell, it leads to a defect in the GPI anchor in the blood cells. Several GPI-anchored proteins in the blood cell, protecting the cell from destruction by the complement system, are deleted by PIG-A mutation. Deletion of these proteins causes a complement activation by some infection, which leads to the destruction of red blood cells.

Part-time proteoglycan Proteoglycans represent glycoproteins to which sulfated glycosaminoglycan is attached covalently. In case of part-time proteoglycan, some portions of core protein without having glycosaminoglycan chains also exist. Amyloide-beta precursor protein (APP), neuroglycan C, and thrombomodulin are typical part-time proteoglycans.

Pattern recognition In case of antibody-dependent pattern recognition, antigen epitope that consists of a couple of amino acid residues binds to antigen-binding site of the antibody (paratope) just like a lock-and-key model. Innate immune responses are initiated by pattern recognition receptors (PRRs), which recognize three-dimensional structures, consisting of a repeating domain of specific low molecular component, such as lipopolysaccharide (LPS) of Gram-negative bacteria.

Peptide mapping In this technique, a particular protein is digested by chemical method or proteolytic enzyme(s) of known specificity, and the peptide fragments are separated by HPLC or electrophoresis. This technique can provide useful information about the amino acid sequence and posttranslational modification of the protein of interest.

Phytosphingosine Sphingosine (2-amino-4-octadecene-1,3-diol) is an 18-carbon amino alcohol with an unsaturated hydrocarbon chain, which forms a primary part of sphingolipids, a class of cell membrane lipids that include sphingomyelin. Sphingosine derived from plants is called phytosphingosin, which has a C-4 hydroxyl group instead of unsaturated carbon and thereby forms a more stable complex structure with additional hydrogen bonds.

PNGase (peptide:N-glycanase) PNGase is also called glycoamidase or N-glycanase. The enzyme cleaves the amide bond between N-glycans and the linkage Asn residues (it should be noted that the enzyme is not a "glycosidase", but "amidase": as a result, the linkage Asn residues are converted to Asp residues because of deamidation reaction). This enzyme, of plant and bacterial origin, has been widely used for structure/function analysis of N-glycans on glycoproteins. Later, this enzyme was found to be ubiquitously observed in the cytosol of eukaryotic cells, and the cytosolic PNGase has been shown to be involved in the process of quality control system for newly synthesized proteins, called ERAD (ER-associated degradation).

Polylactosamine Polylactosamine is a linear polymer of N-acetyllactosamine, Galβ1,4GlcNAc. It is also known as an i antigen, one of the blood group antigens. Polylactosamine is found in diverse glycoproteins and glycolipids, and various functional glycans containing Lewis X antigens are added to its termini. Polylactosamine itself is involved as a ligand of Galectin.

Polypeptide-N-acetylgalactosamine transferase Polypeptide N-acetylgalactosamine transferase is involved in the catalysis of a single glycosidic linkage, GalNAc1-O-Ser/Thr, and collectively the GalNAc-transferase isoforms control the initiation of mucin-type O-glycosylation. It is also called pp-GalAcTase, pp-GalNAc-T, and so on. Polypeptide GalNAc transferase gene family contains 20 genes, of which 15 have been shown to encode functional enzymes. The GalNAc-transferase gene family is the largest mammalian family of glycosyltransferases.

Polysialic acid Polysialic acid (PSA) is a polymer of sialic acid whose degree of polymerization (DP) is 8–200 Sia residues. The most common structure of PSA is the Neu5Ac polymer, whose inter-residual linkage is α2,8. PSA occurs in NCAM, voltage-sensitive sodium channel, capsular polysaccharides of meningitis bacteria, trout polysialoglycoprotein (PSGP), and CD36. PSA has an anti-adhesive effect in cell–cell and cell–

extracellular matrix interactions because of its bulky volume and anionic nature and thereby is involved in the neural cell migration.

Post-translational modification Post-translational modification is the chemical modification of a protein after its translation. Examples are phosphorylation, glycosylation, glycation, and protease cleavage. In general, most of the proteins are synthesized into mature ones by attaching to the functional groups.

Prediction of oligosaccharide modification and interaction Modification of N-linked oligosaccharides can be predicted by the primary and tertiary structures of the proteins. The mode of interaction between oligosaccharide and protein can also be predicted with a computational method using the primary and tertiary structures of the sugar-binding proteins.

Proteasome Proteasomes are gigantic protein complexes inside occurring eukaryotes and archaes, as well as some bacteria. In eukaryotes, they are located in the nucleus and the cytosol. The most common form of the proteasome is known as the 26S proteasome, which is about 2 000 kilodaltons (kDa) in molecular mass and contains one cylinder-shaped 20S core particle, which exhibits various protease activities, and two 19S regulatory subunits (PA700). There is also another regulatory subunit, called 11S regulatory subunit (PA28), and combinations of different regulatory subunits generate various distinct proteasomes. They are involved in not only proteasomal degradation of unwanted proteins, but also in other biological processes such as antigen processing.

Protein O-mannose β-1,2-N-acetylglucosaminyltransferase Protein O-mannose β-1,2-N-acetylglucosaminyltransferase (POMGnT1) catalyzes the formation of GlcNAcβ1-2Man linkage of O-mannosyl glycans. The *POMGnT1* gene is a causative gene of muscle-eye-brain disease (MEB), which is one of the congenital muscular dystrophies.

Protein O-mannose transferase Protein O-mannose transferase catalyzes the transfer of mannose to serine/threonine residues of proteins using dolichol phosphate-mannose (Dol-P-Man) as a donor substrate. This enzyme activity requires two homologs, POMT1 and POMT2. The *POMT1* and *POMT2* genes are the causative genes of Walker-Warburg syndrome (WWS) categorized into congenital muscular dystrophies. Knockout mouse lacking *POMT1* is embryonic lethal.

Proteoglycans Proteoglycans represent a special class of glycoproteins in which more than one glycosaminoglycan (GAG) chain is covalently attached to the core protein. Proteoglycans can be categorized into heparan sulfate-, chondroitin sulfate-, dermatan sulfate-, and keratan sulfate proteoglycans depending on the nature of GAG chains. Proteoglycans are expressed on the cell surface or represent the extracellular matrix components.

P system blood group antigen It is a blood group antigen determined mainly glycan core structures of globo-type glycolipids. Whereas P antigen, globotetrasylceramide (Gb4), is accumulated in P+ blood group, its precursor globotriaosylceramide (Gb3), is accumulated in P- (P-minus) blood group, and is referred to as the pk type. Deficiency in Gb3 synthesis is called little p (p), where P-antigen is absent. Occurrence of pk and p type is rare. On the other hand, there is another type of P-antigen, denoted as P1/P2, and 80% of Caucasians contains P1 antigens. The antigenic structure of P1 has been assumed to be paragloboside (neolactotetraosyl-ceramide), where terminal Galα1-4 structure is observed. It has been found that Gb3 synthetase is also involved in the biosynthesis of P1 antigen, and in P2 the level of expression of this enzyme is found to be at low level.

PTPζ PTPζ is one of the receptor-type protein tyrosine phosphatases of the 21 so far found. Also known as RPTPβ, it is synthesized as a chondroitin sulfate proteoglycan. A protein formed from the extracellular domain of PTPζ through selective splicing is called phosphacan. PTPζ binds to its ligands such as pleiotrophin, a heparin-binding growth factor, or midkine, and mediates diverse signal transduction events.

Pyridylamination (PA sugar) In this pyridylamination method, reducing termini of free oligosaccharide chains are modified with 2-aminopyridin. This modification has advantages of not only enhancement of the sensitivity of detection by HPLC and mass spectrometry analysis, but also better separation of PA-labeled oligosaccharides in the reversed phase HPLC.

Reducing end of sugar chain The reducing end of a sugar chain is the name given to the direction of a sugar chain sequence. One side with an anomeric carbon is called the reducing end; on the other hand, the opposite side is called the non-reducing end. There is more than one non-reducing end in the case of branching sugar chains.

Regenerative medicine Regenerative medicine aims to recover the function of organs or tissues, which cannot be regenerated spontaneously. There are several methods to regenerate organs and tissues: (1) Regenerate the isolated tissues ex vivo and re-inject afterwards. (2) Extract the self-regenerating activities in vivo by providing growth hormone, inducing genes, or giving scaffolds for regeneration. (3) Differentiate the embryonic cells or tissue stem cells to intended organs or tissues.

RNAi RNA interference: the introduction of double-strand RNA (dsRNA) into cells causes its degradation by RNase III-like enzyme, dicer, into 21–25 bases short interfering RNA (siRNA). The siRNA dissociates to single-strand RNA by the action of RNA helicase, and the single-strand RNA is incorporated into the protein complex called RICS (RNA-induced silencing complex). RISC then recognizes the homologous mRNA with the single-strand RNA, degrades them, leading to gene silencing at the mRNA level.

Saccharide primer A saccharide primer is composed of an alkyl chain and a carbohydrate part. Saccharide primers in the culture medium are incorporated into the cells. The primer is a target of glycosyltransferases and the sugar chain is extended on the primer, whose structure is dependent on the biosynthetic pathway of the cell used in the study. Most saccharide primers coordinated with sugar chains are secreted into the medium. The structure of the carbohydrates attached on the primer is similar to that of the oligosaccharides in the cell. It is possible to extend the sugar chain on the primer in a cell-free system.

Scavenger receptor Scavenger receptor is mainly found in macrophages as a receptor of oxidized LDL. Scavenger receptor recognizes negative charges properly arranged on molecular surface, and binds to various molecules including mucins.

Selectin Selectin is a family of cell adhesion molecules that mainly recognize carbohydrates as a ligand. The family consists of three members, E-, P- and L-selectin. E-selectin is expressed in vascular endothelial cells while P-selectin is expressed in vascular endothelial cells and platelets. E- and P-selectins mediate the adhesion between these cells and ligand-positive cells such as leukocytes. L-selectin is expressed on leukocytes and mediates the interaction between leukocytes and ligand-positive cells.

Seminolipid Seminolipid is a sulfoglycolipid expressed specifically in permatocytes and subsequent germ cells. The carbohydrate moiety of seminolipid is galactose-3-sulfate, and the acyl moiety is alkylacylglycerol. The carbohydrate moiety of seminolipid is the same as that of sulfatide and synthesized by common enzymes, ceramide galactosyltransferase and cerebroside sulfotransferase. More than 90% of glycolipids in the testis consists of seminolipid.

Sialic acid Sialic acid is a generic term for N-acetylneuraminic acid and its derivatives, a nine-carbon monosaccharide. The variations of the derivatives are made by the acetylation of hydroxyl group, deletion of N-acetyl group, conversion of N-acetyl group to N-glycolyl group, and so on. It shows acidity because of its carboxyl group. Sialic acid is attached to non-reducing end of the sugar chain, by the action of sialyltransferase using CMP-sialic acid as a donor. Sialic acid is synthesized from N-acetylglucosamine through N-acetylmannosamine.

Sialidase Sialidase is an exo-type α-glycosidase that releases sialic acid from non-reducing terminals of glycans of glycoproteins, glycolipids, and oligosaccharides. It is also called neuraminidase. Sialidase was first discovered in a virus as a receptor-destroying enzyme, and has been found in various tissues of microorganisms (bacteria, protozoa) and vertebrates. Four kinds of animal sialidases are identified with different cellular localizations and zymological properties, and they are known to control a lot of cellular functions.

Sialyl 6-sulfo Lewis X Sialyl 6-sulfo Lewis X is a sugar chain containing NeuAcα2$\rightarrow$3Galβ1$\rightarrow$4[Fucα1$\rightarrow$3]GlcNAc(6-sulfate)β1$\rightarrow$R). Sialyl 6-sulfo Lewis X was first identified as a ligand for L-selectin. The binding affinity of L-selectin is strongly dependent on 6-sulfo group. It also acts as a ligand for E- and P-selectin.

Sialylated glycan Sialylated glycan is a sialic acid-containing glycan and is distributed especially in proteins and gangliosides. Sialylated glycan on the cellular surface is a target for various pathogens including influenza virus. On the other side, it is also a ligand for animal lectins. Several sialylated glycan-binding lectins were reported, such as selectins and siglec proteins. Secletins are involved in the accumulation of leukocytes at the inflammation site. Siglec proteins are known to control immune responses.

Sialyltransferase Sialyltransferase transfers a sialic acid to glycans attached onto proteins and lipids using CMP-NeuAc (or CMP-NeuGc) as a donor. A total of 20 types of sialyltransferases are identified in human and mouse. Sialyltransferases are roughly classified into four groups: ST3Gal-, ST6Gal-, ST8Sia-, and ST6GalNAc-. All sialyltransferases have a common sialyl motif: L, S, and VS. The motif is highly conserved, and is successfully utilized for the PCR cloning of sialyltransferases.

Siglec family Siglec family binds to sialyl oligosaccharides and is a member of the immunoglobulin superfamily. Siglec family is a type I membrane protein. Siglec binds to sialylated glycans using its extracelluer domain, and the signal transduction is mediated by the intracellular domain. Carbohydrate recognition and signal transduction are correlated with each other, but the regulatory mechanism is not clarified completely. The knockout mouse lacking siglec 1–4 is established.

Single molecule fluorescent measurement (fluorescent correlation spectroscopy) There are various methods of fluorescent measurement at a single molecule level by confocal or two photon microscopy, such as fluorescent correlation spectroscopy (FCS), fluorescence intensity distribution analysis (FIDA), or fluorescence polarization (FP) measurement. FCS measures fluorescent intensity fluctuation (translational diffusion time) in an extremely small volume of sample, allowing the determination of the molecular size of the fluorescent molecules. Using this method, real-time, high-speed and high-precision analysis can be achieved for measurement of various molecular parameters, including size, concentration, fluorescence brightness, and the respective amounts of different molecules present, thereby assessing molecular interactions between fluorescent molecules and other molecules in solutions. See also "Fluorescence polarization measurement".

Site-specific glycan structure Glycoproteins having multiple glycosylation sites may contain site-specific glycan structures. Such a difference is because of the fact that processing of glycan chains are affected by protein conformation. Determination of site-specific glycan structure therefore provides us useful information regarding protein structure–function relationship.

Solid-phase carbohydrate synthesis An automatic solid-phase synthesizer for peptide or DNA oligomer is available. In the same way, solid-phase carbohydrate synthesis attempts speeding up by performing the synthesis of sugar chain on a solid-phase support. The requirement of a variety of blocked monosaccharide units and the regulation of stereochemistry of glycosylation are still problems, because all of the oligosaccharides are not linear and thus different from the peptide of DNA. Using enzymatic reaction is also effective. The affinity of solid-phase support for inorganic solvent and the availability of enzyme and nucleotide sugar are keys to the synthesis.

Spermatogenesis Spermatogenesis occurs in the seminiferous tubules in the testis. Spermatogonia, a kind of germline stem cells, are found in the seminiferous tubules. Spermatogonial stem cells differentiate into spermatocytes, and subsequently undergo meiosis twice, differentiating into haploid cells referred to as spermatids, and migrating from the surrounding area to the lumen of the seminiferous tubules on the stroma cells, termed Sertoli cells, with mutual interaction. After that, they mature into spermatozoa via morphogenesis. A series of these steps is called spermatogenesis.

Sphingolipid ceramide N-deacylase Sphingolipid ceramide N-deacylase (SCDase) hydrolyzes the N-acyl linkage between fatty acids and sphingosine bases in the ceramide moiety of various glycosphingolipids and sphingomyelin. It hardly acts on the free ceramide, and the primary structure of SCDase is quite different from the ceramidase that acts only on free ceramide. The enzyme also catalyzes the condensation reaction (reverse hydrolysis) between fatty acids and sphingosine bases to form sphingolipids.

Stable isotope labeling Labeling with stable isotopes (^2H, ^{13}C, ^{15}N, etc) is an important method for the conformational analysis of biomolecules by NMR. Stable isotope labeling is also applied to the structural analysis of oligosaccharides by mass spectrometry. It is performed by: (1) addition of labeled metabolic precursor into the culture medium, (2) chemical synthesis, and (3) in vitro enzymatic reaction.

Stem cell In general, stem cells are defined as cells, which retain the ability to: (1) grow; (2) renew themselves through mitotic division; (3) differentiate into a diverse range of specialized cell types; and (4) repair damage of specific tissues/organs. The two major types of stem cells are: embryonic stem cells (ES cells) that are derived from blastocysts, and somatic stem cells found in adult tissues. Some of the example of somatic stem cells are: hematopoetic stem cells; neuronal stem cells, and hepatic stem cells. They are used for the maintenance and/or repair of respective cells or tissues.

Structural biology Structural biology aims to determine the three-dimensional structures of biological macromolecules at atomic resolution by using X-ray crystallography and nuclear magnetic resonance (NMR) spectrometry and to elucidate the molecular mechanism of functional expression from the structural point of view.

Sugar chain remodeling Sugar chain remodeling is used to change the structure of a sugar chain by artificial manipulation, such as overexpression or gene knockout, of genes of glycosyltransferases. A biological function of sugar chain could be to change dramatically or not after this manipulation.

Sugar chip Sugar chip is prepared by immobilizing the oligosaccharides onto the metal-coated glass plate (typical metal is gold). Sugar chip is used for the surface plasmon resonance (SPR) analysis of the interaction between oligosaccharides and oligosaccharide-binding proteins.

Sugar library Sugar library is a set of oligosaccharides and their derivatives obtained by organic synthesis, enzymatic synthesis, or isolation from natural resources. Sugar library is a powerful tool in the glycomics area. The library can be used as standard samples of structural analysis and also applied to the analysis of sugar–protein interactions combined with the microarray techniques.

Sugar oxazoline Sugar oxazoline is produced by the dehydration and condensation of the acetamido group (C2) with hydroxyl group (C1). Sugar oxazoline is widely used for the chemical and enzymatic glycosylation of 2-acetamido-2-deoxy sugars.

Sugar–protein interaction With structural biology having developed, the mechanism of interaction between proteins, such as lectin and antibody, and sugars has been visualized at the atomic level. Sugar–protein interaction is mainly because of hydrogen bond and van der Waals interactions, etc., which is generally weaker than protein–protein interaction. Sugar–protein interaction has often achieved high affinity and specificity through multivalent binding.

Sulfated glycan Sulfated glycans are any of the groups of glycan-attached sulfate group. Addition of the sulfate group is synthesized with various sulfotransferases. Sulfated glycans have been found in the backbone glycan such as N-linked or O-linked glycan, glycosaminoglycan, and glycolipids. They are involved in significant biological processes such as development and immunity.

Sulfotransferase Sulfotransferase is a transferase enzyme, which acts on a sulfate group. These enzymes utilize activated sulfate, PAPS (3′-phosphoadenosine-5′-phosphosulfate), as a high-energy donor, and act to transfer a sulfate group to a second substance such as protein, glycan, steroid, and tyrosine. These enzymes can be classified into two groups, those localized in the Golgi membrane and those in the cytosol. The enzyme related to sulfonation of carbohydrate is localized in the Golgi membrane.

Surface plasmon resonance In order to excite surface plasmons in a resonant manner, one can irradiate an electron or light beam to the planar metal interface in an appropriate condition, and surface plasmon waves that propagate parallel along a metal interface are resonant with the evanescent wave. Because evanescent waves are affected by the refractive index of the material onto the metal surface, this phenomenon is used in many biosensor applications. For instance, one can measure receptor–ligand interaction by adsorbing the ligand with the metal surface.

Syndecan Syndecans are a family of membrane-bound heparan sulfate proteoglycans. Structurally similar syndecans, syndecan-1 to -4, were reported. Syndecans are involved in a variety of functions, acting as co-receptors of growth factors, integrins, and fibronectins.

T-antigen T-antigen is a cancer-related glycan antigen, Galβ1-3GalNAc-Ser/Thr. It is also called TF antigen, after Thomsen and Friedenreich who discovered this antigen.

Tertiary structure model The molecular structure is described with constituent atomic positions using three-dimensional coordinate system (x, y, and z). Although NMR spectroscopy and X-ray crystallography do not directly provide the atomic position of the molecule, molecular structures determined by them are called "tertiary structure model".

Therapeutic antibody Antibody is produced when exogenous antigen (for example, pathogenic bacteria) invades into the body. Antibody is a component of the immune system that prevents pathogens from entering or damaging the cells. Antibody therapeutics is aimed to utilize the functions of antibody in the immune system. Therapeutic antibodies against TNFα, EGF and Her2 are now used for intractable inflammatory disease and cancer. It has recently been shown that removal of core fucose on human IgG1 oligosaccharide results in 50–100-fold enhancement of antibody-dependent cellular cytotoxicity (ADCC). Hence, control of the IgG fucosylation could be one of the most promising technologies to improve the efficacy of therapeutic antibodies.

Thioester method Conventional solid phase peptide synthesis is limited by the peptide length. In order to prepare long polypeptides, it is necessary to couple short peptides. Thioester method is a technique used for creating long peptides and proteins: (1) a peptide containing thioester bond at its C-terminus is prepared. (2) The thioester is selectively activated by argentite. (3) The activated peptide is coupled to the N-terminal amino group of the other peptide. It is not necessary to protect the functional groups except for the amino groups (Lys, Arg) and thiol group (Cys).

Transferrin Transferrin is an iron-binding glycoprotein composed of 679 residues and possesses two Fe^{3+}-binding sites. Human serum contains 2–3 mg/ml of transferrin. Transferrin expresses biantennary oligosaccharides with N-acetylneuraminic acid at Asn413 and Asn611. Molecular weight of transferrin with oligosaccharides is 79,593. Transferrin is used as a marker molecule to examine the oligosaccharide variants involved in various diseases, such as CDG or chronic rheumatoid arthritis.

Transglycosylation Glycosidase has an ability to release the sugar by hydrolyzing the glycosyl-bond of substrates. The activity is also interpreted as transfer of sugar to water. *Transglycosylation* is a reaction in which sugar is transferred to the compound having hydroxyl group. The catalytic activity of the enzyme is dependent on the tertiary structure of the active site. A variety of oligosaccharides are now synthesized by transglycosylation reaction of exoglycosidases.

Trypanosoma *Trypanosoma* is a group of parasitic protozans. Sleeping sickness and Chagas's disease are caused by *Trypanosoma*. In sleeping sickness, the surface of blood cells is covered with GPI-anchored variant surface glycoprotein. The isoform of variant surface glycoprotein can be switched, dependent on the production of antibodies, resulting in the conversion of antigenicity. The structure and biosynthesis of GPI was first studied by using *Trypanosoma* because the GPI biosynthesis of *Trypanosoma* is highly active.

Tumor marker Tumor markers are produced in tumor cells and can be detected in body fluids such as serum. Examination of tumor marker is useful not only for the diagnosis of cancer and estimation of the primary focus, but also for monitoring therapeutic efficacy and postoperative follow-up.

Ubiquitin ligase A uniquitin ligase, also denoted as E3 enzyme, operates in conjunction with an E1 (ubiquitin-activating enzyme) and an E2 (ubiquitin-conjugating enzyme) to attach ubiquitin to an ε-amino group of lysine side chain(s) on a target protein. The system called "ubiquitin system" comprises E1–E3, and is involved in not only protein degradation, but also in quite diverse biological phenomena. A variety of E3 enzymes with different specificity is known to be responsible for the diverse recognition of substrates by them (e.g., a type of E3 enzyme, Fbs1/2, recognizes N-glycans).

Ultra-high field NMR At present, ultra-high field NMR spectrometers beyond 900 MHz are available owing to the development of new superconducting materials. Ultra-high field NMR spectra with high resolution and sensitivity offer an opportunity for detailed structural analyses of oligosaccharides and glycoproteins.

Verotoxicin/Shiga-like toxin Verotoxin, AB5-type toxin having the A subunit and five B subunits, is produced by the pathogenic *Escherichia coli* (*E. coli*) such as *E. coli* O157:H7. It is named after its ability to kill Vero cells. Shortly after, the verotoxin was referred to as Shiga-like toxin because of its similarities to Shiga toxin. Verotixin specifically binds to globotriaosylceramide (Gb3/CD77).

ZP3 ZP3 is a major glycoprotein component of the zona pellucida, a thick layer surrounding the mammalian egg. A glycan part of ZP3 plays an important role in binding to the sperm and therefore ZP3 serves as a sperm receptor on fertilization.

Index

a

2-aminopyridine 14
α1,2-, α1,3-, α1,3/4-, or α1,6-fucosyltransferases 36
α1,2 fucosyltransferase (α1,2 FT) 263, 366
α1,2 mannosidase Ib (Man Ib) 270
α1,3 fucosyltransferase (α1,3FT) 268
α1,3-fucosyltransferase assay 38
α1,3 galactosyltransferase (α1,3GT) 267
α1,3-N-acetylgalactosamine 363
α1,3/4-fucosyltransferase 367
α1,3GTKO pig 268
α1,4-linkages 50
α1,4-linked GlcNAc residues 227
α1,6 fucosyltransferase (Fut8) 39, 236
α2,3 sialyltransferase (α2,3ST) 268
α2,3Sia (avian-type receptor) 225
α2,3Sia (bird type) to α2,6Sia (human type) 225
α2,6Sia (human-type receptor) 225
α2,8-sialyltransferase 394, 396
α4GlT 51
α4GnT 50–51
α5-deficient Chinese hamster ovary (CHO) cells 358
α-6-D-mannosideβ1,6-N-acetylglucosaminyltransferase V (GnT-V) 267
α- and β-dystroglycan 315
α-dystroglycan (α-DG) 30, 310
α-fetoprotein (AFP) 235
α-Gal epitope (Gal α1-3Galβ1-4 GlcNAc-R) 267
α-galactosylceramide (α-GalCer) 145
α-mannosidase II (MII, Man II) 111, 270
α-mannosidase IIx (MX) 111, 276
Aβ 185
AAL 298

ABO 366
ABO locus 363
acrosome reaction 279
acrosome reaction inducing substance (ARIS) 279
activity assay for MII and MX 113
acute vascular rejection (AVR) 267
AFP-L3 235
ALG8, ALG9 and ALG10 77
ALG11 76
Alzheimer's βsecretase (BACE1) 192
Alzheimer's disease (AD) 185
amino-CTH synthase 53
anaplastic lymphoma kinase (ALK) 355
antibodies 429
"a" series gangliosides 275
asialofetuin–Sepharose 134
assays for nucleotide sugar transporter 105

b

β- and γ-secretases 185
β1,3-glycosyltransferases (β3GT) 24
β1,4-galactosyltranferase-I-deficient (β4GalT-$^{-/-}$)
β1,4-N-acetylgalactosaminyltransferase 396
β1,4-N-acetylglucosaminyltransferase III (GnT-III) 267
β1,4-N-acetylglucosaminyltransferase IV (GnT-IV) 267
β3Gal-T1 24
β3Gal-T2 24
β3Gal-T4 24
β3Gal-T5 24
β3Gn-T6 27
β3Gn-T7 27, 67
β3Gn-T8 27
β4-Galactosyltransferase deficiency 383

491

β4GalNAc T 391
β4GalNAc-T3 22
β4GalNAc-T4 22
β4GalNAc T KO mice 391
β4Gal-T1 19, 383
β4GalT-I KO mice 384
β4Gal-T2 19
β4Gal-T3 19
β4GalT-I$^{-/-}$ mice 343
β4GalT-IV 68
β4GT motif (WGxEDD/V/W) 19
β-dystroglycan 310
β-galactoside α2,3-sialyltransferase family (ST3GalIVI) 42
β-galactoside α2,6-sialyltransferase family 42
bacterial sialyltransferases 43
B cell antigen receptor (BCR) 167
behavioral tests 381
biosynthesis of gangliosides 57
biosynthesis of GPI 73
bisecting GlcNAc 250
bovine submaxillary mucin (BSM) 241
Boxer 306
brefeldin-A (BFA) 213
"b" series gangliosides 275

c
C16S toxin 230
C2-ceramide 300
C2GnT1-deficient mice 244
C5-epimerase (HSepi) 258, 303
CA19-9 36
Ca^{2+}-dependent (C-type) mammalian lectin 162
CabosDB 426
CabosML 426
caenorhabditis elegans 290
calcium-type lectins (MGL/CD301) 158
calnexin 207
calreticulin 207
carbohydrate epitopes 429
carbohydrate microarray 455
carbohydrate–carbohydrate interaction 300
carbohydrate–protein interactions 455
castanospermine (CS) 270
cation-independent mannose-6-phosphate receptor (CI-M6PR) 219
caveolae 333
CAZy (Carbohydrate-Active enZYmes) 9
CAZy Database 423
CD antigens 369
CD1d-restricted Natural Killer T (NKT) 145
CD22 167
CD40L 167
CD44 151, 213
CD57 47, 180

CD59 324
CD72 167
CDG-I 319
CDG-II 319
CDGS 76
cell adhesion molecules (CAMs) 180
cell banks 404
ceramide galactosyltransferase 182, 275
cerebroside sulfotransferase 182, 275
cerebroside sulfotransferase KO mice 389
CGT-null mice 182, 275
chemical syntheses of oligosaccharides 417
Cholera toxin (CTX) 213
chondroitin 4-sulfotransferase (C4ST) 87
chondroitin 6-sulfotransferase (C6ST) 87
chondroitin sulfate (CS) 64, 386
chondroitin sulfate proteoglycans (CSPGs) 188, 195
chondroitin/dermatan sulfate uronyl 2-O-sulfotransferase (CS/DS2ST) 91
chondroitinase ABC 189
chromosomal localization of mouse and sialyltransferases 43
clostridium botulinum 230
cluster designation (CD) 369
CMP-N-acetylneuraminic acid (CMP-NeuAc) hydroxylase 97
CMP-NeuAc 97
CMP-NeuGc 97
CMP-Sia transporter 103
congenital disorders of glycosylation (CDG) 76, 319
congenital disorders of glycosylation (CDG) IIc 295
congenital muscular dystrophy type 1C 317
consensus amino acid sequence for O-fucosylation 33
Consortium for Functional Glycomics (CFG) 435
contact hypersensitivity (CHS) 383
core 1 structure (C1Gal-T) 5, 243
core 2 structure 5
core2 β1,6-N-acetylglucosaminyltransferase-I (C2GnT-I) 228
core 3 structure (β3Gn-T6) 243–244
core-fucose 236
core fucosylation 39
core type 1 278
core type 2 278
CRD 133
cross-sample comparison 448
CSGalNAcT-1 and 2 64
CSS-1, CSS-2, and CSS-3 64
CST-null mice 182–183
CTX/STX 213
cyclooxygenase 2 (COX2) 239
cytokine 145

Index

cytokine receptors 154
cytosolic sialidase 117

d

Dackel, HS co-polymerase 306
DC-sign 22
decay accelerating factors (DAF) 324
delayed-type hypersensitivity (DTH) 383
deoxymannojirimycin (dMM) 270
deoxynojirimycin (dNM) 270
different HNK-1 carbohydrates 49
disaccharide building blocks 418
DLH motif 22
DNA microarray 447
dominant-negative mutants 405
double KO mice 396
D-PDMP 333
DPM1 323
DPM2 323
DPM3 323
drosophila homologs of *EXT* genes 61
Drosophila melanogaster 285
DXD motif 425
dystroglycan 309
dystrophin–glycoprotein complex (DGC) 315

e

E- and P-selectins 148
E3 ubiquitin ligases 204
E-cadherin 351
EDEM (ER-degradation enhancing α-mannosidase-like protein) 207
EGCase I 120
EGCase I, II, and III 119
EGCase II 120
EGCase III (EGALC) 120
eggs 278
endocytosis 217
endoglycoceramidase (EGCase) 119, 120
enzyme replacement therapy 221
epidermal growth factor (EGF) 349
epidermal growth factor (EGF) domains 33
epidermal growth factor receptor (EGFR) 351
ER α-mannosidase I 209
ER-associated degradation (ERAD) 201, 205, 207
ERQC 208
ERT 220
ES cells 263, 265
E-selectin 246
evanescent-field fluorescence 451
expression system of glycosyltransferases 407
EXT genes are 61

EXT-1 and EXT-2, HS 305
EXTL2 and *EXTL3* 60

f

Fabry disease cardiac type 219
Factor VII 35
Fbx6b protein 205
FGF 285
FGF-1 154
FGF-2 154, 197
FGF-signaling 259
fibronectin 254
fluorescence-labeled oligosaccharides 13
fluorescent labeling of glycoprotein 452
Fringe 33
Fringe assay 35
fucose sulfate polymer 279
fucosyltransferase 11
fucosyltransferase (α1,2/ α1,3/ α1,4-fucosyltransferases) 36
fucosyltransferase1 295
fukutin 309
fukutin-related protein 317
Fukuyama congenital muscular dystrophy (FCMD) 309, 315, 316
FUT1 366
FUT1(H) 366
FUT2 366
FUT2(Se) 366
FUT3(Le) 367
FUT4 38
FUT5 36
FUT6 36
FUT7 36, 38
Fut8 359, 379
Fut8 knockout mice 379
FUT9 377
FX 236

g

Galβ1,3GalNAc (TF antigen) 339
GAL4 driver and *UAS-inverted repeat* (*UAS-IR*) 287
GALAXY 413
GalCer 182
galectin 133
galectin-1 263
galectin-9 (Gal-9) 171
GalNAc (Tn antigen) 339
GalNAc 4-sulfate 6-O-sulfotransferase (GalNAc4S-6ST) 87
GalNAc transferase (pp-GalNAc-T) 243
ganglioside clusters 186
GATEWAY®-entry vectors 407
Gaucher disease type 1 219

Gb3 (globotriaosylceramide)   363
Gb3 synthase   363
Gb3/CD77 synthase   53
Gb4 synthase   53
Gb5 synthase   53
GD3   394
GD3 synthase KO mice   394, 396
GDP-4-keto-6-deoxy-D-mannose 3,5-epimerase/4-reductase (Gmer)   296
GDP-fucose   236
GDP-fucose transporter   103, 295
GDP-mannose 4,6 dehydratase (Gmd)   296
germ cells   276
Glc$_3$Man$_9$GlcNAc$_2$ 207
GlcAT-P   48, 180
GlcAT-P and GlcAT-S   47, 381
GlcAT-P KO mice   381
GlcAT-S   48, 180
GlcNAc 6-O-sulfation   151
GlcNAc 6-O-sulfotransferase (GlcNAc6ST)   91
GlcNAcβ:6-O-sulfotransferases   150
GlcNAcβ1–2Man linkage   316
GlcNAc-6-O-sulfotransferase assay   85
GlcNAcT-I   60
glucosylceramide synthase   333
glucuronyltransferase   47, 381
glucuronyltransferase assay   48
glucuronyltransferase assay for glycolipids   48
glycan binding analysis   138
Glycoconjugate Data Bank   435
GlycoEpitope database   429
GlycoGene DataBase (GGDB)   408, 423
Glyco-Net   437
glycopeptides   321
glycoprotein acceptors   48
glycoprotein database   426
glycosaminoglycan sulfotransferase   88
glycosaminoglycan sulfotransferase assay   92
glycosaminoglycan-binding cytokines   154
glycosidase   441
glycosylphosphatidylinositol (GPI)   323
glycosyltransferase   423, 441
glycosyltransferase assays   28
glycosyltransferase evolution   9
glycosyltransferases for the synthesis of glycosphingolipids   54
GM1   330
GM2   330
GM2 activator protein   125, 220
GM2 gangliosidoses   219
GM2 synthase-null mice   276
GM2/GD2 synthase   391
GM2/GD2 synthase-KO mice   276, 396
GM3   333
GnT-III   359
GnT-III, IV, V, and IX   13, 15
GnT-V   359

GPI-anchor glycans binding cytokines   155
GPI-APs   72
GPI-APs and lipid rafts   74
growth retardation   39
GTS-A   10
GTS-B   10
GTS-C   10
GTS-D   11

h

H antigen (O-type antigen)   366
H5N1 influenza A viruses   7, 225
haptoglobin   236
HAS1, HAS2 and HAS3   70, 252, 254
H-D antigen   269
Helicobacter pylori (H. pylori)   227
hemagglutinin (HA)   225
heparan sulfate (HS)   59, 386
heparan sulfate 2-O-sulfotransferase   259
heparan sulfate 2-sulfotransferase (HS2ST)   91, 258
heparan sulfate 3-O-sulfotransferase   289
heparan sulfate 3-sulfotransferase (HS3ST)   92
heparan sulfate 6-O-sulfotransferase   289
heparan sulfate 6-sulfotransferase (HS6ST)   91, 306
heparan sulfate endosulfatase   123
heparan sulfate N-deacetylase/N-sulfotransferase (NDST)   91
heparan sulfate proteoglycans (HSPG)   195, 258, 303
heparan sulfate synthases   59
heparin (HP)   59
hepatocellular carcinoma (HCC)   235
hereditary multiple exostoses (HME)   60
HEXA   219
HEXB   219
Hh   285
hidden Markov model (HMM)   9
HIF-1 (hypoxia inducible factor-1)   249
high endothelial venules   150
high-mannose type glycan binding IL-2   155
high-mannose type N-glycan   112
hinge glycopeptide   341
hinge mucin-type sugar   339
HMGA2 gene   325
HNK-1   47, 180, 381
HNK-1 epitope   386
HNK-1 ST   47
HPLC mapping   413
HS2ST, heparan sulfate 2-O-sulfotransferase   306
HS2ST, HS   305
HS3ST-1, HS   305
HS6ST, heparan sulfate 6-O-sulfotransferase   306
HS6ST-1, HS   305

Index

HS6ST-2, heparan sulfate   306
Hsepi   305
Hsepi A, B, glucuronyl-C5   306
Hsepi, glucuronyl-C5 epimerase   306
human exostosin (EXT)   59
human glycosyltransferases   407
human heparan sulfate/heparin
   glycosyltransferases and epimerase   61
human IgA1   343
human sialyltransferases   43
Hyal-2   216
hyaluronan (HA)   70, 216, 252
hyaluronan synthase   70
hybridoma   404
hyperacute rejection (HAR)   267
hypoxia   249

i

i antigen   27
IgA nephropathy   339, 342, 346, 383
IgA1   339
iGb3   146
iGn-T(β3Gn-T1)   24
IGnT1   367
IGnT2   367
IGnT3   367
Ii antigens   27
Ii blood group systems   366
IL-4   240
immunity   168
immunoglobulin A nephropathy (IgAN)   339, 342, 343, 346
immuno-thin layer chromatography   377
in vivo PNGase assay   203
increased GlcNAcβ1-6 branching   250
inducible RNAi knockdown system   287
influenza   225
inherited GPI deficiency   323
injury repair   196
innate immunity   172
insulin receptor (IR)   333
insulin resistance   333, 336
insulin signaling   330
integrin $α_4β$ 355
integrin $α_5β_1$ 358
integrin $α_6β_1$ 355
Interleukin-2 (IL-2)   155
isoelectric focusing (IEF)   320

k

KegDraw   443
KEGG (Kyoto Encyclopedia of Genes and Genomes)   441
KEGG GLYCAN   441
KEGG PATHWAY   442

KEGG Pathway Database   423
keratan sulfate (KS)   67, 386
keratan sulfate Gal-6-sulfotransferase
   (KSGal6ST)   91
keratan sulfate synthase   19

l

3T3-L1 adipocytes   333
L- and S-sialylmotifs   42
lacto-N-neotetraose   6
lacto-N-tetraose   6
lacto-N-triose II   6
Lactose–Sepharose   134
Lactosyl-Cer synthase   19
lactosylceramide   8
laminin   312
LARGE   310, 317
Lc$_3$Cer synthase   27
LC-MS/MS   177
LDL receptor-related protein (LRP)   355
lectin   432
lectin-affinity database   432
lectin affinity dataBase (LADB)   426
lectin-based glycan profiling   451
lectin-based structural glycomics   433
lectin–glycoprotein interaction   452
lectin (MBP) pathway   163
lectin microarray   451
lentiviral shRNA vector   405
Lewis   366
Lewis x   38, 148
Le^x-glycan(Galβ1–4(Fucα1–3)GlcNAcβ1-)   299
Leydig cells   276
lipid rafts (membrane microdomain)   299
low-density lipoprotein (LDL) receptor-related
   protein-1 (LRP-1)   351, 379
L-selectin   38
L-selectin ligand   386
LTP (long-term potentiation)   381
lung emphysema   39
luteinizing hormone (LH)   22
lymphocyte homing   151
lymphocyte homing assay   388
lysosomal diseases   219
lysosomal sialidase, Neu1   117

m

4D microscopy   292
macrophage galactose-type   158
MALDI TOF   322, 343
mammalian sialidase   116
mannan-binding protein (MBP)   162
mannose 6-phosphate   157

mannose receptor (MR) 219
mass production of monosaccharide derivatives 417
mass spectra database 426
mass spectrometry 319
matrix metalloproteinases 379
MBP-dependent cell-mediated cytotoxicity (MDCC) 164
MBP-ligand 164
MBP-ligand oligosaccharide 165
MDCID 316
MDCIC and limb–girdle muscular dystrophy 2I(LGMD2I) 316
medaka Oryzias latipes 299
membrane microdomains (lipid raft) 299, 333
memory B and T cells 168
methyl β-cyclodextrin (MBCD) 300
MGL1 (CD301a) 158
MGL2 (CD301b) 158
microarray hybridization 448
microdomains 299
midkine (MK) 197, 355
MII 112
MMPs 236
molecular interaction networks 442
monoclonal antibodies 429
mucins 238
mucin-type carbohydrate chains 243
muscle–eye–brain disease (MEB) 315, 316
MX 112
myelin 183, 389
myelin membrane 182
myodystrophy (Large^{myd}) 317

n
N-[2-(2-Pyridylamino)ethyl]-succinamic acid 5-norbornene-2,3-dicarboxyimide ester 41
N-acetylgalactosaminyltransferases (ppGaNTases) 16
N-acetylglucosamine-6-O-sulfotransferase 83
N-acetylglucosamine-6-sulfotransferases 84
N-acetylglucosaminyltransferase I (GnT-I) 270
N-acetylglucosaminyltransferase-II (GnTII) 111
N-acetylglucosaminyltransferases (GnTs) III, IV, V, VI, and IX 13
N-acetylneuraminic acid (Neu5Ac) 225
natural killer (NK) cell 267
N-deacetylase/N-sulfotransferase (NDST) 258, 303, 386
NDST-1 304, 305
NDST-2 305
neoglycolipids 8, 455
NEU 330

Neu1 329
Neu2 329
Neu3 329
Neu4 329
neural cell adhesion molecule (NCAM) 181
neurodegeneration 391, 397
neuroglycan C 195
neuroregeneration 394
neurotoxin (NT) 230
N-glycan 413
N-glycan-binding proteins 204
N-glycolylneuraminic acid (Neu5Gc) 225
N-glycosylation 349, 358
NK cell receptor 270
nocturnal hemoglobinuria (PNH) 324
non-neurological diseases 219
Notch signaling 33, 285, 295
nucleoside diphosphatase (*NDPase*) 104
nucleotide sugar transporters (NSTs) 103

o
O-fucose modification 33
O-fucosyltransferase 1 assay 34
OFUT1 33
Ofut1 34
O-glycan 238, 243, 346
O-glycan structure 346
oligodendrocyte 389
oligosaccharide database 426
O-Mannosylation 30
O-mannosylglycan 309, 315

p
P 366
3′-phosphoadenosine 5′-phosphosulfate (PAPS) 83, 103
4-(2-pyridylamino) butylamine 41
P antigen (Gb4) 364
p individuals (no Gb3) 364
P/pk/p blood group system 363
P1 antigens 364
pancreatic cancer 236
PA-oligosaccharides 413
PAPS transporters 103, 288
PAPST2 289
Paragloboside 364
paranodal junctions 389
PDMP (D-threo-1-phenyl-2-decan-oylamino-3-morpholino-1-propanol) 269
Peptide: N-glycanase (PNGase) 201
PGE$_2$ 239
phosphatidylglucoside (PtdGlc) 177
phosphatidylinositol (PI)-specific phospholipase C (PI-PLC) 72

Index

phylogenetic tree of nucleotide sugar transporters   104
PIGA   323–324
PIGM   323–324
Pk antigen (Gb3)   364
plant sialyltransferases   44
plasma membrane-associated sialidase   117
pleiotrophin (PTN)   197, 355
pluripotent stem (iPS) cells   265
p-methoxyphenyl (*p*MP) glycosides   417
PNG1   201
Pofut1   34
Poly(A)$^+$ RNA preparation   447
polylactosamine   24
polylactosamine synthase   24
poly-*N*-acetyllactosamine   7, 257
POMGnT1   30, 309, 315–316
POMGnT1 assay   31
Pompe disease   219
POMT   30
POMT 1 and 2   289
POMT assay   30
POMT1   316
POMT1/2   315
POMT2   30
porcine endogenous retrovirus (PERV)   267
porcine gastric mucin   230
pp-GalNAc-T2   346
ppGaNTase   16
ppGaNTase assay   17
primary cultures   403
primary hepatomas   39
prosaposin   128
PSA (polysialic acid)   180–181
Purkinje neuron   391
pyridylaminated sugar chain   14
pyridylamination   13

r

regeneration   257
remodeling glycoantigens   269
repeating GlcNAcβ(1 → 4)-GlcAβ(1 → 3) disaccharide   252
retrograde traffic   214
Rib-2, heparan sulfate GlcNAc transferase   306
RNA interference (RNAi)   285
RNAi fly   288

s

Saccharomyces cerevisiae   11
Sandhoff disease   220–221
saposin   125–126
saposin A   127
saposin B   127
saposin C   127
saposin D   127
scavenger receptor (SCR)   239
sCD43 cDNAs   228
sciatic nerve   391
sea urchin egg jelly receptor (suREJ1)   279
selectins   148
seminolipid   94, 389
sensory function   397
Sertoli cells   276
Shiga toxin (STX)   213
SHP-1   169
sialidase assay   115
sialidases   115, 329
sialyl 6-sulfo Lewis x   150, 248
sialyl Lewis A   244, 246–247
sialyl Lewis X   244, 246–247, 384
sialyl Lewis x synthesis   148
sialyl Tn antigen   241
sialylglycopclypeptides   7
sialyl-T (ST) antigen (NeuAcα2-3Galβ1-3GalNAc)   244, 346
sialyltransferase   42
sialyltransferase genes   43
sialyltransferases involved in the biosynthesis of gangliosides   56–57
siglec   136, 240
Siglec 2 (CD22)   167, 240
siglec and glycan binding assay   137
siRNAs   286
skin lesions   397
SL15 (MPDU1)   323
SO$_4$-4GalNAcβ1,4GlcNAcβ-structure   22
software and hardware for 4D microscopy   291
somatic and embryonic stem cells   263
somatic stem cells   263
sperm   278
sperm–egg binding   299
spermatogenesis   275, 389, 391
sphingolipid activator proteins   125
Sphingomona   146
spinal cords   391
SSEA-1 (stage-specific embryonic antigen-1)   38, 265, 377
SSEA-3 (Galβ-globoside)   265
SSEA-4 (Sialyl-Galβ-globoside)   265
SSEA-4 antigens   257
ST3Gal I   244
ST3Gal I, II, III, and IV   193
ST6Gal I   192
ST8SiaII   181
ST8SiaII (STX)   181
ST8SiaIV   181
ST8SiaIV (PST)   181
ST8Sia-I-VI   42
structure search tool   443

sugar-nucleotide synthase   423
sugarnucleotide transporter   423
sulfatide   94, 182–183, 389
sulfatide synthase   94–95
SulfFP1/sulf-1 and SulfFP2   123
sulfotransferase   386, 423
sulfotransferase KO mice   386
swainsonine (SS)   270
synaptic plasticity   381
2-O-sulfotransferases   258, 303
3D structures of carbohydrates   436
3-O-sulfotransferases   258, 303

t
TGF-β1 signaling   379
Th1   240
Th1 cytokine   240
Th2 cytokine   240
thyroid stimulating hormone (TSH)   22
TNFα 333
TRA-1-60   257, 265
TRA-1-81   265
TRA-2-49/6F   265
transcriptome   447
transferrin   319
Trypanosome brucei   72
trypsinogens   379
tunicamycin (TM)   269
type 1 antigens   24

type 2 diabetes   335
type I collagen   254
type I diabetes   269

u
ubiquitin ligase   206
ubiquitin proteasome system   205, 206

v
VEGF   240
viral sialyltransferases   44
viral vector systems   404

w
Walker–Warburg syndrome (WWS)   315, 316
Wg/Wnt   285

x
X-ray crystal structures GlcAT-P and GlcAT-S   49
xylosyltransferase (XylT)   59

z
zona pellucida (ZP)   278

CBS 756
£114.95
(10/83)

A1